Louis C. Guillou
Jean-Jacques Quisquater (Eds.)

Advances in Cryptology – EUROCRYPT '95

International Conference on the
Theory and Application of Cryptographic Techniques
Saint-Malo, France, May 21-25, 1995
Proceedings

 Springer

Series Editors

Gerhard Goos
Universität Karlsruhe
Vincenz-Priessnitz-Straße 3, D-76128 Karlsruhe, Germany

Juris Hartmanis
Department of Computer Science, Cornell University
4130 Upson Hall, Ithaca, NY 14853, USA

Jan van Leeuwen
Department of Computer Science, Utrecht University
Padualaan 14, 3584 CH Utrecht, The Netherlands

Volume Editors

Louis C. Guillou
CCETT
4, rue du Clos Courtel, F-35512 Cesson-Sevigne Cedex, France

Jean-Jacques Quisquater
Department of Electrical Engineering (DICE)
Place du Levant, 3, B-1348 Louvain-la-Neuve, Belgium

CR Subject Classification (1991): E.3-4, G.2.1, D.4.6, F.2.1-2, C.2, J.1

1991 Mathematics Subject Classification:
94A60, 11T71, 11Yxx, 68P20, 68Q20, 68Q25

ISBN 978-3-540-59409-3   ISBN 978-3-540-49264-1 (eBook)
DOI 10.1007/978-3-540-49264-1

CIP data applied for

Originally published by Springer-Verlag Berlin Heidelberg New York in 1995

Typesetting: Camera-ready by author
SPIN: 10485927   06/3142-543210 - Printed on acid-free paper

PREFACE

EUROCRYPT '95. Sponsored by the International Association for Cryptologic Research (IACR), in cooperation with the Centre Commun d'Études de Télévision et Télécommunications (CCETT), a workshop on the theory and applications of cryptographic techniques takes place at the Palais du Grand Large, Saint Malo, France, May 21-25, 1995.

The General Chair of EUROCRYPT '95 is Françoise Scarabin. The Organization Committee was helped by Maryvonne Lahaie and her communication team. Moreover, the CCETT has generously provided the help of a young English lady, Miss Virginia Cooper, for the secretariat of both the Organization and Program Committees. They all did an excellent job in preparing the conference. It is our pleasure to thank them for their essential work.

IACR and EUROCRYPT. According to a very good suggestion expressed during CRYPTO '82, the Association was established at CRYPTO '83. Today, the Association has approximately 600 members and the mailing file managed by its Secretariat consists of more than 2 000 names.

The main goal of the Association is the sponsoring of two annual conferences: CRYPTO, every summer at the University of California, Santa Barbara (UCSB), and EUROCRYPT, every spring in a different European country. Moreover, the Association edits quarterly the Journal of Cryptology (JoC).

After 2 conferences held in 1982 in Burg Feuerstein (Germany) and in 1983 in Udine (Italy), the name EUROCRYPT was used for the very first time in 1984 in Paris (France). Since then, EUROCRYPT has taken place at a variety of venues: in 1985 in Linz (Austria), in 1986 in Linköping (Sweden), in 1987 in Amsterdam (Netherlands), in 1988 in Davos (Switzerland), in 1989 in Houthalen (Belgium), in 1990 in Aarhus (Denmark), in 1991 in Brighton (United Kingdom), in 1992 in Balatonfüred (Hungaria), in 1993 in Lofthus (Norway) and in 1994 in Perugia (Italy). EUROCRYPT '96 is planned to take place in Sarragossa (Spain).

Previous Proceedings. The following 24 proceedings have been published for conferences held at UCSB (CRYPTO) and in Europe (EUROCRYPT).

1. Advances in Cryptology: a Report on CRYPTO 81, ECE Report no. 82–04,
   Allen Gersho, Ed., ECE Dpt, UCSB, Santa Barbara, CA 93106.
2. Cryptography: Proceedings, Burg Feuerstein, 1982,
   T. Beth, Ed., LNCS 149, Springer-Verlag, 1983.
3. Advances in Cryptology: Proceedings of Crypto 82,
   D. Chaum, R. L. Rivest and A. T. Sherman, Eds., Plenum, NY, 1983.
4. Advances in Cryptology: Proceedings of Crypto 83,
   D. Chaum, Ed., Plenum, NY, 1984.
5. Advances in Cryptology: Proceedings of EUROCRYPT 84,
   T. Beth, N. Cot and I. Ingermarsson, Eds., LNCS 209, Springer-Verlag, 1985.
6. Advances in Cryptology: Proceedings of CRYPTO 84,
   R. Blakley and D. Chaum, Eds., LNCS 196, Springer-Verlag, 1985.

7. Advances in Cryptology: Proceedings of EUROCRYPT '85,
   F. Pichler, Ed., LNCS 219, Springer-Verlag, 1986.
8. Advances in Cryptology: CRYPTO '85,
   H. C. Williams, Ed., LNCS 218, Springer-Verlag, 1986.
9. Advances in Cryptology: CRYPTO '86,
   A. M. Odlyzko, Ed., LNCS 263, Springer-Verlag, 1987.
10. Advances in Cryptology: EUROCRYPT '87,
    D. Chaum and W. L. Price, Eds., LNCS 304, Springer-Verlag, 1988.
11. Advances in Cryptology: CRYPTO '87,
    C. Pomerance, Ed., LNCS 293, Springer-Verlag, 1988.
12. Advances in Cryptology: EUROCRYPT '88,
    C. G. Günther, Ed., LNCS 330, Springer-Verlag, 1988.
13. Advances in Cryptology: CRYPTO '88,
    S. Goldwasser, Ed., LNCS 403, Springer-Verlag, 1989.
14. Advances in Cryptology: EUROCRYPT '89,
    J.-J. Quisquater and J. Vandewalle, Eds., LNCS 434, Springer-Verlag, 1990.
15. Advances in Cryptology: CRYPTO '89,
    G. Brassard, Ed., LNCS 435, Springer-Verlag, 1990.
16. Advances in Cryptology: EUROCRYPT '90,
    I. B. Damgard, Ed., LNCS 473, Springer-Verlag, 1991.
17. Advances in Cryptology: CRYPTO '90,
    A. J. Menezes and S. A. Vanstone, Eds., LNCS 537, Springer-Verlag, 1991.
18. Advances in Cryptology: EUROCRYPT '91,
    D. W. Davies, Ed., LNCS 547, Springer-Verlag, 1991.
19. Advances in Cryptology: CRYPTO '91,
    J. Feigenbaum, Ed., LNCS 576, Springer-Verlag, 1992.
20. Advances in Cryptology: EUROCRYPT '92,
    R. A. Rueppel, Ed., LNCS 658, Springer-Verlag, 1993.
21. Advances in Cryptology: CRYPTO '92,
    E. F. Brickell, Ed., LNCS 740, Springer-Verlag, 1993.
22. Advances in Cryptology: EUROCRYPT '93,
    T. Helleseth, Ed., LNCS 765, Springer-Verlag, 1994.
23. Advances in Cryptology: CRYPTO '93,
    D. R. Stinson, Ed., LNCS 773, Springer-Verlag, 1994.
24. Advances in Cryptology: CRYPTO '94,
    Y. G. Desmedt, Ed., LNCS 839, Springer-Verlag, 1994.

No proceedings were published for the conferences held in 1983 in Udine (Italy) and in 1986 in Linköping (Sweden). Moreover at the time of writing this preface, the proceedings of EUROCRYPT '94 held in Perugia (Italy) are still waiting for publication. A careful examination of the list induces the following five remarks.

- The words *'Advances in cryptology'* appeared on the first proceedings.
- Since 1984, CRYPTO and EUROCRYPT are written in capitals.
- Since EUROCRYPT '85, the number of the year is preceded by '.
- Since CRYPTO '85, the words *'Proceedings of'* have disappeared.
- Among these 24 proceedings, 21 were published by Springer Verlag.

Lecture Notes in Computer Science 921

Edited by G. Goos, J. Hartmanis and J. van Leeuwen

Advisory Board: W. Brauer   D. Gries   J. Stoer

Springer-Verlag Berlin Heidelberg GmbH

Submissions, Program, Proceedings. CRYPTO '94 and EUROCRYPT '95 are the first two IACR conferences where the proceedings are available at the conference; the subsequent advance of the submission deadlines by two months explains the slight decrease in the number of submissions: 135 at CRYPTO '92, 117 at EUROCRYPT '93, 136 at CRYPTO '93, 137 at EUROCRYPT '94, 114 at CRYPTO '94, 113 at EUROCRYPT '95.

This outcome does not appear to be long term, there being 150 submissions for CRYPTO '95. Equally the Board of Directors of the IACR is currently looking at solutions to address this problem for later conferences.

Thus the Program Committee of EUROCRYPT '95 received 113 submissions among which one was withdrawn by the author and one by the Program Chair for double submission. The editors would like to thank everyone who submitted a paper. The success of a conference depends ultimately upon the quality of the contributions. EUROCRYPT and CRYPTO have been and remain leading conferences in cryptology due to the high quality of the submissions.

Each paper was submitted for evaluation and comments to at least 4 members of the Program Committee. The process was anonymous, as it has been since 1989. The Program Committee has selected 33 papers among the 111 remaining submissions, i.e., slightly less than one third.

The rule, introduced in 1991, whereby a member of the Program Committee can be the author of at most one accepted paper, has been respected. Moreover, a new rule states that the status of Program Chair is not compatible with that of author.

The Program Chair is very grateful to all the members of the Program Committee for their hard work. It was a pleasure working with all of them.

Several experts helped the Program Committee members in reaching their decisions. In the name of the Program Committee, the Program Chair would also like to express here his appreciation for their efforts and their expertises.

The editors thank the authors for providing them in due time with the final versions of their papers. The availability of the proceedings at the conference is a significant progress, appreciated by the editors and also, by each participant.

The Author Index at the end of this book consists of 60 names. We know the date of birth of 30 peoples in this list: 7 are in their forties; 11 in their thirties; 12 in their twenties, four of them being only 24 years old! The youngest one will be 24 on the last day of the conference. The significant percentage of young authors is an encouraging sign of vitality of the IACR conferences.

Rump Session. The rump session is now an established tradition at IACR conferences. It aims at presenting the most recent results and at establishing the constestation of results presented in the other sessions. The publication of the proceedings at the conference seriously reduces the possibility of publishing the rump talks in the book. However, one contestation has been presented in due time and the corresponding rump talk is provided as the last paper of this book. As long as fair play is respected, such a contestation is another proof of the vitality of the IACR conferences. Of course, each author bears the full responsibility for his or her paper.

Special Session. In the program, a special session is dedicated to the introduction of arithmetic co-processors in the security self-programmable one-chip microcomputers (SPOMs), such as those used in smart cards. Allowing an efficient use of PK and ZK techniques, such arithmetic co-processors will deeply modify the use of smart cards in their various applications.

With the agreement of the Program Committee, the Program Chair set up a Special Committee chaired by Pascal Chour (AQL) and Marc Girault (SEPT). With the help of Guy Monnier (SGS Thomson) and David Arditti (CNET-Paris), the Special Committee has done an admirable job in orienting and focusing the preparation of the three invited talks of the special session and in organizing a corresponding illustrative exhibition.

David Naccache (Gemplus), Michel Ugon (Bull CP8) and Peter Landrock (Cryptomathic) have agreed to draft and to talk respectively on the following three aspects: hardware (architectural principles, trade-offs, performances, provisional calendars of the silicon founders); software (possible security mechanisms for functional aspects, such as digital signature, entity authentication, key management, file management, card issuing); applications (estimated consequences in major applications such as banking, telephone, television, health care, network security, electronic purse, transportation ...). A copy of the three talks is available for each participant as a special pre-publication.

The subject is particularly hot if we consider the major work of Europay International, MasterCard International and Visa International in drafting the so-called EMV specifications. The goal of the three organizations is a general worldwide use of SPOMs in credit cards. The present production of SPOMs for smart cards is about 30 million pieces per year, approximately one half of which are for banking purposes. The needs of the banks which are members of the three international organizations are evaluated around 300 million pieces per year.

Ten years ago, EUROCRYPT '84 held a special session on smart cards; at that time, we were at the very beginning of a general French development with the publication of specifications, in January 1984, by the GIE des Cartes Bancaires, the French interbank association; today, we are on the verge of a general worldwide development with the publication of the EMV specifications.

However the EMV phenomena should not hide all the other emerging applications. Let us quote Gustavus J. Simmons: "Smart cards *will put a sophisticated information-integrity device in the wallet or purse of practically every person in the industrialized world, and will therefore probably be the most extensive application ever made of cryptographic schemes*" (Preface of *Contemporary Cryptology, The Science of Information Integrity*, IEEE Press, 1992).

Louis Claude Guillou, Program Chair

CCETT, Cesson Sévigné, France

Jean-Jacques Quisquater, Co-Editor

March 1995

EUROCRYPT '95

EUROCRYPT '95
Saint-Malo, France
May 21–25, 1995

Sponsored by the

International Association for Cryptologic Research (IACR)

in cooperation with the

Centre Commun d'Études de Télévision et Télécommunications (CCETT)

General Chair

Françoise Scarabin, CCETT, France

Program Chair

Louis C. Guillou, CCETT, France

Program Committee

Mihir Bellare .... IBM T. J. Watson Research Center, USA
Johannes Buchmann .... U. Saarland, Germany
Mike Burmester .... Royal Holloway, U. London, UK
Paul Camion .... INRIA, France
Donald Davies .... Fair Winds, UK
Amos Fiat .... U. Tel Aviv-ARL, Israel
Hideki Imai .... U. Tokyo, Japan
Lars R. Knudsen .... U. Aarhus, Denmark
Ueli Maurer .... ETH, Switzerland
Birgit Pfitzmann .... U. Hildesheim, Germany
Jean-Jacques Quisquater .... UCL-Math RiZK, Belgium
Ronald L. Rivest .... MIT, USA
Jacques Stern .... ENS, France
Douglas Stinson .... U. Nebraska, USA
Moti Yung .... IBM T. J. Watson Research Center, USA
Gideon Yuval .... Microsoft, USA

CONTENTS

Cryptanalysis

Signatures

Number Theory

Protocol Aspects

Secret Sharing

Electronic Cash

Shift Registers and Boolean Functions

Authentication Codes

New Schemes

Complexity Aspects

Implementation Aspects

Rump Session

Attacking the Chor–Rivest Cryptosystem by Improved Lattice Reduction

C.P. Schnorr and H.H. Hörner
Johann Wolfgang Goethe–Universität Frankfurt
Fachbereich Mathematik/Informatik
Postfach 111932, D–60054 Frankfurt a.M., Germany

Abstract. We introduce algorithms for lattice basis reduction that are improvements of the famous L^3-algorithm. If a random L^3-reduced lattice basis $b_1, \ldots, b_n$ is given such that the vector of reduced Gram–Schmidt coefficients $(\{\mu_{i,j}\}\ 1 \leq j < i \leq n)$ is uniformly distributed in $[0, 1)^{\binom{n}{2}}$, then the pruned enumeration finds with positive probability a shortest lattice vector. We demonstrate the power of these algorithms by solving random subset sum problems of arbitrary density with 74 and 82 many weights, by breaking the Chor–Rivest cryptoscheme in dimensions 103 and 151 and by breaking Damgård's hash function.

1 Introduction and Summary

We address the challenging problem whether it is possible to find, for a given integer lattice basis $b_1, \ldots, b_n \in \mathbb{Z}^m$, in polynomial time a nonzero lattice vector of length $n^{O(1)}\lambda_1$, where λ_1 is the minimal length of nonzero lattice vectors. The L^3-algorithm of Lenstra,Lenstra, Lovász [LLL82] finds in polynomial time a lattice vector of length $2^{\frac{n}{2}}\lambda_1$. Schnorr [S87, S94] has extended this algorithm from block size $\beta = 2$ to arbitrary block sizes $2 \leq \beta \leq n$. Roughly speaking, this extension goes as follows. Whereas the L^3-algorithm iteratively swaps two consecutive basis vectors b_i, b_{i+1} if this decreases the length of $\widehat{b}_i$, the orthogonal projection of b_i in $\text{span}(b_1, \ldots, b_{i-1})^\perp$, block reduction with block size β iteratively transforms blocks $b_i, b_{i+1}, \ldots, b_{i+\beta-1}$ of β consecutive basis vectors as to minimize $\widehat{b}_i$. The first vector of a block reduced basis satisfies $\|b_1\| \leq \gamma_\beta^{\frac{n-1}{\beta-1}}\lambda_1$, where $\gamma_\beta \sim \frac{\beta}{\pi e}$ is the Hermite constant of dimension β. For an implementation of block reduction, see the algorithm BKZ of [SE94]. With block size $\beta = 20$ it is only 10 times slower than L^3-reduction but for large block sizes β the delay factor is about $\beta^{O(\beta)}$. This delay factor is the time to construct a shortest vector $\widehat{b}_i$ for a block of size β using complete enumeration of all short lattice vectors. A shortest vector of the entire lattice can be found by the algorithm of Kannan [KA87] in exponential time $n^{O(n)}$.

In this paper we present and analyse a new rule for pruning the enumeration of short lattice vectors. This pruning very likely finds a shortest lattice vector, and is exponentially faster than complete enumeration. It is based on the Gaussian volume heuristic that estimates the number of points of lattice L in nice subsets $S \subset \text{span}(L)$ as $\text{vol}(S)/\det L$. If a random L^3-reduced lattice basis $b_1, \ldots, b_n$ is given such that the vector of reduced Gram–Schmidt coefficients $(\{\mu_{i,j}\}\ 1 \leq j < i \leq n)$ is uniformly distributed in $[0, 1)^{\binom{n}{2}}$, then the pruned

L.C. Guillou and J.-J. Quisquater (Eds.): Advances in Cryptology - EUROCRYPT '95, LNCS 921, pp. 1-12, 1995.

enumeration finds with positive probability a shortest lattice vector. We let $\{r\}$ denote the residue modulo 1 of the real number r in the interval $[0,1)$.

Pruning the enumeration by the Gaussian volume heuristic is more powerful and more flexible than the previous pruning rule of [SE94]. We combine the new pruning with the block reduction algorithm BKZ of [SE94]. This pruned block reduction is the most powerful lattice reduction algorithm so far. It solves almost all subset sum problems of dimension 74 and 82 for all densities, it breaks the Chor–Rivest cryptosystem in dimensions 103 and 151, and it easily breaks Damgård's knapsack hash function [DA89]. Our experiments raise new hope that almost shortest lattice vectors can be found in polynomial time.

Lagarias and Odlyzko [LO85] have been the first to solve subset sum problems by lattice reduction. Their attack on subset sum problems of low density was improved by [RK88]. Since then the main progress came from block reduction [SE94], [S87], [S94] and by introducing a superior lattice basis [CJLOSS92]. Kaib and Ritter [KR94] propose an alternative approach based on lattice reduction in the l_∞-norm.

2 Basic concepts for efficient lattice reduction

Let $\mathbb{R}^n$ be the m-dimensional real vector space with ordinary inner product $\langle\,,\,\rangle$ and Euclidean length $\|y\| = \langle y,y\rangle^{1/2}$. A discrete, additive subgroup $L \subset \mathbb{R}^m$ is called a *lattice*. Every lattice is generated by some set of linearly independent vectors $b_1,\ldots,b_n \in L$, called a *basis* of L, $L = \{t_1 b_1 + \cdots + t_n b_n \mid t_1,\ldots,t_n \in \mathbb{Z}\}$. Let $L(b_1,\ldots,b_n)$ denote the lattice with basis $b_1,\ldots,b_n$. Its *rank* or *dimension* is n and its *determinant* is $\det L = \det[\langle b_i, b_j\rangle_{1\leq i,j\leq n}]^{1/2}$.

With an ordered lattice basis $b_1,\ldots,b_n \in \mathbb{R}^m$ we associate the Gram-Schmidt orthogonalisation $\widehat{b}_1,\ldots,\widehat{b}_n \in \mathbb{R}^m$ which can be computed together with the Gram-Schmidt coefficients $\mu_{i,j} = \langle b_i,\widehat{b}_j\rangle/\langle\widehat{b}_j,\widehat{b}_j\rangle$ by the recursion $\widehat{b}_1 = b_1$, $\widehat{b}_i = b_i - \sum_{j=1}^{i-1}\mu_{i,j}\widehat{b}_j$ for $i = 2,\ldots,n$. We let π_i denote the orthogonal projection $\pi_i : \mathbb{R}^m \to \operatorname{span}(b_1,\ldots,b_{i-1})^\perp$ for $i = 1,\ldots,n$, $\pi_i(b_j) = \sum_{s=i}^{j}\mu_{s,j}\widehat{b}_s$. Then $\pi_i(L)$ is a lattice of rank $n - i + 1$.

An ordered basis $b_1,\ldots,b_n \in \mathbb{R}^m$ is L^3-*reduced*, according to A.K. Lenstra, H.W. Lenstra and L. Lovász [LLL82], with $\delta \in [1/4,1)$ if (1) and (2) hold:

(1) $$|\mu_{i,j}| \leq 1/2 \quad \text{for} \quad 1 \leq j < i \leq n$$

(2) $$\delta \cdot \|\widehat{b}_{k-1}\|^2 \leq \|\widehat{b}_k + \mu_{k,k-1}\widehat{b}_{k-1}\|^2 \quad \text{for} \quad k = 2,\ldots,n.$$

A basis satisfying (1) is called *size-reduced*. The L^3-algorithm of Lovász [LLL82] transforms an integer lattice basis in polynomial time into an L^3-reduced basis of the same lattice. Schnorr, Euchner [SE94] propose a floating point version L^3FP of the L^3-algorithm. This algorithm is used whenever we apply L^3-reduction.

A lattice basis $b_1,\ldots,b_n$ is *block reduced* with block size β if it is size reduced and if $\widehat{b}_i$, for $i = 1,\ldots,n$, is the shortest nonzero vector of the lattice $\pi_i L(b_i,\ldots,b_{\min(i+\beta-1,n)})$. Block reduction has been analysed in [S87], [S94].

We consider the following function c_t with integer entries $u_t, \ldots, u_n$

$$c_t(u_t, \ldots, u_n) := \|\pi_t \sum_{i=t}^{n} u_i b_i\|^2 = \sum_{j=t}^{n} \left(\sum_{i=j}^{n} u_i \mu_{i,j} \right)^2 \|\widehat{b}_j\|^2 \text{ for } t = 1, \ldots, n.$$

We present the core of the procedure ENUM of [SE94] that generates a shortest lattice vector by complete enumeration in depth first order.

Algorithm ENUM
INPUT $\|\widehat{b}_i\|^2$, $\mu_{i,t}$ for $1 \leq t \leq i \leq n$.
OUTPUT a minimal nonzero place $(u_1, \ldots, u_n)$ and a minimal value $\bar{c}_1$ for the function c_1.

1. FOR $i = 1, \ldots, n$ DO $\widetilde{c}_i := u_i := \widetilde{u}_i := y_i := 0$
   $\widetilde{u}_1 := u_1 := 1$, $t := 1$, $\bar{c}_1 := \widetilde{c}_1 := \|\widehat{b}_1\|^2$.
   (we always have $\widetilde{c}_t = c_t(\widetilde{u}_t, \ldots, \widetilde{u}_n)$, $\bar{c}_1$ is the current minimum of c_1)
2. WHILE $t \leq n$
   $\widetilde{c}_t := \widetilde{c}_{t+1} + (y_t + \widetilde{u}_t)^2 \|\widehat{b}_t\|^2$
   IF $\widetilde{c}_t < \bar{c}_1$
   THEN IF $t > 1$
         THEN $t := t - 1$, $y_t := \sum_{i=t+1}^{t_{max}} \widetilde{u}_i \mu_{i,t}$, $\widetilde{u}_t := \lceil -y_t \rfloor$
         ELSE $\bar{c}_1 := \widetilde{c}_1$, $u_i := \widetilde{u}_i$ for $i = 1, \ldots, n$
   ELSE $t := t + 1$
   $\widetilde{u}_t := \begin{cases} \widetilde{u}_t + 1 & \text{if } t = t_{max} \\ \text{next}(\widetilde{u}_t, -y_t) & \text{otherwise} \end{cases}$
   END while

Here $\lceil r \rfloor \stackrel{def}{=} \lceil r - 0.5 \rceil$, t_{max} is the maximal previous value of t. We define $a' = \text{next}(a, r)$ to be the integer which is, next to $a \in \mathbb{Z}$, nearest to $r \in \mathbb{R}$. We have $|a - r| \leq |a' - r| \leq |a - r| + 1$, $sign(a' - r) \neq sign(a - r)$, $|a - r| = |a' - r| \Rightarrow a < r < a'$.

Correctness. The algorithm ENUM enumerates in depth first order all nonzero integer vectors $(\widetilde{u}_t, \ldots, \widetilde{u}_n)$ for $t = 1, \ldots, n$ that satisfy $c_t(\widetilde{u}_t, \ldots, \widetilde{u}_n) < \bar{c}_1$ where $\bar{c}_1$ is the actual minimal value for the function c_1. All enumerated vectors satisfy $\widetilde{u}_i > 0$ for the largest i with $\widetilde{u}_i \neq 0$. For fixed $\widetilde{u}_{t+1}, \ldots, \widetilde{u}_n$, the sequence of values $\widetilde{u}_t$, generated by iterating the function $\text{next}(*, -y_t)$, makes the sequence $c_t(\widetilde{u}_t, \ldots, \widetilde{u}_n)$ non decreasing. Therefore, if the test $\widetilde{c}_t < \bar{c}_1$ fails for the current vector $(\widetilde{u}_t, \ldots, \widetilde{u}_n)$, the subsequent increment of stage t has the effect to *discard* all vectors $(u, \widetilde{u}_{t+1}, \ldots, \widetilde{u}_n)$ where $\widetilde{u}_t$ preceeds u in the iteration of $\text{next}(*, -y_t)$. The discarded vectors can not lead to the minimum of the function c_1.

3 Pruning the enumeration

We prune the enumeration of vectors $(\widetilde{u}_t, \ldots, \widetilde{u}_n)$ in ENUM by tightening up the test "IF $\widetilde{c}_t < \bar{c}_1$". We cut off the depth first search at $(\widetilde{u}_t, \ldots, \widetilde{u}_n)$ if the probability that $(\widetilde{u}_t, \ldots, \widetilde{u}_n)$ can be completed as to satisfy $c_1(\widetilde{u}_1, \ldots, \widetilde{u}_n) < \bar{c}_1$ is less than a chosen threshold 2^{-p}.

The Gaussian volume heuristic. A general principle, dating back to Gauss, estimates the number of points of lattice L in nice subsets $S \subset \text{span}(L)$ as $\text{vol}(S)/\det L$.

How to apply it. Suppose we have chosen integers $\widetilde{u}_t, \ldots, \widetilde{u}_n$ and we search for $\widetilde{u}_1, \ldots, \widetilde{u}_{t-1}$ as to satisfy $c_1(\widetilde{u}_1, \ldots, \widetilde{u}_n) < \bar{c}_1$. We let $\bar{L}$ denote the lattice $\bar{L} = L(b_1, \ldots, b_{t-1})$. So we want to add to the given lattice vector $b = \sum_{i=t}^{n} \widetilde{u}_i b_i$ a vector $\bar{b} = \sum_{i=1}^{t-1} \widetilde{u}_i b_i$ in $\bar{L}$ as to satisfy $\|b + \bar{b}\|^2 < \bar{c}_1$. We decompose b into orthogonal parts $b = y - z$ with $z = -\sum_{j=1}^{t-1} \sum_{i=t}^{n} \widetilde{u}_i \mu_{i,j} \widehat{b}_j \in \text{span}(\bar{L}), y \in \text{span}(\bar{L})^{\perp}$, $\widetilde{c}_t = \|y\|^2$. This means, we search for a point in

$$(b + \bar{L}) \cap S(\sqrt{\bar{c}_1 - \widetilde{c}_t}, y) = \bar{L} \cap S(\sqrt{\bar{c}_1 - \widetilde{c}_t}, z)$$

where $S(r, y)$ is the $(t-1)$-dimensional sphere with radius r and center y in $y + \text{span}(\bar{L})$. Here the equality holds since $z = y - b$. Now we apply the volume heuristic to the lattice $\bar{L}$ and the sphere $S(\sqrt{\bar{c}_1 - \widetilde{c}_t}, z) \subset \text{span}(\bar{L})$. Hence the expected number of vectors $(\widetilde{u}_1, \ldots, \widetilde{u}_{t-1}) \in \mathbb{Z}^{t-1}$ satisfying $c_1(\widetilde{u}_1, \ldots, \widetilde{u}_n) \leq \bar{c}_1$ is $\text{vol} \, S(\sqrt{\bar{c}_1 - \widetilde{c}_t}, z)/\det \bar{L}$. We propose to cut off the enumeration of $(\widetilde{u}_1, \ldots, \widetilde{u}_{t-1})$ if this ratio is less than 2^{-p} for a fixed chosen p.

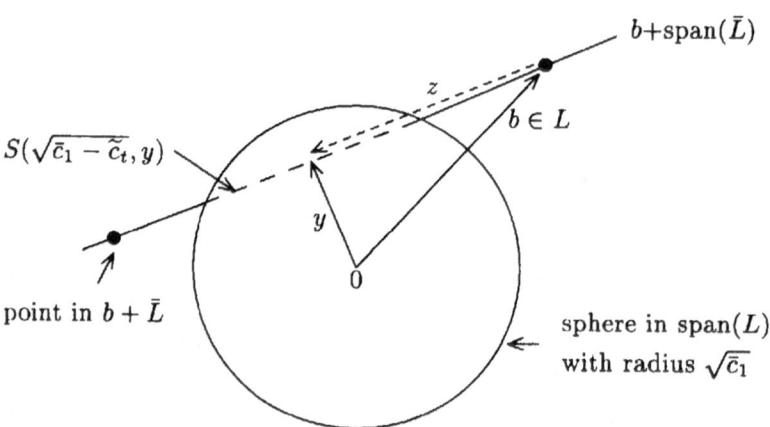

Figure: the volume heuristic

GAUSS–ENUM. We replace in ENUM the condition "IF $\tilde{c}_t < \bar{c}_1$" by "IF vol $S(\sqrt{\bar{c}_1 - \tilde{\tilde{c}}_t}, z)/\det \bar{L} < 2^{-p}$". We call the new procedure GAUSS–ENUM. The parameter p controls the pruning. Large values p correspond to weak pruning, $p = \infty$ corresponds to complete enumeration (no pruning). The inequality vol $S(\sqrt{\bar{c}_1 - \tilde{\tilde{c}}_t}, z)/\det \bar{L} < 2^{-p}$ is equivalent to $\tilde{c}_t < \bar{c}_1 - \eta$ where

$$\eta = \frac{1}{\pi}\left(\frac{t-1}{2}\right)!^{\frac{2}{t-1}}\left(2^{-p}\prod_{i=1}^{t-1}\|\hat{b}_i\|\right)^{\frac{2}{t-1}}.$$

If GAUSS–ENUM cuts off the depth first search at $(\tilde{u}_t, \ldots, \tilde{u}_n)$ the probability, that $(\tilde{u}_t, \ldots, \tilde{u}_n)$ can be completed as to satisfy $c_1(\tilde{u}_1, \ldots, \tilde{u}_n) < \bar{c}_1$, is at most 2^{-p}. In the analysis of GAUSS ENUM we disregard that GAUSS–ENUM discards, in addition to the vectors $(\tilde{u}_1, \ldots, \tilde{u}_n)$, also the vectors $(\tilde{u}_1, \ldots, \tilde{u}_{t-1}, u, \ldots, \tilde{u}_n)$ where $\tilde{u}_t$ precedes u in the iteration of next$(*, -y_t)$. This can be repaired by a slight change in GAUSS–ENUM. However this yields a reduction algorithm that is less efficient in practice.

Justification of the volume heuristic. The Gaussian principle does not hold in general. MAZO and ODLYZKO [MO90] show that it fails even in the case of spheres and the lattice $L = \mathbb{Z}^n$ for particular choices of the center z. However the principle holds if the center of the sphere is "uniformly distributed *(u.d.)* modulo the lattice".

Definition. *For a lattice L with basis $b_1, \ldots, b_n$ a probability distribution of points $\sum_{i=1}^n t_i b_i$ in $\operatorname{span}(L)$ is called u.d. modulo L if the reduced vector $(\{t_i\}\ i = 1, \ldots, n)$ is u.d. in $[0, 1)^n$.*

This notion does not depend on the choice of the basis. If $b_1, \ldots, b_n$ and $\bar{b}_1, \ldots, \bar{b}_n$ are two bases of lattice L there is a matrix $U \in \operatorname{GL}_n(\mathbb{Z})$ satisfying $[\bar{b}_1, \ldots, \bar{b}_n] = [b_1, \ldots, b_n]\, U$. Since $|\det U| = 1$ the linear transformation by U transforms the uniform distribution on $\sum_{i=1}^n b_i\, [0, 1)$ into the uniform distribution on $\sum_{i=1}^n \bar{b}_i\, [0, 1)$. Alternatively we can express the uniformity modulo L in terms of the Gram-Schmidt orthogonalization $\hat{b}_1, \ldots, \hat{b}_n$ associated with the basis $b_1, \ldots, b_n$. The vector $\sum_{i=1}^n t_i'\, \hat{b}_i$ in $\operatorname{span}(L)$ is u.d. modulo L if and only if the vector $(\{t_i'\}\ i = 1, \ldots, n)$ is u.d. in $[0, 1)^n$.

Lemma 1. *Let L be a lattice and $S(r, z) \subset \operatorname{span}(L)$ the sphere with fixed radius r and random center z that is u.d. modulo L. Then $E_z \#(S(r, z) \cap L) =$ vol $S(r, z)/\det L$ holds for the expectation E_z.*

Proof. For two points $z, \bar{z} \in \operatorname{span}(L)$ that coincide modulo L, i.e. $z = \bar{z} \bmod L$, we have $\#(S(r, z) \cap L) = \#(S(r, \bar{z}) \cap L)$. The average number of lattice points in $S(r, z)$ is the average number of lattice points per volume part vol $S(r, z)$. Hence the expected value of $\#(S(r, z) \cap L)$ is vol $S(r, z)/\det L$. $\square$

We apply Lemma 1 to the situation in GAUSS-ENUM with $\widetilde{u}_t, \ldots, \widetilde{u}_n$ being fixed, $\widetilde{c}_t = c_t(\widetilde{u}_t, \ldots, \widetilde{u}_n)$, $\bar{c}_1 > \widetilde{c}_t$ and a lattice point of $\bar{L}$ is searched in the sphere $S(\sqrt{\bar{c}_1 - \widetilde{c}_t}, z)$ with center $z = -\sum_{j=1}^{t-1} \sum_{i=t}^{n} \widetilde{u}_i \mu_{i,j} \widehat{b}_j$.

Theorem 2. *If the vector* $(\{\mu_{i,j}\}\ 1 \le j < i \le n)$ *is u.d. in* $[0,1)^{\binom{n}{2}}$ *then for every fixed nonzero* $(\widetilde{u}_t, \ldots, \widetilde{u}_n) \in \mathbb{Z}^{n-t+1}$ *the center z is u.d. modulo the lattice* $\bar{L} = L(b_1, \ldots, b_{t-1})$. *Moreover*

$$E_z \ \#[\ (\widetilde{u}_1, \ldots, \widetilde{u}_{t-1}) \in \mathbb{Z}^{t-1} : c_1(\widetilde{u}_1, \ldots, \widetilde{u}_n) \le \bar{c}_1] = \text{vol } S(\sqrt{\bar{c}_1 - \widetilde{c}_t}, z)/\det \bar{L}.$$

Proof. We can assume that $\widetilde{u}_n \ne 0$ since otherwise we can decrease n. We see that the vectors $(\{\widetilde{u}_n \mu_{n,j}\}\ j = 1, \ldots, t-1)$ and $(\{\sum_{i=t}^{n} \widetilde{u}_i \mu_{i,j}\}\ j = 1, \ldots, t-1)$ are u.d., in $[0,1)^{t-1}$. This shows that z is u.d. modulo $\bar{L}$. Since

$$\#[(\widetilde{u}_1, \ldots, \widetilde{u}_{t-1}) \in \mathbb{Z}^{t-1} : c_1(\widetilde{u}_1, \ldots, \widetilde{u}_n) \le \bar{c}_1] = \#(S(\sqrt{\bar{c}_1 - \widetilde{c}_t}, z) \cap \bar{L})$$

the expression for the expectation E_z follows from Lemma 1. $\square$

Success rate of GAUSS–ENUM. Suppose a distribution of L^3–reduced lattice bases so that the vector $(\{\mu_{i,j}\}\ 1 \le j < i \le n)$ is u.d. in $[0,1)^{\binom{n}{2}}$ and let $p > \log_2 n$. Whenever the depth first search is cut off at a fixed vector $(\widetilde{u}_t, \ldots, \widetilde{u}_n) \in \mathbb{Z}^{n-t+1}$ then, by theorem 2, the event that a lattice vector shorter than $\sqrt{\bar{c}_1}$ gets lost, has probability at most 2^{-p}. Therefore the probability of missing the shortest lattice vector is at most 2^{-p} times the average number of cutoffs. While the number of cutoffs can be arbitrarily large for badly reduced bases statistical experiments show that, for random L^3–reduced basis, the average number of cutoffs is proportional to $c_{p,n} 2^p$ where the factor $c_{p,n}$ decreases to 0 as p increases. E.g. for $n < 30$ and $p = 7$ the probability of success is at least 0.1 .

Expected time bound for GAUSS–ENUM. Using even more heuristic arguments we can show for $p > \log_2 n$: Given a random basis $b_1, \ldots, b_n$ and $\bar{c}_1 \le \|b_1\|^2$, GAUSS-ENUM performs on the average only $O(n^2 2^p)$ arithmetic steps to find a lattice vector b with $\|b\|^2 < \bar{c}_1$, respectively to terminate if such b does not exist.

4 Solving subset sum problems

Given positive integers $a_1, \ldots, a_n, s$ we wish to solve the equation $\sum_{i=1}^{n} a_i x_i = s$ with $x_1, \ldots, x_n \in \{0,1\}$. We assume that we are also given $q = \sum_{i=1}^{n} x_i$, the number of 1–entries of the solution. So we search for a $\{0,1\}$–solution $(x_1, \ldots, x_n)$ of the two equations $\sum_{i=1}^{n} a_i x_i = s$, $\sum_{i=1}^{n} x_i = q$. Following [CJLOSS92] we associate to this problem the following lattice basis $b_0, \ldots, b_n \in \mathbb{Z}^{n+3}$

$$
\begin{aligned}
b_0 &= (\ 1,\, q,\, q,\, \ldots\, q,\ \ n^2 s,\ \ n^2 q\ ) \\
b_1 &= (\ 0,\, n,\, 0,\, \ldots 0,\ \ n^2 a_1,\ \ n^2\ ) \\
b_2 &= (\ 0,\, 0,\, n,\, \ldots 0,\ \ n^2 a_2,\ \ n^2\ ) \\
\vdots\ \ &= \ \ \vdots\ \ \vdots\ \ \vdots\ \ \ddots\ \vdots\ \ \ \ \vdots\ \ \ \ \ \vdots \\
b_n &= (\ 0,\, 0,\, 0,\, \ldots n,\ n^2 a_n,\ \ n^2\ )
\end{aligned}
$$

(3)

According to [CJLOSS92] the shortest vector z of the lattice $L(b_0, \ldots, b_n)$ solves via (4) almost all subset sum problems of density less then 0.9408, where the density is $n / \max_i \log_2 a_i$. Even beyond this density threshold, solutions of the problems in this paper are associated with very short lattice vectors.

With a $\{0,1\}$-solution $x = (x_1, \ldots, x_n)$ of the subset sum problem we associate lattice vectors $z = (z_0, \ldots, z_{n+2}) = \pm (-b_0 + \sum_{i=1}^{n} x_i b_i)$ that satisfy $|z_0| = 1, z_{n+1} = z_{n+2} = 0, z_i/z_0 \in \{q, q-n\}$ for $i = 1, \ldots, n$. Conversely every such lattice vector $z = (z_0, \ldots, z_{n+2})$ induces a subset sum solution

(4) $x_i := [\text{ IF } z_i/z_0 = q - n \text{ THEN } 1 \text{ ELSE } 0]$ for $i = 1, \ldots, n$

We have tested the following algorithm for general subset sum problems with $n = 74$ and $n = 82$ many weights and for the Chor–Rivest subset sum problem with $n = 103$.

Algorithm PRUNED SUBSET SUM
INPUT lattice basis $b_0, \ldots, b_n \in \mathbb{Z}^{n+3}$ as in (3).
Perform four successive stages of reduction :

1. L^3–reduction.

2. block reduction with block size 20.

3. pruned block reduction with block size 50 and $p = 10$.

4. pruned block reduction with block size 70 and $p = 12$.

Algorithmic details. 1. For L^3–reduction we use the algorithm L^3FP of [SE94]. We set $\delta = 0.99$, we apply the deep insertion rule of [SE94] for the first basis vector.

2. Block reduction is done by the algorithm BKZ of [SE94] with $\delta = 0.99$ resulting in a basis $b_0, \ldots, b_n$ satisfying for $i = 0, \ldots, n$

(5) $0.99 \|\hat{b}_i\| \leq \|\pi_i(b)\|$ for all nonzero $b \in L(b_i, \ldots, b_{\min(i+\beta-1,n)})$.

3. Pruned block reduction is done the same way as block reduction except that we use instead of algorithm ENUM the algorithm GAUSS–ENUM with an appropriate pruning parameter p. The resulting basis may occasionally fail the inequalities (5).

4. Test for solution and early termination. Subsequent to every size–reduction of a basis vector b_j it is always tested whether b_j solves the subset sum problem, i.e. whether (4) induces a solution x for $z = b_j$. Also for each stage of the reduction, the vectors of the reduced basis are tested for solution. The algorithm terminates as soon as a solution has been found.

5. *Reduction to the sublattice* $\widetilde{L} = \{(z_0, \ldots, z_{n+2}) \in L(b_0, \ldots, b_n) : z_{n+1} = z_{n+2} = 0\}$. After the L^3–reduction in stage 1 we construct a basis of the lattice $\widetilde{L}$ and we continue the reduction process with this basis. Working with the lattice $\widetilde{L}$ simplifies subsequent reductions since $rank(\widetilde{L}) = rank(L) - 2$. To construct a basis of $\widetilde{L}$ we linearly transform the L^3–reduced basis $b_0, \ldots, b_n$ of L so that $b_{i,j} = 0$ holds for $i = 1, \ldots, n - 2$ and $j = n + 1, n + 2$. Then we eliminate the vectors b_{n-1}, b_n from the basis and we remove from the vectors b_i $i = 0, \ldots, n-2$ the last two coordinates $b_{i,n+1}, b_{i,n+2}$. Upon entry of stage 2 we randomly permute the basis so that it starts with the vectors b_i that have a nonzero coordinate $b_{i,0}$. This enhances the generation of short lattice vectors z which induce via (4) a subset sum solution.

5 Attacks on the Chor–Rivest cryptosystem

Chor, Rivest present a public key encryption method for which deciphering has the form of a subset sum problem of high density, for details see [CR88]. Chor, Rivest propose examples of their scheme with $n = 197$ and $n = 211$ many weights. For testing possible attacks they also designed a small example with $n = 103$ many weights and subset sum problems of density 1.271. The Lagarias–Odlyzko method which is based on L^3–reduction completely failed for the $n = 103$ subset sum problems.

Interestingly, block reduction with pruned enumeration solves the Chor–Rivest subset sum problems with $n = 103$ many weights in only 1.5 hours average time with 42% success rate. Thus the widespread believe that subset sum problems with density greater than 1 cannot be solved via lattice reduction is outright wrong. The Chor–Rivest scheme with $n = 103$ and density 1.271 is even less difficult than random subset sum problems with $n = 82$ and density 1.

Generation of the Chor–Rivest subset sum problems. We take the particular weights $a_1, \ldots, a_{103}$ of the example constructed by Chor, Rivest. We generate 50 random vectors $(x_1, \ldots, x_{103}) \in \{0,1\}^{103}$ so that $\sum_{i=1}^{103} x_i = 12$, and we set $s := \sum_{i=1}^{103} x_i a_i$. In the corresponding subset sum problem we are given $a_1, \ldots, a_{103}, s$ and have to solve the equations $\sum_{i=1}^{103} x_i a_i = s$, $\sum_{i=1}^{103} x_i = 12$ with $x_1, \ldots, x_{103} \in \{0,1\}$. (The number 12 arises from the particular construction of the weights a_i starting from the field $\mathbb{F} = GF(103^{12})$, a generator g for the group of units $\mathbb{F}^\star$, an element $t \in \mathbb{F}$ that is algebraic of degree 12 over $GF(103)$, a random permutation π in $Sym(n)$ and a random number d with $0 \le d < 103^{12} - 2$, and setting $a_i := \log_g(t + \pi(i)) + d$ for $i = 1, \ldots, 103$.) We solve these 50 subset sum problems by applying the algorithm PRUNED SUBSET SUM to the lattice basis (3) with $n = 103$, $q = 12$.

The first table shows, for each of the stages $i = 1, 2, 3, 4$, in column 4 the number of successes on stage i, in column 5 the number of successes up to stage i, in column 6 the average time (with respect to all 50 problems) of stage i, in column 7 the total time up to stage i and in column 8 the maximal time of stage i. The last column contains the total time for all 50 problems divided by the number of successes. All times are in minutes for a HP 715/50 workstation under HP–UX 9.05 .

stage	block size	p	# successes		time in minutes			total time per success
			on stage	up to stage	average	av. total	maximal	
1	2	∞	0	0	0.6	0.6	0.7	∞
2	20	∞	3	3	7.6	8.2	16.9	163.5
3	50	10	18	21	86.8	95.0	247.3	226.1
4	70	12	14	35	173.4	268.4	938.3	383.5

Stage 1 which performs L^3–reduction does not find any solution. This confirms the previous results of Odlyzko showing that L^3–reduction is too weak even if the CJLOSS basis (3) is used which is much stronger than the Lagarias-Odlyzko basis used in the experiments of Odlyzko.

Stage 4 by itself is quite inefficient. It takes a total of 619 minutes per success. This suggests to replace stage 4 by a repetition of stages 1,2,3 with a randomly permuted input basis. The next table shows the results for two repetitions of stages 1,2,3.

	# successes		time in minutes		total time per success
	in round	total	average	av. total	
stages 1,2,3	21	21	95.0	95.0	226.1
1. repetition	11	32	65.3	160.3	250.5
2. repetition	6	38	33.5	193.8	255.0

With two repetitions the success rate is 76% with an average time of 3.2 hours. It may be of interest that an alternative algorithm of Ritter, see [KR94], solves all $n = 103$ Chor-Rivest problems in about 7 hours maximal time.

Chor–Rivest subset sum problems with more weights. A limited number of first experiments have been carried out by H.H. Hörner in attacking a Chor-Rivest cryptosystem with $n = 151$ many weights and $q = 16$ [H94]. So far he could solve 5 out of 50 random problems with an average time of 195 hours for the solved problems.

6 Attacks on Damgård's knapsack hash function

In [DA89] a hash function h his proposed based on the subset sum problem. Choose random numbers $a_1, \ldots, a_{256}$ in the interval $[1, 2^{120} - 1]$ and hash a message m consisting of the bits $m_1, \ldots, m_{256}$ into the integer $h(m_1, \ldots, m_{256}) = \sum_{i=1}^{256} a_i m_i$.

To construct a collision for h it is sufficient to find a nonzero $\{\pm 1, 0\}$–solution $(x_1, \ldots, x_{256})$ of the equation $\sum_{i=1}^{256} a_i x_i = 0$. This yields messages m, m' with bits $m_i = \max\{0, x_i\}$, $m_i' = -\min\{x_i, 0\}$ for $i = 1, \ldots, 256$ satisfying $h(m) = h(m')$.

Following an analysis of Joux, Stern [JS94] collisions exist almost surely even for the restricted problem with 80 out of the 256 weights a_i. We construct nonzero $\{\pm 1, 0\}$–solutions of the equation $\sum_{i=1}^{100} a_i x_i = 0$. We associate to this problem the following lattice basis $b_1, \ldots, b_n \in \mathbb{Z}^{n+1}$ with $b_i = (0, \ldots, 1^{(i)}, \ldots, 0, na_i)$ for $i = 1, \ldots, n$ and $n = 100$.

A nonzero lattice vector $z = (z_1, \ldots, z_{n+1})$ yields a collision if $z_{n+1} = 0$ and $(z_1, \ldots, z_n) \in \{\pm 1, 0\}^n$. We apply to this basis a two–stage reduction consisting of an L^3–reduction and a single pruned block reduction with block size 50 and alternative p–values $8, 9, \ldots, 12$. We test after each size–reduction whether the reduced vector z yields a collision. (The more powerful reduction algorithm PRUNED SUBSET SUM is less efficient since the shortest lattice vector is most likely not in $\{\pm 1, 0\}^n$. This follows from the analysis in [JS94].)

Each row in the following table corresponds to 20 random vectors $(a_1, \ldots, a_{100}) \in [1, 2^{120} - 1]^{100}$. We report the number of successes, the average running time in minutes, the minimal and maximal size of the detected collision (the size of a collision $(x_1, \ldots, x_n) \in \{\pm 1, 0\}^n$ is $\#\{i : x_i \neq 0\}$), and the pruning parameter p.

block size	p	# successes	av. time in minutes	min size	max size
50	8	7	235.04	48	58
50	9	16	261.98	44	62
50	10	16	365.84	45	59
50	11	19	388.05	37	61
50	12	20	386.65	44	60

A first collision for Damgård's hash function has been constructed in [JG94] using pruned block reduction via the pruning of [SE94]. They report one success for ten problems. The new results demonstrate the superiority of pruning via the volume heuristic.

7 General subset sum problems

We report on solving random subset sum problems of arbitrary density in dimensions $n = 74$ and 82. The previously most powerful algorithm [SE94] could solve almost all problems in dimension $n = 66$ by combining block reduction with some sort of pruning. The new algorithm PRUNED SUBSETSUM prunes the enumeration of short lattice vectors by the volume heuristic. It solves for $n = 74, 82$ a substantial fraction of all random subset sum problems of arbitrary density.

In the following table, every row with entries n, b corresponds to 20 random input bases (3) that are generated as follows. Pick random integers $a_1, \ldots, a_n$ in the interval $[1, 2^b]$, pick a random subset $I \subset \{1, \ldots, n\}$ of size $n/2$ and put $s = \sum_{i \in I} a_i$. To solve the corresponding subset sum problem $\sum_{i=1}^n a_i x_i = s$ we apply the algorithm PRUNED SUBSET SUM to the lattice basis (3) with $q = n/2$. The numbers in columns S, S1, S2, S3, S4 denote the total number of successes, and the number of successes in stages 1, 2, 3, 4.

n	b	# successes					average time in minutes				
		S	S1	S2	S3	S4	stage 1	stage 2	stage 3	stage 4	total
74	26	20	19	1	0	0	0.08	0.00			0.08
74	34	20	5	13	2	0	0.13	0.23	0.05		0.40
74	42	19	1	4	13	1	0.15	1.05	0.62	0.07	1.88
74	50	17	0	0	11	6	0.20	1.43	4.08	2.75	8.47
74	58	15	0	0	4	11	0.23	2.12	10.98	14.67	28.00
74	66	12	0	0	6	6	0.28	2.72	21.82	27.67	52.48
74	74	12	0	0	6	6	0.32	3.68	27.28	31.72	63.00
74	82	20	0	0	19	1	0.38	5:70	11.42	0.37	17.87
74	90	20	0	8	12	0	0.45	4.07	2.17		6.68
74	98	20	0	15	5	0	0.55	3.00	0.38		3.93
82	34	20	2	16	2	0	0.17	0.37	0.05		0.06
82	42	20	0	7	13	0	0.20	1.30	0.88		2.38
82	50	18	0	1	13	4	0.27	1.77	4.78	1.32	8.13
82	58	8	0	0	4	4	0.28	2.85	10.25	26.65	40.05
82	66	5	0	0	0	5	0.33	3.20	30.25	65.93	99.73
82	74	4	0	0	1	3	0.37	3.73	60.67	171.37	236.13
82	82	5	0	0	1	4	0.45	4.73	104.87	172.53	282.06
82	90	14	0	0	9	5	0.53	6.00	60.95	73.72	141.20
82	98	20	0	0	17	3	0.61	7.55	33.90	7.65	49.72
82	106	20	0	3	17	0	0.68	9.87	9.28		19.83

PRUNED SUBSET SUM is remarkably efficient for densities less than 0.9408 where the shortest lattice vector most likely yields a solution, see lines $n = 74, b \geq 82$ and $n = 82, b \geq 90$. This gives new hope that shortest, or near shortest lattice vectors can be found in polynomial time.

Random subset sum problems with $n = 82$ and density 1 are harder than the Chor–Rivest scheme with $n = 103$ and density 1.271. Here stage 4 of the algorithm PRUNED SUBSET SUM is necessary for the generation of solutions. Only 5 out of 20 problems for $n = 82, b = 82$ are solved in 282 minutes. The Chor–Rivest problems are easier because the problem solution yields a shortest lattice vector with no further vector being nearly as short.

References

[CJLOSS92] M.J. Coster, A. Joux, B.A. LaMacchia, A.M. Odlyzko, C.P. Schnorr and J. Stern: Improved Low–Density Subset Sum Algorithms; comput. complexity 2, Birkhäuser–Verlag Basel (1992), 111–128.

[CR88] B. Chor and R.L. Rivest: A knapsack–type public key cryptosystem based on arithmetic in finite fields; IEEE Trans. Inform. Theory, vol IT–34 (1988), 901–909.

[DA89] I. B. Damgård: A Design Principle for Hash Functions; Advances in Cryptology, Proc. Crypto 89, Springer LNCS 435 (1990), 416–427.

[H94] H.H. Hörner: Verbesserte Gitterbasenreduktion; getestet am Chor–Rivest Kryptosystem und an allgemeinen Rucksack–Problemen. Diplomarbeit, Universität Frankfurt (August 1994).

[JG94] A. Joux and L. Granboulan: A Practical Attack against Knapsack based Hash Functions; Prodeedings EUROCRYPT'94, Springer LNCS (1994).

[JS94] A. Joux and J. Stern: Lattice Reduction: a Toolbox for the Cryptanalyst, TR DGA/CELAR, ENS (1994).

[KA87] R. Kannan: Minkowski's convex body theorem and integer programming; Math. Oper. Res. 12 (1987), 415–440.

[KR94] M. Kaib and H. Ritter: Block Reduction with Respect to Arbitrary Norms; TR U. Frankfurt (1994).

[LO85] J.C. Lagarias and A.M. Odlyzko: Solving low–density subset sum problems; J. Assoc. Comp. Mach. 32(1) (1985), 229–246.

[LLL82] A.K. Lenstra, H.W. Lenstra Jr. and L. Lovász: Factoring polynomials with rational coefficients; Math. Ann. 261 (1982), 515–534.

[MO90] J.E. Mazo and A.M. Odlyzko: Lattice Points in high–dimensional spheres; Monatsh. Math. 110 (1990), 47–61.

[RK88] S. Radziszowski and D. Kreher: Solving subset sum problems with the L^3 algorithm; J. Combin. Math. Combin. Comput. 3 (1988), 49–63.

[S87] C.P. Schnorr: A hierarchy of polynomial time lattice basis reduction algorithms; Theoretical Computer Science 53 (1987), 201–224.

[S94] C.P. Schnorr: Block reduced lattice bases and successive minima; Combinatorics, Probability and Computing 3 (1994), 507–522.

[SE94] C.P. Schnorr and M. Euchner: Lattice Basis Reduction: Improved Practical Algorithms and Solving Subset Sum Problems; Mathematical Programming 66 (1994), 181–199.

Convergence in Differential Distributions

Luke O'Connor[*,1,2]

[1] Distributed Systems Technology Centre, Australia
[2] Information Security Research Centre, QUT, Australia

Abstract. Differential cryptanalysis is a general attack based on the notion of differences. The success of the attack is derived from the probability of a *differential*. While it has been observed that the distribution of differentials can be modeled as a Markov chain, there have been few analyses that take advantage of this observation because of the prohibitive computations involved. In this paper we apply the Markov approach to the differentially 2-uniform mappings, and show that they converge exponentially fast with high probability.

1 Introduction

Differential cryptanalysis is a general attack based on the distribution of differences in a cipher [3, 2]. The notion of difference can be defined arbitrarily, but in this paper we will assume it to mean the XOR (exclusive-or) of two binary strings. The probability of an r-round differential $\Delta P, \Delta C_r$ is the probability that a pair of plaintexts of difference ΔP have a ciphertext difference of ΔC_r after r-rounds. A cipher is called *iterated* if there is a function $\mathbf{F}$, the round function, such that the cipher operates by applying $\mathbf{F}$ repeatedly. Lai, Massey and Murphy [10] have observed that it is possible in some ciphers to model the distribution of differentials in an iterated cipher as a homogeneous Markov chain $\mathbf{P}$ when the subkeys are assumed to be independent. The states of the chain correspond to the set of nonzero differences. Such ciphers are called *Markov ciphers*, examples of which include DES [14] and IDEA [9]. If $\mathbf{P}^{(r)} = [P_{ij}^{(r)}]$ is the rth power of $\mathbf{P}$, then the probability of the differential $\Delta P = i, \Delta C_r = j$ is given as $P_{ij}^{(r)}$. If it can be shown that all entries of $\mathbf{P}^{(r)}$ are tending to some small value ϵ as r becomes large, this is taken as *strong evidence* that product ciphers built from the round function $\mathbf{F}$ will be resistant to differential cryptanalysis.

A typical analysis of a Markov chain would then proceed to classify the states so as to determine the asymptotic behaviour of $\mathbf{P}^{(r)}$. State classification is usually performed by inspection but this is not possible when $\mathbf{P}$ is large, as is the case for DES and IDEA with 2^{128} entries each. However, some general properties of the $\mathbf{P}$ matrix suggest an approach to approximate its asymptotic

* The work reported in this paper has been funded in part by the Cooperative Research Centres program through the Department of the Prime Minister and Cabinet of Australia. Correspondence should be sent to DSTC, ITE Building, QUT GP, GPO Box 2434, Brisbane Q 4001, Australia. Email: oconnor@dstc.edu.au.

L.C. Guillou and J.-J. Quisquater (Eds.): Advances in Cryptology - EUROCRYPT '95, LNCS 921, pp. 13-23, 1995.
© Springer-Verlag Berlin Heidelberg 1995

behaviour. It is known [9] that **P** is doubly stochastic when **F** is bijective, and **P**$^{(r)}$ tends to the uniform distribution when **P** is *ergodic* (defined below). This means that if **P** could be shown to be ergodic, all entries of **P**$^{(r)}$ would be tending towards the value $1/2^n$ where n is the block size. We are then confronted with the following two problems: (a) demonstrate that **P** is ergodic, and (b) determine the rate at which **P** approaches the unifrom distribution if it is ergodic. Note that (a) will determine if the differences are distributed uniformly, while (b) will determine an appropriate number of rounds for the cipher. Hornauer, Stephan, and Wernsdorf [8] were able to prove that certain round functions **F** yield ergodic transition matrices **P** by examining the group of mappings that could be formed by iterating the round function. Results for more general chains are reported by O'Connor and Golić [13, 12] where a combinatorial approach is used to show that if **F** is selected uniformly then **P** is ergodic with probability tending to one. On the other hand, there are no known results related to the rate of convergence of specific chains. However, Lai [9] has performed experiments on 'mini' verions of IDEA (8-bit block length) and shown the convergence to be rapid.

1.1 Results

In this paper we examine round functions that are differentially 2-uniform [11], meaning that the XOR table entries for nonzero input differences are either zero or two. As we will show, the answer to (a) depends on the *density* of nonzero entries in **P**, while (b) depends on the *magnitude* of the nonzero entries in **P**, both of which are known when **F** is differentially 2-uniform. Our main result is then to show that transition matrices **P** derived from differentially 2-uniform mappings **F** are ergodic with high probability and are expected to converge exponentially fast to the uniform distribution. This is proven by bounding the second largest eigenvalue of **P**.

The paper proceeds as follows. In section 2 we review concepts related to finite Markov chains, and show that differentially 2-uniform mappings are highly likely to have ergodic transition matrices **P**. In section 3 we introduce the concept of a rapidly mixing Markov chain, and in section 3.1 show that **P** rapidly approaches the uniform distribution with high probability.

Throughout the paper we will use bold notation to refer to objects related to the round function **F**, such as **P** and **G**. As many definitions and concepts will apply to all Markov chains we will use $P = [p_{ij}]$ to denote an arbitrary N-state chain, referring to it generally using normal (not bold) notation.

2 Finite Markov Chains

General definitions of Markov chains can be found in Feller [6], but we will review some concepts briefly. A chain $P = [p_{ij}]$ is *ergodic* if it is finite, aperiodic and irreducible. A sufficient condition for aperiodicity of an N-state chain P is that $p_{ii} > 0$ for some i, $1 \leq i \leq N$, while P is irreducible if for all i, j there exists an r such that $p_{ij}^{(r)} > 0$, $1 \le i, j \le N$. If P is ergodic then there exists a unique distribution $\Pi = (\pi_1, \pi_2, \ldots, \pi_N)$ such that

$$\pi_j = \lim_{r \to \infty} p_{ij}^{(r)}. \tag{1}$$

The distribution Π is said to be the *limiting distribution* for P and is known to be the uniform distribution for P that are doubly stochastic.

Let $\mathbf{F} : Z_2^n \to Z_2^n$, a bijective mapping, be the round function of an r-round iterated cipher E, such that at round i, the round mapping is $C_{i+1} = \mathbf{F}(C_i + K_i)$ where C_i is the ciphertext at round i and K_i is the subkey at round i, $1 \le i \le r$. It can be verified that E is a Markov cipher when '+' denotes XOR and the K_i are assumed to be independent. The differential transition matrix $\mathbf{P} = [P_{ij}]$ is obtained from the XOR table of $\mathbf{F}$ as follows. For each input difference $\Delta X = i$ and output difference $\Delta Y = j$, $1 \le i, j \le 2^n - 1$, P_{ij} is defined as

$$P_{ij} = 2^{-n} \cdot \sum_{\substack{X, X' \in Z_2^n \\ \Delta X = X + X'}} [\mathbf{F}(X) + \mathbf{F}(X') = \Delta Y] \tag{2}$$

where $[\cdot]$ is a boolean predicate evaluating to 0 or 1. Then $\mathbf{P} = [P_{ij}]$ is an $N \times N$ matrix where $N = 2^n - 1$ since the degenerate cases where $i = 0$ or $j = 0$ are excluded. The transition matrix $\mathbf{P}$ is doubly stochastic as $\mathbf{F}$ is bijective [9].

Since $\mathbf{P}$ is clearly finite, to prove ergodicity we must demonstrate that the chain is aperiodic and irreducible. Note that $\mathbf{P}$ would be ergodic if all N^2 entries were nonzero since $P_{ii} > 0$ implies that $\mathbf{P}$ is aperiodic, and irreducibility follows trivially as $P_{ij} > 0$ for all i, j. We will argue that when the number of nonzero entries in $\mathbf{P}$ exceeds some bound $B < N^2$, then $\mathbf{P}$ is ergodic with high probability. In particular, using results from random graph theory, we will show that B is approximately $N \log N$.

2.1 Differentially uniform mappings

A mapping $\mathbf{F}$ is differentially δ-uniform [11] if each entry in the XOR table for $\mathbf{F}$ is at most δ. Since each entry in an XOR table is even, a differentially 2-uniform mapping $\mathbf{F}$ has an XOR table that consists entirely of zeros and twos, with exactly 2^{n-1} nonzero entries in each row of the table. Thus the XOR table for differentially 2-uniform mappings has the maximum number of nonzero entries for a bijective mapping, as does the corresponding transition matrix $\mathbf{P}$, which is $(2^n - 1) \cdot 2^{n-1} = N(N/2 + 1/2)$.

Example 1. Let $\rho : GF(2^3) \to GF(2^3)$ be a bijective mapping defined as $\rho(x) = x^3 \bmod f(x)$ where $f(x) = x^3 + x + 1$ is irreducible over $GF(2)$. The mapping ρ then corresponds to $0 - 0, 1 - 1, 2 - 3, 3 \to 4, 4 - 5, 5 \to 6, 6 - 7, 7 - 2$, and is known to be differentially 2-uniform [11]. The XOR table for ρ and the corresponding transition matrix $\mathbf{P}$ are then

$$XOR_\rho = \begin{bmatrix} 8 & 0 & 0 & 0 & 0 & 0 & 0 & 0 \\ 0 & 2 & 0 & 2 & 0 & 2 & 0 & 2 \\ 0 & 0 & 2 & 2 & 2 & 2 & 0 & 0 \\ 0 & 2 & 2 & 0 & 2 & 0 & 0 & 2 \\ 0 & 0 & 0 & 0 & 2 & 2 & 2 & 2 \\ 0 & 2 & 0 & 2 & 2 & 0 & 2 & 0 \\ 0 & 0 & 2 & 2 & 0 & 0 & 2 & 2 \\ 0 & 2 & 2 & 0 & 0 & 2 & 2 & 0 \end{bmatrix} \qquad \mathbf{P} = \frac{1}{8} \cdot \begin{bmatrix} 2 & 0 & 2 & 0 & 2 & 0 & 2 \\ 0 & 2 & 2 & 2 & 2 & 0 & 0 \\ 2 & 2 & 0 & 2 & 0 & 0 & 2 \\ 0 & 0 & 0 & 2 & 2 & 2 & 2 \\ 2 & 0 & 2 & 2 & 0 & 2 & 0 \\ 0 & 2 & 2 & 0 & 0 & 2 & 2 \\ 2 & 2 & 0 & 0 & 2 & 2 & 0 \end{bmatrix}. \qquad (3)$$

Note that $\mathbf{P}$ is obtained by deleting the first row and column of XOR_ρ and dividing by 8. Observe that $\mathbf{P}$ is aperiodic since $P_{11} > 0$. □

2.2 Random Graph Theory

As has been observed by many authors, a transition matrix P can be considered as the adjacency matrix for a directed graph $G = (V, E)$, where $V = \{v_1, v_2, \ldots, v_N\}$ and there is a directed edge from v_i to v_j if and only if $p_{ij} > 0$. We will call G the *underlying graph* of P. A directed graph G is strongly connected if for all v_i, v_j, there is a directed path from vertex v_i to vertex v_j. Also, an edge in G is called a loop if it connects a vertex to itself.

Lemma 1. The matrix P is irreducible if and only if G is strongly connected. The matrix P is ergodic if G is strongly connected and has a loop.

Proof. If vertices v_i and v_j in G are connected by the path $v_i v_1' v_2' \cdots v_r' v_j$ then by construction $p_{ij}^{(r)} > 0$. When G is strongly connected this is true for all vertex pairs v_i, v_j which implies that $p_{ij}^{(r)} > 0$ for some r, and P must be irreducible. Further, if G has a loop then there must exist an i for which $p_{ii} > 0$, and hence P is aperiodic. The lemma now follows. □

Example 2. The underlying graph $\mathbf{G}$ corresponding to the $\mathbf{P}$ matrix in (3) is shown in Figure 1. It can be verified that $\mathbf{G}$ is strongly connected, and since $\mathbf{G}$ contains 4 loops, it follows that $\mathbf{P}$ is ergodic. □

Clearly the probability of G having a loop and being strongly connected increases as the number of edges in G increases. We will say that almost all graphs with N vertices and m edges possess a certain property if the fraction of graphs with the property tends to one when $N \to \infty$. Using simple combinatorial arguments, it can be shown that almost all directed graphs G with N vertices and m edges have a loop when $m = N \cdot \gamma_N$, $\gamma_N \to \infty$. Further, Palásti [15] has shown that almost all directed graphs G with m edges are strongly connected when $m = N(\log N + \gamma_N)$, $\gamma_N \to \infty$. Here γ_N is any function that tends to infinity as N does such as $\log \log N$.

Theorem 2. Let $m = N(\log N + \gamma_N)$ where $\gamma_N \to \infty$. Then almost all directed graphs G have a loop and are strongly connected.

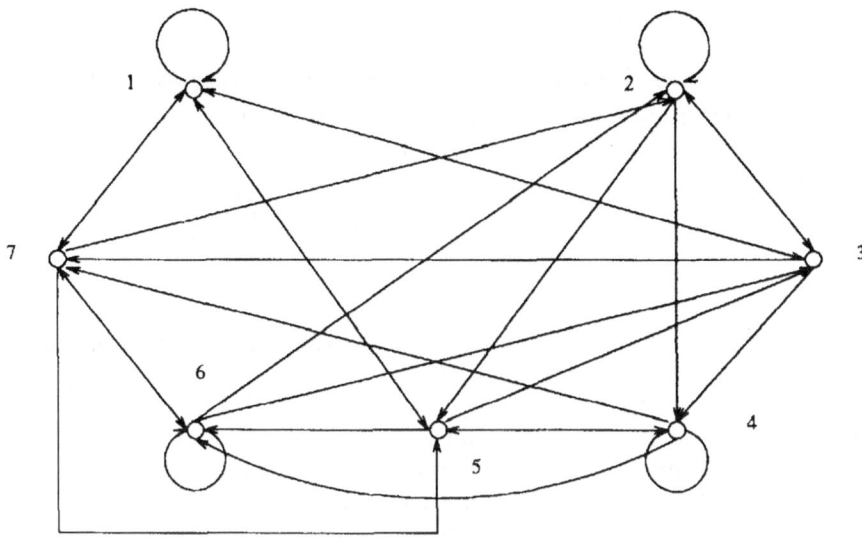

Fig. 1. The underlying graph of ρ, $0 \to 0, 1 \to 1, 2 \to 3, 3 \to 4, 4 \to 5, 5 \to 6, 6 \to 7, 7 \to 2$,

Proof. For any two events α_1, α_2, $\Pr(\alpha_1, \alpha_2) \geq \Pr(\alpha_1) + \Pr(\alpha_2) - 1$. If the events of interest are 'G has a loop' and 'G is strongly connected', then the joint probability tends to one when $m = N(\log N + \gamma_N)$ and $\gamma_N \to \infty$. $\square$

Recall that the $\mathbf{P}$ matrix associated with a differentially 2-uniform mapping $\mathbf{F} : Z_2^n \to Z_2^n$ has $(2^n - 1) \cdot 2^{n-1} = N(N/2 + 1/2) > N^2/2$ nonzero entries. Theorem 2 states that a transition matrix with at least $N \log N$ *randomly distributed* nonzero edges is ergodic with high probability. Even though the nonzero entries of $\mathbf{P}$ are *not* randomly distributed, it is still highly likely that $\mathbf{P}$ is ergodic since the number of edges in $\mathbf{G}$ exceeds the required $N \log N$ bound of Theorem 2. We therefore make the following assumption

Proposition 2.1 Let $\mathbf{F} : Z_2^n \to Z_2^n$ be a bijective differentially 2-uniform round function. Then the transition matrix $\mathbf{P}$ derived from $\mathbf{F}$ is assumed to be ergodic.

To support this assumption we have verified that from the 40,320 bijective mappings $\mathbf{F} : Z_2^3 \to Z_2^3$, 10,752 are differentially 2-uniform, and each has an ergodic transition matrix $\mathbf{P}$. There are only 7 distinct transition matrices $\mathbf{P}$, with 1536 mappings $\mathbf{F}$ yielding the same $\mathbf{P}$ matrix. (We note that 3 is the smallest n for which differentially 2-uniform bijective mappings exist).

3   Convergence and Rapid Mixing

Let $P = [p_{ij}]$ be ergodic with $\Pi = (\pi_1, \pi_2, \ldots, \pi_N)$ its limiting distribution. Also let $P_i^{(r)}$ be the distribution of the states after r steps when started in state i. The *variation* [5] between $P_i^{(r)}$ and Π is defined as

$$\| P_i^{(r)} - \Pi \| = \frac{1}{2} \cdot \sum_{j=1}^{N} |p_{ij}^{(r)} - \pi_j| = \frac{1}{2} \cdot \sum_{j=1}^{N} e_{ij}^{(r)}. \tag{4}$$

A chain is *rapidly mixing* if the error terms $e_{ij}^{(r)}$ in (4) converge to zero as a fast function of r (see [16] for a survey). Initially the results related to rapid mixing only applied to special chains, such as those that were *time reversible*. A chain P is time reversible if $\pi_i p_{ij} = \pi_j p_{ji}$ for all states i, j. If P is nonreversible then define $M(P) = P \cdot \tilde{P}$ where $\tilde{P} = [\tilde{p}_{ij}]$ and $\tilde{p}_{ij} = \pi_j p_{ji}/\pi_i$. It can be shown [7] that $M(P)$ is time reversible, is ergodic if P is ergodic, and has limiting distribution Π if Π is the limiting distribution of both P and $\tilde{P}$. The usefulness of $M(P)$ is that it is possible to bound $\| P_i^{(r)} - \Pi \|$ by considering its eigenvalues.

For an ergodic chain P the Perron-Frobenius theorem [1] states that the largest eigenvalue is 1 while all other eigenvalues are less than 1 in modulus. In particular, the N eigenvalues of $M(P)$ are real and nonnegative [7]. Consequently, the convergence of the chain is determined by the magnitude of the second largest eigenvalue. Let the N eigenvalues of $M(P)$ be $1 = \beta_1 > \beta_2 \geq \cdots \geq \beta_N > 0$.

Theorem 3 Fill [7]. For any state i, $4\pi_i \cdot \| P_i^{(r)} - \Pi \|^2 \leq (\beta_2)^r$.   □

There are several methods to bound β_2 when $M(P)$ is time reversible, and the method we will use is based on the Poincaré inequality [5] and canonical paths [16]. One result from the investigation of rapidly mixing Markov chains has been to show that the convergence of the chain P depends on the geometric properties of G_M, the underlying graph of $M(P)$. Let $M(P) = [q_{ij}]$, and from the time reversible property define

$$d_{ij} \stackrel{\text{def}}{=} \pi_i q_{ij} = \pi_j q_{ji}. \tag{5}$$

For $G_M = (V, E)$, let $\delta(v_i, v_j)$ be a (directed) path between vertices v_i and v_j with no repeated edges. Let Γ be a collection of paths $\delta(v_i, v_j)$ containing one path for each vertex pair v_i, v_j in G_M. At least one such path set Γ will exist since P, and hence $M(P)$, is irreducible. Now define the length of the path $\delta(v_i, v_j)$ to be

$$|\delta(v_i, v_j)| \stackrel{\text{def}}{=} \sum_{e.e \in \delta(v_i, v_j)} d(e)^{-1} \tag{6}$$

where the sum is over all edges e in the path $\delta(v_i, v_j)$ and $d(e) = d_{ij}$. Finally define κ as

$$\kappa \stackrel{\text{def}}{=} \kappa(\Gamma) = \max_e \sum_{\delta(v_i, v_j): e \in \delta(v_i, v_j)} \pi_i \pi_j \cdot |\delta(v_i, v_j)| \tag{7}$$

where the maximum is taken over all directed edges in the graph and the sum is over all paths that traverse the edge e. Note that κ is essentially a measure of 'bottlenecks'. A bottleneck S in a graph G is a set of vertices for which there are relatively few edges directed in or out of S as compared to $|S|$. Intuitively, if the chain enters a state corresponding to a vertex in S then the process gets 'stuck' in S and does not mix rapidly. Consequently, in any path set Γ, the edges joining S to the rest of the graph will be traversed frequently. We are now ready to state the Poincaré inequality.

Proposition 3.1 (Poincaré inequality) For an ergodic time reversible chain

$$\beta_2 \leq 1 - \frac{1}{\kappa}.$$

$\square$

Since in general the transitions matrices $\mathbf{P}$ describing differential distributions are not time reversible, our goal is to show that the convergence of $\mathbf{P}$ can be bound using Theorem 3 and the Poincaré inequality.

3.1 Bounding eigenvalues

We begin by showing that the $M(\mathbf{P})$ matrix derived from a differentially 2-uniform mapping $\mathbf{F}$ has the *complete graph* (all vertex pairs connected by an edge) as its underlying graph.

Lemma 4. $M(\mathbf{P}) = \mathbf{P}\mathbf{P}^T$.

Proof. Recall that $M(\mathbf{P}) = \mathbf{P}\check{\mathbf{P}}$ is dervied from $\mathbf{P} = [P_{ij}]$ by defining $\check{\mathbf{P}} = [\check{P}_{ij}]$ where $\check{P}_{ij} = \pi_j P_{ji}/\pi_i$. But since $\pi_i = \pi_j$ for all states i, j it follows that $\check{\mathbf{P}} = \mathbf{P}^T$, the transpose of $\mathbf{P}$. $\square$

Corollary 3.1 $M(\mathbf{P})$ has no zero entries. Equivalently, the underlying graph $\mathbf{G}_M$ of $M(\mathbf{P})$ is complete.

Proof. Let $M(\mathbf{P}) = [q_{ij}]$ and observe that

$$q_{ij} = \sum_{k=1}^{N} P_{ik} \cdot \check{P}_{kj} = \sum_{k=1}^{N} P_{ik} \cdot P_{jk}. \tag{8}$$

But by construction, each row $P_{i1}, P_{i2}, \ldots, P_{iN}$ of $\mathbf{P}$ has $2^{n-1} = N/2 + 1/2$ nonzero entries. Since the majority of entries in each row are nonzero, there must be at least one k in (8) for which $P_{ik} \cdot \check{P}_{kj} > 0$, implying $q_{ij} > 0$. Since this is true for all pairs of states i, j, it follows that $M(\mathbf{P})$ has no nonzero entries. $\square$

Now consider applying the Poincaré inequality to bound the variation in $\mathbf{P}$, requiring a path set Γ for $\mathbf{G}_M$. But since $\mathbf{G}_M$ is complete we simply take the path between v_i and v_j to be the directed edge connecting these vertices so that

$\delta(v_i, v_j) = e_{ij}$. Obviously each edge is used only once in the path set which implies that the summation on the RHS of (7), the equation defining κ, will only have one term. To determine κ we need only determine the length $|\delta(v_i, v_j)|$ of each path (i.e. edge).

Recall that the length of an edge $e = e_{ij}$ is the inverse of $d_{ij} = \pi_i q_{ij}$ where $1/\pi_i = N$ and q_{ij} is given as in (8). Since $P_{ij} = 2/2^n$ if $P_{ij} > 0$ then q_{ij} can be written as

$$q_{ij} = \frac{4}{2^{2n}} \cdot \sum_{k=1}^{N} [P_{ik} > 0] \cdot [P_{jk} > 0] \qquad (9)$$

where $[\cdot]$ is a boolean predicate evaluating to either 1 or 0. When $i = j$ the value of the summation in (9) is $2^{n-1} = N + 1/2$. However when $i \neq j$, we will model the summation as a random variable α_e distributed binomially as $b(p, N)$, where $p = (\frac{1}{2} + \frac{1}{2N})^2$ is the probability that two rows from $\mathbf{P}$ are nonzero in the same entry. There are $N^2 - N$ random variables α_e to consider, corresponding to q_{ij}, $i \neq j$.

Given the restrictions on $M(\mathbf{P})$, the criterion (7) for κ reduces to

$$\kappa = \kappa(\Gamma) = \max_e \frac{1}{N^2} \cdot \frac{N \cdot 2^{2n}}{4 \cdot \alpha_e} = \max_e \frac{(N+1)^2}{4N \cdot \alpha_e}. \qquad (10)$$

So the maximum is obtained when α_e is minimized. Observe that $\beta_2 \leq 2^{-t}$ when $\kappa = \frac{2^t}{2^t - 1}$, implying that there is some α_e for which

$$\alpha_e = \frac{2^t - 1}{2^t} \cdot \frac{(N+1)^2}{4N} = (1 - 2^t) \cdot pN. \qquad (11)$$

It follows that a good bound on β_2 is obtained if the smallest α_e is just slightly less than the mean pN of its distribution $b(p, N)$.

Example 3. For the $\mathbf{P}$ matrix defined in (3), it can be verified that $M(\mathbf{P})$ is a 7×7 matrix with 1 on the main diagonal and $1/8$ elsewhere. In this case $pN = 16/49$, and $\alpha_e = (1 - 2^{-3}) \cdot pN$ for all q_{ij}, $i \neq j$, implying that $\| P_i^{(r)} - \Pi \|$ is the same value for all i. The Poincaré inequality states that $\beta_2 \leq 2^{-3}$, and Table 1 shows that $(\beta_2)^r = 2^{-3r}$ bounds $4/7 \cdot \| P_i^{(r)} - \Pi \|$ as predicted from Theorem 3.

□

For larger n we will not be able to compute $M(\mathbf{P})$ explicitly as we did in the example above, and some probabilistic statement must be made. The next lemma (adapted from [4]) gives a bound on the probability that $|\alpha_e - pN| \geq 2^{-t} \cdot pN$.

Lemma 5. If $0 < p < \frac{1}{2}$, and $(pN)^{-\frac{1}{2}} < \epsilon < \frac{1}{6}$, then

$$\Pr(|\alpha_e - pN| \geq \epsilon pN) < e^{-\frac{\epsilon^2 pN}{3(1-p)} + \frac{\epsilon}{1-p}} \qquad (12)$$

where e is the base of the natural logarithm.

□

r	$\| P_i^{(r)} - \Pi \|$	$4/7 \cdot \| P_i^{(r)} - \Pi \|^2$	2^{-3r}
1	0.10714	0.45918	0.875
2	0.53571	0.11479×10^{-1}	0.10937
3	0.13392×10^{-1}	0.71747×10^{-3}	0.13671×10^{-1}
4	0.66964×10^{-2}	0.17936×10^{-3}	0.17089×10^{-2}
5	0.16741×10^{-2}	0.11210×10^{-4}	0.21362×10^{-3}

Table 1. Convergence bounds for **P** .

If $2^{2t} = o(N)$ and letting $p = 1/4$, the probability that all $N^2 - N$ α_e deviate from pN by less than a factor of $(1 - 2^{-t})$ is $\left(1 - e^{-\frac{N}{9 \cdot 2^{2t}} + \frac{4}{3 \cdot 2^t}} \right)^{N^2 - N}$ which reduces to

$$1 - e^{\ln N + \ln(N-1) - \frac{N}{9 \cdot 2^{2t}} + \frac{4}{3 \cdot 2^t}} + O\left(e^{2 \ln N - \frac{2N}{2^{2t}}} \right) . \tag{13}$$

Using (13) we are able to argue convergence results for ciphers with large given parameters such as block size and number of rounds.

Example 4. Consider a 16-round cipher of block size $n = 64$, or $N = 2^{64} - 1$, for which we wish to show that $\beta_2 \leq \frac{1}{32}$. In this case $t = 5$ and (13) bounds the deviation probability as greater than $1 - e^{128 - 2^{50}}$ which for all practical purposes is 1. Then if at least one α_e is less than the mean, Theorem 3 states that

$$4\pi_i \cdot \| P_i^{(16)} - \Pi \|^2 \leq (\beta_2)^{16} < 2^{64} \cdot 2^{-5 \cdot 16} = 2^{-16}. \tag{14}$$

Since the variation has N terms, the average squared error per state is then approximately $\frac{1}{4} \cdot 2^{-64} \cdot 2^{-16} = 2^{-82}$, giving the average error value $e_{ij}^{(16)}$ to be 2^{-41}. □

4 Conclusion

The main aim of the paper was to show that bounds on the convergence of Markov chains describing differential distributions can be obtained for differential 2-uniform mappings. The convergence is expected to be exponential since there is a large separation between β_1 and β_2, the two largest eigenvalues of the $M(\mathbf{P})$ matrix. This separation is due to the fact that $M(\mathbf{P})$ contains no zero entries for a differential 2-uniform mapping.

The analysis has shown that the density of zero entries in the XOR table for the round function **F** will determine if it is ergodic and also the rate of convergence, since $M(\mathbf{P})$ will mostly be nonzero if **P** is mostly nonzero. It is then possible to extend the analysis to mappings other than those that are differentially 2-uniform, if the density of zeros can be approximated and a bound is known on the largest entry in the XOR table so that α_e can be approximated.

However, we conjecture that the rate of convergence in differential 2-uniform mappings is optimal (fastest) given that $M(\mathbf{P})$ is totally nonzero and all nonzero entries in $\mathbf{P}$ are bounded by the constant 2.

Strictly speaking, the convergence result states that the probability of a differential can be made arbitrarily close to $1/N = \frac{1}{2^n-1}$. However since each nonzero XOR table entry is even, the lowest nonzero probability of a differential is $\frac{2}{2^n} = \frac{1}{2^{n-1}}$. The discrepancy is introduced by modeling differentials using Markov chains. In practice, if the probability of the most likely differential is at most $3/2^n$, then all 2^n plaintext-ciphertext pairs must be examined, which renders key search redundant [9]. A rapidly converging Markov chain for differential cryptanalysis strongly suggests that all differentials will have a probability approaching the practical minimum.

Acknowledgement

I would like to thank Jovan Golić for many helpful discussions.

References

1. U. Bhat. *Elements in applied stochastic processes*. John Wiley and Sons, 1972.
2. E. Biham and A. Shamir. Differential cryptanalysis of DES-like cryptosystems. *Journal of Cryptology*, 4(1):3–72, 1991.
3. E. Biham and A. Shamir. *Differential cryptanalysis of Data Encryption Standard*. Springer-Verlag, 1993.
4. B. Bollobás. *Random graphs*. Academic Press, 1985.
5. P. Diaconis and D. Stroock. Geometric bounds for eigenvalues of Markov chains. *Annals of Applied Probability*, 1(1):37–61, 1991.
6. W. Feller. *An Introduction to Probability Theory and its Applications*. New York: Wiley, 3rd edition, Volume 1, 1968.
7. J. Fill. Eigenvalue bounds on convergence to stationarity for nonreversible Markov chains, with an application to the exclusion process. *Annals of Applied Probability*, 1(1):62–87, 1991.
8. G. Hornauer, W. Stephan, and R. Wernsdorf. Markov ciphers and alternating groups. *Advances in Cryptology, EUROCRYPT 93, Lecture Notes in Computer Science, vol. 765, T. Helleseth ed., Springer-Verlag*, pages 453–460, 1994.
9. X. Lai. *On the design and security of block ciphers*. ETH Series in Information Processing, editor J. Massey, Hartung-Gorre Verlag Konstanz, 1992.
10. X. Lai, J. Massey, and S. Murphy. Markov ciphers and differential analysis. In *Advances in Cryptology. EUROCRYPT 91, Lecture Notes in Computer Science, vol. 547. D. W. Davies ed., Springer-Verlag*, pages 17–38, 1991.
11. K. Nyberg. Differentially uniform mappings for cryptography. *Advances in Cryptology, EUROCRYPT 93, Lecture Notes in Computer Science, vol. 765, T. Helleseth ed., Springer-Verlag*, pages 55–64, 1994.
12. L. J. O'Connor. Designing product ciphers using Markov chains. *proceedings of the Workshop on Selected Areas in Cryptography, Kingston. Canada, May 1994*, pages 2–13. 1994.

13. L. J. O'Connor and J. Dj Golić. A unified markov approach to differential and linear cryptanalysis. to be presented at Asiacrypt, November 1994.
14. National Bureau of Standards. Data Encryption Standard. FIPS PUB 46, Washington, D. C. (January 1977).
15. I. Palásti. On the strong connectedness of random graphs. *Studia Sci. Math. Hungar.*, 1:205–214, 1966.
16. U. Vazirani. Rapidly mixing Markov chains. In B. Bollobás, editor, *Probabilistic combinatorics and its applications, proceedings of Symposia in Applied Mathematics, volume 44*, pages 99–121, 1991.

A Generalization of Linear Cryptanalysis and the Applicability of Matsui's Piling-up Lemma

Carlo Harpes, Gerhard G. Kramer, James L. Massey

Swiss Federal Institute of Technology,
Signal and Info. Proc. Lab., CH-8092 Zürich
email: harpes@isi.ee.ethz.ch

Abstract. Matsui's linear cryptanalysis for iterated block ciphers is generalized by replacing his linear expressions with I/O sums. For a single round, an I/O sum is the XOR of a balanced binary-valued function of the round input and a balanced binary-valued function of the round output. The basic attack is described and conditions for it to be successful are given. A procedure for finding effective I/O sums, i.e., I/O sums yielding successful attacks, is given. A cipher contrived to be secure against linear cryptanalysis but vulnerable to this generalization of linear cryptanalysis is given. Finally, it is argued that the ciphers IDEA and SAFER K-64 are secure against this generalization.

Keywords. Linear cryptanalysis, differential cryptanalysis, piling-up lemma, IDEA, SAFER K-64.

1 Introduction

Linear cryptanalysis, which was introduced by Matsui in [5] to attack DES, is an attack that applies to any iterated block cipher. In this paper, we develop a generalized version of linear cryptanalysis that widens somewhat the class of ciphers for which the attack will be successful and that provides additional insight into Matsui's attack.

In Section 2, we define an I/O sum for one round as the XOR of a balanced binary-valued function of the round input and a balanced binary-valued function of the round output. This generalizes Matsui's "linear expressions". We also introduce key-dependent imbalance and average-key imbalance as measures for the usefulness of an I/O sum.

In Section 3, we adapt Matsui's linear cryptanalysis to the use of I/O sums. We describe a basic attack that exploits a multi-round I/O sum for the entire cipher excluding the last round and tries to find the last-round key. In Section 4, we formulate the hypothesis of wrong-key randomization, which states that using a wrong key in the last round to estimate an I/O sum decreases its key-dependent imbalance. The generalized attack succeeds if it is based on an I/O sum satisfying this hypothesis and if enough plaintext/ciphertext pairs are available.

Section 5 treats the case where the average-key imbalance of an I/O sum is unknown. To handle this case, we introduce a threefold sum as an I/O sum XOR-ed with a binary-valued function of the key and show that the imbalance of the

L.C. Guillou and J.-J. Quisquater (Eds.): Advances in Cryptology - EUROCRYPT '95, LNCS 921, pp. 24-38, 1995.

threefold sum is a lower bound on the average-key imbalance of the parent I/O sum. In practice, finding effective I/O sums is done by finding effective threefold sums whose imbalance is much easier to compute.

In Section 6, we develop a procedure for finding effective "homomorphic" threefold sums. This procedure relies on Matsui's piling-up lemma and applies to ciphers whose round function is a cascade of a keyed group operation and a possibly-keyed bijective function. We argue that ciphers that insert keys by certain modulo operations, such as IDEA and SAFER K-64, are generally resistant to this procedure, and we show that, after a slight modification, the procedure can be applied to DES-like ciphers too.

Section 7 defines QRweak, a mini-cipher vulnerable to the generalization of linear cryptanalysis, but secure against differential and linear cryptanalysis. We also argue that the cipher IDEA is secure against the generalization of linear cryptanalysis by showing that the presented procedure for finding effective homomorphic threefold sums finds no effective threefold sum for one round of either IDEA(8) or IDEA(16). We also show that SAFER K-64 has this desirable property.

Section 8 summarizes the main results.

2 Preliminaries

An *r-round iterated block cipher of block-size n* (Fig. 1) consists of r successive applications of a *keyed round function*, with a different key in each round. The *full key* is $K^{(1..r)} := (K^{(1)}, \ldots, K^{(r)})$, where $K^{(i)}$ is the *round key* applied in the i-th *round* for $i = 1, 2, \ldots, r$. The round keys take on values in a set $\mathcal{K}$, the *round key space*. The plaintext X and ciphertext Y take values in $\mathcal{X}$, the set of binary n-tuples. For each round key k, the keyed round function F_k is a bijection on $\mathcal{X}$. Let $Y^{(i)}$ denote the output n-tuple of the i-th round so that $Y = Y^{(r)}$, and let $Y^{(0)} := X$.

Throughout this paper, capital letters such as X, Y, $Y^{(1)}$, $\tilde{Y}^{(r-1)}$, $K^{(1)}$, etc. will denote random variables and the corresponding lowercase letters will denote specific values of these random variables, e.g., fixed keys. A superscript will specify the round(s) to which a variable is associated, e.g., $Y^{(r-1)}$ is the output of the $(r-1)$-th round, $K^{(1..r-1)}$ is the tuple of round keys from the first to the $(r-1)$-th round, etc.

We always assume that the plaintext and all keys used within the cipher are independent and uniformly random over the appropriate spaces, except when we explicitly fix the keys by specifying, e.g., $K^{(1..r)} = k^{(1..r)}$. This assumption defines the random experiment on which linear cryptanalysis is formalized and for which all probabilities are calculated. A binary-valued function is *balanced* if it takes on the value 0 for exactly half of its possible arguments and the value 1 otherwise.

In [5], Matsui exploits a cipher's weakness that he expresses in terms of "linear expressions". In Matsui's terminology, a linear expression for one round is an "equation" for a certain modulo-two sum of round input bits and round output bits as a sum of round key bits. The expression should be satisfied with

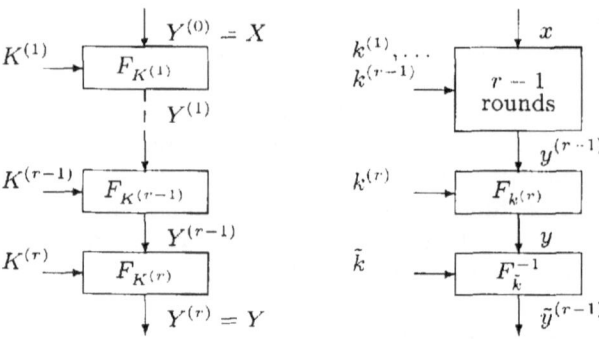

Fig. 1. Structure and notation for an iterated block cipher (left) and notation used in the attack (right).

probability much more (or much less) than 0.5 to be useful. Our generalization of linear cryptanalysis resides in replacing Matsui's linear expressions by the more general notion of I/O sums.

Definition 1. An *I/O sum* $S^{(i)}$ for the i-th round is a modulo-two sum of a balanced binary-valued function f_i of the round input $Y^{(i-1)}$ and a balanced binary-valued function g_i of the round output $Y^{(i)}$, that is,

$$S^{(i)} := f_i(Y^{(i-1)}) \oplus g_i(Y^{(i)}) , \qquad (1)$$

where $\oplus$ denotes modulo-two addition, i.e., the XOR operation.

The functions f_i and g_i will be called the *input function* and the *output function*, respectively, of the I/O sum $S^{(i)}$.

I/O sums for successive rounds are *linked* if the output function of each round before the last coincides with the input function of the following round (i.e., $f_i = g_{i-1}$). When ρ successive $S^{(i)}$ are linked, their sum,

$$S^{(1..\rho)} := \bigoplus_{i=1}^{\rho} S^{(i)} = g_0(Y^{(0)}) \oplus g_\rho(Y^{(\rho)}) \qquad (2)$$

will be called a *multi-round I/O sum*.

As a measure for the "effectiveness" of a linear expression in an attack, Matsui uses the magnitude of the difference between $\frac{1}{2}$ and the probability that the expression is satisfied. We will instead use "imbalances", which are similarly defined but with an extra factor of two so that the imbalance will lie between 0 and 1 inclusive.

Definition 2. The *imbalance* $I(V)$ of a binary-valued random variable V (whose values are the real numbers 0 and 1) is the non-negative real number $|2P[V = 0] - 1|$ or, equivalently, $|E[2V - 1]|$, where $P[V = 0]$ is the probability that V takes on the value 0 and $E[.]$ denotes expectation.

The *key-dependent imbalance* $I(S^{(1..\rho)} \mid k^{(1..\rho)})$ of the I/O sum $S^{(1..\rho)}$ is the imbalance of this sum conditioned on the event that $K^{(1..\rho)} = k^{(1..\rho)}$. The *average-key imbalance* of the I/O sum $S^{(1..\rho)}$ is the expectation of these key-dependent imbalances and will be denoted as $\overline{I}(S^{(1..\rho)})$. An I/O sum is *effective* if it has a large average-key imbalance, and is *guaranteed* if its average-key imbalance is 1, the maximum possible.

As an example, suppose that $S^{(1)} = f(X) \oplus g(Y^{(1)}) = h(K^{(1)})$ where h is a balanced function. Then $S^{(1)}$ has imbalance $I(S^{(1)}) = I(h(K^{(1)})) = 0$. However, because $S^{(1)} = h(k^{(1)})$, a constant, when $K^{(1)} = k^{(1)}$, the key-dependent imbalance of $S^{(1)}$ is 1 for all keys $k^{(1)}$ and hence the average-key imbalance is also 1. Thus $S^{(1)}$ is a guaranteed I/O sum.

3 Attacks by the Generalization of Linear Cryptanalysis

The *basic* attack by the generalization of linear cryptanalysis exploits an effective I/O sum $S^{(1..r-1)} = g_0(X) \oplus g_{r-1}(Y^{(r-1)})$ for the first $r-1$ rounds with the intention of finding the last-round key. It is assumed that the attacker has access to N plaintext/ciphertext pairs (hereafter called *p/c-pairs*) with *uniformly randomly* chosen plaintexts [although experience suggests that *any* N different p/c-pairs will do just as well]. The basic attack proceeds as follows (Fig. 1).

0. Set up a counter $c[\tilde{k}]$ for each possible last-round key $\tilde{k}$ and initialize all counters to 0.
1. Choose a p/c-pair (x, y).
2. For each possible $\tilde{k}$, evaluate $\tilde{y}^{(r-1)} := F_{\tilde{k}}^{-1}(y)$ and, if $g_0(x) \oplus g_{r-1}(\tilde{y}^{(r-1)}) = 0$, increment $c[\tilde{k}]$ by 1.
3. Repeat Step 1 and 2 for all N available p/c-pairs.
4. Output all keys $\tilde{k}$ that maximize $|c[\tilde{k}] - \frac{N}{2}|$ as candidates for the key actually used in the last round.

The quantity $c[\tilde{k}]$ is proportional to an obvious estimate of the key-dependent imbalance of the I/O sum under the assumption that $\tilde{k}$ is the right key. Under suitable statistical assumptions, Step 4 implements the maximum-likelihood decision rule for the last-round key when the counts are considered to be the observation [8].

The basic attack must in practice be speeded up by exploiting "key equivalence". Two keys $k, k' \in \mathcal{K}$ are *equivalent* if $g_{r-1}(F_{k'}^{-1}(y)) = g_{r-1}(F_k^{-1}(y)) \oplus c$ for some c and for all $y \in \mathcal{X}$. The basic attack can never distinguish between equivalent keys. Therefore, we consider in Step 2 only one representative of each *key (equivalence) class*. Just as differential and conventional linear cryptanalysis determine only some portion of the last-round key, the generalization of linear cryptanalysis determines only the (equivalence) class in which the true key lies. The key class containing the actual key used in the last round is the *right class* and its representative is the *right key*. The other key classes are *wrong classes* and their representatives are *wrong keys*. In practice, the number of key classes must be reasonably small, since the computation in the attack is proportional to that number.

The *success probability* p_{GLC} of the attack is the probability of the event that the output list contains only the right class. The *conditional success probability* $p_{GLC|k^{(1..r)}}$ is the probability of this event when the key $K^{(1..r)} = k^{(1..r)}$. Matsui considers in [6] an improvement of linear cryptanalysis similar to "list decoding" of error-detecting codes [3]. Applied to our generalization, this improvement consists of trying out all keys in all equivalence classes in order of decreasing apparent imbalance $|c[\tilde{k}] - \frac{N}{2}|$ until the true key is found. The basic attack can also be speeded up (as was done in [6]) by first classifying all p/c-pairs in Step 2 into text classes – each consisting of p/c-pairs that cause the same set of counters to be incremented – and then incrementing the counters once for each text class. Matsui also improved his attack by determining the key to the first and the last round simultaneously [6]. We can use an $(r-2)$-round I/O sum $S^{(2..r-1)}$ instead of $S^{(1..r-1)}$ for a similar improvement of our basic attack.

4 Success of the Generalization of Linear Cryptanalysis

Theorem 3 below states that using enough p/c-pairs in the basic generalization of linear cryptanalysis reveals information on the last-round key provided that the following hypothesis holds.

Hypothesis of wrong-key randomization for an (r-1)-round I/O sum.
Let $S^{(1..r-1)} = g_0(X) \oplus g_{r-1}(Y^{(r-1)})$ be an effective I/O sum for the cipher in Fig. 1. Then, for virtually all possible full keys $k^{(1..r)}$ and for all wrong keys $\tilde{k}$ for the last round, the key-dependent imbalance $I(S^{(1..r-1)} \mid k^{(1..r-1)})$ is substantially reduced if the output of the $(r-1)$-th round is replaced by the estimate $\tilde{Y}^{(r-1)}$ computed from the ciphertext Y and a wrong key $\tilde{k}$ for the last round. That is, for all wrong keys $\tilde{k}$,

$$\frac{I(\tilde{S}^{(1..r-1)} \mid k^{(1..r)}\tilde{k})}{I(S^{(1..r-1)} \mid k^{(1..r-1)})} << 1 \tag{3}$$

where $\tilde{S}^{(1..r-1)} = g_0(X) \oplus g_{r-1}(\tilde{Y}^{(r-1)})$ and $\tilde{Y}^{(r-1)} = F_{\tilde{k}}^{-1}(Y)$.

$\tilde{S}^{(1..r-1)}$ can be considered as a kind of $(r+1)$-round I/O sum where the $(r+1)$-th round has round function F^{-1} and fixed round key $\tilde{k}$ (Fig. 1, right), but it coincides with either $S^{(1..r-1)}$ or its complement if $\tilde{k}$ is a representative of the right key class (as this implies $g_{r-1}(Y^{(r-1)}) = g_{r-1}(\tilde{Y}^{(r-1)}) \oplus c$). For a good cipher, the key-dependent imbalance of multi-round I/O sums can be expected to decrease with an increasing number of rounds.

Theorem 3. *Suppose that $S^{(1..r-1)}$ is an effective $(r-1)$-round I/O sum for which the hypothesis of wrong-key randomization in the basic attack holds. Then, for virtually all keys, the generalization of linear cryptanalysis with I/O sum $S^{(1..r-1)}$ finds the key class in which the true key of the last round lies as reliably as desired provided that sufficiently many (randomly chosen) p/c-pairs are available.*

By using similar arguments, Matsui showed that, for a fixed key $k^{(1..r)}$, the success probability of linear cryptanalysis is approximately proportional (in our notation) to $(I(S^{(1..r-1)} \mid k^{(1..r)}))^2$ where $S^{(1..r-1)}$ is the considered I/O sum [6, 5]. The crucial point is that the success probability is an increasing function of the key-dependent imbalance of the considered I/O sum, which suggests that this imbalance is a robust measure for the usefulness of such a sum.

5 Random Keys and Threefold Sums

The success probability p_{GLC} of an attack exploiting the I/O sum $S^{(1..r-1)}$ depends on the average-key imbalance $\overline{I}(S^{(1..r-1)})$ in approximately the same manner as $p_{\mathrm{GLC}|k^{(1..r)}}$ depends on $I(S^{(1..r-1)} \mid k^{(1..r)})$. This approximation is virtually exact when the key-dependent imbalances for all keys are virtually equal. We state this as a hypothesis, which is analogous to the hypothesis of stochastic equivalence for differential cryptanalysis [2].

Hypothesis of fixed-key equivalence for an I/O sum. *The key-dependent imbalance of an effective I/O sum $S^{(1..r-1)}$ is virtually independent of the key $k^{(1..r-1)}$; more precisely,*

$$I(S^{(1..r-1)} \mid k^{(1..r-1)}) \approx \overline{I}(S^{(1..r-1)}) \tag{4}$$

is satisfied for virtually all keys $k^{(1..r-1)}$ that can result from the cipher's key scheduling algorithm.

In fact, the average-key imbalance gives us valuable information even without the hypothesis of fixed key equivalence. For example, if $\overline{I}(S^{(1..r-1)}) = 2^{-100}$, we know that at most a fraction 2^{-40} of the keys will give a key-dependent imbalance greater than 2^{-60}. Such an argument allows one to bound the number of "weak keys" for a cipher and suggests that average-key imbalance is a robustly good measure for the usefulness of an I/O sum.

The cryptanalyst needs to find I/O sums that are effective for many (or virtually all) keys. One possibility is to assume that the given hypothesis holds for any I/O sum, fix the key, calculate the key-dependent imbalance by Monte Carlo methods for virtually all possible I/O sums, and then select the most effective ones. We describe below an alternative procedure that requires far less computation. To formulate this procedure requires us to introduce the notion of threefold sums.

Definition 4. A *threefold sum* $T^{(i)}$ for the i-th round is a modulo-two sum of three terms: the first, a balanced binary-valued function f_i of the round input $Y^{(i-1)}$; the second, a balanced binary-valued function g_i of the round output $Y^{(i)}$; and the third, some binary-valued function h_i of the round key $K^{(i)}$; i.e.,

$$T^{(i)} := f_i(Y^{(i-1)}) \oplus g_i(Y^{(i)}) \oplus h_i(K^{(i)}) \ . \tag{5}$$

The function h_i is the *key function* of the threefold sum. Note that the first part of the expression for $T^{(i)}$ is the I/O sum $S^{(i)}$ (cf. (1)). We will call $S^{(i)}$ the *parent* I/O sum for $T^{(i)}$. The imbalance of a threefold sum is calculated under our universal assumption that the arguments of the input function and of the key function are independent and uniformly distributed.

We now analyze the relation between threefold sums and their parent I/O sums. We begin by lower bounding the average-key imbalance of the parent I/O sum $S^{(1..\rho)}$ by the imbalance of the threefold sum $T^{(1..\rho)} = S^{(1..\rho)} \oplus h(K^{(1..\rho)})$ in the manner

$$
\begin{aligned}
\bar{I}(S^{(1..\rho)}) &= E\left[\left|2P[S^{(1..\rho)}=0 \mid K^{(1..\rho)}] - 1\right|\right] \\
&= E\left[\left|2P[T^{(1..\rho)}=0 \mid K^{(1..\rho)}] - 1\right|\right] \\
&\geq \left|2E[T^{(1..\rho)}] - 1\right| = I(T^{(1..\rho)}) ,
\end{aligned}
\tag{6}
$$

where we have used Jensen's inequality and the convexity of the absolute-value function. Furthermore, equality holds if and only if $2P[T^{(1..\rho)} = 0|K^{(1..\rho)} = k^{(1..\rho)}] - 1$ has the same sign for all $k^{(1..\rho)}$. When equality holds, we will call the key function h a *maximizing key function* of $T^{(1..\rho)}$. We thus have proved the following proposition.

Proposition 5. [Threefold sums with maximizing key function] *Let $S^{(1..\rho)}$ be a multi-round I/O sum. Then the function $h_{\max}$ on $\mathcal{K}^\rho$ defined as*

$$
h_{\max}(k^{(1..\rho)}) = \begin{cases} 0 & \text{if } P[S^{(1..\rho)} = 0 \mid K^{(1..\rho)} = k^{(1..\rho)})] \geq \frac{1}{2} \\ 1 & \text{otherwise} \end{cases}
\tag{7}
$$

is a function h which maximizes the imbalance of the multi-round threefold sums $S^{(1..\rho)} \oplus h(K^{(1..\rho)})$, i.e., which upper bounds the imbalance of any other threefold sum with the same parent. Furthermore, this maximum imbalance is the average-key imbalance of the parent I/O sum, i.e.,

$$
\bar{I}(S^{(1..\rho)}) = I(S^{(1..\rho)} \oplus h_{\max}(K^{(1..\rho)})) .
\tag{8}
$$

The example below indicates that the $(r-1)$-round threefold sums used in Matsui's linear cryptanalysis of DES are not likely to have a maximizing key function - thus their imbalances provide only a lower bound on the average-key imbalance of the parent I/O sum. Matsui's approximated success probability is then a pessimistic estimate of the true success probability. In fact, Matsui has noted that his attacks perform better than predicted.

We show in Section 6 how to find the imbalance of threefold sums $S \oplus h(K)$ for a particular family $\mathcal{H}$ of key functions h. Obviously $\bar{I}(S) \geq \max_{h \in \mathcal{H}} I(S \oplus h(K))$, but the right side is often a good approximation to $\bar{I}(S)$. As it is generally infeasible to compute $\bar{I}(S)$ exactly, one has to rely on such an approximation when trying to find effective threefold sums. Section 6 describes such families of key functions for which we are able to compute this approximate imbalance.

Example 1. This example illustrates that if a threefold sum is the sum of three-fold sums that are linked, in the sense that their parent I/O sums are linked, and that have maximizing key functions, then this threefold sum can still have a key function that is not maximizing. Consider the cipher in Fig. 2 consisting of a cascade of an MA-box for four input bits ($\mathcal{X} = \{0,1\}^4$) as defined for the cipher IDEA [2], an XOR operation with a four-bit key, and another MA-box.

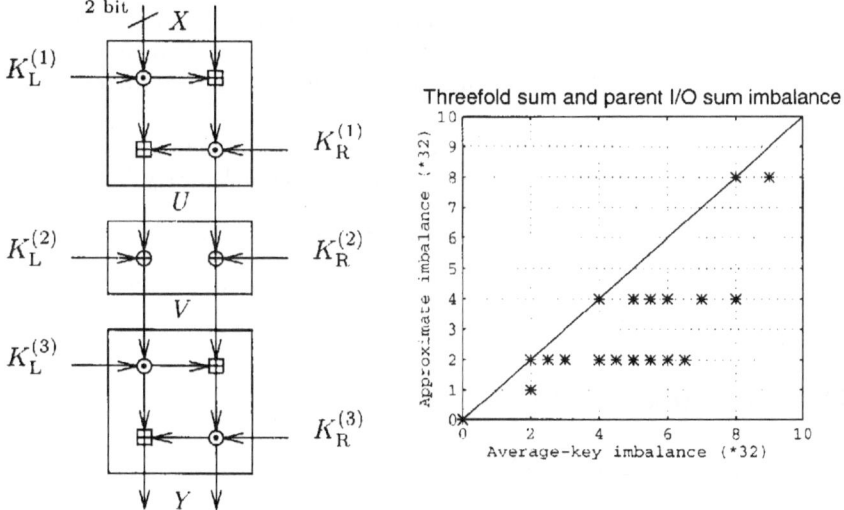

Fig. 2. Cipher using the MA-structure of IDEA(8).

Consider the threefold sum

$$T_{a,b,c} := (\overset{\bullet}{a} \bullet X) \oplus (\overset{\bullet}{c} \bullet Y) \oplus (h_{\max,a,b}(K^{(1)}) \oplus (\overset{\bullet}{b} \bullet K^{(2)}) \oplus h_{\max,b,c}(K^{(3)}))$$

where $h_{\max,a,b}$ and $h_{\max,b,c}$ are the maximizing key functions of

$$T^{(1)}_{\max,a,b} := (\overset{\bullet}{a} \bullet X) \oplus (b \bullet U) \oplus h_{\max,a,b}(K^{(1)}) \text{ and}$$

$$T^{(3)}_{\max,b,c} := (\overset{\bullet}{b} \bullet V) \oplus (\overset{\bullet}{c} \bullet Y) \oplus h_{\max,b,c}(K^{(3)}) ,$$

respectively, where $\bullet$ denotes the bitwise scalar product, and where $a, b, c \in \mathcal{X}$. Let $T^{(2)}_{\max,b,b} := (\overset{\bullet}{b} \bullet U) \oplus (\overset{\bullet}{b} \bullet V) \oplus (\overset{\bullet}{b} \bullet K^{(2)})$, so that $T_{a,b,c} = T^{(1)}_{\max,a,b} \oplus T^{(2)}_{\max,b,b} \oplus T^{(3)}_{\max,b,c}$. We show in the next section that $I(T_{a,b,c}) = I(T^{(1)}_{\max,a,b}) \cdot I(T^{(3)}_{\max,b,c})$ as $I(T^{(2)}_{\max,b,b}) = 1$.

In Fig. 2, we compare the approximate imbalance $\max_{b \in \mathcal{X}} I(T_{a,b,c})$ and the average-key imbalance of the parent I/O sum $S_{a,c} := (\overset{\bullet}{a} \bullet X) \oplus (c \bullet Y)$. For each of the 225 pairs (a, c) for which $a \neq 0$ and $c \neq 0$, we plot one star, which may overlap with other stars. To compute the average-key imbalances of the I/O sum, we have to find the cipher output for each key and input combination. For such a small cipher, this is still feasible.

We observe that the average-key imbalances of the I/O sum may be strictly larger than the approximate imbalances, even for a pair yielding the greatest approximate imbalance. It follows that the sum of threefold sums each with maximizing key function does not always have a maximizing key function, even if the imbalance of this sum is the largest possible. There are eight threefold sums with the greatest average-key imbalance 9/32, e.g., the one with $(a, c) = (2, 10)$.[1] If one of these threefold sums is used in an attack, the estimation of the success probability based on the approximate imbalance 8/32 will be pessimistic. Nonetheless, the highest approximate imbalance is quite close to the true average-key imbalance.

Finally, we analyze one of the most effective I/O sums, namely the one with $(a, c) = (2, 10)$, more closely. The key-dependent imbalances vary between

$$I(S_{2,10}(k^{(1..3)} = (5, 2, t))) = 0 \text{ and } I(S_{2,10}(k^{(1..3)} = (5, 0, t))) = 0.5$$

for $t \in \{0, 1\}^4$. This means that the hypothesis of fixed-key equivalence is not valid.

6 Finding Effective Threefold Sums

6.1 Applicability of Matsui's "Piling-Up Lemma"

In the language of threefold sums, Matsui's piling-up lemma becomes the statement that the imbalance of a sum of threefold sums is the product of their imbalances, i.e.,

$$I\left(\bigoplus_{i=1}^{\rho} T^{(i)}\right) = \prod_{i=1}^{\rho} I(T^{(i)}) , \tag{9}$$

provided that these threefold sums are *independent*.

Example 2. Consider a cascade of two "two-bit"-adders – $Y^{(1)} = K^{(1)} \boxplus X$ and $Y^{(2)} = K^{(2)} \boxplus Y^{(1)}$, where $\boxplus$ denotes addition modulo $2^n = 4$ – and the linked threefold sums

$$T^{(1)} = \text{MSB}(X) \oplus \text{MSB}(Y^{(1)}) \oplus \text{MSB}(K^{(1)}) ,$$
$$T^{(2)} = \text{MSB}(Y^{(1)}) \oplus \text{MSB}(Y^{(2)}) \oplus \text{MSB}(K^{(2)}) ,$$

where the function MSB gives the most significant bit of its argument. It is easy to check that $I(T^{(1)}) = I(T^{(2)}) = \frac{1}{2}$ (the threefold sums are equal to 0 if there is no carry bit, i.e., with probability $\frac{1}{4}$) and yet $I(T^{(1)} \oplus T^{(2)}) = 0 \neq I(T^{(1)}) \cdot I(T^{(2)})$.

---

[1] In this paper, the usual radix two representation of integers as n-tuples of $\mathcal{X}$ is considered, except sometimes when we consider multiplication, where the all zero n-tuple denotes the integer 2^n.

Thus, the piling-up formula does not hold and hence we can conclude that $T^{(1)}$ and $T^{(2)}$ are *not* independent. We also note that $T^{(1)} \oplus T^{(2)}$ does not have a maximizing key function and that the average-key imbalance of the parent I/O sum is $\frac{1}{2}$, which also does not satisfy the piling-up formula.

The reason that Matsui's piling-up lemma is of interest is that, in actual ciphers, it is infeasible to evaluate a multi-round imbalance directly, as this would involve evaluating the multi-round output for all input and key combinations. One is forced to find imbalances of one-round threefold sums and then use Matsui's piling-up lemma to find the imbalance of their sum. If these one-round threefold sums are linked, we thus get the multi-round threefold sum imbalance, which gives a lower bound on the average-key imbalance of the parent I/O sum. The above example indicates the desirability of conditions guaranteeing the independence of one-round threefold sums, since the piling-up formula (9) applied to dependent threefold sums can suggest misleading results. The following lemma specifies such a condition.

Lemma 6. *For an iterated cipher as in Fig. 1 with independent round keys, let $T^{(i)}$ be a threefold sum for the i-th round. If for each $i = 2, \ldots, \rho$, $T^{(i)}$ is independent of the round input $Y^{(i-1)}$, then the threefold sums $T^{(1)}, \ldots, T^{(\rho)}$ are independent.*

6.2 A Procedure for Finding Effective "Homomorphic" Threefold Sums

The independence of a one-round threefold sum and its input can be assured when a group operation occurs at the beginning of each round. This fact is fundamental for Theorem 8. Hereafter, we denote the left and the right part of a key $K^{(i)}$ by $K_{\mathrm{L}}^{(i)}$ and $K_{\mathrm{R}}^{(i)}$, respectively.

Definition 7. An I/O sum is *homomorphic* if the input and the output functions are homomorphisms for some considered group operation(s). A threefold sum is homomorphic if the parent I/O sum is homomorphic.

For example, a one-round homomorphic threefold sum, independent of its input, for a cipher that inserts the key $K_{\mathrm{L}}^{(i)}$ with the group operation "$*_i$" at the entry of the i-th round, is

$$T^{(i)} := f_i(Y^{(i-1)}) \oplus g_i(Y^{(i)}) \oplus (f_i(K_{\mathrm{L}}^{(i)}) \oplus h_i(K_{\mathrm{R}}^{(i)})) , \qquad (10)$$

where f_i and g_i are homomorphic binary functions for $*_i$ and $*_{i+1}$, respectively, i.e., for all $U, V \in \mathcal{X}$, $f_i(U *_i V) = f_i(U) \oplus f_i(V)$ and $g_i(U *_{i+1} V) = g_i(U) \oplus g_i(V)$.

Theorem 8. *Consider a cascade of ρ rounds with keyed round functions $F^{(1)}, \ldots F^{(\rho)}$, for which*

$$Y^{(i)} = F_{K^{(i)}}^{(i)}(Y^{(i-1)}) = \phi_i(Y^{(i-1)} *_i K_{\mathrm{L}}^{(i)}, K_{\mathrm{R}}^{(i)}) \qquad (11)$$

where "$*_i$" denotes a group operation in $\mathcal{X}$, $\phi_i(., k_R^{(i)})$ is a bijection on $\mathcal{X}$ for all $k_R^{(i)}$, $T^{(i)}$ is a homomorphic threefold sum for the i-th round, and $T^{(1)}, \ldots, T^{(\rho)}$ are linked. Then the imbalance of the ρ-round threefold sum $T^{(1..\rho)} := \bigoplus_{i=1}^{\rho} T^{(i)}$ is given by Matsui's piling-up formula (9). This means that for the parent I/O sums,

$$\overline{I}(S^{(1..\rho)}) \geq \prod_{i=1}^{\rho} \overline{I}(S^{(i)}) \geq \prod_{i=1}^{\rho} I(T^{(i)}) \ . \tag{12}$$

It follows that one can find an effective ρ-round threefold sum for a cipher whose round functions have a group operation at the entry (cf. 11) as follows:

1. For $i = 1, \ldots, \rho + 1$, find the set $\mathcal{H}_i$ of all binary functions on $\mathcal{X}$ that are homomorphisms for "$*_i$".
2. For $i = 1, \ldots, \rho$, find the imbalance of all i-th-round homomorphic threefold sums with input function $g_{i-1} \in \mathcal{H}_i$ and output function $g_i \in \mathcal{H}_{i+1}$.
   Discard the threefold sums with small imbalance.
3. Consider each possible list of ρ linked threefold sums containing one threefold sum found in Step 2 for each round.
   Use Theorem 8 to find the imbalance of the ρ-round threefold sum that can be written as the sum of all threefold sums in the same list.
   Find the ρ-round threefold sum with the largest imbalance.

6.3 Discussion of the Given Procedure

The complexity of the above procedure depends mainly on the number of homomorphisms onto $(\{0, 1\}, \oplus)$ for the group operations. If "$*_i$" is the bitwise XOR operation in $\mathcal{X}$, the only such homomorphisms are the *linear* functions defined by $l_a(x) = a \bullet x$ for all $x \in \mathcal{X}$, where a is a non-zero n-tuple. An I/O sum (or a threefold sum) whose input and output functions are l_a and l_b, respectively, is called *linear* with *linear-mask* (a, b). If all group operations "$*_i$" are the XOR operation, the given procedure considers only threefold sums whose component functions are linear. Thus for DES and other ciphers using XOR, the given procedure leads to no improvement of Matsui's method for finding effective linear expressions and to no real generalization of his linear cryptanalysis.

For the two groups $(\{0, 1\}^n, \odot)$ (multiplication modulo $2^n + 1$ with $n = 2, 4, 8$ or 16, and 0 representing 2^n) and $(\{0, 1\}^n, \boxplus)$ (addition modulo 2^n) of order 2^n used in IDEA [1, 2], there exists only one homomorphism, viz. the quadratic residue function QR for $\odot$ and the parity function (i.e., the least significant bit function LSB) for $\boxplus$. For ciphers using these operations, there are only very few possible linked threefold-sums, so that there is little chance that one of the corresponding threefold sums is effective. Thus the procedure for finding effective homomorphic threefold sums and the generalization of linear cryptanalysis is not very powerful against most ciphers using such operations to insert the key.

It is generally infeasible to analyze the imbalances of all possible threefold sums and even infeasible to find the imbalance of a single $(r-1)$-round threefold sum if one cannot deduce it from the imbalances of smaller sections such as rounds or S-boxes by using Matsui's piling-up lemma. The only threefold sums we know to which Matsui's piling-up lemma applies are homomorphic. We are aware of no practical alternative for finding imbalances. The procedure given in this section considers promising candidates for the most effective threefold sum, but it never guarantees that the threefold sum found is the most effective possible.

Example 3. To show that such a guarantee does not exist, we consider the 3-bit round function F defined by $Y = F_K(X) = \phi(X \boxplus K)$ where the function table of ϕ is $\underline{\phi} := (0, 1, 3, 5, 2, 4, 7, 6)$. The only homomorphic function l for $\boxplus$ is given by $\underline{l} = (0, 1, 0, 1, 0, 1, 0, 1)$. Since $\overline{I}(l(X) \oplus l(Y)) = 0$, but $\overline{I}(f(X) \oplus l(Y)) = \frac{1}{4}$ for $\underline{f} := (1, 1, 1, 1, 0, 0, 0, 0)$, the threefold sum with input function f has higher imbalance than the only homomorphic threefold sum.

6.4 Application to DES

A procedure for finding effective homomorphic threefold sums for DES has been implemented in [7]. It is similar to our procedure, but it requires more ingenuity to link threefold sums efficiently as there exist many guaranteed one-round threefold sums. The following example illustrates how one-round threefold sums that are independent of their input are constructed. By Lemma 6, this guarantees that Matsui's piling-up lemma is applicable when we cascade them.

Example 4. Let U_i denotes the i-th bit of a random variable U, where we number the bits from left to right starting with 1. This differs from Matsui's numbering (right to left starting with 0). For example, what we call the 2nd input bit to an S-box, Matsui calls the 4-th input bit; our 3rd plaintext bit X_3 is his bit $P_H[29]$ and Y_{64} corresponds in his notation to $C_L[0]$.

Let U denote the 6-bit input to the fifth S-box S5, and $V := S5(U)$ the 4-bit output. The threefold sum $U_2 \oplus V_1 \oplus V_2 \oplus V_3 \oplus V_4$ has imbalance $\frac{5}{8}$. By considering the permutation and the expansion in DES, one can find relations that enable us to transform the threefold sum for S5 into the threefold sum

$$T^{(1)} = (X_3^{(1)} \oplus X_8^{(1)} \oplus X_{14}^{(1)} \oplus X_{25}^{(1)} \oplus X_{49}^{(1)}) \oplus (Y_{35}^{(1)} \oplus Y_{40}^{(1)} \oplus Y_{46}^{(1)} \oplus Y_{57}^{(1)}) \oplus K_{26}^{(1)},$$

which has the same imbalance $\frac{5}{8}$ and is independent of the input $X^{(1)}$. Similarly,

$$T^{(3)} = (X_3^{(3)} \oplus X_8^{(3)} \oplus X_{14}^{(3)} \oplus X_{25}^{(3)}) \oplus (Y_{17}^{(3)} \oplus Y_{35}^{(3)} \oplus Y_{40}^{(3)} \oplus Y_{46}^{(3)} \oplus Y_{57}^{(3)}) \oplus K_{26}^{(3)}.$$

Since the I/O sum $S^{(2)}$ linked to both $T^{(1)}$ and $T^{(3)}$ is guaranteed, $T^{(1..3)}$ (with $T^{(2)} = S^{(2)}$) has imbalance $I(T^{(1)}) \cdot I(T^{(3)}) = \frac{25}{64}$, which is quite effective.

7 Some Examples

7.1 The QRweak Cipher

We now contrive a cipher, QRweak, to be weak against our generalization of linear cryptanalysis, but secure against linear and differential cryptanalysis. QRweak is a four round iterated block cipher of block-size eight (Fig. 1 with $r = 4$, $n = 8$) whose round function is defined as $F_k(x) = \phi(x \odot k)$, where $\odot$ denotes multiplication modulo 257 and where the function ϕ changes the bit order of the argument in the manner $t_1 t_2 t_3 t_4 t_5 t_6 t_7 t_8 \mapsto t_6 t_4 t_8 t_3 t_1 t_5 t_2 t_7$ and then XORs the result with the integer 34. The function ϕ of the last round can be omitted. Our aim is to find the key of the last round, given that all p/c-pairs are known. The only homomorphic one-round threefold sum that is independent of the input is $\mathrm{QR}(X) \oplus \mathrm{QR}(Y) \oplus \mathrm{QR}(K)$ and has imbalance $I^{\mathrm{QR}} = \frac{21}{64}$, where QR is the quadratic residues modulo $2^8 + 1$ function. The parent I/O sum of the "two and a half"-round threefold sum

$$\mathrm{QR}(X) \oplus \mathrm{QR}(Y^{(2)} \odot K^{(3)}) \oplus \mathrm{QR}(K^{(1)}) \oplus \mathrm{QR}(K^{(2)}) \oplus \mathrm{QR}(K^{(3)})$$

with imbalance $(\frac{21}{64})^2 = 10.77\%$ is used in our attack. The success probability is $p_{\mathrm{GLC}} \approx 5.5\%$, whereas linear and differential cryptanalysis as well as random key guessing yield a success probability of only about 0.39%.

7.2 Cryptanalysis of IDEA

We now apply the procedure for finding effective homomorphic threefold sums to the cipher IDEA [1], earlier called IPES [2]. The round function is a function on 64-bit words, consisting of a group operation denoted by $\otimes$, a keyed involution In and a permutation P_I. Each 64-bit word X can be considered as a concatenation of four 16-bit words $X1, X2, X3, X4$ and denoted as a 4-tuple. The group operation is defined as $X \otimes K := (X1 \odot K1, \ X2 \boxplus K2, \ X3 \boxplus K3, \ X4 \odot K4)$, where $\odot$ denotes multiplication modulo $2^{16} + 1$ with 0 representing 2^{16}, and $\boxplus$ addition modulo 2^{16}. As there exists only one non-constant homomorphism for $\boxplus$ and only one for $\odot$, there exist $2^4 - 1$ homomorphisms for $\otimes$, namely the functions

$$f_i(X) = (i1 \cdot \mathrm{QR}(X1)) \oplus (i2 \cdot \mathrm{LSB}(X2)) \oplus (i3 \cdot \mathrm{LSB}(X3)) \oplus (i4 \cdot \mathrm{QR}(X4)) \quad (13)$$

for all binary four-tuples $i = i1\,i2\,i3\,i4$ different from 0000, where QR is the quadratic-residue modulo $2^{16} + 1$ function.

A homomorphic I/O sum for IDEA can be characterized by the *IDEA-mask* (a, b) where the non-zero 4-tuple a is the mask of the input function and the non-zero 4-tuple b the mask of the output function.

We try to find effective homomorphic one-round I/O sums. For some of the 225 IDEA-masks, e.g., $(1110, 0100)$, we can show that the average-key imbalance, and thus all key-dependent imbalances, are zero. For other IDEA-masks, it is computationally infeasible to evaluate the key-dependent imbalances exactly. For the mini-cipher IDEA(8), the average-key imbalance of all one-round homomorphic I/O sums are zero. For IDEA(16), the one-round I/O sums with IDEA-masks $(1111, 1011)$, $(1101, 1111)$, $(1011, 1001)$, and $(1001, 1101)$ have average-key imbalance 0.002441, the four with $(1111, 1001)$, $(1001, 1111)$, $(1101, 1101)$, and $(1011, 1011)$ have average-key imbalance 0.00122, and all other I/O sum average-key imbalances are zero. Moreover, the number of p/c-pairs that must be analyzed in the generalization of linear cryptanalysis is about the square of the key-dependent imbalance and is here far larger than the total number of p/c-pairs. We conclude that the procedure for finding effective homomorphic threefold sums does not find any effective threefold sum for IDEA(8) and IDEA(16). Furthermore, the maximum key-dependent homomorphic I/O sum imbalance is only 0.00586. As this is only slightly larger than the maximum I/O sum average-key imbalance, there are no weak keys for the MA-box with respect to our attack. These conclusions doubtlessly hold true for (full-sized) IDEA as well. Thus IDEA seems secure against the generalization of linear cryptanalysis.

7.3 Cryptanalysis of SAFER K-64

SAFER K-64 is an iterated block-cipher, presented by Massey in [4]. The round function of SAFER K-64 consists of two half-rounds, each consisting of a keyed group operation and an unkeyed bijection either consisting of exponential and logarithm functions modulo 257 or a "Pseudo-Hadamard Transform". We have been able to prove that the procedure for finding effective homomorphic threefold sums for "one and a half" or more rounds of SAFER K-64 does not find a homomorphic threefold sum with non-zero imbalance. We conclude that SAFER K-64 is secure against the procedure for finding effective homomorphic threefold sums.

8 Conclusion

We have defined a generalization of linear cryptanalysis of iterated block ciphers and focused on its basic attack, which exploits an effective $(r - 1)$-round I/O sum to find information about the key of the last round. We have given sufficient conditions for a successful basic attack. These results can be extended to non-basic attacks in a manner similar to Matsui's improvements on basic linear cryptanalysis [6].

We have given a careful analysis of the applicability of Matsui's piling-up lemma. For the family of ciphers that insert keys by group operations, we have developed a procedure for finding some (arguably the best, but not necessarily all) effective multi-round threefold sums. This procedure requires finding homomorphisms for the used group operations. For ciphers using XOR (such as DES), the procedure finds only linear threefold sums, which are the same as Matsui's linear expressions. For ciphers using modular addition and multiplication with large moduli (such as IDEA), the choice of homomorphic sums is severely limited so that such ciphers tend to be immune to our generalization of linear cryptanalysis.

Finally, we argued that IDEA is secure against the generalization of linear cryptanalysis by showing that the presented procedure for finding effective homomorphic threefold sums finds no effective threefold sum for IDEA(8) or for IDEA(16). Similarly, the procedure for finding effective homomorphic threefold sums finds no effective threefold sum for "one and a half" (or more) rounds of SAFER K-64.

9 Acknowledgments

It is a pleasure to thank Richard De Moliner, Thomas Jakobsen, Kenneth Paterson, and Christian Waldvogel for helpful discussions. Thanks also to Patrick Berny, Steve Perkins, and Tobias Zürcher for their enthusiastic assistance in the cryptanalysis of IDEA and SAFER K-64.

References

1. X. Lai, *On the Design and Security of Block Ciphers*, vol. 1 of *ETH Series in Information Processing*. Hartung-Gorre Verlag Konstanz, J. L. Massey ed., 1992. ISBN 3-89191-573-X.
2. X. Lai, J. L. Massey, and S. Murphy, "Markov ciphers and differential cryptanalysis," in *Advances in Cryptology – Eurocrypt'91*, LNCS 574, pp. 17–38, Springer, 1991.
3. S. K. Langford and M. E. Hellman, "Differential-linear cryptanalysis," in *Advances in Cryptology – Crypto'94*, LNCS 839, pp. 17–25, Springer, 1994.
4. J. L. Massey, "SAFER K-64: A byte-oriented block-ciphering algorithm," in *Fast Software Encryption* (R. Anderson, ed.), LNCS 809, pp. 1–17, Springer, Dec. 1993.
5. M. Matsui, "Linear cryptanalysis method for DES cipher," in *Advances in Cryptology – Eurocrypt'93*, LNCS 765, pp. 386–397, Springer, 1993.
6. M. Matsui, "The first experimental cryptanalysis of the data encryption standard," in *Advances in Cryptology – Crypto'94*, LNCS 839, pp. 1–11, Springer, 1994.
7. M. Matsui, "On correlation between the order of S-boxes and the strength of DES," in *Advances in Cryptology – Eurocrypt'94*, 1994.
8. S. Murphy, F. Piper, M. Walker, and P. Wild, "Likelihood estimation for block cipher keys," submitted for publication, 1994.

On the Efficiency of Group Signatures Providing Information-Theoretic Anonymity

Lidong Chen[1] and Torben P. Pedersen[2]

[1] Mathematics Department, Texas A & M University, College Station, TX 77843 – 3368, USA, email: Lily.Chen@math.tamu.edu[***]

[2] Department of Comp. Sci., Aarhus University, Ny Munkegade, DK-8000 Århus C, Denmark, email: tppedersen@daimi.aau.dk[†]

Abstract. Group signatures, introduced by Chaum and van Heijst at Eurocrypt'91, allow members of a group to make signatures on behalf of the group while remaining anonymous. Furthermore, in case of disputes a designated group authority, who is given some auxiliary information, can identify the signer. Chaum and van Heijst presented four schemes, one of which protects the anonymity of the signer information-theoretically. However, this scheme as well as subsequent schemes with this property requires that the signer basically needs a new secret key for each signature and that the group authority secretly stores a very long string.

This paper analyses such group signature schemes and obtains lower bounds on the length of both the secret keys of the group members and the auxiliary information of the authority depending on the number of signatures each is allowed to make and the number of group members. These bounds are optimal as they are met by the scheme suggested by Chaum and van Heijst.

1 Introduction

Group signatures, introduced in [CH91], allow members of a group (e.g. a company or family) to make signatures on behalf of the group in such a way that

- only members can make signatures, and
- the actual member who made a given signature remains anonymous except that
- in case of dispute a designated authority (who is given some extra information) can identify the signer.

Such a signature scheme can, for example, be used in invitations to submit tenders. All companies submitting a tender then form a group and each company signs its tender anonymously using the group signature. Later, when the preferred tender has been selected, the winner can be identified, whereas the signers of all other tenders will remain anonymous. All submitters are bound to their tender by the signature, as the signer can be identified without his cooperation.

[***] Work done while visiting Aarhus University

[†] Supported by Carlsbergfondet

L.C. Guillou and J.-J. Quisquater (Eds.): Advances in Cryptology - EUROCRYPT '95, LNCS 921, pp. 39-49, 1995.

1.1 Related Work

Group signatures should not be confused with the related notion of group oriented signatures first suggested in [Boy89b] and [CH89]. Here certain subsets of a group of people are allowed to sign on behalf of the group. Such schemes do not provide a method for identifying the (subset of the) members who actually made the signature (see [D93] for an overview). Another related concept is that of multi-signatures which require a digital signature from many persons (see [O88] and [OO93]).

Chaum and van Heijst present in [CH91] (see also [H92]) four group signature schemes: one protects the anonymity of the signer unconditionally, whereas the other three only give computational anonymity. The scheme giving information-theoretic anonymity is very simple and works as follows given any digital signature scheme. Each member chooses a pair of keys for the signature scheme and sends the public key secretly to the authority. When the authority has received all public keys, it forms the public key of the group as a list of all the individual public keys in random order. Each member can now sign a message using the secret key. The receiver verifies that the signature is valid with respect to one of the keys in the public key of the group. Only the authority, who knows the correspondence between public keys and group members, will be able to identify the originator of a given signature.

A serious drawback of this scheme is that if a member wants to sign many messages he needs a new key pair for each message (otherwise the signatures can be linked). The secret key of each group member, the auxiliary information and the public key of the group therefore get longer the more signatures a member is allowed to make. Other disadvantages are

- Each member, P, is only protected against framing (i.e., other persons making a signature for which the authority will identify P as the signer) under a cryptographic assumption. This problem can be remedied by replacing the digital signature scheme by unconditionally secure signatures (see [CR91]) or fail-stop signatures (see [WP90]).
- If a new person wants to be a member, all other members must select a new key-pair. Otherwise, it is easy to identify the public key of the new member. This problem has been solved in [CP94a].

A scheme solving both of these problems, while retaining unconditional anonymity is presented in [CP94b].

1.2 Results and Contents

Although, the scheme of Chaum and van Heijst has been improved in some ways, it has not been possible to construct a scheme which is more efficient in terms of the length of secret keys and auxiliary information. This paper explains that lack of efficiency by giving lower bounds on the sizes of these (see Section 3). These bounds say that the length of the secret key of each member grows as $T \log_2 n$, if each member can make T signatures and n is the number of members. Similarly, the length of the auxiliary information of the authority grows as $Tn \log_2 n$.

In order to obtain these lower bounds a definition of group signatures with information-theoretic anonymity is needed. This is given in Section 2.

2 Definitions

This section defines secure group signatures giving information-theoretic anonymity. Throughout this paper $\mathcal{M}$ denotes the message space.

As computational model we use Turing machines and interactive Turning machines as defined in [GMR89]. In protocols (specifically, in *gen* in Definition 1 below) it is assumed that each pair of participants can send secret messages to each other.

Definition 1. A group signature scheme for a group of n members $P_1, \ldots, P_n$ and an authority A is a tuple $(n, k, gen, sign, test, iden)$. Here k is the security parameter, *gen* is a protocol involving $n + 1$ polynomially bounded participants and *sign*, *test*, *iden* are all polynomial time (in k) algorithms.

- On secret inputs (k, n, ρ_0) from A and ρ_i from P_i, $(i = 1, 2, \ldots, n)$, where each of $\rho_0, \rho_1, \ldots, \rho_n$ is a random bit-string, *gen* produces a common output, *pk*, secret output *aux* to A and secret output s_i to each P_i. Here, *pk* is the public key of the group, s_i is the secret key of P_i $(i = 1, 2, \ldots, n)$ and *aux* is the auxiliary information for A.
- *sign* is a *probabilistic* algorithm which on input s_i and $m \in \mathcal{M}$ outputs $sign(s_i, m)$. A string $\sigma \in \{0, 1\}^*$ is called a *correct signature* on $m \in \mathcal{M}$, if there exists $i \in \{1, 2, \ldots, n\}$ such that $\sigma = sign(s_i, m)$.
- *test* is used to test signatures. On input *pk*, m, and a possible signature on m, it outputs *true* or *false*. A string σ is called an *acceptable signature* on m with respect to *pk* if $test(pk, m, \sigma) = true$.
- *iden* is used by A to identify the signer. On input *aux*, $m \in \mathcal{M}$ and an acceptable signature on m, it outputs $i \in \{1, 2, \ldots, n\} \cup \{?\}$ (the output ? indicates that *iden* could not identify the signer).

For any $i \in \{1, 2, \ldots, n\}$, and any $m \in \mathcal{M}$, the scheme must satisfy

$$test(pk, m, sign(s_i, m)) = true \qquad (1)$$

and

$$iden(aux, m, sign(s_i, m)) = i \qquad (2)$$

Remark. From (2) it immediately follows that different secret keys produce different signatures:

$$\forall i, j \in \{1, 2, \ldots, n\} \, \forall m \in \mathcal{M} : i \neq j \Rightarrow sign(s_i, m) \neq sign(s_j, m).$$

Remark. From (1) it follows that a correct signature is also acceptable (but an acceptable signature is not necessarily correct).

According to the informal description in the introduction a group signature scheme must

- be secure against forgeries;
- provide anonymity of the signer; and
- enable the authority to identify the signer.

Each of these properties will be defined in the following.

2.1 Security Against Forgeries

It must be infeasible to forge signatures in adaptively chosen message attacks (see [GMR88]). Let $\mathcal{F}$ be a polynomial time algorithm, which on input pk and possibly aux, works as follows.

1. Repeat the following:
    (a) Generate a message $m \in \mathcal{M}$ and $i \in \{1, 2, \ldots, n\}$;
    (b) Get $sign(s_i, m)$.
2. Output a message $m_0 \in \mathcal{M}$ different from all m's generated above and $\tilde{\sigma}(m_0)$.

Definition 2. Let a group signature scheme $(n, k, gen, sign, test, iden)$ and T, polynomial in k, be given. The scheme is *secure against forgeries* after signing T messages if the following holds: For any polynomial time $\mathcal{F}$ as above getting at most T signatures from each P_i, for all but a negligible fraction of the keys (distributed according to gen),

$$\forall c > 0, \exists k_0, \forall k > k_0$$
$$Prob[test(pk, m_0, \tilde{\sigma}(m_0)) = true] \leq k^{-c},$$

where $(m_0, \tilde{\sigma}(m_0))$ is the output of $\mathcal{F}$. The probability is over the random coins of signatures and the random coins of $\mathcal{F}$.

2.2 Anonymity

Every group member should be able to make signatures on behalf of the group without leaking any (Shannon) information about his identity — only the group authority must be able to link signatures to members. Thus in the definition of anonymity, the authority can be trusted, but some group members may try to identify other members. As we are interested in information-theoretic anonymity, these curious members may have unlimited computing power.

Let a non-empty subset $J \subseteq \{1, 2, \ldots, n\}$ be given. The members in J are assumed to be honest, but the members outside J may deviate from the prescribed methods (no assumption about computing power) — these members are denoted by $\tilde{P}_j$ for $j \notin J$. Consider an execution of *gen* by A, $(P_j)_{j \in J}$ and $(\tilde{P})_{j \notin J}$, and assume that this protocol ends with a public key, *pk*, of the group and secret keys sk_j of P_j for $j \in J$ (otherwise the group is not set up properly). Let $view_J$ denote the view (including *pk*) of the faulty participants (see [GMR89]), and let $SK(view_J)$ denote the set of possible secret keys of $(P_j)_{j \in J}$ given this view. This set is equipped with a distribution induced from the random coins of the authority and $(P_j)_{j \in J}$. In the following let $J = \{a_1, a_2, \ldots, a_{|J|}\}$. Then $SK(view_J)$ is a set of tuples $(sk_{a_1}, sk_{a_2}, \ldots, sk_{a_{|J|}})$.

For all positive integers t and L, $0 < L \leq |J|t$, define a subset of $J^L = J \times J \times \ldots \times J$ (L times) by

$$\mathcal{I}_J(t, L) = \{(i_1, \ldots, i_L) \in J^L \mid \forall j \in J : |\{l \in \{1, \ldots, L\} \mid i_l = j\}| \leq t\}.$$

Thus each $j \in J$ appears at most t times in $\underline{i} = (i_1, \ldots, i_L) \in \mathcal{I}_J(t, L)$. For $J = \{1, 2, \ldots, n\}$, $\mathcal{I}_J(t, L)$ will be denoted $\mathcal{I}(t, L)$.

If $\sigma(m_i)$ is a correct signature on $m_i \in \mathcal{M}$ for $i = 1, \ldots, L$, then $\sigma(\underline{m})$ denotes $(\sigma(m_1), \sigma(m_2), \ldots, \sigma(m_L))$. For every $\underline{i} \in \mathcal{I}_J(t, L)$, "$\sigma(\underline{m}) \Leftarrow \underline{i}$" denotes the event that there exists $(sk_{a_1}, sk_{a_2}, \ldots, sk_{a_{|J|}}) \in SK(view_J)$ such that for all $j \in \{1, 2, \ldots, L\}$:

$$sign(sk_{i_j}, m_j) = \sigma(m_j).$$

Definition 3. Let a group signature scheme $(n, k, gen, sign, test, iden)$ and T, polynomial in k, be given. The scheme provides *anonymity* for signing T messages if for any non-empty $J \subseteq \{1, 2, \ldots, n\}$ and any $\tilde{P}_{j \notin J}$ in the scenario described above, and for any $L \leq |J|T$ different messages

$$\underline{m} = (m_1, m_2, \ldots, m_L)$$

the following holds. Given correct signatures on these messages made by $(P_j)_{j \in J}$

$$\sigma(\underline{m}) = (\sigma(m_1), \sigma(m_2), \ldots, \sigma(m_L)),$$

where each P_j has made at most T signatures, then for any $\underline{i} \in \mathcal{I}_J(T, L)$,

$$Prob[\sigma(\underline{m}) \Leftarrow \underline{i}] = \frac{1}{|\mathcal{I}_J(T, L)|}.$$

The probability is over the choice of $(sk_{a_1}, sk_{a_2}, \ldots, sk_{a_{|J|}}) \in SK(view_J)$ and the random coins used in the signatures.

This definition can be generalised to allow the same message to be signed several times. However, we have chosen this definition as messages in practice will be unique (e.g., contain a time stamp) in order to prevent replay attacks.

2.3 Signer Identification

For any subset J of $\{1, 2, \ldots, n\}$, let $\mathcal{F}_J$ be a polynomial time algorithm, which works as follows:

1. Execute *gen* — the members not in J may deviate from the prescribed protocol.
2. Repeat the following:
   (a) Generate a message $m \in \mathcal{M}$, and a number $i \in J$;
   (b) Get $sign(s_i, m)$.
3. Output a message $m_0 \in \mathcal{M}$ different from all m's in 2 and an acceptable signature $\sigma(m_0)$ on m_0.

Definition 4. Let a group signature scheme $(n, k, gen, sign, test, iden)$ and T, polynomial in k, be given. The scheme provides *signer identification* for signing T messages if the following holds: For any subset J of $\{1, 2, \ldots, n\}$, and for any polynomial time algorithm $\mathcal{F}_J$ as above getting at most T signatures from each P_i $(i \in J)$,

$$\forall d > 0, \exists k_0, \forall k > k_0$$
$$Prob[iden(aux, m_0, \sigma(m_0)) \in \{1, 2, \ldots, n\} \setminus J] \geq 1 - k^{-d},$$

where $(m_0, \sigma(m_0))$ is the output of $\mathcal{F}_J$. The probability is over the random coins of $\mathcal{F}_J$, and the random coins used in *gen* and *sign*.

There are two aspects of this definition. Firstly, if $|J| = n - 1$ the signer must be identified by the authority with overwhelming probability. Secondly, it says that no subset of (polynomially bounded) group members can frame a member outside this subset.

Remark. If the dishonest members (i.e., those outside J) are allowed unlimited computing power, this definition gives unconditional security against framing.

2.4 Secure Group Signatures

The preceding three definitions give

Definition 5. A group signature scheme is *secure* for signing T messages, if it is secure against forgery and provides both anonymity and signer identification after each member has made at most T signatures.

Remark. The definition easily generalises to let P_i sign T_i messages, $i = 1, 2, \ldots, n$.

3 Lower Bounds

Based on the definition given above this section shows that the length of the secret keys and auxiliary information grows by the number of signatures and group members. In the following it is assumed that all members and the authority participates honestly when generating the keys, since we want to give lower bounds on correct keys.

3.1   Secret Keys

The main idea in the proof of the lower bound of the secret keys is to partition the set of possible secret keys of each member into nonempty, disjoint subsets. Then the number of possible secret keys is bounded by the number of subsets.

A public key pk, produced by gen, corresponds to a set of possible secret keys defined as

$$SK(pk) = \{(sk_1, sk_2, \ldots, sk_n) \mid \exists aux, \rho_0, \rho_1, \ldots, \rho_n :$$
$$gen(n, k, \rho_0, \rho_1, \ldots, \rho_n) = (pk, (sk_1, sk_2, \ldots, sk_n), aux)\}.$$

We will omit pk in the following. The set $SK^{(i)}$ is defined as all possible secret keys of P_i, $i = 1, 2, \ldots, n$, i.e. $SK^{(i)}$ is the projection of SK on the i'th coordinate. If $s_i \in SK^{(i)}$ denotes the actual secret key of P_i, then

$$(s_1, s_2, \ldots, s_n) \in SK.$$

For a t-tuple $\underline{i} = (i_1, i_2, \ldots, i_t) \in \{1, 2, \ldots, n\}^t$ and t different messages $\underline{m} = (m_1, m_2, \ldots, m_t)$ define for every $r, 1 \leq r \leq n$

$$SK_{\underline{i}}^{(r)}(\underline{m}) = \{sk \in SK^{(r)} \mid sign(sk, m_j) = sign(s_{i_j}, m_j), j = 1, 2, \ldots, t\},$$

where s_i is the secret key of P_i ($i = 1, 2, \ldots, n$). $SK_{\underline{i}}^{(r)}(\underline{m})$ is the set of possible keys of P_r which will give P_{i_j}'s signature on m_j for $j = 1, 2, \ldots, t$.

Lemma 6. *If a group signature scheme $(n, k, gen, sign, test, iden)$ provides anonymity for signing T messages, then for any $t \leq T$, the following holds: For all $\underline{i} = (i_1, i_2, \ldots, i_t)$, and any t different messages $\underline{m} = (m_1, m_2, \ldots, m_t)$,*

$$SK_{\underline{i}}^{(r)}(\underline{m}) \neq \emptyset \qquad for \ r = 1, 2, \ldots, n.$$

Proof. Assume there exist $t \leq T$ different messages $\underline{m} = (m_1, \ldots, m_t)$, and $\underline{i} = (i_1, i_2, \ldots, i_t)$, such that

$$SK_{\underline{i}}^{(r_0)}(\underline{m}) = \emptyset,$$

for some r_0. Let $\sigma(m_j) = sign(s_{i_j}, m_j)$, $j = 1, 2, \ldots, t$ and $\underline{i_0} = (r_0, r_0, \ldots, r_0)$. Then

$$Prob[\sigma(\underline{m}) \Leftarrow \underline{i_0}] = 0,$$

which contradicts the definition of anonymity. $\qquad\qquad\square$

Theorem 7. *Let a group signature scheme $(n, k, gen, sign, test, iden)$ be given. If it provides anonymity for signing T messages, then for any $r \in \{1, 2, \ldots, n\}$,*

$$|SK^{(r)}| \geq n^T.$$

Proof. First, for any $t \leq T$ different messages $\underline{m} = (m_1, m_2, \ldots, m_t)$, if

$$\underline{i} = (i_1, i_2, \ldots, i_t) \neq (i'_1, i'_2, \ldots, i'_t) = \underline{i'},$$

then

$$SK_{\underline{i}}^{(r)}(\underline{m}) \cap SK_{\underline{i'}}^{(r)}(\underline{m}) = \emptyset.$$

Otherwise there exists

$$sk \in SK_{\underline{i}}^{(r)}(\underline{m}) \cap SK_{\underline{i'}}^{(r)}(\underline{m}),$$

such that for some $j \in \{1, 2, \ldots, n\}$, $i_j \neq i'_j$,

$$sign(sk, m_j) = sign(s_{i_j}, m_j) \quad \text{and} \quad sign(sk, m_j) = sign(s_{i'_j}, m_j),$$

which contradicts Definition 1 (see the remark following that definition).

Second, by Lemma 6, for any t different messages $\underline{m} = (m_1, \ldots, m_t)$, and any t-tuple $\underline{i} = (i_1, i_2, \ldots, i_t) \in \{1, 2, \ldots, n\}^t$,

$$|SK_{\underline{i}}^{(r)}(\underline{m})| \geq 1.$$

Finally, for any t different messages $\underline{m} = (m_1, m_2, \ldots, m_t)$

$$|SK^{(r)}| \geq \sum_{\underline{i} \in \{1, 2, \ldots, n\}^t} |SK_{\underline{i}}^{(r)}(\underline{m})| \geq n^t,$$

for any $t \leq T$. $\qquad\qquad\qquad\qquad\qquad\qquad\qquad\qquad\qquad\qquad\qquad$ $\square$

Thus each member must have a secret key chosen from a set of at least n^T possible secret keys. In other words, at least $T \log n$ bits are needed to represent some of the secret keys of each group member. Thus, the length of secret keys grows linearly in the number of signatures.

3.2 Auxiliary Information

In this section, we consider the length of the auxiliary information held by the authority. Let $(n, k, gen, sign, test, iden)$, T and an integer L, $0 < L \leq nT$ be given. Consider the following experiment given L different messages $m_1, m_2, \ldots, m_L$:

1. Generate $(pk, (s_1, \ldots, s_n), aux)$ correctly using gen.
2. Choose $(i_1, i_2, \ldots, i_L) \in \mathcal{I}(T, L)$ uniformly at random.
3. Let an (L, T)-history $hist_L(\underline{m}) = (pk, (m_1, \sigma_1), \ldots, (m_L, \sigma_L))$ be defined by

$$\sigma_j = sign(s_{i_j}, m_j) \qquad \text{for } j = 1, \ldots, L.$$

Let AUX be the random variable of the authority's auxiliary information (defined on the probability space induced by gen). Let ID be the uniformly distributed random variable taking the value $(i_1, i_2, \ldots, i_L) \in \mathcal{I}(T, L)$.

The following lemma follows immediately from the definition of unconditional anonymity.

Lemma 8. *If the group signature scheme $(n, k, gen, sign, test, iden)$ provides anonymity for signing T messages, then for any (L, T)-history $hist_L(\underline{m})$, ID is uniformly distributed on $\mathcal{I}(T, L)$. Especially, the conditional entropy of ID given $hist_L(\underline{m})$ is*

$$H(ID \mid hist_L(\underline{m})) = \log_2 |\mathcal{I}(T, L)| = \log_2 \left(\frac{(Tn)!}{(T!)^n} \right).$$

Theorem 9. *If the group signature scheme $(n, k, gen, sign, test, iden)$ provides anonymity for signing T messages and signer identification, then*

$$H(AUX) \geq Tn(\log n - 1).$$

Proof. Let $L = Tn$, and consider an (L, T)-history, $h = hist_L(\underline{m})$. The entropy of AUX can be written

$$
\begin{aligned}
H(AUX) &\geq H(AUX \mid h) \\
&= H(AUX, ID \mid h) - H(ID \mid AUX, h) \\
&= H(AUX|ID, h) + H(ID \mid h) - H(ID|AUX, h).
\end{aligned}
$$

Requirement (2) of Definition 1 implies that $H(ID|AUX, h) = 0$ and thus

$$H(AUX) \geq H(ID \mid h) + H(AUX|ID, h) \geq H(ID \mid h).$$

From the lemma above,

$$H(ID \mid h) = \log \frac{(Tn)!}{(T!)^n}.$$

Stirlings Formula

$$n! \approx e^{-n} n^n \sqrt{2\pi n}$$

gives

$$\log \frac{(Tn)!}{(T!)^n} \approx Tn \log n + \log \sqrt{2\pi Tn} - n \log \sqrt{2\pi T} \geq Tn(\log n - 1).$$

This completes the proof. ☐

This bound can be interpreted as follows. The authority needs some information corresponding to each signature that each member is allowed to make — in total nT pieces. Each of these must be unique for the member — this requires $\log n$ bits.

3.3 Comparison with Upper Bound

In the construction of Chaum and van Heijst the length of the auxiliary information is

$$\log\left(\frac{(Tn)!}{(T!)^n}\right).$$

This is exactly the bound which was obtained above.

Furthermore, if a secret key of the digital signature scheme used in this construction requires K bits then the length of the secret key of each member is KT bits. This should be compared with the bound $T\log_2 n$. Since all secret keys must be different, $K \geq \log_2 n$. Thus, except for the length of the secret keys of the given digital signature scheme (which we have only bounded by $\log_2 n$) the upper and lower bounds meet.

4 Conclusion

A detailed definition of group signatures with information-theoretic anonymity has been given, and it has been shown that in such schemes the length of the secret keys and the auxiliary information grows linearly in the number of signatures. These bounds only require anonymity and that the authority can identify signers of correct signatures, but the definitions of security against forgery and signer identification are not used.

On the one hand these bounds say that the scheme of Chaum and van Heijst is optimal, on the other they imply that such group signature schemes have some limits which might make them less attractive in some applications.

Some group signature schemes offering only computational anonymity have been suggested (e.g., see [CH91] and [CP94a]), but it is still an open problem to construct efficient such schemes, which can be proved secure under a "common cryptographic assumption".

References

[Boy89b] C. Boyd. Digital Multisignatures. In *Cryptography and Coding*, pages 241 – 246, 1989.

[CH91] D. Chaum and E. van Heijst. Group Signatures. In *Advances in Cryptology - proceedings of EUROCRYPT 91*, Lecture Notes in Computer Science #547, pages 257-265. Springer-Verlag, 1991.

[CR91] D. Chaum and S. Roijakkers Unconditionally Secure Digital Signatures In *Advances in Cryptology - proceedings of CRYPTO 90*. Lecture Notes in Computer Science #537, pages 206–214, 1991.

[CP94a] L. Chen and T. P. Pedersen New Group Signature Schemes. In *Advances in Cryptology - Proceedings of Eurocrypt '94*.

[CP94b] L. Chen and T.P. Pedersen. Group Signatures: Unconditional Security for Members. Technical Report DAIMI PB – 481, Aarhus University, September 1994.

[CH89]   R. A. Croft and S. P. Harris. Public-Key Cryptography and Reusable Shared Secrets. In *Cryptography and Coding*, pages 189 – 201, 1989.

[D93]   Y. Desmedt. Threshold Cryptosystems. In *Advances in Cryptology - proceedings of AUSCRYPT 92*, Lecture Notes in Computer Science #718, pages 3–14, 1993.

[GMR88] S. Goldwasser, S. Micali, and R. L. Rivest. A Digital Signature Scheme Secure Against Adaptive Chosen Message Attack. *SIAM Journal on Computing*, 17(2):281 – 308, April 1988.

[GMR89] S. Goldwasser, S. Micali, and C. Rackoff. The Knowledge Complexity of Interactive Proof-Systems. *SIAM Journal of Computation*, 18(1):186–208, 1989.

[H92]   E. van Heijst. *Special Signature Schemes*. PhD thesis, CWI, 1992.

[O88]   T. Okamoto. A Digital Multisignature Scheme Using Bijective Public-Key Cryptosystems. *ACM Trans. on Comp. Sys.*, 6(8):432 – 441, 1988.

[OO93]   K. Ohta and T. Okamoto. A Digital Multisignature Scheme Based on the Fiat-Shamir Scheme. In *Advances in Cryptology - proceedings of ASIACRYPT 91*, Lecture Notes in Computer Science #739, pages 139 – 148. Springer-Verlag, 1993.

[WP90]   M. Waidner and B. Pfitzmann. The Dining Cryptographers in the Disco: Unconditional Sender and Recipient Untraceability with Computationally Secure Serviceability In *Advances in Cryptology - proceedings of Eurocrypt 89*. Lecture Notes in Computer Science #434, page 690, 1990.

Verifiable Signature Sharing

Matthew K. Franklin   Michael K. Reiter

AT&T Bell Laboratories, Holmdel, New Jersey, USA
franklin,reiter@research.att.com

Abstract. We introduce Verifiable Signature Sharing (VΣS), a cryptographic primitive for protecting digital signatures. VΣS enables the holder of a digitally signed document, who may or may not be the original signer, to share the signature among a set of proxies so that the honest proxies can later reconstruct it. We present efficient VΣS schemes for exponentiation based signatures (e.g., RSA, Rabin) and discrete log based signatures (e.g., ElGamal, Schnorr, DSA) that can tolerate the malicious (Byzantine) failure of the sharer and a constant fraction of the proxies. We also describe our implementation of these schemes and evaluate their performance. Among the applications of VΣS is the incorporation of digital cash into multiparty protocols, e.g., to enable cash escrow and secure distributed auctions.

1 Introduction

In this paper we introduce a new cryptographic primitive for protecting digital signatures, called Verifiable Signature Sharing (VΣS). VΣS enables the holder of a digitally signed document, who may or may not be the original signer, to *share* the signature among a set of *proxies* so that the honest proxies can later *reconstruct* it. At the end of the sharing phase, each proxy can verify whether a valid signature for the document can be reconstructed, even if the original signature holder and/or some proxies are malicious. In addition, malicious proxies gain no information about the signature held by an honest sharer prior to reconstruction (but do see the document itself).

While VΣS can be solved in theory using known cryptographic techniques, the approach taken here is to focus on *practical* solutions to this problem. In fact, our study of VΣS was motivated by an effort to implement common financial trading vehicles in computer systems. For instance, VΣS provides an elegant way to escrow digital cash [5, 7]: by verifiably sharing the bank's signature of a bank note, the moneyholder can escrow (in the true sense) the monetary value of the note among the proxies. That is, each proxy can verify that the proxies collectively possess the cash, but the cooperation of sufficiently many proxies is required to spend it. This ability, in turn, facilitates secure distributed financial services, such as sealed-bid auctions. By having bidders escrow cash bids with a set of auction servers, the winner's payment can be guaranteed while the losers' payments are protected. A separate paper reports on the implementation of an auction service using these techniques [14].

L.C. Guillou and J.-J. Quisquater (Eds.): Advances in Cryptology - EUROCRYPT '95, LNCS 921, pp. 50-63, 1995.
© Springer-Verlag Berlin Heidelberg 1995

More generally, VΣS has applications whenever a signed document should become valid only under certain conditions (e.g., a will, a "springing power of attorney" [17], or an exchange of contracts). Verifiably sharing the document's signature, with "trigger" instructions given to all proxies, ensures that the signature will not be released until the honest proxies believe that the triggering events have occurred.

We have developed VΣS schemes for many digital signature schemes, including RSA [24] (with public exponent 3), Rabin [22], ElGamal [12], Schnorr [25], and the Digital Signature Standard [11]. Our protocols are simple and efficient. Sharing requires a single broadcast from the signature holder to the proxies, followed by a single round of broadcasts among the proxies. Reconstruction requires no interaction, beyond a single message sent from each proxy to the reconstructor. Our protocols can tolerate a malicious sharer and a constant fraction of malicious proxies. Almost all of our protocols ensure the secrecy of the signature in a strong sense (related to simulatability), but we consider a weaker, heuristic notion of secrecy as well. Some of our protocols exploit precomputation by the proxies to improve performance. We summarize our results in Table 1, which includes performance data in milliseconds for our protocols on a network of SPARCstation 10s, for 512-bit moduli and the minimum number of proxies that can tolerate one faulty proxy.

scheme	faults	precomp?	secrecy	test size	share time	recon time
RSA ($e = 3$)	$\lfloor(n-1)/5\rfloor$	yes	strong	$n = 6$	330 ms	6 ms
Rabin	$\lfloor(n-1)/4\rfloor$	yes	strong	$n = 5$	251 ms	3 ms
ElGamal	$\lfloor(n-1)/3\rfloor$	no	strong	$n = 4$	862 ms	321 ms
Schnorr	$\lfloor(n-1)/3\rfloor$	no	strong	$n = 4$	514 ms	117 ms
DSA	$\lfloor(n-1)/3\rfloor$	no	weak	$n = 4$	587 ms	116 ms

Table 1. Summary of results (n = number of proxies)

The rest of this paper is structured as follows. In Section 2, we place VΣS in the context of related work. Section 3 presents our system model and definitions. VΣS for exponentiation-based signature schemes is presented in Section 4, and for discrete log-based schemes in Section 5. VΣS schemes with heuristic secrecy are discussed in Section 6. Section 7 addresses the performance of the schemes, including a discussion of the numbers in Table 1. We conclude in Section 8.

2   Related Work

In this section, we describe previous cryptographic research to which Verifiable Signature Sharing is connected, and from which we have borrowed many of our techniques.

VΣS is related to, but different from, the idea of threshold signature schemes (see survey [10]). Such schemes enable any subset of sufficiently many members of an organization to sign documents on behalf of the organization. Some of the functionality of our VΣS schemes could be achieved by the sharer distributing the ability to sign the document among the proxies using a threshold signature scheme. One important difference is that the sharer must be the original signer of the document. A second difference is that a too-large coalition gains the ability to forge any document, rather than just reconstruct those signatures that are shared. We can state this second difference another way. Threshold signature schemes—and, more generally, function sharing schemes [9]—enable the efficient computation of a secret function (signing) on a public input (the document). VΣS enables the efficient computation of a public function (signature verification) on a shared secret input (the signature).

VΣS is related to a distributed verification protocol for undeniable signatures, due to Pedersen [21]. In an undeniable signature scheme, each signed document has associated with it some secret information, distinct from the signer's private key. To verify an undeniable signature of a document requires interaction with the possessor of the corresponding secret information. Pedersen shows how the original signer of a document can distribute the ability to verify it. In this case, the signed document is public, the secret information for that signature is shared among the verifiers, and the private key of the signer is uninvolved in the verification. Some of his methods, which in turn are based on ideas from Feldman [13], are central to our VΣS schemes for ElGamal, Schnorr, and DSA.

VΣS is related to "fair public-key cryptosystems" [20]. These schemes also enable certain public predicates, related to encryption and decryption keys, to be evaluated efficiently on shared secret inputs. The shared secret input in this case is a private decryption key. The public predicate is that the decryption key bears the appropriate inverse relationship to the public key.

The notion of secure distributed computation, or "mental games" [15, 2, 6], is also closely related. That work builds on secret sharing [26, 4] and verifiable secret sharing [8, 13, 21] techniques. General and powerful "completeness theorems" for securely evaluating arbitrary boolean or arithmetic circuits yield VΣS as a special case. However, the message complexity of these general solutions (and the number of encryptions) is typically a large constant times the size of the circuit. The circuit size of known public-key signature schemes is too large for these methods to be feasible. By adapting some of the techniques of earlier schemes in novel ways, we achieve VΣS with low communication complexity.

3   Model and Definitions

A VΣS scheme is a pair of protocols (for "sharing" and "reconstructing") involving a dealer D (called the "sharer" in the Introduction), n proxies $P_1, \ldots, P_n$, and a reconstructor R. The dealer begins with a document m and a signature $\sigma(m)$ for the document, created by itself or by another. At the end of the sharing protocol, the honest proxies either *accept* or *reject*. If the honest proxies accept, then they know m (but not $\sigma(m)$) at the end of the sharing protocol, and the reconstructor will know $\sigma(m)$ (and m if desired) at the end of the reconstruction protocol. The reconstructor does not participate in the sharing protocol, and the dealer does not participate in the reconstruction protocol.

Faulty parties (proxies and/or the dealer) are assumed to be controlled by a single adversary, who can see the contents of their memories, and can cause them to deviate from the protocol in an arbitrary manner (Byzantine faults). The reconstructor is assumed to be honest throughout the reconstruction protocol. Since our reconstruction protocols are non-interactive, the only effect of a faulty reconstructor is to prevent its own success.

We assume that the dealer can send an authenticated message privately to any proxy, and that any proxy can send an authenticated message privately to the reconstructor. We also assume that the dealer or any proxy can reliably broadcast a message to the set of proxies (Byzantine Agreement [19]). More precisely, each reliable broadcast in our VΣS protocols results in the same message being delivered to all honest proxies (and thus is *terminating* in the sense of [16]), and this message is the same as the message broadcast if the sender is honest.

Finally, the dealer and the proxies are assumed to know the public key of the original signer, which is assumed to be a valid public key.

3.1 t-Resilience

We say that our VΣS scheme is t-resilient if the following conditions hold:

Completeness: If the dealer is honest, and at most t proxies are faulty, then each honest proxy will accept.

Soundness: If at most t proxies are faulty and any honest proxy accepts at the end of the sharing protocol, then every honest proxy accepts at the end of the sharing protocol and reconstruction will be successful (regardless of whether the dealer is honest or faulty).

Secrecy: Anything that can be computed by an adversary that controls up to t faulty proxies, after participating in k sharing protocols with an honest dealer, can also be computed by the adversary without participating in any sharing protocols (i.e., from the k unsigned documents and the public keys of the original signer).

We will say that our VΣS scheme is t-resilient with "heuristic secrecy" if only completeness and soundness can be proven, while heuristic evidence supports some weaker version of secrecy, e.g., that no "useful" information can be computed from what is seen by a computationally bounded adversary.

3.2 Secret Sharing

Our protocols make extensive use of the polynomial based secret sharing scheme due to Shamir [26]. A dealer wishes to share a secret s, from a finite field F, among n parties. The dealer chooses $a_1, \ldots, a_t$ from the uniform distribution on F; this is denoted $a_1, \ldots a_t \in_R F$. Let $g(x) = a_t x^t + \cdots + a_1 x + s$. Party i is given the share $(i, g(i))$, $1 \le i \le n$. The parties $S \subseteq \{1 \ldots n\}$, $|S| = t + 1$, can recover the secret using the Lagrange formula for interpolation ($z = 0$): $g(z) = \sum_{i \in S} c_i g(i)$, where $c_i = \prod_{j \in S, j \ne i} (z - j)(i - j)^{-1}$. No subset of up to t shares yields any information about the secret.

Our VΣS schemes for exponentiation based signatures use polynomial based secret sharing over the ring Z_N, where N is the product of two large primes. The properties of secret sharing still hold in this case (e.g., see [9]). In particular, the Lagrange interpolation formula is well-defined as long as n is smaller than the prime factors of N (and thus $(i - j)^{-1} \bmod N$ always exists).

4   VΣS for Exponentiation Based Signatures

In this section, we present VΣS schemes for exponentiation based signature schemes, i.e., RSA with public exponent 3, and Rabin. The following is an informal description of how the protocol works for RSA. The signature is the cube root of the hash of a document. The dealer uses Shamir's polynomial based secret sharing scheme to share the signature among the n proxies, using some degree t polynomial. The dealer then broadcasts the corresponding points of the cube of this polynomial. The proxies convince themselves that (1) a value has been shared using a degree t polynomial; and (2) the cube of this polynomial shares the cube of the signature, i.e., the hash of the document itself. This is enough to ensure that reconstruction will be successful.

To improve the communication complexity of the sharing protocol, we will assume in this section that the proxies have performed some precomputation before participating in any sharing or reconstruction. Specifically, the ith proxy knows (only) the value at i of a large number of random degree t polynomials. One such "pre-shared" polynomial will be used per execution of the sharing protocol. The proxies can then convince themselves that the signature-sharing polynomial has degree t by adding one of the pre-shared polynomials to it, and verifying the degree of the sum. The proxies can precompute these pre-shared polynomials using standard multi-party secure computation protocols. The computation for all shared polynomials can proceed in parallel, requiring only a constant number of rounds of communication. Alternatively, a single trusted source can provide the proxies with such values, and then destroy itself.

4.1   RSA Signatures

Suppose that the signature scheme is ("hashed") RSA with encrypting exponent $e = 3$, decrypting exponent d, and modulus N. The dealer begins with $m, \sigma(m)$, where $\sigma(m) = (h(m))^d \bmod N$ for some publicly known one-way hash function h. Assume that $p_1(x), p_2(x), \ldots$ are a supply of random degree t polynomials shared among the proxies, i.e., each proxy P_i holds $p_1(i), p_2(i), \ldots$. No proxy knows anything further about these polynomials. The polynomial $p_j(x)$ will be used in the jth sharing protocol. We do not consider concurrent executions of the sharing protocol.

Sharing Protocol

1. D sends messages privately and by reliable broadcast to the proxies:
   a. D chooses $a_t, \ldots, a_1 \in_R Z_N$, and finds $f(x) = a_t x^t + \cdots + a_1 x + \sigma(m)$.
   b. $D \to P_i$ privately: $y_i = f(i) \bmod N$ (for all i, $1 \leq i \leq n$).
   c. $D \to P_1, \ldots P_n$ by reliable broadcast: $m, y_1^3 \bmod N, \ldots, y_n^3 \bmod N$.

2. Let the broadcast values (from 1c) be denoted $m, z_1, \ldots, z_n$. Assume that this is the jth sharing protocol in which the proxies are participating. Each proxy P_i computes $r_i = p_j(i) + y_i \bmod N$, and makes the following reliable broadcast to all proxies:
   a. $(r_i, \text{COMPLAIN})$, if $z_i \neq y_i^3 \bmod N$.
   b. (r_i, ALLOW), otherwise.

3. Without further communication, each proxy ACCEPTS if the following conditions are met (and REJECTS otherwise):
   a. The values $z_1, \ldots, z_n$ lie on a polynomial g of degree at most $3t$.
   b. $g(0) = h(m)$.
   c. The values $r_1, \ldots, r_n$ that were reliably broadcast (in 2) lie on a polynomial f^* of degree t, with at most t errors. (Any nonsensical response is counted as an error.)
   d. At most t proxies COMPLAINED (in 2a) or contributed an error to f^* (in 3c).

Reconstruction Protocol

1. Each proxy P_i sends to R the following information:
   a. The value y_i that was privately sent to it (in 1b of the sharing protocol);
   b. The values $z_1, \ldots, z_n$ that were reliably broadcast (in 1c of the sharing protocol);
   c. The error locations (proxy identities) in the degree t polynomial f^* computed at the end of the sharing protocol.

2. Without further communication, R reconstructs the signature as follows:
   a. R finds $z_1, \ldots, z_n$ by majority vote of the lists received in 1b of this protocol.
   b. R finds ERROR-LOCS, the majority vote of the lists of error locations received in 1c of this protocol.
   c. R discards every private share y_i (received in 1a of this protocol) such that $z_i \neq y_i^3 \bmod N$ or such that i is in ERROR-LOCS.
   d. R interpolates the remaining private shares (received in 1a of this protocol) to find a degree t polynomial (with *no* errors). The signature is taken to be the value of the resulting polynomial at zero.

Theorem 1 *This VΣS scheme is t-resilient whenever $n \geq 5t + 1$.*

Proof. It suffices to show that the three conditions for t-resilience are satisfied. Suppose at most t proxies are faulty.

Completeness: If the dealer is honest, then $z_1, \ldots, z_n$ lie on the polynomial $g(x) = (f(x))^3$, which is of degree at most $3t$ (since the degree of f is at most t). So, each honest proxy will recover g through interpolation. Only a faulty proxy will complain or give a bad r_i in step 2, so at most t are seen to do so by each honest proxy. Thus all honest proxies will accept.

Soundness: We claim that the reconstruction protocol will be successful, and all honest proxies will accept, whenever any honest proxy accepts at the end of the sharing protocol. Let $\hat{y}_i$ be the value sent privately by the (possibly faulty) dealer $\hat{D}$ to P_i, let $m, \hat{z}_1, \ldots, \hat{z}_n$ be the values broadcast by $\hat{D}$ to the proxies, and let $\hat{r}_i$ be the value broadcast by P_i to the other proxies. If any honest proxy accepts during the sharing protocol, then every honest proxy will accept (since the information on which they base their decisions was all received by reliable broadcast). It suffices to show that the reconstruction will be successful when all honest proxies accept.

If all honest proxies accept, then all found the same degree t polynomial through $\hat{r}_1, \ldots, \hat{r}_n$ with at most t errors. Since all honest proxies agree on the polynomial, all sent the same error locations to R, and so ERROR-LOCS will consist of exactly these locations (by majority vote). R will also recover $\hat{z}_1, \ldots, \hat{z}_n$ exactly as they were reliably broadcast by $\hat{D}$, again by majority vote from the honest proxies. Since a degree t polynomial passes through $\hat{r}_1, \ldots, \hat{r}_n$ minus points at ERROR-LOCS, a (different) degree t polynomial $\hat{f}$ passes through $\hat{y}_1, \ldots, \hat{y}_n$ minus points at ERROR-LOCS and faulty proxies (except with negligible probability). The points $\hat{y}_1, \ldots, \hat{y}_n$ minus points at complainer locations and faulty proxies have cubes that agree with $\hat{z}_1, \ldots, \hat{z}_n$.

Let $\hat{g}$ be the unique degree $3t$ polynomial that passes through $\hat{z}_1, \ldots, \hat{z}_n$. Then $\hat{f}^3$ and $\hat{g}$ agree at the points of all proxies minus complainers, error locations, and faulty proxies. Thus they have at least $n - 2t \geq 3t + 1$ points in common. Since both are degree $3t$, this means that $\hat{f}^3 = \hat{g}$. Thus it only remains to be shown that R recovers $\hat{f}$. The points from all honest proxies, minus error locations and complainers, are included in the interpolation by R, and a point from a faulty proxy is included only when it lies on $\hat{f}$ (since otherwise its cube will be inconsistent). Hence at least $t + 1$ points are included, and so $\hat{f}$ is recovered.

Secrecy: For every sharing protocol that is executed with an honest dealer, the adversary sees a random degree $3t$ polynomial g that passes through the hash of the document, a random degree t polynomial f^* (together with "allow" messages from the honest proxies), and at most t points on the cube root of g. This probability distribution can be sampled by an adversary who knows only the document that is being signed: choose f^* at random; choose t random values to be the cube roots, cube them, and then find a random degree $3t$ polynomial g that passes through the hash of the document and the cubes ($t + 1$ points). For any computation that involves participation in sharing protocols, the adversary could perform an equivalent computation by simulating its participation instead.

4.2 Rabin Signatures

Using Rabin's signature scheme, $\sigma(m) = \sqrt{h(m)} \bmod N$, where h is a one-way hash function and the factorization of N is the private key of the signer. The same scheme as for RSA signatures now works (with squares instead of cubes). Protection against t faulty proxies requires $n \geq 4t + 1$.

5 $V\Sigma S$ for Discrete Log Based Signatures

In this section, we present $V\Sigma S$ schemes for discrete log based signature schemes, illustrating our methods with ElGamal and Schnorr signature schemes.

5.1 Verifiable Discrete Log Sharing

Central to our $V\Sigma S$ schemes is a scheme to verifiably share the discrete log of a public value. This scheme is essentially the Verifiable Secret Sharing scheme due to Pedersen [21], although small differences arise from differences in security models. We assume that p, q, g are known to all parties before the start of the protocol, where p is a large prime, q is a large prime factor of $p - 1$, and where g has order q in Z_p^*, i.e., $\langle g \rangle = \{g^1 \bmod p, \ldots, g^{p-1} \bmod p\}$ has q elements. The dealer broadcasts α and shares a value β, claiming that (1) $\alpha \in \langle g \rangle$, and (2) β is the discrete log of α modulo p with respect to the base g. The sharing protocol ends with the proxies convinced of this claim.

Sharing Protocol

1. D sends messages privately and by reliable broadcast to the proxies:
   a. D chooses $a_t, \ldots, a_1 \in_R Z_q$, and finds $f(x) = a_t x^t + \cdots + a_1 x + \beta$.
   b. $D \to P_i$ privately: $\beta_i = f(i) \bmod q$ (for all i, $1 \leq i \leq n$).
   c. $D \to P_1, \ldots, P_n$ by reliable broadcast: $\alpha, g^{a_1} \bmod p, \ldots, g^{a_t} \bmod p$.
2. Let the broadcast values (from 1c) be denoted $\alpha, u_1, \ldots, u_t$. Each proxy P_i makes the following reliable broadcast to all proxies:
   a. COMPLAIN if $g^{\beta_i} \not\equiv \alpha \prod_{j=1}^{t} (u_j)^{i^j} \bmod p$.
   b. ALLOW otherwise.
3. Without further communication, each proxy ACCEPTS if the following are true (and REJECTS otherwise):
   a. At most t proxies COMPLAINED (in 2a). (Any nonsensical response is counted as a COMPLAIN.)
   b. $\alpha^q \equiv 1 \bmod p$.

Reconstruction Protocol

1. Each proxy P_i sends to R the following information:
   a. The value β_i that was privately sent to it (in 1b of the sharing protocol);
   b. The values $\alpha, u_1, \ldots, u_t$ that were reliably broadcast (in 1c of the sharing protocol);

2. Without further communication, R finds the discrete log of α as follows:

    a. R finds $\alpha, u_1, \ldots, u_t$ by majority vote of the lists received in step 1b of this protocol.

    b. R discards every private share β_i (received in 1a of this protocol) that is inconsistent with $\alpha, u_1, \ldots, u_t$, i.e., if $g^{\beta_i} \not\equiv \alpha \prod_{j=1}^{t}(u_j)^{i^j} \bmod p$.

    c. R interpolates the remaining private shares (received in 1a of this protocol) to find a polynomial in $Z_q[x]$ of degree at most t (with *no* errors). The discrete log is taken to be the value at zero of this polynomial.

The definition of t-resilience for VΣS from Section 3.1 can be extended in the obvious way to log sharing schemes. For the proof (of soundness) of the following theorem, it will be useful to define the "g-log" of an element of Z_p^*. Choose $g_1, \ldots, g_k$ such that every element $y \in Z_p^*$ can be written as $g^z \prod_{i=1}^{k} g_i^{z_i} \bmod p$ for exactly one z, $0 \leq z \leq q-1$. Call z the "g-log" of y mod p.

Theorem 2 *This log sharing scheme is t-resilient whenever $n \geq 3t + 1$.*

Proof. The conditions for t-resilience are met.

Completeness: No honest proxy complains in step 2a when the dealer is honest, and so at most t complaints are broadcast. Moreover, $\alpha^q \equiv 1 \bmod p$ for any $\alpha \in \langle g \rangle$, and so each honest proxy accepts.

Soundness: If any honest proxy accepts at the end of the sharing protocol, then all honest proxies accept (since the decision is based solely on information sent to all proxies by reliable broadcast). It suffices to show that reconstruction will be successful whenever all honest proxies accept. Let the broadcast from step 1c of the sharing protocol (and of step 2a of the reconstruction protocol) be $\hat{\alpha}, \hat{u}_1, \ldots, \hat{u}_t$. Let the private share sent to each proxy P_i in step 1b be $\hat{\beta}_i$. Let $\hat{a}_j$ be the "g-log" of $\hat{u}_j$ modulo p for all j, $1 \leq j \leq t$, i.e., $g^{\hat{a}_j} \equiv \hat{u}_j \prod_{i=1}^{k} g_i^{z_{ji}} \bmod p$ (for some $z_{j1}, \ldots, z_{jk}$). Let $\hat{\beta}$ be the g-log of $\hat{\alpha}$ modulo p, i.e., $g^{\hat{\beta}} \equiv \hat{\alpha} \prod_{i=1}^{k} g_i^{z_i} \bmod p$ (for some $z_1, \ldots, z_k$). Let $\hat{f}(x) = \hat{a}_t x^t + \cdots + \hat{a}_1 x + \hat{\beta}$. Suppose that $(i, \hat{\beta}_i)$ passes the test in step 2b of the reconstruction protocol, and thus is used by R in its interpolation in step 2c. Then $g^{\hat{\beta}_i} \equiv \hat{\alpha} \prod_{j=1}^{t}(\hat{u}_j)^{i^j} \bmod p$, and so $g^{\hat{\beta}_i} \equiv (\prod_{i=1}^{k} g_i^{z_i'}) g^{\hat{\beta}} \prod_{j=1}^{t} g^{\hat{a}_j i^j} \bmod p$ (for some $z_1', \ldots, z_k'$), and thus $g^{\hat{\beta}_i} \equiv (\prod_{i=1}^{k} g_i^{z_i'}) g^{\hat{f}(i)} \bmod p$. By uniqueness of g-logs, $\hat{\beta}_i \equiv \hat{f}(i) \bmod q$. Thus, in step 2c of reconstruction, R will interpolate only points that lie on $\hat{f}(x)$. Furthermore, the interpolation will include all points from honest proxies who did not complain in step 2a of the sharing protocol. Since there are at least $n - 2t \geq t + 1$ such points, R will successfully recover $\hat{f}(x)$, and thus find $\hat{f}(0) \bmod q$, the g-log of $\hat{\alpha}$. Since $\hat{\alpha}$ passed step 3b of the sharing protocol, $\hat{\alpha} \in \langle g \rangle$, i.e., the g-log of $\hat{\alpha}$ is in fact the discrete log of $\hat{\alpha}$.

Secrecy: For every sharing protocol that is initiated by an honest dealer, the adversary controlling t faulty proxies sees (1) the element $\alpha \in \langle g \rangle$ whose discrete log is to be shared, (2) $g^{a_1} \bmod p, \ldots, g^{a_t} \bmod p$, for random $a_1, \ldots, a_t \in_R Z_q^*$, and (3) t points on $f(x) = a_t x^t + \cdots + a_1 x + \beta \bmod p$ where β is the discrete log of

α. Suppose that the faulty proxies are $P_1, \ldots, P_t$. The probability distribution of the adversary's view of the protocol can be sampled by an adversary who knows only α, as follows. Choose the t points to be $(1, \beta_1), \ldots, (t, \beta_t)$, where $\beta_1, \ldots, \beta_t \in_R Z_q$. Let b_{ij} be the i, j entry of the inverse of the Van der Monde matrix whose i, j entry is i^j, $0 \leq i, j \leq t$. Then each $u_i = g^{a_i} \bmod p$ can be computed by $u_i = (\alpha)^{b_{i0}} \prod_{j=1}^{t} (g^{\beta_j b_{ij}}) \bmod p$.

5.2 ElGamal Signatures

Using the log sharing scheme from Section 5.1 (or simple variants), we can construct VΣS schemes for many variants of ElGamal signature [12]. For example, let the public key be g, p, y where p is a large prime with large prime factor q, g has order q in Z_p^*, and $y = g^x \bmod p$ for some x. Let the private key be x. The signature of a document m is given by $\sigma(m) = [r, s]$, where $r = g^k \bmod p$ for $k \in_R Z_q$, and where $s = k^{-1}(m - xr) \bmod q$. A signature can be publicly verified by checking that $g^m \equiv y^r r^s \bmod p$. This is a slight modification of the original scheme proposed by ElGamal (computing s modulo q instead of modulo $p - 1$).

Using the log sharing scheme from Section 5.1, we can construct a VΣS scheme for this signature scheme. The dealer D reliably broadcasts m, r to all proxies. Each proxy can now compute $g^m y^{-r} \equiv g^{m-xr} \equiv r^s \bmod p$. D now uses the scheme from Section 5.1 to verifiably share s to the proxies, i.e., to verifiably share the log of $r^s \bmod p$ with respect to the base r. Instead of step 3b of the sharing protocol, the proxies verify that $r^q \equiv 1 \bmod p$.

The reconstruction protocol for the VΣS scheme is essentially the same as the reconstruction protocol for the discrete log sharing scheme. The only modification is that each proxy also sends r to the reconstructor, who determines the actual r by majority vote.

Theorem 3 *This VΣS scheme is t-resilient whenever $n \geq 3t + 1$.*

Proof. The conditions for t-resilience are met.

Completeness: When the dealer is honest, each honest proxy accepts the log sharing protocol, and verifies that $r^q \boxminus 1 \bmod p$, and thus accepts.

Soundness: Suppose that the (possibly faulty) dealer broadcasts $m, \hat{r}$ to the proxies. If $\hat{r}^q \equiv 1 \bmod p$, then $\hat{r}$ has order q in Z_p^*, and thus the assumptions for the discrete log sharing scheme are met. The old step 3b of the sharing protocol of the discrete log scheme is unnecessary, since $\hat{r}^s \boxminus g^m y^{-\hat{r}} \bmod p$ is known to have order q by construction. If the honest proxies accept the log sharing protocol, then reconstruction will successfully recover $\hat{s}$ such that $\hat{r}^s \boxminus g^m y^{-\hat{r}} \bmod p$. Thus $g^m \equiv y^{\hat{r}} \hat{r}^{\hat{s}} \bmod p$, i.e., $[\hat{r}, \hat{s}]$ is a valid signature of m.

Secrecy: The adversary chooses a random $r \in \langle g \rangle$, and then simulates the log sharing protocol for $g^m y^{-r} \bmod p$, as described in part 3 of the proof of resilience for log sharing (Theorem 2).

5.3 Schnorr Signatures

In a Schnorr signature, the public key is g, p, y where p is a large prime, q is a large prime factor of $p - 1$, g has order q in Z_p^*, and $y = g^x \bmod p$ for some x. The private key is x. The signature of a document m is given by $\sigma(m) = [c, z]$ where $c = h(g^r \bmod p, m)$ for random r and one-way hash function h, and where $z = cx + r \bmod q$. A signature can be publicly verified by checking that $c = h(g^z y^{-c} \bmod p, m)$.

Using the log sharing scheme from Section 5.1, we can construct a VΣS scheme for Schnorr signatures. The dealer D reliably broadcasts m, c, u to all proxies, where $u = g^z \bmod p$. D shares z with the proxies so that they are convinced they hold shares of the log of $u \bmod p$ with respect to the base g. The proxies accept if they accept the log sharing protocol and $c = h(u y^{-c} \bmod p, m)$. In the reconstruction protocol, R finds u, c by majority vote from the proxies, and reconstructs z as in the log sharing scheme.

Theorem 4 *This VΣS scheme is t-resilient whenever $n \geq 3t + 1$.*

Proof. The conditions for t-resilience are met.

Completeness: When the dealer is honest, each honest proxy accepts the log sharing protocol, and sees that $c = h(u y^{-c} \bmod p, m)$.

Soundness: Suppose that the (possibly faulty) dealer broadcasts $m, \hat{u}, \hat{c}$ to all proxies. If the honest proxies accept, then (1) $\hat{c} = h(\hat{u} y^{-\hat{c}} \bmod p, m)$ and (2) R will recover $\hat{z}$ as the log of $\hat{u} \bmod p$ with respect to the base g. Thus $\hat{c} = h(g^{\hat{z}} y^{-\hat{c}} \bmod p, m)$, i.e., $[\hat{c}, \hat{z}]$ is a valid signature of m.

Secrecy: The adversary chooses a random r, finds $w = g^r \bmod p$, finds $c = h(w, m)$, and finds $u = w y^c \bmod p$. The adversary then follows the simulation from part 3 of the proof of resilience for log sharing (Theorem 2).

6 VΣS for the DSA with Heuristic Secrecy

In this section, we consider VΣS schemes for which the secrecy requirement is weakened. Previously, we have required that the adversary learns *nothing* from participating in these protocols, in the strong sense that its view could be simulated without participation. We replace this with a heuristic condition that the adversary *appears to learn nothing useful*. Of course, it is impossible to prove such an informal condition.

In the DSA, the public key is g, y, p, q where p is a large prime, q is a large prime factor of $p - 1$, g is a generator of Z_q^*, and $y = g^x \bmod p$ for some private key $x \in Z_q^*$. The signature of a document m is $\sigma(m) = [r, s]$, where $r = (g^k \bmod p) \bmod q$ for some $k \in_R Z_q^*$, and where $s = k^{-1}(h(m) + xr) \bmod q$ for a one-way hash function h into Z_q^*. To verify a signature, anyone can check that $r = (g^{u_1} y^{u_2} \bmod p) \bmod q$, where $u_1 = h(m) s^{-1} \bmod q$, and where $u_2 = rs^{-1} \bmod q$.

A VΣS scheme with heuristic secrecy for DSA proceeds as follows. D reliably broadcasts m, s, α to all proxies, where $\alpha = y^{u_2} \bmod p$. D does a verifiable share of u_2 to the proxies so that they are convinced it is the log of $\alpha \bmod p$ with respect to the base y. The proxies accept if they accept the log sharing protocol, and if $y^v \equiv \alpha^s \bmod p$, where $v = (g^{u_1}\alpha \bmod p) \bmod q$. The signature will be easy to reconstruct since $r = u_2 s \bmod q$.

Theorem 5 *This VΣS scheme is t-resilient with heuristic secrecy whenever $n \geq 3t + 1$.*

Proof. The conditions for t-resilience with heuristic secrecy are met.

Completeness: When the dealer is honest, each honest proxy accepts the log sharing protocol. Furthermore, the test $y^v \equiv \alpha^s \bmod p$ will succeed for every honest proxy, since it compares values that were reliably broadcast by the dealer.

Soundness: Suppose that the (possibly faulty) dealer broadcasts $m, \hat{s}, \hat{\alpha}$ to the proxies. If the honest proxies accept the log sharing protocol, then reconstruction will successfully recover $\hat{u}_2$, the y-log of $\hat{\alpha} \bmod p$. R will reconstruct the signature to be $[\hat{u}_2\hat{s} \bmod q, \hat{s}]$. We claim that this is a valid signature of m: $y^{\hat{u}_2\hat{s}} \equiv \hat{\alpha}^{\hat{s}} \equiv y^{\hat{v}} \equiv y^{(g^{\hat{s}_1}\hat{\alpha} \bmod p) \bmod q} \bmod p$, and thus $\hat{u}_2\hat{s} \bmod q = (g^{\hat{u}_1}\hat{\alpha} \bmod p) \bmod q = (g^{\hat{u}_1}y^{\hat{u}_2} \bmod p) \bmod q$. Thus $[\hat{u}_2\hat{s} \bmod q, \hat{s}]$ passes the test for being a valid signature of m.

Heuristic Secrecy: An adversary controlling t faulty proxies sees m, s, α and t shares of u_2. The t shares reveal no useful information, since they come from a distribution that is uniformly random. There appears to be no way to recover r from m, s, α, y without an ability to take discrete logs, but it is difficult to prove this because of the complex dependencies among these values.

7 Performance

As described in Section 1, this work was motivated by a practical effort to experiment with a number of financial trading vehicles in real systems. We have implemented the VΣS schemes described here as part of this effort. Our present implementation uses the arbitrary precision arithmetic package of Cryptolib [18] and the reliable multicast protocol of Rampart [23]. This reliable multicast protocol, which incorporates timeouts into its methods of fault detection and recovery, satisfies the "reliable broadcast" portion of our communication model, under the assumption that messages from honest parties induce timeouts in other honest parties sufficiently infrequently.

A brief summary of the performance for our implementation, in the case of no failures, is shown in Table 1 (see Section 1). The tests described in Table 1 were performed among user processes on a network of moderately loaded, single processor SPARCstation 10s running SunOS 4.1.3. The moduli N (for RSA and Rabin) and p (for ElGamal, Schnorr, and DSA) were 512 bits long in these tests. The modulus q was 160 bits long for Schnorr and DSA, and 511 bits long for ElGamal. The column titled "share time" describes the mean latency in milliseconds (ms) of the sharing protocol for each VΣS scheme presented, beginning when the dealer initiates the protocol and ending when the proxies accept. This cost includes the computational costs incurred by the dealer and the proxies, as well as the communication latency of the $n + 1$ reliable multicasts. The private messages from the dealer to the proxies are piggybacked on the dealer's reliable multicast, each encrypted so that only the intended proxy can decipher it (using symmetric key encryption in our implementation). The column titled "recon time" shows the computational latency incurred by the reconstructor in the reconstruction protocol of each VΣS scheme. The latency of the communication from the proxies to the reconstructor (and the accompanying voting at the reconstructor) is not included in these times, because for some applications that we envision, proxies may not communicate to the reconstructor simultaneously.

The experience of implementing these schemes revealed a number of opportunities to exploit concurrency and precomputation to improve the latency of the protocols. For instance, in each scheme a proxy can perform many of the tests for determining acceptance or rejection in parallel with reliably multicasting its ALLOW/COMPLAIN message, because many of these tests depend only on information multicast from the dealer. In addition, since polynomial interpolation is used heavily in these protocols, it is useful to compute the interpolation coefficients (i.e., the quantities c_i; see Section 3.2) prior to executing the protocol, if the modulus is known in advance. The numbers in Table 1 reflect the use of both of these optimizations. Moreover, the noisy interpolation steps in the RSA and Rabin schemes were optimized for the case $t = 1$; see [3] for an algorithm for arbitrary t. The column labeled "precomp?" in Table 1 refers only to the pre-sharing of polynomials in the RSA and Rabin schemes, and not to the local precomputation of interpolation coefficients in all of the implementations.

8 Conclusions

We have identified a new cryptographic primitive for protecting digital signatures, called Verifiable Signature Sharing, and have provided practical implementations of this primitive for RSA, Rabin, ElGamal, Schnorr, and DSA signatures. Our experimental data confirms that these techniques perform sufficiently well to be useful in a wide range of applications, including the integration of digital cash into secure protocols. In a separate paper [14], we describe the use of these techniques to construct a secure distributed auctioning system.

For some of our VΣS schemes, further speed-ups are possible by making small modifications to the underlying signature scheme. It is an intriguing open question to find new signature schemes that best balance the efficiencies of signature construction, signature verification, and verifiable signature sharing.

References

1. D. Beaver, S. Micali, and P. Rogaway, "The round complexity of secure protocols," ACM STOC 1990, 503–513.
2. M. Ben-Or, S. Goldwasser, and A. Wigderson, "Completeness theorems for non-cryptographic fault-tolerant distributed computation," ACM STOC 1988, 1–9.

3. E. Berlekamp and L. Welch, "Error correction of algebraic block codes," U.S. Patent Number 4,633,470.

4. G. Blakely "Safeguarding cryptographic keys," AFIPS National Computer Conference 48 (1979), 313–317.

5. D. Chaum, "Security without identification: transaction systems to make big brother obsolete," CACM 28 (1985), 1030–1044.

6. D. Chaum, C. Crépeau, and I. Damgård, "Multiparty unconditionally secure protocols," ACM STOC 1988, 11–19.

7. D. Chaum, A. Fiat, and M. Naor, "Untraceable electronic cash," Crypto 1988, 319–327.

8. B. Chor, S. Goldwasser, S. Micali, and B. Awerbuch, "Verifiable secret sharing and achieving simultaneity in the presence of faults," IEEE FOCS 1985, 383–395.

9. A. DeSantis, Y. Desmedt, Y. Frankel, and M. Yung, "How to share a function securely," ACM STOC 1994, 522–533.

10. Y. Desmedt, "Threshold cryptography," European Transactions on Telecommunications and Related Technologies 5 (1994), 449–457.

11. NIST FIPS PUB 181, "Digital signature standard," U.S. Department of Commerce/National Institute of Standards and Technology.

12. T. ElGamal, "A public key cryptosystem and a signature scheme based on discrete logarithms," IEEE Trans. Information Theory IT-31 (1985), 469–472.

13. P. Feldman, "A practical scheme for non-interactive verifiable secret sharing," IEEE FOCS 1987, 4427–437.

14. M. K. Franklin and M. K. Reiter, "The design and implementation of a secure auction service," IEEE Symposium on Security and Privacy, Oakland, CA, 1995 (to appear).

15. O. Goldreich, S. Micali, and A. Wigderson, "How to play any mental game," ACM STOC 1987, 218–229.

16. V. Hadzilacos and S. Toueg, "Fault-tolerant broadcasts and related problems," In *Distributed Systems* (2nd edition), Chapter 5, Addison-Wesley, 1993.

17. J. Hoffman, "New power-of-attorney form is introduced," The New York Times, October 1, 1994.

18. J. Lacy, D. Mitchell, and W. Schell, "CryptoLib: cryptography in software," 4th USENIX Security Workshop, pp. 1–17, 1993.

19. L. Lamport, R. Shostak, and M. Pease, "The Byzantine generals problem," ACM TOPLAS 4 (1982), 382–401.

20. S. Micali, "Fair public-key cryptosystems," Crypto 1992, 113-138.

21. T. Pedersen, "Distributed provers with applications to undeniable signatures," Eurocrypt 1991, 221–242.

22. M. Rabin, "Digitalized signatures and public key functions as intractable as factorization," Technical Report MIT/LCS/TR-212, Laboratory for Computer Science, Massachusetts Institute of Technology, 1979.

23. M. K. Reiter, "Secure agreement protocols: Reliable and atomic group multicast in Rampart," 2nd ACM Conf. Computer and Comm. Security, 68–80, 1994.

24. R. Rivest, A. Shamir, and L. Adleman, "A method for obtaining digital signatures and public-key cryptosystems," CACM 21 (1978), 120–126.

25. C. Schnorr, "Efficient signature generation by smart cards," J. Cryptology 4 (1991), 161–174.

26. A. Shamir, "How to share a secret," CACM 22 (1979), 612–613.

Server(Prover/Signer)-Aided Verification of Identity Proofs and Signatures

Chae Hoon Lim and Pil Joong Lee

Department of Electrical Engineering
Pohang University of Science and Technology (POSTECH)
Pohang, 790-784, KOREA
E-mail : lch@baekdu.postech.ac.kr ; pjl@cipher.postech.ac.kr

Abstract. Discrete log based identification and signature schemes are well-suited to identity proof and signature generation, but not suitable for verification, by smart cards, due to their highly asymmetric computational load between the prover/signer and the verifier. In this paper, we present very efficient and practical protocols for fast verification in these schemes, where the verifier with limited computing power performs its computation fast with the aid of the powerful prover/signer. The proposed protocols require very small amounts of computation and communication. The prover/signer only needs to perform a few modular exponentiations in real-time and the two interacting parties only need to communicate a few long numbers. Using the proposed prover-aided verification (PAV) protocol, the verifier can perform the Schnorr-like identification scheme almost as fast as the Guillou-Quisquater scheme. We generalize the PAV protocol into the signer-aided verification (SAV) protocol, which can be used for verification of any public function.

1 Introduction

Based on zero-knowledge proof techniques, a lot of identification and digital signature schemes have been developed [1-6]. Among them, Schnorr-like schemes [4-6] are particularly attractive for use in smart cards or other environments with limited computing power, since the prover/signer needs almost no8 fax. real-time computation with preprocessing/precomputation techniques [4,7-9]. However, verification requires exponentiation involving a lot of multiplications, which is disadvantageous compared to Fiat-Shamir-like schemes [1-3]. This asymmetric computational load may restrict applications of these schemes, when implemented on a weak power device such as a smart card, into environments where only one-way proofs are sufficient. Thus what is further desired for these schemes

L.C. Guillou and J.-J. Quisquater (Eds.): Advances in Cryptology - EUROCRYPT '95, LNCS 921, pp. 64-78, 1995.
© Springer-Verlag Berlin Heidelberg 1995

would be that proofs in the other way are also efficient for smart card implementations. This motivated us to develop methods for speeding up the computation by the verifier in some way or another.

We first considered the applicability of the server-aided approach to secret computation, first proposed by Matsumoto et al. [10] and since then widely studied by many researchers [11-15], to the public verification of identity proofs and signatures. And we found a related work performed by Yen and Laih [16], but unfortunately their protocol can be easily shown to be insecure. Furthermore, this kind of protocols seems to need too much amount of communication to be practical for smart card applications. There are fundamentally different requirements for the server-aided secret computation protocol and the server-aided public verification protocol. The former has to guarantee the security of the involved secret information, while the latter requires the assurance of the integrity of computation results returned from the server. This difference of requirements seems make the latter protocol needlessly complicated and hard to guarantee the correctness of the required verification.

In this paper we propose efficient protocols for speeding up the verification of identity proofs and signatures. The key idea is to use the precomputation based on a fixed base element and then mirror the action of the prover/signer. In the proposed protocols, we assume that the proving/signing terminal is much more powerful than the verifying device so that it can perform several exponentiations in real-time. This situation will commonly arise in a smart card based system when a powerful terminal proves to a smart card or when signature verification is performed on a smart card. Thus the resulting protocols may be called as prover/signer-aided verification (PAV/SAV) protocols since the verifier performs the required verification fast by borrowing the computing power of the prover/signer.

Compared to the protocol for server-aided RSA computation, our proposed protocols are much more efficient and practical since only a few long numbers need to be exchanged and only a few modular exponentiations need to be performed by the prover/signer. For example, using the proposed prover-aided verification (PAV) protocol, the verifier can execute the Schnorr-like identification scheme [4-6] almost as fast as the Guillou-Quisquater scheme [2], only with exchange of one long number and without loss of security. This will make Schnorr-like identification schemes much more attractive for smart card implementations since now smart cards can also perform the required verification fast. By generalizing the PAV protocol, we also present a signer-aided verification (SAV) protocol with which the verifier can check the validity of signatures with any desired convincing probability. For example, if a convincing probability of $1 - 2^{-t}$ is acceptable in a real-time protocol, a signature generated by Schnorr's scheme can be verified in about $3t$ multiplications on average. Finally we show that the proposed techniques can be used for verification of any public function by presenting a fully generalized version of the server-aided verification protocol.

2 Prover-Aided Verification of Identity

Throughout this paper, we will use the following conventions, unless otherwise stated. Let p and q be two large public primes such that q divides $p-1$ and g be an element of order q in Z_p. We denote the bit-length of p (q, resp.) by n (l, resp.) (i.e., $|p| = n, |q| = l$). Let (s, v) be the secret and public key pair of the prover/signer, where $v = g^{-s} \bmod p$ with $s \in Z_q$. We assume that precomputation of random powers to the fixed base g is performed in advance and thus does not take time during the protocol execution.

The computation of $a^x b^y \bmod p$ with $|x| = l$ and $|y| = t$ is assumed to be carried out by the square-and-multiply algorithm with a precomputed value of $ab \bmod p$ (see [4]). We assume that the most significant bits of the exponents, x and y, are always one for completeness, though their effect on the performance is negligible. Then the above computation can be completed in $1.5l + 0.25(t-1)$ multiplications for $l > t$, and $1.75l - 0.75$ for $l = t$, on average. Multiplication will always denote multiplication mod p and multiplication mod q will be neglected when counting the number of multiplications.

2.1 PAV Protocol for Schnorr's Identification Scheme

The following is the five-move protocol for prover-aided verification in Schnorr's identification scheme [4]. Here t is a parameter that determines the security level of the identification scheme, usually lying between 20 and 40.

0) (Preprocessing) The prover picks a random number $r \in Z_q$ and computes $x = g^r \bmod p$. Similarly the verifier computes $z = g^{-K} \bmod p$ with randomly chosen K over Z_q.
1) The prover sends x to the verifier.
2) The verifier randomly picks an integer $e \in [0, 2^t)$ and sends it to the prover.
3) The prover computes and sends $y = r + se \bmod q$.
4) The verifier randomly picks an integer $k \in [0, 2^t)$, computes and sends $u = (K + y)k^{-1} \bmod q$.
5) The prover computes $w = g^u \bmod p$ and sends it back to the verifier.
6) Finally the verifier checks if the following equation holds :

$$x = w^k v^e z \bmod p \tag{1}$$

Note that for security the precomputed value z should not be revealed to the prover at least until the protocol is completed. This must be observed in every protocol presented in this paper. If desired, the computation of $k^{-1} \bmod q$ in step 4) may be performed in the preprocessing stage. Steps 1) - 3) exactly correspond to the original Schnorr scheme whose verification equation equals

$$x = g^y v^e \bmod p. \tag{2}$$

On the other hand, steps 4) - 6) correspond to the protocol in which the verifier computes $g^y \bmod p$ with the aid of the prover. Thus we can see that by borrowing the prover's computing power the verifier can reduce the computational load of l bit exponentiation to that of t bit exponentiation.

Security : Equation (1) shows that the values of x, v and e, which are determined in the first half of the protocol, cannot be modified, without knowledge of k, in the latter half. On the other hand, the value of u, which is the only data available to the prover for extracting information on the secret number k, releases no information on k since even z is not available at this point. As a result, the prover may guess k but has no way to verify its guess. The above two facts show that the PAV protocol is unconditionally sound since no information on k is released in the Shannon-theoretic sense and since without knowing k the dishonest prover cannot convince the verifier with more than guessing probability.

There may be a slight advantage on the prover's side. Throughout the whole protocol, the prover is given two chances of cheating the verifier : either by guessing e in step 1) as in the original Schnorr scheme or by guessing k in step 5). The latter guess can be successful independently of the former guess since, once the former is turned out to be wrong from the response of step 2), the prover knows how to manipulate w to pass the verification of step 6), of course, under the assumption that its guess at k is correct. Thus the added steps 4) and 5) only gives the prover another chance of random guessing. This will be of little value to the (dishonest) prover. Consequently, we conclude that the prover-aided approach to fast verification preserves almost the same security level of the original scheme.

Efficiency : The verifier can check the verification equation (1) in about $1.75t + 0.25$ multiplications on average. This is almost the same amount of computation as is required in the GQ scheme. Note that with the original verification equation (2), about $1.5l + 0.25(t - 1)$ multiplications are required. For example, with $l = 160$ and $t = 20$, the equality of equation (1) can be checked in 35.25 multiplications, while validating equation (2) requires 244.75 multiplications, on average. Thus about 210 multiplications can be saved in this case using the proposed verification protocol.

The above efficiency is obtained only by increasing the number of communication bits by $n+l$. The computational complexity imposed on the prover is also very small, just one exponentiation ($1.5(l - 1)$ multiplications on average). No restriction on the computing power of the prover will be necessary due to this increase of computational amount, since such computation can be carried out in real-time even on the PC (personal computer). Therefore, in typical smart card-based systems, we will be able to obtain great computational advantage using the PAV protocol only with a small increase of communication.

2.2 PAV Protocol for Brickell-McCurley's Scheme

Brickell and McCurley [5] modified the Schnorr scheme in order to enhance the security at the cost of more computation and communication. The basic differences are that all exponents are selected and computed modulo $p-1$ rather than modulo q and that q is kept secret from the users (so the modulus p should be chosen such that $p-1$ is hard to factor). The resulting protocol can be proven to be secure, assuming that $p-1$ is hard to factor, and remains as secure as the Schnorr scheme even if $p-1$ is factored.

The PAV protocol for the Brickell-McCurley (BM) scheme is the same as that for the Schnorr scheme, except that all arithmetics on exponents should be done modulo $p-1$. Thus the performance improvement by the PAV protocol is much more drastic in this scheme. For example, with $n = 512$ and $t = 20$, the original verification requires 772.75 multiplications on average, while the prover-aided verification still requires 35.25 multiplications. This amounts to more than a twenty four-fold improvement. Since main disadvantage of the BM scheme can be eliminated with the PAV protocol, the BM scheme may be preferred to the Schnorr scheme in view of security.

We finally would like to mention that the PAV protocol does not affect the provable security of the original scheme since no additional information on the secret key of the prover is involved in the prover-aided verification part. Note that a three-move identification scheme is said to be secure (in the sense of Feige-Fiat-Shamir [17]) if the protocol execution releases no useful information on the prover's secret.

2.3 PAV Protocol for Okamoto's Scheme

Okamoto [6] has proposed another modification of Schnorr's scheme with the feature of provable security. Since it is somewhat different from the Schnorr scheme in basic construction, we describe his scheme together with the proposed verification protocol. Let p and q be as before and g_1 and g_2 be elements of order q in Z_p. The public key of the prover in the Okamoto scheme is $v = g_1^{-s_1} g_2^{-s_2} \bmod p$, where s_1 and s_2 in Z_q are his secret keys. The PAV protocol for Okamoto's scheme is as follows.

0) (Preprocessing) The prover randomly picks $r_1, r_2 \in Z_q$ and computes $x = g_1^{r_1} g_2^{r_2} \bmod p$. Similarly the verifier computes $z = g_1^{-K_1} g_2^{-K_2} \bmod p$ with $K_1, K_2 \in Z_q$.

1) The prover sends x to the verifier.

2) The verifier randomly picks an integer $e \in [0, 2^t)$ and sends it to the prover.

3) The prover computes $y_1 = r_1 + s_1 e \bmod q$ and $y_2 = r_2 + s_2 e \bmod q$ and sends them to the verifier.

4) The verifier randomly picks $k \in [0, 2^t)$, computes $u_1 = (K_1 + y_1)k^{-1} \bmod q$ and $u_2 = (K_2 + y_2)k^{-1} \bmod q$, and sends them back to the prover.

5) The prover computes $w = g_1^{u_1} g_2^{u_2} \bmod p$ and sends it to the verifier.
6) Finally the verifier checks if the following equation holds :

$$x = w^k v^e z \bmod p \qquad (3)$$

Though the Okamoto scheme is somewhat different from the Schnorr scheme, we can see that the performance of the PAV protocol remains almost the same. Compare the above equation (3) with the original verification equation :

$$x = g_1^{y_1} g_2^{y_2} v^e \bmod p \qquad (4)$$

The only difference is that in the above the verifier computes $g_1^{y_1} g_2^{y_2} \bmod p$ as $w^k z \bmod p$ with the aid of the prover. Note that it is unnecessary to use different values of k to compute u_1 and u_2 due to the involvement of distinct random secrets, K_1 and K_2, of l bit size (In any case, knowing one small random secret will be sufficient to cheat the verifier).

Table 1 below summarizes the performance of three identification schemes and their PAV versions. The certificate for the public key v is not taken into account when counting the number of communication bits and the computational amounts for preprocessing are also excluded. The number of multiplications is counted for the average case. Finally, note that we are using the parameters n, l and t as $n = |p|, l = |q|$ and $t = |e| = |k|$, respectively.

			Schnorr	Brickell et al.	Okamoto
Orig	Mul	P	almost 0	almost 0	almost 0
		V	$1.5l + 0.25(t-1)$	$1.5n + 0.25(t-1)$	$1.75l + 0.125t + 2.37$
	Comm		$2n + l + t$	$3n + t$	$2n + 2l + t$
PAV	Mul	P	$1.5(l-1)$	$1.5(n-1)$	$1.75l + 0.25$
		V	$1.75t + 0.25$	$1.75t + 0.25$	$1.75t + 0.25$
	Comm		$3n + 2l + t$	$5n + t$	$3n + 4l + t$

Table 1. Performance of PAV protocols for three identification schemes

Finally we note that the proposed PAV protocol can also be adapted for identification schemes with composite moduli. For example, in Girault's modification of the Schnorr scheme based on composite discrete logarithms [18], the order of the based element g is made public and thus the PAV protocol for Schnorr's scheme can be applied directly. On the other hand, in the similar protocol using the self-certified public key [19], the based element g has a maximal order modulo a composite and the signature component y is not reduced modulo any number. Thus it is not feasible to compute multiplicative inverses of exponents. For this scheme, the verifier may first raise both sides of the verification equation to the k-th power and then apply the PAV protocol (or it may use the protocol to be presented in section 3). Of course, the performance will be somewhat degraded in this case.

3   Signer-Aided Verification of Signatures

There exists the same asymmetry of computational load in digital signature schemes derived from identification schemes based on the discrete logarithm problem. Thus these signatures are easy to generate but hard to verify with smart cards. This section is devoted to developing an efficient protocol for signer-aided verification of signatures. Of course, the role of the powerful server need not be assumed by the signer itself in this case. Since typical application of this protocol will be signature verification on the smart card, the server may be a powerful terminal with which the smart card interacts.

We only explain the proposed SAV protocol with Schnorr's signature scheme, but it can be used for verification of other signature schemes based on the discrete logarithm problem as well (e.g., see [20-23] for generalized ElGamal-type signature schemes and their message recovery variants). In fact, the proposed technique can be applied to server-aided verification of any public function, as will be illustrated in the next section.

3.1   SAV Protocol for Schnorr's Signature Scheme

For the moment, let us suppose that the signer's public key $v = g^{-s} \bmod p$ is globally known and frequently used (this may be the case if we have to frequently verify signatures of some central authorities). Then we can adapt the PAV protocol into the SAV (signer-aided verification) protocol as follows, where h denotes a one-way hash function producing randomly and uniformly c-bit digests (see below).

0) (Preprocessing) The verifier computes $z = g^{-K_1} v^{-K_2} \bmod p$ with $K_1, K_2 \in Z_q$.

1) The signer sends the signature $\{x, y, m\}$ to the verifier, where $x = g^r \bmod p$ and $y = r + se \bmod q$ with $e = h(x, m)$.

2) The verifier computes $e = h(x, m)$. Then it randomly picks an integer $k \in (0, 2^t]$, computes $u_1 = (K_1 + y)k^{-1} \bmod q$ and $u_2 = (K_2 + e)k^{-1} \bmod q$, and sends them to the signer.

3) The signer computes and sends $w = g^{u_1} v^{u_2} \bmod p$.

4) The verifier then checks if $x = w^k z \bmod p$ holds. If the check succeeds, the verifier accepts and stores $\{e, y\}$ as a valid signature for message m.

We first want to note that the length of hash-values used in any signature schemes should be at least 128 bits, contrary to the minimal length of 64 or 72 bits that many researchers (e.g., see [1,2,4]) suggested. This is because the signer can find two different messages with the same signature using the birthday paradox if short hash-values are used. If such a thing is feasible, then the signer may deny later the signature of one message by presenting the other message with the same signature. This situation is essentially the same, as far as the legality of signature is concerned, as the case where an outside attacker finds two different messages with the same hash-value, obtains a signature for the message favorable to the signer and then claims that the signer signed the other message favorable to himself.

A slight modification may achieve the same effect that can be obtained by the use of longer hash-values without increasing the computational load of the verifier, but this does not matter in the current SAV protocol. From now on, we will assume that hash-values are randomly distributed over Z_q (i.e., $c = l = |q|$) as in the DSS [24].

The above SAV protocol achieves a security level of 2^{-t}. The signer cannot use in step 3) a value of v different from the one publicly known or sent in step 1), due to its involvement in the computation of z. Other security considerations are the same as in the PAV protocol. Thus, a fake signature can be made to be accepted only when the guess of k is correct. If a false acceptance with probability of 10^{-6} can be tolerated in a real-time protocol, then the signature can be verified in 29.5 multiplications on average. However, this protocol seems not practical in general, since the precomputation using the signer's public key is not possible in most cases. Thus the above SAV protocol needs to be augmented by somewhat different technique.

The problem we are faced with is to compute the part $v^e \bmod p$ of the verification equation $x = g^y v^e \bmod p$ with the aid of the signer, where the signer's public key v is assumed to vary in every run of the protocol. Our solution is to blind the public key v by raising to the k-th power and then multiplying by a random power of g, i.e., form $u = g^K v^k \bmod p$ ($K \in Z_q, k \in (0, 2^t])$, so that the signer, no matter how powerful it is, cannot deduce k from u (and thus cannot modify v) with more than the guessing probability of 2^{-t}. For this, the verifier must compute $v^k \bmod p$ before beginning the signer-aided verification, which increases the verifier's computational load almost twice compared to the above case. The following is the final SAV protocol for Schnorr's signature scheme.

0) (Preprocessing) The verifier computes $z_1 = g^{-K_1} \bmod p$ and $z_2 = g^{-K_2} \bmod p$ with $K_1, K_2 \in Z_q$.

1) The signer sends the signature $\{x, y, m\}$ to the verifier, where $x = g^r \bmod p$ and $y = r + se \bmod q$ with $e = h(x, m)$.

2) The verifier randomly picks an integer $k \in (0, 2^t]$ and computes $u_1 = z_1 v^k \bmod p$ using the signer's public key v. The verifier also computes $u_2 = (K_2 + ky + K_1 e) \bmod q$ with $e = h(x, m)$ and sends u_1 and u_2 to the signer.

3) The signer computes and sends $w = u_1^e g^{u_2} \bmod p$.

4) Finally the verifier checks that $x^k = w z_2 \bmod p$. If the equation holds, the verifier accepts and stores $\{e, y\}$ as a valid signature for message m.

The above verification is based on the following identity :

$$x^k = (g^{-K_1} v^k)^e \cdot g^{K_2 + K_1 e + ky} \cdot g^{-K_2} \bmod p \tag{5}$$

Note that since the value of e computed as $e = h(x, m)$ by the verifier is embedded in u_2, it is of no use for the signer to use a different value of e when computing w in step 3). The on-line computational load for the verifier is about $3t - 1$ multiplications on average. Thus, with a convincing probability of $1 - 10^{-6}$, the verifier can validate a signature in 59 multiplications on average. This is a substantial improvement over direct verification requiring about 279.25 multiplications, if a small probability of false acceptance can be tolerated. If more strict verification is required, we may choose $t = 30$, in which case the signature can be verified in 89 multiplications with probability of false acceptance of 10^{-9}.

It is interesting to note that the SAV protocol may be viewed as an interactive proof system for language membership [25], though the proof is trivial, where the language L consists of a set of valid signatures generated with the Schnorr scheme, i.e.

$$L = \{(x, y, m, v) | x = g^y v^e \bmod p \text{ with } e = h(x, m)\}. \tag{6}$$

In the SAV protocol, the verifier with limited computing power wants to be convinced that a given instance belongs to L. The above discussion shows that the SAV protocol satisfies the two conditions of an IP system, completeness and soundness.

The following table shows the performance of the proposed SAV protocol for Schnorr's signature scheme. Here we assume that the hash-value e is of l bit size. The message m and the public key certificate are not included in the number of communication bits.

		Original	SAV
Mul	Signer	almost 0	$1.75l - 0.75$
	Verifier	$1.75l - 0.75$	$3t - 1$
Commun		$n + 2l$	$4n + 2l$

Table 2. Performance of SAV protocol for Schnorr's signature scheme

3.2 Batch SAV Protocol for Schnorr's Scheme

A collection of signatures can be verified more efficiently by processing in a batch. Naccache et al. [26] presented (interactive and probabilistic) batch verification protocols for DSA at Eurocrypt'94, together with several other useful techniques to improve the performance of DSA (but the interactive batch verification protocol was shown to be insecure [27]).

Let $\{x_i, y_i\}$, for $i = 1, 2, \cdots, N$, be Schnorr's signatures for messages m_i signed by the same signer, where $x_i = g^{r_i} \bmod p$ and $y_i = r_i + se_i \bmod q$

with $e_i = h(x_i, m_i)$. Then the verifier can check the validity of the signatures by batch-processing with the equation

$$\prod_{i=1}^{N} x_i^{k_i} = g^{\sum_{i=1}^{N} k_i y_i} \cdot v^{\sum_{i=1}^{N} k_i e_i} \bmod p, \qquad (7)$$

where k_i's are random numbers of t-bit size chosen by the verifier. The parameter t determines the level of confidence for batch verification.

We first explain a method for efficiently evaluating the left-hand side of equation (6) using the idea from [8]. It can be computed by arranging the N terms of small powers into a groups consisting of b terms, preparing all products of possible combinations among b terms in each group and then applying the square-and-multiply algorithm. We can then show that the required computation can be completed in $\frac{2^b-1}{2^b}(t-1)a + t + (2^b - b)a - 2$ multiplications on average. For this, we also need a storage for $(2^b - 1)a$ values.

Table 3 below summarizes, for some selected parameters, the numbers of multiplications and storage required for the computation of the left-hand side of equation (6) using this method. From the table, we can see that if the verifying device is equipped with sufficient storage, a number of signatures can be verified with great efficiency. Batch verification on the PC may be such a case. For example, 16 signatures generated by the same signer can be validated in about 464 multiplications on average, where $t = 30$ is assumed and 279.25 multiplications for computing the right-hand side of equation (6) are included.

N	2	3	4	6		8		12		16
(a,b)	(1,2)	(1,3)	(2,2)	(3,2)	(2,3)	(4,2)	(2,4)	(4,3)	(3,4)	(4,4)
Storage	3	7	6	9	14	12	30	28	45	60
Mul $t = 20$	34.3	39.6	50.5	66.8	61.3	83.0	77.6	104.5	107.4	137.3
$t = 30$	51.8	58.4	75.5	99.3	88.8	123.0	106.4	149.5	145.6	184.8

Table 3. Resource requirements for computing the left-hand side of equ. (6)

Now, let us consider the batch verification on the smart card. Since typical smart cards under current technology do not have much storage, a relatively small number of signatures can be processed at a time. In this case, the computation of the right-hand side of equation (6) seems a quite heavy load to the smart card. Thus we may use a batch SAV protocol for this computation. Let us consider the following protocol.

0) (Preprocessing) The verifier computes $z_1 = g^{-K_1} \bmod p$ and $z_2 = g^{-K_2} \bmod p$ with $K_1, K_2 \in Z_q$.

1) The signer sends $\{x_i, y_i, m_i\}$ $(1 \leq i \leq N)$ to the verifier.

2) The verifier first computes $e_i = h(x_i, m_i)$. Then it randomly picks $N + 1$ integers k_i, for $i = 0, 1, \cdots, N$, over $(0, 2^t]$, and then computes $u_1 = z_1 v^{k_0}$ mod p and u_2, u_3 as

$$u_2 = k_0^{-1} \sum_{i=1}^{N} k_i e_i \bmod q, \ u_3 = K_2 + K_1 u_2 + \sum_{i=1}^{N} k_i y_i \bmod q.$$

The verifier then sends u_1, u_2 and u_3 to the signer.

3) The signer computes and sends $w = u_1^{u_2} g^{u_3} \bmod p$.

4) Finally the verifier checks if the following equation holds :

$$\prod_{i=1}^{N} x_i^{k_i} = w z_2 \bmod p \tag{8}$$

If it holds, the verifier accepts and stores $\{e_i, y_i\}$ as valid signatures for messages m_i for $i = 1, 2, \cdots, N$.

The above batch verification is based on the following identity :

$$\prod_{i=1}^{N} x_i^{k_i} = (g^{-K_1} v^{k_0})^{k_0^{-1} \sum_{i=1}^{N} k_i e_i} \cdot g^{k_0^{-1} K_1 \sum_{i=1}^{N} k_i e_i + \sum_{i=1}^{N} k_i y_i + K_2} \bmod p \tag{9}$$

Using the above batch SAV protocol, the verifier can compute the right-hand side of equation (6) in $1.5t - 0.5$ multiplications on average if we neglect the arithmetics mod q. Therefore, we can verify, for example, four signatures in about 80 multiplications on the smart card, with a convincing probability of $1 - 10^{-6}$ ($t = 20$), if the smart card has a scratch pad memory for ten values or so. Note that if different signers are involved, each signer's public key must be blinded individually and thus the performance will be degraded. But this is also the case for direct verification.

The batch SAV protocol has one undesirable property, compared to the SAV protocol of the previous subsection, in the sense that its security is dependent upon the computing power of the signer. That is, for small N, the signer may try to find the random secret numbers k_i's from the value of u_2 by an exhaustive search using the birthday paradox. This is clearly undesirable but seems inevitable due to the involvement of secret numbers in the exponent of v.

From the equation $u_2 k_0 + \sum_{i=1}^{N/2} k_i e_i = \sum_{i=1+N/2}^{N} k_i e_i \bmod q$ where we assume that N is even, k_i's can be computed in $L \log_2 L$ operations with $L = 2^{t(1+N/2)}$. For example, for $N = 2$ and $t = 20$, we have $L = 2^{40}$. However, such an attack can be mounted only after u_2 is given. Thus it is unlikely that this attack makes any practical threat to the protocol even for the above minimal parameters, since it is infeasible to perform 2^{40} operations in a second or so. Other security considerations are the same as in the SAV protocol.

4   Server-Aided Verification of General Functions

We now present a fully generalized version of server-aided verification protocols which can be used for verification of any public function. Suppose that the verifier, with the aid of a powerful server, wants to check the equality of the following general equation defined over a finite group G :

$$y^\beta = \prod_{i=1}^{N} x_i^{\alpha_i} \tag{10}$$

All involved elements are assumed to be public and variable. The following protocol allows the verifier to test the equality of the above equation with a convincing probability of $1 - 2^{-t}$.

0) (Preprocessing) The verifier randomly picks an element $g \in G$ and computes $z_i = g^{K_i}$ with $K_i, \in G$ for $i = 0, 1, \cdots, N$.
1) The verifier randomly picks an integer $k \in (0, 2^t]$ and then computes the following values :

$$u_0 = z_0 y^k, \quad u_i = z_i x_i^k \ (1 \le i \le N), \quad u_{N+1} = K_0\beta - \sum_{i=1}^{N} K_i\alpha_i + K_{N+1}$$

   Then the verifier sends $\{g, u_i, \alpha_i, \beta\}$ to the server.
2) The server computes and sends the following value :

$$w = g^{u_{N+1}} u_0^{-\beta} \prod_{i=1}^{N} u_i^{\alpha_i}$$

3) Finally the verifier checks if $z_{N+1} = w$ holds.

The above server-aided verification is based on the following identity :

$$g^{K_{N+1}} = g^{K_0\beta - \sum_{i=1}^{N} K_i\alpha_i + K_{N+1}} \cdot (g^{K_0} y^k)^{-\beta} \cdot \prod_{i=1}^{N} (g^{K_i} x_i^k)^{\alpha_i} \tag{11}$$

The element g may be globally fixed and, if the group order $|G|$ is known, all the exponents can be reduced modulo $|G|$. The protocol achieves a security level of 2^{-t} since the only way to cheat the verifier is to guess k and manipulate y and/or x_i. The number of group multiplications required of the verifier is around $(1.5t - 0.5)(N + 1)$ on the average. If there are M fixed elements in equation (9), this quantity can be reduced to $(1.5t - 0.5)(N - M + 1)$.

All the protocols presented so far are special cases of the above protocol. Note that with t-bit randomizers (blinding factors), signature schemes involving a fixed base element can be verified in $3t - 1$ multiplications while the other schemes such as Guillou-Quisquater [2] and Ohta-Okamoto [3] can be verified in about $4.5t - 1.5$ multiplications on average. Even for the GQ scheme, this is a considerable improvement over direct verification in case where a moderate level of confidence is sufficient (e.g., 88.5 vs 223.25 for $t = 20$ and 128 bit hash-values).

The above server-aided approach to fast verification will be useful for most public key cryptographic schemes when executed between two parties with asymmetric computing power. Typical applications may be found in the interactive protocols between smart cards and terminals. Since the proposed protocol is independent of the size of exponents and its security level is independent of the server's power, the advance of cryptanalytic methods (based either on software or on hardware) will never adversely affect its performance. Rather, the performance may be further improved in case that the size of group order is increased.

5 Summary and Conclusion

We have presented an elegant way to speed up the computation by the verifier in discrete logarithm-based identification schemes (Schnorr, Brickell-McCurley, Okamoto, etc.), with the aid of the powerful prover. The proposed prover-aided verification (PAV) protocol is secure and efficient : Only with a small amount of additional communication and with almost the same level of security as the original scheme, the verifier can perform the Schnorr-like identification scheme almost as fast as the Guillou-Quisquater scheme. In particular, the efficiency of the proposed protocol is independent of the size of exponents and thus Brickell-McCurley's scheme may be preferred to the Schnorr scheme due to its enhanced security. The proposed PAV protocol will make Schnorr-like identification schemes much more attractive for smart card implementations since now smart cards can also perform the required verification fast.

By generalizing the PAV protocol, we have also presented a signer-aided verification (SAV) protocol that can be adapted for verification of any public function. The proposed SAV protocol is also quite efficient in both computation and communication. With a convincing probability of $1 - 2^{-t}$, the validity of a signature can be checked in about $3t$ multiplications on average for discrete logarithm-based schemes and in about $4.5t$ multiplications on average for the GQ scheme. The batch SAV protocol enables more efficient verification of a collection of signatures.

The proposed server-aided verification protocol will be useful for many public key cryptographic schemes carried out between users with asymmetric computing powers. Smart card verification of identity proofs and signatures will be one of the most attractive application areas of the protocol. Another important application can be found in designing efficient protocols for authenticated key exchange between smart cards and servers (computers) (see [28]).

Finally we would like to mention that if the communication cost is relatively low, we can considerably reduce the computational complexity for the SAV protocol by adapting the server-aided approach for RSA computation (e.g., see [29]). Of course, in this case, its security relies on the computing power of the server as in the batch SAV protocol presented in this paper.

References

1. A.Fiat and A.Shamir : 'How to prove yourself : Practical solution to identification and signature problems', *Advances in Cryptology-Crypto'86*, Springer-Verlag, pp.186-194 (1988).
2. L.C.Guillou and J.J.Quisquater : 'A practical zero-knowledge protocol fitted to security microprocessor minimizing both transmission and memory', *Advances in Cryptology-Eurocrypt'88*, Springer-Verlag, pp.123-128 (1988).
3. K.Ohta and T.Okamoto : 'A modification of the Fiat-Shamir scheme', *Advances in Cryptology-Crypto'88*, Springer-Verlag, pp.232-243 (1990).
4. C.P.Schnorr : 'Efficient signature generation by smart cards', *Journal of Cryptology*, 4(3), pp.161-174 (1991).
5. E.F.Brickell and K.S.McCurley : 'An interactive identification scheme based on discrete logarithm and factoring', *Journal of Cryptology*, 5(1), pp.29-39 (1992).
6. T.Okamoto : 'Provably secure and practical identification schemes and corresponding signature schemes', *Advances in Cryptology-Crypto'92*, Springer-Verlag, pp.31-53 (1993).
7. E.F.Brickell, D.M.Gordon, K.S.McCurley and D.B.Wilson : 'Fast exponentiation with precomputation', *Advances in Cryptology-Eurocrypt'92*, Springer-Verlag, pp.200-207 (1993).
8. C.H.Lim and P.J.Lee : 'More flexible exponentiation with precomputation', *Advances in Cryptology-Crypto'94*, Springer-Verlag, pp.95-107 (1994).
9. P.de Rooij : 'Efficient exponentiation using precomputation and vector addition chains', In *Pre-proceedings of Eurocrypt'94*, pp.403-416 (1994).
10. T.Matsumoto, K.Kato and H.Imai : 'Speeding up secret computations with insecure auxiliary devices', *Advances in Cryptology-Crypto'88*, Springer-Verlag, pp.497-506 (1990).
11. J.J.Quisquater and M.De Soete : 'Speeding up smart card RSA computation with insecure coprocessors', In *Proc. Smart Card 2000*, North-Holland, 191-197 (1991).
12. B.Pfitzmann and M.Waidner : 'Attacks on protocols for server-aided RSA computation', *Advances in Cryptology-Eurocrypt'92*, Springer-Verlag, pp.153-162 (1993).
13. T.Matsumoto, H.Imai, C.S.Laih and S.M.Yen : 'On verifiable implicit asking protocols for RSA computation', *Advances in Cryptology-Auscrypt'92*, Springer-Verlag, pp.296-308 (1993).
14. S.Kawamura and A.Shimbo : 'Fast server-aided secret computation protocols for modular exponentiation', *IEEE J. Selected Areas in Commun.*, 11(5), 778-784 (1993).
15. J.Burns and C.J.Mitchell : 'Parameter selection for server-aided RSA computation schemes', *IEEE Trans. Computers*, 43(2), 163-174 (1994).
16. S.M.Yen and C.S.Laih : 'Server-aided honest computation for cryptographic applications', *Computers Math. Applic.*, 26(12), pp.61-64 (1993).
17. U.Feige, A.Fiat and A.Shamir : 'Zero-knowledge proofs of identity', *J. Cryptology*, 1(2), pp.77-94 (1988).
18. M.Girault : 'An identity-based identification scheme based on discrete logarithms modulo a composite number', *Advances in Cryptology-Eurocrypt'90*, Springer-Verlag, pp.481-486 (1991).

19. M.Girault : 'Self-certificated public keys', *Advances in Eurocrypt'91*, Springer-Verlag, pp.490-497 (1991).

20. K.Nyberg and R.Rueppel : 'Message recovery for signature schemes based on the discrete logarithm problem', submitted to *Designs, Codes and Cryptography* (also appears in *Pre-proceedings of Eurocrypt'94*).

21. P.Horster, H.Petersen and M.Michels : 'Meta-ElGamal signature schemes', In *Proceedings of 2nd ACM Conference on Computer and Communication Security* (1994).

22. P.Horster, H.Petersen and M.Michels : 'Meta message recovery and meta blinded signature schemes based on the discrete logarithm problem and their applications', In *Pre-Proceedings of Asiacrypt'94*, pp.185-196 (1994).

23. L.Harn and Y.Xu : 'Design of generalized ElGamal type digital signature schemes based on discrete logarithm', *Electronics Letters*, 30(24), pp.2025-2026 (1994).

24. NIST : 'Digital signature standard', *FIPS PUB 186* (1994).

25. S.Goldwasser, S.Micali and C.Rackoff : 'The knowledge complexity of interactive proof systems', *SIAM J. Comput.*, 18(1), pp.186-208 (1989).

26. D.Naccache, D.M'raihi, D.Raphaeli and S.Vaudenay : 'Can D.S.A. be improved ? -Complexity trade-offs with the digital signature standard', In *Pre-proceedings of Eurocrypt'94* (1994).

27. C.H.Lim and P.J.Lee : 'Security of interactive DSA batch verification', *Electronics Letters*, 30(19), pp.1592-1593 (1994).

28. C.H.Lim and P.J.Lee : 'Fast authenticated key exchange with the aid of the communicating partner', in preparation (available from the authors by e-mail).

29. C.H.Lim and P.J.Lee : 'Signer-aided probabilistic verification of digital signatures using random decomposition', in preparation (available from the authors by e-mail).

Counting the number of points on elliptic curves over finite fields: strategies and performances

Reynald Lercier[1] and François Morain[*][2][**]

[1] CELAR/SSIG, Route de Laillé, F-35170 Bruz
Email: lercier@polytechnique.fr
[2] LIX, École Polytechnique, F-91128 Palaiseau CEDEX, FRANCE
Email: morain@polytechnique.fr

Abstract. Cryptographic schemes using elliptic curves over finite fields require the computation of the cardinality of the curves. Dramatic progress have been achieved recently in that field by various authors. The aim of this article is to highlight part of these improvements and to describe an efficient implementation of them in the particular case of the fields $GF(2^n)$, for $n \leq 600$.

1 Introduction

Elliptic curves have been used successfully to factor integers [26, 36], and prove the primality of large integers [6, 15, 4]. Moreover they turned out to be an interesting alternative to the use of $\mathbf{Z}/N\mathbf{Z}$ in cryptographical schemes [33, 21]. Elliptic curve cryptosystems over finite fields have been built, see [5, 30]; some have been proposed in $\mathbf{Z}/N\mathbf{Z}$, N composite [23, 12, 42]. More applications were studied in [19, 22]. The interested reader should also consult [31].

In order to perform key exchange algorithms using an elliptic curve E over a finite field K, the cardinality of E must be known. The first suggestions in that direction were to use supersingular curves for which the cardinality is easy to compute [33, 21, 18, 5, 30]. But these curves turned out to be disastrous, since the discrete logarithm problem can be reduced to the discrete logarithm problem over an extension field of K of small degree [29]. For non supersingular curves, no reduction algorithm is known in general and the only known attack on such schemes is to use a variant of Pollard's algorithm [16] and this algorithm has exponential running time. Hence, it appears promising to use these curves since we can achieve the same level of confidence one has with $\mathbf{Z}/N\mathbf{Z}$ with much shorter keys.

Two types of finite fields $GF(q)$ have been suggested. The first one considers curves over $GF(p)$ where p is a large prime, the second one curves defined over

---

[*] On leave from the French Department of Defense, Délégation Générale pour l'Armement.

[**] Part of this study was done under contract n⁰0044193 with DGA/CELAR.

L.C. Guillou and J.-J. Quisquater (Eds.): Advances in Cryptology - EUROCRYPT '95, LNCS 921, pp. 79-94, 1995.

$GF(2^n)$ where n is some integer. It is possible to use the properties of complex multiplication as stated in [4] to build an elliptic curve with cardinality satisfying some properties [38, 22, 34, 35, 24, 8]. On the other hand, one can use random curves and try to compute its cardinality. It was not until recently that Schoof's polynomial time algorithm for solving this problem could be efficiently implemented and give satisfactory results. The aim of this paper is to give some hints on how this was made possible and to give some precise timings on randomly selected curves.

Since there are industrial applications for elliptic curves over $GF(2^n)$ [16, 31], we will focus on this case. We will briefly compare the running time of our implementation with that of the case $GF(p)$, p a large prime.

The structure of this paper is as follows. Section 2 recalls basic facts on elliptic curves. Section 3 describes Schoof's algorithm in a synthetic way using the contributions of Atkin, Elkies, Couveignes–Morain and the decisive ideas of Couveignes for the computation of isogenies in characteristic 2. We will present some strategies combining these ideas. Some details of the implementation are given in Section 4; precise timings on random curves for various fields are also given.

Throughout the paper, we let $K = GF(q) = GF(p^n)$ be a finite field of characteristic p.

2 Elliptic curves over finite fields

We recall well known properties of elliptic curves. All these can be found in [46] (see also [31]).

The general equation of an elliptic curve E is given as:

$$\mathcal{F}(X, Y, Z) := Y^2 Z + a_1 XYZ + a_3 YZ^2 - (X^3 + a_2 X^2 Z + a_4 XZ^2 + a_6 Z^3) = 0$$

where the a_i's are in K and the discriminant Δ defined by

$$d_2 = a_1^2 + 4a_2, d_4 = 2a_4 + a_1 a_3, d_6 = a_3^2 + 4a_6, d_8 = a_1^2 a_6 + 4a_2 a_6 - a_1 a_3 a_4 + a_2 a_3^2 - a_4^2,$$

$$c_4 = d_2^2 - 24d_4, \Delta = -d_2^2 d_8 - 8d_4^3 - 27d_6^2 + 9d_2 d_4 d_6$$

is invertible in K. The j-invariant of the curve is $j(E) = c_4^3 / \Delta$.

It is possible to define on the set of points $E(K)$ of E

$$E(K) = \{(x, y) \in K^2, \mathcal{F}(x, y, 1) = 0\} \cup \{O_E\}$$

an Abelian law using the so-called *tangent-and-chord* method, O_E being the neutral element $(0, 1, 0)$. We refer to the references given above for the precise equations of the law.

Let m denote the cardinality of the set $E(K)$ of points on E. Then, it is well known that $m = q + 1 - t$ where t is an integer satisfying $|t| \le 2\sqrt{q}$.

3   Counting the number of points

3.1   Torsion points

Let E be an elliptic curve and let N be an integer. Define $E[N]$ as the set of points of $E(\overline{K})$ of order N. When N is prime to p, then $E[N]$ is isomorphic to $(\mathbf{Z}/N\mathbf{Z}) \times (\mathbf{Z}/N\mathbf{Z})$ and when $N = p^e$, it is either $\{O_E\}$ or $(\mathbf{Z}/p^e\mathbf{Z})$.

It can be shown that there exists a polynomial $f_N(X)$ in $\mathbf{Q}[a_1, a_2, a_3, a_4, a_6][X]$ of degree

$$d_N = \begin{cases} (N^2 - 1)/2 & \text{if } (N,p) = 1, N \text{ odd,} \\ (N^2 - 4)/2 & \text{if } (N,p) = 1, N \text{ even,} \\ (p^{2e} - p^e)/2 & \text{if } N = p^e, \end{cases}$$

such that $P = (X, Y, 1)$ is in $E[N]$ if and only if $f_N(X) = 0$ in $\overline{K}$. The polynomial f_N is called *division polynomial*.

3.2   Schoof's algorithm

Schoof's algorithm [43] uses the properties of the Frobenius π_E which maps $E(\overline{K})$ onto itself and which sends a point $(X, Y, 1)$ to $(X^q, Y^q, 1)$. It is known that this endomorphism has characteristic equation

$$\pi^2 - t\pi + q = 0 \tag{1}$$

where t is related to the cardinality m of $E(K)$ via $m = q + 1 - t$.

Let ℓ be a prime number. Equation (1) is still valid when π_E is restricted to the group $E[\ell]$, and equivalently

$$\pi^2 - t\pi + q \equiv 0 \bmod \ell. \tag{2}$$

We can find $t_\ell \equiv t \bmod \ell$ by finding which value of τ, $0 \le \tau < \ell$, satisfies

$$(X^{q^2}, Y^{q^2}) + q(X, Y) = \tau(X^q, Y^q)$$

in $GF(q)[X, Y]/(\mathcal{F}(X, Y, 1), f_\ell(X))$. If we know $t \bmod \ell$ for enough ℓ such that

$$\prod \ell > 4\sqrt{q}$$

then we can determine t using the Chinese remaindering theorem.

3.3 An overview of the improvements of Atkin and Elkies

Though Schoof's algorithm has polynomial running time, its implementation was rather inefficient, due to the size of the polynomials involved. However, Atkin first and then Elkies devised theoretical and practical improvements. We suppose from now on that we want to compute $t_\ell \equiv t \bmod \ell$, ℓ a prime number different from p (see below for the particular case $\ell = p$).

Firstly, Atkin [2] explained how to use the properties of the modular polynomial $\Phi_\ell(X, Y)$ modulo p to get a list of possible values of t_ℓ. The polynomial $\Phi_\ell(X, Y)$ is symmetric in X and Y and has degree $\ell + 1$. The polynomial $\Phi(X) = \Phi_\ell(X, j(E))$ describes the cyclic subgroups of $E[\ell]$. It can have basically two splitting in K: $(11r \ldots r)$ with $\ell - 1 = rs$ or $(r \ldots r)$ with $\ell + 1 = rs$ (there are two particular cases described in the paper which are rare and we omit the relevant details for the sake of simplicity). In the first case, ℓ is said to be an *Elkies prime* and an *Atkin prime* in the second. In each case r is the order of α/β where α and β are the roots of

$$\pi^2 - t\pi + q \equiv 0 \bmod \ell$$

and lie in $GF(\ell)$ if ℓ is an Elkies prime (and thus $t^2 - 4q$ must be a square modulo ℓ) and in $GF(\ell^2)$ otherwise (implying that $t^2 - 4q$ is not a square modulo ℓ). Once r is known, there are $\varphi(r)$ possible values of t_ℓ and in many cases, this value is much less than ℓ; we denote by $c(\ell)$ the number of possible values of $t \bmod \ell$. It remains to combine these values in a clever way, using a *match and sort* technique described in [2]. This paper contains also many ideas concerning the alternative use of other modular equations, that turn out to be essential in practice, but that we do not want to describe here (for this see also [40]).

Elkies [14] remarked that when $t^2 - 4q$ is a square modulo ℓ, then $f_\ell(X)$ has a factor $g_\ell(X)$ of degree $(\ell - 1)/2$. Moreover, π_E has an eigenspace associated with g_ℓ, which means that we now look for some k, $1 \le k < \ell$ such that

$$(X^q, Y^q) = k(X, Y)$$

in $GF(q)[X, Y]/(\mathcal{F}(X, Y, 1), g_\ell(X))$; then we recover $t_\ell = (k^2 + q)/k \bmod \ell$. This change was crucial, because it was then possible to use polynomials of degree $(\ell - 1)/2$ rather than of degree $(\ell^2 - 1)/2$. Elkies gave an algorithm to compute g_ℓ using further properties of modular equations. Another approach was given in [9].

Atkin [3] gave his own solution to the problem of computing $g_\ell(X)$ using more modular equations and modular forms. Though rather tricky to implement, his approach is very fast in practice.

Recently, Couveignes and Morain showed how to use powers of small Elkies primes [11].

All these ideas are also described in [44] and were implemented [3, 25, 40, 39]. The results are striking, the record being that of the computation of the cardinality of a curve modulo a prime p of 500 digits (see the end of this article).

The only remaining problem was that these ideas could not work when $p = 2$. As a matter of fact, the theory of Atkin and Elkies remains valid, but one could not use the ordinary parameterization of elliptic curves via Weierstrass' $\wp$-functions to get a suitable way of computing g_ℓ. Couveignes solved this problem in his thesis [10], using formal groups as a powerful tool. The first successful implementation of these ideas is due to Lercier and Morain [27].

3.4 Couveignes's algorithm

Couveignes's algorithm [10] works in any characteristic $p > 0$. We simplify the exposition in the case $p = 2$.

When ℓ is an Elkies prime, we known that the initial curve

$$E : y^2 + xy = x^3 + a_6, \tag{3}$$

is isogenous to a curve

$$E^* : y^2 + xy = x^3 + a_6^* \tag{4}$$

that can be easily computed from the modular equation Φ_ℓ. The difficulty lies in the computation of the isogeny I from E to E^* defined by

$$I(x, y) = (U(x), V(x, y)) = \left(\frac{g(x)}{h^2(x)}, \frac{k(x)}{y h^3(x)} \right). \tag{5}$$

Then $h(x)$ is the factor of the division polynomial we are looking for.

Setting $t = -x/y$ and $s = -1/y$, the formal groups defined by (3) is the set of pairs (t, s) satisfying

$$t^3 + ts + a_6 s^3 = s$$

where t and s are formal series in $K((\tau))$. A morphism M from E to E^* satisfies the equality

$$M((t_1(\tau), s_1(\tau)) + (t_2(\tau), s_2(\tau))) = M(t_1(\tau), s_1(\tau)) + M(t_2(\tau), s_2(\tau)) \tag{6}$$

in $K((\tau))$. This equation is not sufficient to get I since there are much more morphisms than isogenies.

Since I can be written as (5), letting $z(\tau) = s(\tau)/t(\tau)$, U is a formal series such that

$$U(\tau) = U(z(\tau)) = z(\tau) \frac{\hat{h}^2(z(\tau))}{\hat{g}(z(\tau))} \tag{7}$$

with $\hat{h}$, a polynomial of degree $(\ell - 1)/2$ and $\hat{g}$, a polynomial of degree ℓ. We write $U(\tau) = \tau + \sum_{i=2}^{\infty} u_i \tau^i$ and we find the u_i's coefficient by coefficient. If i is not a power of 2, we look for u_i such that the equality

$$U((\tau, s(\tau)) + (A\tau, s(A\tau))) = (U(\tau), s^*(U(\tau))) + (U(A\tau), s^*(U(A\tau))), \tag{8}$$

holds up to τ^{i+1}, A being a constant in the field chosen as described in [10]; when i is a power of 2, we do the same thing using

$$U((\tau, s(\tau)) + (\tau, s(\tau))) = (U(\tau), s^*(U(\tau))) + (U(\tau), s^*(U(\tau))). \tag{9}$$

We have to compute $4\ell + 1$ terms of $U(\tau)$ in order to get $4\ell + 2$ terms once substituted in $z^*(\tau) = s^*(t)/t$ and finally have $2\ell + 1$ terms as a series in $z(\tau) = s(\tau)/\tau$, to be able to recognize

$$U(z) = \frac{zg(z)^2}{h(z)}$$

with the Massey–Berlekamp algorithm [28].

3.5 A synthetic description of the algorithm

The general algorithm for computing the cardinality of $E(K)$ runs as follows: we use two variables M_u and M_l which contain respectively the product of primes ℓ for which t_ℓ is known and for which t_ℓ is in some subset of possible values. Typically, M_u contains Elkies primes and M_l Atkin primes. The variable M will contain the current number of combinations to be tried; a bound on M is given as a constant $\mathcal{M}$ (more details on its choice will be given later).

The general procedure is:

procedure SEA(K, E)

1. $\ell := 1$; $M_u := 1$; $M_l := 1$; $M := 1$;
2. **while** $(M_u \times M_l < 4\sqrt{q})$ **or** $(M > \mathcal{M})$ **do**
   - (a) $\ell := \text{nextprime}(\ell)$;
   - (b) **if** $\ell = p$ **then** LEqualPCase(ℓ);
   - (c) compute $\Phi(X) = \Phi_\ell(X, j(E))$ and find the number ν of roots of Φ in K;
   - (d) **if** $\nu = 2$ (ℓ is an Elkies prime) **then** ElkiesCase(ℓ);
   - (e) **if** $\nu = 0$ (ℓ is an Atkin prime) **then** AtkinCase(ℓ);
3. use the match and sort technique to finish the computations.

The core of the computations consists of the two procedures:

procedure ElkiesCase(ℓ)

1. compute a factor $g_\ell(X)$ of $f_\ell(X)$ of degree $(\ell - 1)/2$ using Atkin's algorithm if p is large and Couveignes's otherwise;
2. find an eigenvalue of π_E related to g_ℓ, i.e., $1 \leq k < \ell$ such that $(X^q, Y^q) = k(X, Y)$ in $GF(q)[X, Y](\mathcal{F}(X, Y, 1), g_\ell(X))$; deduce from this that $t \equiv (k^2 + q)/k \bmod \ell$;
3. $M_u := M_u \times \ell$.

procedure AtkinCase(ℓ)

1. find the least r such that $X^{q^r} \equiv X \bmod \Phi$ and set $c(\ell) = \varphi(r)$; $M := M \times c(\ell)$;
2. $M_l := M_l \times \ell$.

Details concerning these two procedures are given in [3, 14, 9, 40, 25]. Recent improvements are due to Müller [41] (see also [45]) and Dewaghe [13] and have been incorporated in our programs.

In the case $\ell = p$, covered by procedure LEqualPCase, the polynomial $f_{p^e}(X)$ can be written as $P(X)^{p^{e-1}}$ where $P(X)$ is of degree $(p^{e+1} - p)/2$. Schoof's original algorithm or sometimes Elkies' algorithm can be used (see [31] for the case $p = \ell = 2$; the general case will be dealt with in [27]).

3.6   A more elaborate strategy

Let us give a variant of the algorithm we described above using four more constants $\mathcal{A}$, $\mathcal{E}$, $\mathcal{S}$ and $\mathcal{C}$ that will reflect the choice of possible strategies:

(c) if $\nu = 2$ and $\ell \leq \mathcal{E}$ then
      1. ElkiesCase(ℓ); compute the semi-order d of the eigenvalue k mod ℓ,
         i.e., the smallest d such that $k^d \equiv \pm 1$ mod ℓ;
      2. for $n := 2$ while $\ell^{n-1}d \leq \mathcal{C}$ do compute t mod ℓ^n; $M_u := M_u \times \ell$;
else if $(\ell^2 - 1)/2 \leq \mathcal{S}$ then for $n := 1$ while $(\ell^{2n} - \ell^{2n-2})/2 \leq \mathcal{S}$ do
      compute t mod ℓ^n using Schoof's original algorithm;
     else if $\ell \leq \mathcal{A}$ then AtkinCase(ℓ);

In the above description, the quantity $\ell^{n-1}d$ represents the degree of a factor of $f_{\ell^n}(X)$, see [11]; $(\ell^{2n} - \ell^{2n-2})/2$ is the degree of a factor of $f_{\ell^n}(X)$.

This presentation captures many possible strategies. First of all, setting $\mathcal{E} = \mathcal{A} = 0$ yields Schoof's original algorithm. Setting $\mathcal{E} = 0$ gives Atkin's first algorithm [2]. Introducing $\mathcal{C}$ makes it possible to use the ideas of [11]. We will detail the constants of our implementation in the next section.

4   Implementation and results

4.1   General remarks

We note that almost all the ideas (and tricks) of Atkin are still valid when the characteristic is 2. The first implementation of part of the above ideas is described in [32], which contains many interesting details.

4.2   Basic arithmetic

Our implementation is based on the library GFM written by F. Chabaud [7] (on top of BigNum – cf. [17]), and improved by the authors. It represents $GF(2^n)$ as the residue class ring $GF(2)[T]/(T^n + f(T))$ where $f(T)$ is a polynomial of degree smaller than n such that $T^n + f(T)$ is irreducible over $GF(2)$. In practice – in the range $1 \leq n \leq 600$ – we were always able to find a suitable f of degree less than 15.

The algorithm spends most of the time doing multiplications of elements in the field. To speed up this operation, we first perform the multiplication of two polynomials with coefficients in $GF(2)$ using a table storing all the products PQ of polynomials P and Q of degree at most 7 (at the expense of a storage of 128 kilo-bytes). Then we reduce this polynomial of degree at most $2n - 2$ modulo $T^n + f(T)$ using a second table storing the coefficients of $q(T)f(T)$ for all $q(T)$ of degree smaller than 15 (at the expense of a storage of 256 kilo-bytes too). Inversion in the field is done as in [31, Chap. 6, pp. 85].

We give in Table 1 containing precise timings (in seconds) for performing 10^6 operations. All benchmarks have been done on a DecAlpha 3000/500.

Table 1. Benchmarks for field arithmetic in $GF(2^n)$

n	Squaring	Multiplication	Inversion
65	5.2	27.4	567
89	5.3	30.8	834
105	5.8	33.6	994
155	7.3	63.8	1963
196	8.5	101.6	2835

4.3 Polynomial arithmetic

One of the main costs of the algorithm is the computation of $X^{2^n} \bmod P(X)$ where $P(X) \in GF(2^n)[X]$. As squaring of polynomials of degree d can be performed in $O(d)$ squarings in $GF(2^n)$, we have to improve the reduction of a polynomial $g(X)$ (of degree at most $2d$) modulo a polynomial $P(X)$ (of degree d). This usually costs $O(d^2)$ multiplications in K (cf. [20]), but can be improved using Newton's method and Karatsuba's algorithm as described for instance in [1] (see also [37]). In the graph given in Figure 1, we plotted the time needed to square a polynomial modulo another polynomial in $GF(2^{105})$ for all degree less than 300.

4.4 Timings

In [16], the authors give running times for curves defined over $GF(2^{65})$, $GF(2^{89})$ and $GF(2^{105})$. We used these fields as benchmarks for our implementation. We took the 50 curves defined as $y^2 + xy = x^3 + a_6$ where $a_6 \in GF(2)[T]$ and $2 \leq a_6(2) \leq 51$ (none of such coefficient a_6 belongs to a smaller extension of $GF(2^{65})$, $GF(2^{89})$ and $GF(2^{105})$).

We give: ℓ_{max}, the maximal prime used; the number of U (resp. L) primes; M, the number of combinations; the cumulated time for X^q, X^{q^r}, Schoof's algorithm; computing g_ℓ and k when ℓ is Elkies; the time for the match and sort program; the total time. For each category, we give the minimal, maximal and average values.

Fig. 1. Time for squaring a polynomial over $GF(2^{105})$

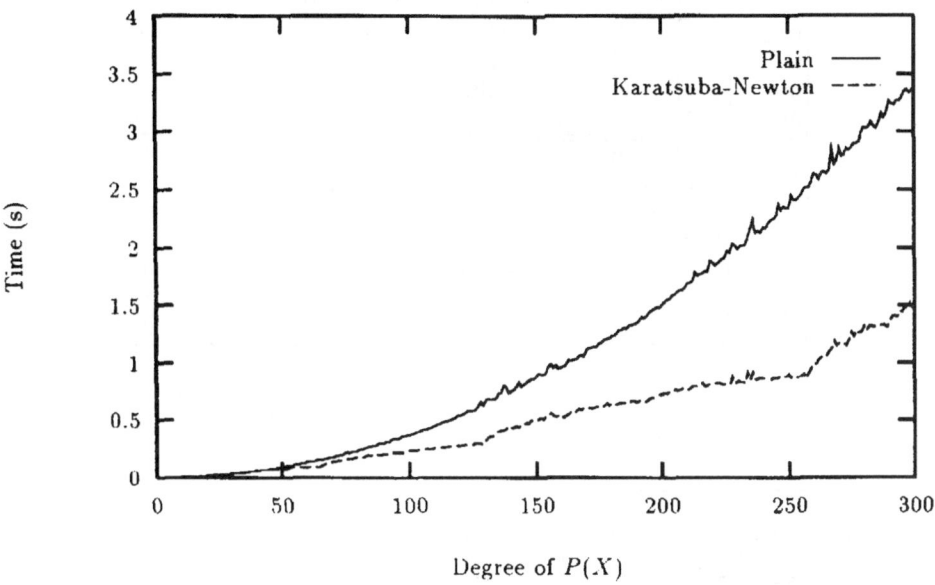

Degree of $P(X)$

Consider first $K = GF(2^{65})$. In this case, $\mathcal{M} = \infty$. Results are put in Table 2. These tables show immediately that throwing away Atkin primes is really a bad idea. Playing with the different parameters finally yields the best results for our three fields in Table 3. In each case, one has $\mathcal{A} = \infty$.

A dynamic strategy. When ℓ is an Elkies prime, the cost of `ElkiesCase` turns out to be greater than that of `AtkinCase`. In order to have a program as fast as possible, it is sometimes better to treat an Elkies prime as an Atkin prime. This motivates our *dynamic* strategy. Let L denote the least prime such that $\prod_{\ell \leq L} \ell > 4\sqrt{q}$ and denote by $\tilde{c}(\ell)$ an upper bound on $c(\ell)$. An upper bound for the number of combinations is then $\prod_{\ell < L} \tilde{c}(\ell)$. The program runs as above and as soon as for the current prime ℓ, one has

$$\left(\prod_{l < \ell} c(l) \right) \left(\prod_{\ell \leq l \leq L} \tilde{c}(l) \right) < \mathcal{M}$$

one decides to treat the remaining primes as Atkin primes.

We can compute an upper bound $\tilde{c}(\ell)$ for $c(\ell)$ as $\tilde{c}(\ell) = \max\{\varphi(r)\}$ where $r \mid \ell - \varepsilon$ and $(q/\ell) = (-1)^{(\ell-\varepsilon)/r}$ for all choices of ε in $\{\pm 1\}$. (Note that $\varepsilon = +1$ if ℓ is an Elkies prime and -1 otherwise.)

The results of this strategy are as follows, with $\mathcal{A} = \infty$ in all cases: This shows that this strategy is useful in the last case only.

Table 2. Different parameters for $GF(2^{65})$

	$\mathcal{E}=\infty, \mathcal{C}=64, \mathcal{S}=84, \mathcal{A}=0$				$\mathcal{E}=0, \mathcal{C}=0, \mathcal{S}=0, \mathcal{A}=\infty$		
	min	max	avg		min	max	avg
ℓmax	19	53	35	ℓmax	31	31	31
#U	6	9	7	#U	1	3	2
#L	0	10	4	#L	8	10	9
M	1	1	1	M	$1.02\,10^3$	$1.47\,10^6$	$2.79\,10^5$
X^q	1.5	18.9	7.2	X^q	4.7	4.9	4.8
X^{q^r}	0.0	0.0	0.0	X^{q^r}	1.1	3.9	3.0
Schoof	0.0	71.5	17.4	Schoof	0.0	0.0	0.0
g	17.4	337.0	106.0	g	0.0	0.0	0.0
k	5.1	27.3	14.7	k	0.0	0.0	0.0
M-S	0.0	0.1	0.1	M-S	0.1	2.1	1.3
Total	26.9	410.0	146.0	Total	7.2	10.5	9.1

Table 3. Best parameters for $GF(2^n)$

	$GF(2^{65})$ $\mathcal{E}=2, \mathcal{C}=2, \mathcal{S}=0$			$GF(2^{89})$ $\mathcal{E}=3, \mathcal{C}=4, \mathcal{S}=0$			$GF(2^{105})$ $\mathcal{E}=3, \mathcal{C}=4, \mathcal{S}=24$		
	min	max	avg	min	max	avg	min	max	avg
ℓmax	29	29	29	37	41	39	43	43	43
#U	1	4	2	1	6	3	4	6	5
#L	6	9	7	6	12	9	8	10	9
M	10^3	$3.7\,10^5$	$5.8\,10^4$	$7.7\,10^2$	$2.8\,10^7$	$2.4\,10^6$	$1.5\,10^5$	$7.1\,10^8$	$6.6\,10^7$
X^q	3.8	4.0	3.9	10.2	14.9	12.5	22.4	24.8	23.3
X^{q^r}	1.3	3.2	2.2	3.8	12.0	8.1	11.5	18.3	14.7
Schoof	0.0	0.0	0.0	0.0	0.0	0.0	0.0	30.0	12.9
g	0.0	1.1	0.4	0.0	2.5	1.0	0.0	2.4	1.0
k	0.1	0.2	0.1	0.3	0.6	0.5	0.3	0.8	0.6
M-S	0.1	1.7	1.1	0.2	5.9	2.5	0.5	18.8	5.7
Total	6.1	8.8	7.7	17.9	32.2	24.6	43.0	73.9	58.1

4.5 Comparison with the case $GF(p)$

For the sake of comparisons, we give some timings when using the field $GF(p)$ where p is the least prime greater than 2^{65} (resp. 2^{89}, etc.). We considered the 50 random curves of equation $y^2 = x^3 + x + b$ for $1 \le b \le 50$ for each of these primes. In all cases, one has $\mathcal{A} = \infty$ and $\mathcal{C} = 0$. Results are given in Table 5.

Table 4. Best parameters for the dynamic strategy

	$GF(2^{65})$ $\mathcal{C}=16, \mathcal{S}=4, \mathcal{M}=10^6$			$GF(2^{89})$ $\mathcal{C}=16, \mathcal{S}=4, \mathcal{M}=10^8$			$GF(2^{105})$ $\mathcal{C}=32, \mathcal{S}=12, \mathcal{M}=10^{10}$		
	min	max	avg	min	max	avg	min	max	avg
ℓ_{max}	29	29	29	37	41	37	41	43	42
$\#U$	2	4	2	2	5	3	2	5	3
$\#L$	6	8	80	7	10	9	9	12	10
M	10^3	$3.7\,10^5$	$5.3\,10^4$	$3\,10^3$	$2.7\,10^7$	$2.9\,10^6$	$7.4\,10^4$	$9.4\,10^8$	$8.3\,10^7$
X^q	3.3	3.5	3.4	9.0	12.3	9.4	15.6	22.2	20.2
X^{q^r}	1.1	2.7	1.9	3.2	8.6	5.3	8.7	15.9	12.3
Schoof	0.0	3.3	1.8	0.0	3.4	1.7	0.0	11.0	2.5
g	0.0	0.2	0.1	0.0	0.9	0.2	0.0	2.1	0.7
k	0.1	0.2	0.2	0.3	0.9	0.6	0.4	2.2	0.9
M-S	0.1	1.5	1.0	0.2	5.7	2.5	0.6	27.3	6.7
Total	6.0	10.8	8.3	15.4	25.0	19.6	32.0	65.2	43.3

Table 5. Best running times for $GF(p)$

	$GF(2^{65}+131)$ $\mathcal{E}=7, \mathcal{S}=0$			$GF(2^{89}+29)$ $\mathcal{E}=\infty, \mathcal{S}=3$			$GF(2^{105}+39)$ $\mathcal{E}=\infty, \mathcal{S}=3$		
	min	max	avg	min	max	avg	min	max	avg
ℓ_{max}	29	31	30	41	43	41	43	47	46
$\#U$	2	6	3	5	13	7	3	13	8
$\#L$	5	9	7	1	8	5	2	11	6
M	$1.7\,10^3$	$6.6\,10^5$	$1.\,10^5$	8	$2.0\,10^6$	$6.1\,10^4$	32	$1.1\,10^7$	$6.0\,10^05$
X^q	1.1	2.2	1.8	0.3	4.2	2.3	0.3	10.9	4.7
X^{q^r}	1.1	2.2	1.8	0.3	4.2	2.3	0.3	10.9	4.7
Schoof	0.0	0.0	0.0	0.0	0.0	0.0	0.0	0.0	0.0
g	0.0	0.0	0.0	0.1	0.9	0.4	0.2	1.8	0.9
k	0.0	0.1	0.0	0.7	8.1	3.7	1.1	13.5	7.7
$M-S$	0.3	1.3	0.6	0.4	2.4	0.6	0.5	6.9	1.1
Total	6.3	10.0	8.8	23.5	32.9	25.3	40.1	60.4	51.6

4.6 Records

In [32, 31], the authors gave timings for larger fields $GF(2^{155})$ and $GF(2^{195})$. For these fields and for larger fields (the last one being the current record, as of February 1995), our implementation gave the following timings, for the curve:

$$E_X : y^2 + xy = x^3 + T^{16} + T^{14} + T^{13} + T^9 + T^8 + T^7 + T^6 + T^5 + T^4 + T^3$$

(the coefficient was chosen as the binary expression of 91128 – our zip code – converted to a polynomial if $GF(2^n)$). Table 6 corresponds to the first version of our implementation, using an incremental search for the isogeny relying on a lot of composition of series, but in such a way that we could not use fast algorithms for series computations. The gap between $GF(2^{400})$ and $GF(2^{500})$ is due to a simplification of the formulas used by Couveignes in his work.

Table 7 refers to the new implementation that uses another approach, enabling one to use fast algorithms for doing series computations. The comparison for the case $GF(2^{300})$ is striking. More details will be given in [27]. We also add Table 8 which gives timings for fields $GF(p)$ with p a large prime. It seems that the program for $GF(2^n)$ is slower than the program for large prime characteristic around $n = 150$.

The record for the large prime case (as of February 1995) is our computation of the cardinality of $E : Y^2 = X^3 + 4589X + 91228$ modulo $p = 10^{499} + 153$ (4589 is the extension number of one of us). It took roughly 4200 hours of DEC 3000 - M300X including 2900 hours for the computation of various X^p. We had to consider 163 primes less than 1000, out of which 71 were of type L.

Table 6. Records for the first implementation

	$GF(2^{155})$	$GF(2^{196})$	$GF(2^{300})$	$GF(2^{400})$	$GF(2^{500})$	$GF(2^{601})$
ℓ_{max}	59	73	109	173	179	241
$\#U$	8	12	20	26	27	29
$\#L$	9	9	9	13	11	23
M	$1.21\,10^6$	$1.06\,10^8$	$1.2\,10^{10}$	$2.5\,10^9$	$1.3\,10^7$	$2.1\,10^{10}$
X^q	128.7	453.5	15886	92643	29137	109708
X^{q^r}	40.2	195.9	47562	94965	6106	52885
Schoof	0	3.7	12194	186607	65799	240091
g	334.5	1714.7	672994	1119077	518697	3139250
k	46.8	116.6	40655	774895	492213	1392113
M-S	59	698	7183	27088	3609	1728
Total	609	3183	796474	2511000	1112093	4935776

5 Conclusion

It should be apparent from the preceding tables that the implementation of Schoof's algorithm in characteristic 2 is somewhat faster than in large characteristic, at least for small fields. As a matter of fact, asymptotically, the large prime case is faster. One of the reasons for this is that polynomial arithmetic is faster for $GF(2^n)$, since X^q consists of squarings only, and squaring is an easy operation in characteristic 2. As n increases, the cost of computing the isogeny takes much more time than in the large prime case. There is still room for many improvements in that direction. We think that the situation might evolve very rapidly soon.

Table 7. Records for the second implementation

	$GF(2^{155})$	$GF(2^{196})$	$GF(2^{300})$
ℓ_{max}	59	73	109
$\#U$	6	11	18
$\#L$	11	10	11
M	$2\,10^7$	10^8	$3\,10^8$
X^q	121	440	3221
X^{q^r}	42	127	356
Schoof	0	69	0
g	24	580	22974
k	19	141	3613
M-S	10	23	56
Total	217	1381	30221

Table 8. Comparisons with $GF(p)$

	$GF(2^{155}+15)$	$GF(2^{196}+21)$	$GF(2^{300}+157)$
ℓ_{max}	67	79	113
$\#U$	11	16	20
$\#L$	8	6	10
M	$1.5\,10^6$	$6.1\,10^5$	$6.6\,10^8$
X^q	126.2	307.5	1575.3
X^{q^r}	16.6	6.8	97.3
Schoof	0.1	0.1	0.2
g	1.8	7.1	19.4
k	25.1	113.1	491.9
M-S	3.3	2.8	104.4
Total	179.1	443.7	2350.1

Acknowledgments. First of all, the authors want to express their gratitude to J.-M. Couveignes, without whom this work could not have been possible. We also thank one of the referees for suggesting a title corresponding more closely to the content of this article.

References

1. Aho, A. V., Hopcroft, J. E. , and Ullman, J. D. *The design and analysis of computer algorithms.* Reading. Addison–Wesley, 1974.
2. Atkin, A. O. L. The number of points on an elliptic curve modulo a prime. Draft, 1988.
3. Atkin, A. O. L. The number of points on an elliptic curve modulo a prime (ii). Draft, 1992.

4. ATKIN, A. O. L., AND MORAIN, F. Elliptic curves and primality proving. *Math. Comp. 61*, 203 (July 1993), 29–68.

5. BENDER, A., AND CASTAGNOLI, G. On the implementation of elliptic curve cryptosystems. In *Advances in Cryptology* (1989), G. Brassard, Ed., vol. 435 of *Lecture Notes in Comput. Sci.*, Springer-Verlag, pp. 186–192. Proc. Crypto '89, Santa Barbara, August 20–24.

6. BOSMA, W. Primality testing using elliptic curves. Tech. Rep. 85-12, Math. Instituut, Universiteit van Amsterdam, 1985.

7. CHABAUD, F. Sécurité des crypto-systèmes de McEliece. Mémoire de DEA, École polytechnique, 1993.

8. CHAO, J., TANADA, K., AND TSUJII, S. Design of elliptic curves with controllable lower boundary of extension degree for reduction attacks. In *Advances in Cryptology - CRYPTO '94* (1994), Y. Desmedt, Ed., vol. 839 of *Lecture Notes in Comput. Sci.*, Springer-Verlag, pp. 50–55. Proc. 14th Annual International Cryptology Conference, Santa Barbara, Ca, USA, August 21–25.

9. CHARLAP, L. S., COLEY, R., AND ROBBINS, D. P. Enumeration of rational points on elliptic curves over finite fields. Draft, 1991.

10. COUVEIGNES, J.-M. *Quelques calculs en théorie des nombres*. Thèse, Université de Bordeaux I, July 1994.

11. COUVEIGNES, J.-M., AND MORAIN, F. Schoof's algorithm and isogeny cycles. In preparation, February 1995. Preliminary version appeared in *ANTS-I* (1994), L. Adleman and M.-D. Huang, Eds., vol. 877 of *Lecture Notes in Comput. Sci.*, Springer-Verlag, pp. 43–58. 1st Algorithmic Number Theory Symposium - Cornell University, May 6-9, 1994.

12. DEMYTKO, N. A new elliptic curve based analogue of RSA. In *Advances in Cryptology - EUROCRYPT '93* (1994), T. Helleseth, Ed., vol. 765 of *Lecture Notes in Comput. Sci.*, Springer-Verlag, pp. 40–49. Workshop on the Theory and Application of Cryptographic Techniques, Lofthus, Norway, May 23–27, 1993.

13. DEWAGHE, L. Remarques sur l'algorithme SEA. In preparation, Dec. 1994.

14. ELKIES, N. D. Explicit isogenies. Draft, 1991.

15. GOLDWASSER, S., AND KILIAN, J. Almost all primes can be quickly certified. In *Proc. 18th STOC* (1986), ACM, pp. 316–329. May 28–30, Berkeley.

16. HARPER, G., MENEZES, A., AND VANSTONE, S. Public-key cryptosystems with very small key length. In *Advances in Cryptoloy - EUROCRYPT '92* (1993), R. A. Rueppel, Ed., vol. 658 of *Lecture Notes in Comput. Sci.*, Springer-Verlag, pp. 163–173. Workshop on the Theory and Application of Cryptographic Techniques, Balatonfüred, Hungary, May 24-28, 1992, Proceedings.

17. HERVÉ, J.-C., SERPETTE, B., AND VUILLEMIN, J. BigNum: A portable and efficient package for arbitrary-precision arithmetic. Tech. Rep. 2, Digital Paris Research Laboratory, May 1989.

18. KALISKI, JR., B. S. A pseudo-random bit generator based on elliptic logarithms. In *Proc. Crypto 86* (1986), vol. 263 of *Lecture Notes in Comput. Sci.* Proceedings Crypto '86, Santa Barbara (USA), August 11–15, 1986.

19. KALISKI, JR., B. S. One-way permutations on elliptic curves. *Journal of Cryptology 3*, 3 (1990), 187–199.

20. KNUTH, D. E. *The Art of Computer Programming: Seminumerical Algorithms.* Addison-Wesley, 1981.

21. KOBLITZ, N. Elliptic curve cryptosystems. *Math. Comp. 48*, 177 (Jan. 1987), 203–209.

22. KOBLITZ, N. Elliptic curve implementation of zero-knowledge blobs. *Journal of Cryptology 4*, 3 (1991), 207–213.

23. KOYAMA, K., MAURER, U. M., OKAMOTO, T., AND VANSTONE, S. A. New public-key schemes based on elliptic curves over the ring Z_n. In *Advances in Cryptology* (1991), vol. 576 of *Lecture Notes in Comput. Sci.*, Springer-Verlag, pp. 252–266. Proc. Crypto '91, Santa Barbara, August 12–15.

24. LAY, G.-J., AND ZIMMER, H. G. Constructing elliptic curves with given group order over large finite fields. In *ANTS-1* (1994), L. Adleman and M.-D. Huang, Eds., vol. 877 of *Lecture Notes in Comput. Sci.*, Springer-Verlag, pp. 250–263. 1st Algorithmic Number Theory Symposium - Cornell University, May 6-9, 1994.

25. LEHMANN, F., MAURER, M., MÜLLER, V., AND SHOUP, V. Counting the number of points on elliptic curves over finite fields of characteristic greater than three. In *ANTS-I* (1994), L. Adleman and M.-D. Huang, Eds., vol. 877 of *Lecture Notes in Comput. Sci.*, Springer-Verlag, pp. 60–70. 1st Algorithmic Number Theory Symposium - Cornell University, May 6-9, 1994.

26. LENSTRA, JR., H. W. Factoring integers with elliptic curves. *Annals of Math. 126* (1987), 649–673.

27. LERCIER, R., AND MORAIN, F. Counting the number of points on elliptic curves over finite fields of characteristic 2. In preparation, Oct. 1994.

28. MASSEY, J. L. Shift-register and BCH decoding. *IEEE Trans. on Information Theory IT-15*, 1 (Jan. 1969), 122–127.

29. MENEZES, A., OKAMOTO, T., AND VANSTONE, S. A. Reducing elliptic curves logarithms to logarithms in a finite field. In *Proceedings 23rd Annual ACM Symposium on Theory of Computing (STOC)* (1991), ACM Press, pp. 80–89. May 6–8, New Orleans, Louisiana.

30. MENEZES, A., AND VANSTONE, S. A. The implementation of elliptic curve cryptosystems. In *Advances in Cryptology* (1990), J. Seberry and J. Pieprzyk, Eds., no. 453 in Lecture Notes in Comput. Sci., Springer-Verlag, pp. 2–13. Proceedings Auscrypt '90, Sysdney (Australia), January 1990.

31. MENEZES, A. J. *Elliptic curve public key cryptosystems*. Kluwer Academic Publishers, 1993.

32. MENEZES, A. J., VANSTONE, S. A., AND ZUCCHERATO, R. J. Counting points on elliptic curves over F_{2^m}. *Math. Comp. 60*, 201 (Jan. 1993), 407–420.

33. MILLER, V. Use of elliptic curves in cryptography. In *Advances in Cryptology* (1987), A. M. Odlyzko, Ed., vol. 263 of *Lecture Notes in Comput. Sci.*, Springer-Verlag, pp. 417–426. Proceedings Crypto '86, Santa Barbara (USA), August11–15, 1986.

34. MIYAJI, A. On ordinary elliptic curve cryptosystems. In *Advances in Cryptology - ASIACRYPT '91* (1991), vol. 739 of *Lecture Notes in Comput. Sci.*, Springer-Verlag, pp. 50–55.

35. MIYAJI, A. Elliptic curves over F_p suitable for cryptosystems. In *Advances in cryptology - AUSCRYPT '92* (1993), J. Seberry and Y. Zheng, Eds., vol. 718 of *Lecture Notes in Comput. Sci.*, Springer-Verlag, pp. 479–491. Workshop on the theory and application of cryptographic techniques, Gold Coast, Queensland, Australia, December 13-16, 1992.

36. MONTGOMERY, P. L. Speeding the Pollard and elliptic curve methods of factorization. *Math. Comp. 48*, 177 (Jan. 1987), 243–264.

37. MONTGOMERY, P. L. *An FFT extension of the Elliptic Curve Method of factorization*. PhD thesis, University of California – Los Angeles, 1992.

38. MORAIN, F. Building cyclic elliptic curves modulo large primes. In *Advances in Cryptology - EUROCRYPT '91* (1991), D. Davies, Ed., vol. 547 of *Lecture Notes in Comput. Sci.*, Springer–Verlag, pp. 328–336. Proceedings of the Workshop on the Theory and Application of Cryptographic Techniques, Brighton, United Kingdom, April 8–11, 1991.

39. MORAIN, F. Implantation de l'algorithme de Schoof-Elkies-Atkin. Preprint, January, 1994.

40. MORAIN, F. Calcul du nombre de points sur une courbe elliptique dans un corps fini : aspects algorithmiques. To appear in the Actes des Journées Arithmétiques 1993, Feb. 1995.

41. MÜLLER, V. Looking for the eigenvalue in Schoof's algorithm. In preparation, Oct. 1994.

42. OKAMOTO, T., FUJIKODA, A., AND FUJISAKI, E. An efficient digital signature scheme based on an elliptic curve over the ring Z_n. In *Advances in Cryptology - CRYPTO '92* (1992), vol. 740 of *Lecture Notes in Comput. Sci.*, Springer-Verlag, pp. 54–65.

43. SCHOOF, R. Elliptic curves over finite fields and the computation of square roots mod p. *Math. Comp. 44* (1985), 483–494.

44. SCHOOF, R. Counting points on elliptic curves over finite fields. To appear in Proc. Journées Arithmétiques 93, Jan. 1995.

45. SHOUP, V. A new polynomial factorization algorithm and its implementation. Preprint, 1994.

46. SILVERMAN, J. H. *The arithmetic of elliptic curves*, vol. 106 of *Graduate Texts in Mathematics.* Springer, 1986.

An Implementation of the
General Number Field Sieve
to Compute Discrete Logarithms mod p

Damian Weber

FB Informatik
Universität des Saarlandes
Postfach 151150
66041 Saarbrücken
Germany

Abstract. There are many cryptographic protocols the security of which depends on the difficulty of solving the discrete logarithm problem ([8], [9], [14], etc.). In [10] and [18] it was described how to apply the number field sieve algorithm to the discrete logarithm problem in prime fields. This resulted in the asymptotically fastest known discrete log algorithm for finite fields of p elements. Very little is known about the behaviour of this algorithm in practice. In this report we write about our practical experience with our implementation of their algorithm whose first version was completed in October 1994 at the Department of Computer Science at the Universität des Saarlandes.

1   Introduction

The importance of the Discrete Logarithm Problem has its roots in its cryptographic significance. Many protocols in cryptography, for example the Digital Signature Standard [14], are secure if the underlying Discrete Logarithm Problem is difficult to solve.
A lot of algorithms have already been created to find a solution to it and therefore to break one cryptosystem associated with it.

There is an early method which proves to be quite successful for groups of smooth orders, i.e. which have no large prime factor. It was published by Pohlig and Hellman [17] and independently by Silver. It can be improved by an idea of Shanks [19]. We actually use the very practical improvement of Pollard [16].
The first of the class of index calculus algorithms to which the algorithm we discuss belongs to was published by Kraitchik and Cunningham and later rediscovered and analyzed by Adleman, Merkle and Pomerance [5]. It has a running time of $L_p[\frac{1}{2}, \delta]$ for some $\delta > 0$. With $L_p[\nu, \delta]$ we mean the commonly used expression

$$L_p[\nu, \delta] = \exp(\delta(\log p)^\nu \cdot (\log \log p)^{1-\nu}).$$

There are variations of this index calculus algorithm, discovered by Coppersmith, Odlyzko and Schroeppel [4] with conjectured running time $L_p[\frac{1}{2}, \delta]$.

L.C. Guillou and J.-J. Quisquater (Eds.): Advances in Cryptology - EUROCRYPT '95, LNCS 921, pp. 95-105, 1995.

The first algorithm with expected running time $L_p[\frac{1}{3}, 3^{\frac{2}{3}}]$ was detected by Dan Gordon [10] in 1992. This was improved by Oliver Schirokauer [18] in 1993 achieving an expected running time of $L_p[\frac{1}{3}, (\frac{64}{9})^{\frac{1}{3}}]$.

It is based on the Number Field Sieve, a method which has already been used to factor integers ([1],[2]).

If this algorithm proves to be valuable in practice, the security parameters of much implemented cryptosystems have to be thought over. In this report we write about our practical experience with our first implementation of their algorithm whose first version was completed in October 1994 at the Department of Computer Science at the Universität des Saarlandes.

For our implementation we used the methods described in [10] and [18]. After a short description of their algorithm we show how the general problem can be treated conveniently. Furthermore we consider the running time in practice, and give some impressive results concerning the comparison with our implementation of the algorithm of Pohlig and Hellman, which has an expected running time of $L_p(1, \frac{1}{2})$. Clearly, as the algorithm of Pohlig and Hellman is not an index calculus method, there is need for a comparison with an implementation of the methods of Coppersmith, Odlyzko and Schroeppel [4].

2 The Discrete Logarithm Problem in $\mathbb{F}_p$

We consider $\mathbb{F}_p(\cdot)$, the cyclic multiplicative group of the prime fields of p elements, p prime, which has order $p - 1$.
Let $a, b \in \mathbb{F}_p$.
If there exists $x \in \mathbb{N}_0$ such that

$$a^x = b,$$

we define the least such $x \in \mathbb{N}_0$ as the *discrete logarithm* of b to the base a.

3 The General Number Field Sieve (GNFS)

With $a, b \in \mathbb{F}_p$, we determine the discrete logarithm x of b to the base a modulo $q \in \mathbb{N}$ where q is a prime divisor of $p - 1$. Then we combine the results for every q dividing $p - 1$ via the Chinese remainder algorithm. In order to determine the discrete logarithm x modulo q, we use the GNFS to construct q-th powers in $\mathbb{F}_p$.
If we are able to find integers s, t with the property

$$a^s \cdot b^t \equiv w^q \bmod p$$

for some $w \in \mathbb{F}_p$, and a rational integer q coprime to t, then we have computed $x \bmod q$. If this is the case, writing $b \equiv a^x \bmod p$ leads to

$$x \equiv -st^{-1} \bmod q.$$

So the task is to construct a q-th power in $\mathbb{F}_p$, written as a nontrivial product of powers of a and b.

First, choose an irreducible polynomial

$$f(X) = X^n + a_{n-1}X^{n-1} + \ldots + a_1 X + a_0,$$

an integer m and a rational factorbase $\mathcal{F}\mathcal{B}_1$.

We denote by K the field $\mathbb{Q}[\alpha]$. We will work in the ring of integers $\mathcal{O}_K \subset K$, α a zero of f in $\mathbb{C}$.

We choose an algebraic factor base $\mathcal{F}\mathcal{B}_2$ consisting of first degree prime ideals with norm less than some bound.

For a set M of integers and an integer l we say that l is M-smooth, if all the prime divisors of l lie in M. We call l to be m-smooth for an integer m, l is $\{1, \ldots, m\}$-smooth.

After the choice of the polynomial f and the factor bases $\mathcal{F}\mathcal{B}_1$ and $\mathcal{F}\mathcal{B}_2$, the following conditions on f must be satisfied:

- $f(m) \equiv 0 \mod p$,
- $m = h \cdot b$ where h is a $\mathcal{F}\mathcal{B}_1$ smooth integer,
- p does not ramify in $\mathcal{O}_K$,
- the constant term of f is a $\mathcal{F}\mathcal{B}_2$-smooth integer,
- p does not divide the discriminant of f,
- q does not ramify in $\mathcal{O}_K$ for each divisor q of $p - 1$ we want to apply the algorithm to.

Because of the first condition the map

$$\varphi : \mathbb{Z}[\alpha] \longrightarrow \mathbb{F}_p$$
$$\alpha \longmapsto m$$

is a ring homomorphism.

The algorithm determines a non empty set S of pairs (c, d) with the following property:

- $\prod_S (c + dm)^{e_{c,d}}$ is only divisible by a and b and

- $\prod_S (c + d\alpha)^{e_{c,d}}$ is a q-th power in $\mathbb{Z}[\alpha]$.

Therefore, the following congruence holds.

$$\prod (c + dm)^{e_{c,d}} = \prod \varphi(c + d\alpha)^{e_{c,d}}$$
$$= \varphi(\omega^q)$$
$$\equiv w^q \mod p$$

It is clear immediately that $\prod_S (c + dm)^{e_{c,d}}$ is a q-th power in $\mathbb{F}_p$.

4 The Sieving Stage

In the sieving stage we collect pairs (c, d) of integers for which

- $c + dm$ is $\mathcal{F}\mathcal{B}_1$-smooth
- $c + d\alpha$ is $\mathcal{F}\mathcal{B}_2$-smooth.

Each pair which satisfies these conditions we call a *hit*.

If we have more than $|\mathcal{F}\mathcal{B}_1| + |\mathcal{F}\mathcal{B}_2|$ hits collected, a solution of a linear system mod q leads to an equation

$$\prod(c + dm) = \prod \varphi(c + d\alpha)$$
$$= \varphi(\omega^q)$$
$$\equiv w^q \bmod p$$

where the left side is only divisible by a and b.

In the case of a, b being primes, we have $a^s b' \equiv w^q \bmod p$ as desired. If a, b are not primes, we get a small linear system modulo q. It is convenient to avoid this by using the reduction we describe in section 6.

5 Constructing q-th powers in $\mathcal{O}_K$

In the previous section the construction of a q-th power in $\mathcal{O}_K$ is required. The details of this construction are to be found in [18]. In the following we give a brief overview.

Let

$$q\mathcal{O}_K = \prod_{\rho=1}^{r} \pi_\rho$$

be the decomposition of q into prime ideals of $\mathcal{O}_K$ and

$$\epsilon = \operatorname{lcm}_\rho \{N(\pi_\rho) - 1\}.$$

It follows that

$$\gamma^\epsilon \equiv 1 \bmod \mathcal{O}_K/\pi_\rho \qquad \text{for } \gamma \in \mathcal{O}_K/\pi_\rho \text{ and } 1 \leq \rho \leq r.$$

Define λ to be the following map.

$$\lambda : (\mathcal{O}_K, \cdot) \longrightarrow q\mathcal{O}_K/q^2\mathcal{O}_K(+)$$
$$\gamma \longrightarrow \gamma^\epsilon - 1$$

Because of q not being ramified in $\mathcal{O}_K$, this is actually a homomorphism of semi groups and a homomorphism on the group of units of $\mathcal{O}_K$.

We consider a special case of the main result of [18].

Proposition 1. *Let γ be an element of $\mathcal{O}_K$ whose norm is not divisible by q. Let U be the group of units of $\mathcal{O}_K$. Let*

$$U' = \{\eta \in U \mid \eta \equiv 1 \bmod q\mathcal{O}_K\}.$$

Then γ is a q-th power in $\mathcal{O}_K$, if

i) the class number of K is not divisible by q,
ii) $U' \subset U^q$,
iii) $\mathrm{ord}_Q(\gamma) \equiv 0 \bmod q$ for all prime ideals Q of $\mathcal{O}_K$,
iv) $\lambda(\gamma) = 0$.

For each pair (c, d) which we have detected as a hit we compute the image under λ modulo q^2 using the α-power basis of $\mathbb{Z}[\alpha]/q^2\mathbb{Z}[\alpha]$.

$$\lambda(c + d\alpha) = \sum_{j=0}^{n-1} b_j \alpha^j \bmod q^2\mathcal{O}_K.$$

We aim

$$\sum \lambda(c + d\alpha) = 0 \bmod q^2.$$

But all the $\lambda(c + d\alpha)$ are multiples of q. Therefore we can divide each b_j by q and then take the sum $\bmod q$ instead of computing $\bmod q^2$.

With this argument we supply to the exponent vector of the prime ideals the coefficients b_j of the image under λ. This means the exponent vector gets extended by n entries.

This concludes the construction.

6   A Reduction of the General Problem

It is convenient to transform the original discrete logarithm problem into an easier one, which means the numbers a, b are prime and smaller than a given bound. We require the following conditions on a and b and we will show how to change the original task appropriately.

Condition 1: a shall be a prime $\in \mathcal{FB}_1$
Condition 2: b shall be a prime $\leq \sqrt[v]{p}$

We give a brief description how this can be achieved.

1. We factor $p - 1 = \prod_{i=1}^{r} q_i^{e_i}$ using the elliptic curve method because this method is fast enough for numbers having less than 40 digits.
2. We find a generator g modulo p, which is in our rational factor base $\mathcal{FB}_1$.
3. We find l such that $a^l \cdot b \equiv c \bmod p$ is $\sqrt[v]{p}$-smooth

$$c = \prod s_i^{e_i'}.$$

4. For every i we solve $g^{r_i} \equiv s_i \bmod p$:

(a) using the GNFS ($m = h \cdot s_i$, h $\mathcal{FB}_1$–smooth) we find a relation

$$g^{y_j} \cdot s_i^{y'_j} \equiv d_i^{q_j^{e_j}} \bmod p$$

for $1 \le j \le r$,

(b) it follows that $x_i \equiv -\frac{y_j}{y'_j} \bmod q_j^{e_j}$.

5. We solve $g^z \equiv a \bmod p$, $z \equiv z_j \bmod q_j^{e_j}$ and compute

$$\log_a b \equiv \frac{x_i}{z_i} - l \bmod q_i^{e_i}$$

7 A 25–digit example

The first example which could not be done with our implementation of the Pohlig–Hellman procedure was the following.

We solved

$$7^x = 17 \bmod p,$$

where p is the 25-digit number 1234567890123456789000421.

Factoring $p - 1$ by trial division equals

$$1234567890123456789000421 - 1 = 2^2 \cdot 3 \cdot 5 \cdot q,$$

where q is the 23-digit prime number 20576131502057613150007.

Since $7^{\frac{p-1}{q'}} \not\equiv 1 \bmod p$ for $q' \in \{2, 3, 5, q\}$, 7 is a generator of $\mathbb{F}_p$. Because of the 23-digit prime factor q in the factorization of $p - 1$, our implementation of the methods of Pohlig–Hellman Pollard was not successful within 96 hours.

So we started our GNFS algorithm by using the polynomial

$$f(X) = X^3 + 57007X^2 - 27942X - 31727$$

and set $m = 107257599$ with $f(m) \equiv 0 \bmod p$.

As $b = 17$ is an element of our rational factor base there is no need of satisfying condition 3 of section 6.

The primes of both factor bases were the first-degree primes with norm less than 2400; no large primes were used.

The sieving interval for $c + dm$, $c + d\alpha$ was chosen as follows:

$$-4000000 \le c \le 4000000$$
$$1 \le d \le 500.$$

The sieving procedure was performed on a Sparc ELC workstation with 16 MB RAM within 12 hours.

The solution of the 728×703 linear system $\bmod\ q$ was done on a Sparc ELC workstation with 64 MB RAM within 8 hours. The solutions of $x \bmod 2^2, 3, 5$

were easily to obtain by using the Pohlig–Hellman–Shanks procedure. In the final step we had to combine the results by using the Chinese–Remainder–Algorithm and

$$x = 256351350915151146893061 \bmod 1234567890123456789000420$$

was established.

8   A 40–digit example

At February 2, 1995, the solution of our second interesting example has been achieved with the aid of Zayer's implementation of the General Number Field Sieve, which has already been used successfully to factorize a 70–digit number and to sieve in the case of a 107–digit number [2].

Here we solved

$$23^r = 29 \bmod p,$$

where p is the 40-digit number $3108193812051968080419611909199224122909$.

Factoring $p - 1$ by trial division equals

$$3108193812051968080419611909199224122909 - 1 = 2^2 \cdot 3^2 \cdot q, \text{ where}$$

q is the 38–digit prime number $86338717001443557789433664144422892303$.

Again $23^{\frac{p-1}{q'}} \not\equiv 1 \bmod p$ for $q' \in \{2, 3, q\}$, and 23 is a generator of $\mathbb{F}_p$.
The GNFS algorithm has been started by using the polynomial

$$f(X) = X^3 + 67025431X^2 + 3599000298704X - 6293411590817$$

and the integer $m = 14593810375959$ with $f(m) \equiv 0 \bmod p$.
Again $b = 29$ is an element of our rational factor base, so there is no need to satisfy condition 3 of section 6.
The primes of the rational factor base are the 1493 primes $p \leq 12503$.
The primes of the algebraic factor base are the 1978 first-degree prime ideals with norm less than 17321. The sieving interval for $c + dm, c + d\alpha$ was chosen as follows

$$-5000000 \leq c \leq 5000000$$
$$1 \leq d \leq 5000.$$

The sieving procedure was done on a Sparc ELC workstation with 16 MB RAM within 21 hours. The solution of the 3500×3477 linear system $\bmod q$ was obtained on a Paragon massively parallel system at KFA in Jülich within 40 minutes on 60 nodes.
The task was carried out by a structured Gauss implementation of Thomas Denny [6], which uses the LIP package of Arjen Lenstra [13]. The solutions of

$x \bmod 2^2, 3, 5$ were easily obtained by using the Pohlig–Hellman–Shanks procedure. In the final step we had to combine the results using the Chinese-Remainder-Algorithm and

$$x = 1761149741453474132304575201715643940920$$
$$\bmod 31081938120519680804196119091199224122908$$

was established.

9 Experimental Results of the Pohlig–Hellman Algorithm

The algorithm of Pohlig and Hellman works well in the case of q being small. We want to know how small the biggest prime factor q should be so that compared with the GNFS implementation this algorithm is faster. Our experimental results show that there is no reason to work with it if $q \geq 10^{12}$.

As above, the computations have been done on a Sparc ELC workstation with 16 MB RAM (21 Mips).

As the running time of both algorithms depends on the largest prime factor of $p - 1$, we have used the hardest primes p, namely those primes where $\frac{p-1}{2}$ is a prime, too. So the amount of time needed to solve the whole problem can be viewed as the time to determine the solution in the subgroup of quadratic residues $\bmod\ p$ which has order $\frac{p-1}{2}$.

We consider twenty discrete log problems, ten with primes of thirteen decimal digits and ten with primes of fifteen decimal digits.

DL–Problem				Running time		
$2^x \equiv$	5	mod	2000000000123	6	min	25 sec
$2^x \equiv$	5	mod	2000000001443	45	min	58 sec
$13^x \equiv$	17	mod	2000000001599	36	min	40 sec
$5^x \equiv$	7	mod	2000000002487	19	min	10 sec
$11^x \equiv$	13	mod	2000000003879	99	min	7 sec
$2^x \equiv$	5	mod	2000000004347	18	min	2 sec
$2^x \equiv$	5	mod	2000000007107	22	min	2 sec
$2^x \equiv$	5	mod	2000000007683	41	min	14 sec
$5^x \equiv$	10	mod	2000000007767	15	min	33 sec
$5^x \equiv$	7	mod	2000000008367	24	min	56 sec
$2^x \equiv$	5	mod	100000000005083	84	min	22 sec
$5^x \equiv$	10	mod	100000000005527	307	min	34 sec
$5^x \equiv$	7	mod	100000000007807	354	min	38 sec
$5^x \equiv$	10	mod	100000000008863	215	min	15 sec
$13^x \equiv$	19	mod	100000000010279	186	min	33 sec
$2^x \equiv$	5	mod	100000000012307	117	min	59 sec
$2^x \equiv$	5	mod	100000000013027	262	min	48 sec
$17^x \equiv$	23	mod	100000000015439	557	min	8 sec
$2^x \equiv$	5	mod	100000000016747	197	min	43 sec
$2^x \equiv$	6	mod	100000000017899	250	min	30 sec

This is an average running time of 32 min. 55 sec. for the 13-digit primes and an average running time of 253 min. 27 sec for the 15-digit primes.

10 GNFS versus Pohlig–Hellman–Algorithm

We have solved the problems of the section before with our Number field sieve implementation.

We start with our choice for the 13-digit primes p. Here we have chosen a polynomial of degree 2 and $m = 1388297$.

If one takes $m = \sqrt{p}$ instead, the average norm of $c + \alpha$ is $3.38 \cdot 10^9$, which is slightly larger compared to our choice of m, where we got $1.21 \cdot 10^9$.

Each factor base is bounded by a value between 200 and 300, so we expect to get between 100 and 140 factor base elements totally.

So the polynomials are determined by m

$$f(X) = X^2 + 52317X + (p - m^2 - 52317m).$$

We have allowed one large prime for each of the two factor bases. The large prime bound for $\mathbb{Z}$ is 2000000 and for $\mathcal{O}_K$ it is 4000000. A description of the large prime variation can be found in [11].

The choice of the parameters in the case of the 15 digit primes was quite similar. With $m = 9964218$ we have an average norm of $9.9 \cdot 10^8$ instead of $3.4 \cdot 10^9$ with $m = \sqrt{p}$.

Here the polynomials are

$$f(X) = X^2 + 71692X + (p - m^2 - 52317m).$$

The large prime bounds are chosen as above.

In the following table we list the number of full relations, i.e. relations without large primes as well as the number of single large prime relations with either a large rational prime or a prime ideal with large norm and the number double large prime relations, i.e. a rational single large prime and an algebraic single large prime.

We have listed the running times of the sieve and the solution of the linear system mod q.

Prime	full	single	double	Sieve	LS	total
2000000000123	63	95	48	02:51	01:48	04:39
2000000001443	133	0	0	04:29	02:14	06:43
2000000001599	98	105	63	03:07	01:46	04:53
2000000002487	51	70	56	03:08	01:53	05:01
2000000003879	35	42	26	03:47	01:19	05:06
2000000004347	33	70	37	05:05	02:20	07:25
2000000007107	40	67	39	05:42	02:46	08:27
2000000007683	174	303	122	05:24	03:46	09:10
2000000007767	61	96	41	07:03	02:22	09:25
2000000008367	55	85	28	05:24	03:36	09:00
100000000005083	72	123	53	05:32	04:18	09:50
100000000005527	111	148	50	03:34	03:38	07:12
100000000007807	80	87	49	03:24	08:17	11:41
100000000008863	124	87	54	03:57	08:41	12:37
100000000010279	114	143	51	03:14	03:43	06:57
100000000012307	68	145	48	06:38	06:04	12:42
100000000013027	95	86	54	05:02	06:04	11:06
100000000015439	104	110	64	04:38	04:20	08:58
100000000016747	137	103	0	04:01	03:47	07:48
100000000017899	50	102	35	07:42	02:45	10:27

This is an average running time of 6 min. 59 sec. for the 13-digit primes and an average running time of 9 min. 56 sec for the 15-digit primes.

All the integer computations were performed by using the libI package of Ralf Dentzer [7], which we have embedded in a C++ class library called LiDIA [15].

References

1. D. Bernstein, A. K. Lenstra, *A general Number Field Sieve Implementation* , in [11], 1991

2. J. Buchmann, J. Loho, J. Zayer, *An implementation of the general number field sieve* , Advances in Cryptology Crypto (1993) Lecture Notes in Computer Science 773, pp. 159–165

3. J. P. Buhler, H. W. Lenstra, C. Pomerance, *Factoring integers with the number field sieve*, in [11], 1992

4. D. Coppersmith, A. Odlyzko, R. Schroeppel, *Discrete Logarithms in GF(p)* , Algorithmica 1, 1986, pp. 1–15

5. K. Mc Curley, *The Discrete Logarithm Problem*, Cryptology and Computational Number Theory, Proc. Symp. in Applied Mathematics, American Mathematical Society, 1990

6. Th. Denny, *A Structured Gauss Implementation for GF(p)*, Universität des Saarlandes, to appear

7. R. Dentzer, *libI: eine lange ganzzahlige Arithmetik*, IWR Heidelberg, 1991

8. W. Diffie, M. Hellman, *New directions in Cryptography*. IEEE Trans. Inform. Theory 22 (1976), pp. 472-492

9. T. ElGamal, *A public key cryptosystem and a signature scheme based on discrete logarithms*, IEEE Trans. Inform. Theory 31 (1985), pp. 469-472

10. D. Gordon, *Discrete Logarithms in GF(p) using the Number Field Sieve*, University of Georgia, preprint 1992

11. A. K. Lenstra, H. W. Lenstra, *The development of the number field sieve*, Springer-Verlag, 1993

12. A. K. Lenstra, H. W. Lenstra, M. S. Manasse, J. M. Pollard, *The number field sieve*, Abstract: Proc. 22nd Ann. ACM Symp. on Theory of Computing (STOC)(1990),564-572

13. A. K. Lenstra, *lip: A long integer package*, Bellcore, 1989

14. National Institute of Standards and Technology, *The Digital Signature Standard, proposal and discussion*, Comm. of the ACM, 35 (7), pp. 36-54, 1992

15. I. Biehl, J. Buchmann, Th. Papanikolaou *LiDIA − A library for computational number theory*, Universität des Saarlandes, submitted to ISAAC 1995

16. J. M. Pollard, *Monte Carlo Methods for Index Computation (mod p)*, Math. Comp. 32, 918 924, 1978

17. S. Pohlig, M. Hellman, *An improved algorithm for computing logarithms over GF(p) and its cryptographic significance*, IEEE Trans. on Inform. Theory 24, 106–110, 1978

18. O. Schirokauer, *Discrete Logarithms and Local Units*, Phil. Trans. R. Soc. Lond. A (1993) 345, 409-423

19. D. Shanks, *Class Number, a Theory of Factorization and Genera*, Proc. Symposium Pure Mathematics Vol. 20, American Mathematical Society, Providence, R. I., 1970, pp. 415-440

A Block Lanczos Algorithm for Finding Dependencies over GF(2)

Peter L. Montgomery

780 Las Colindas Road, San Rafael, CA 94903-2346 USA.
Work performed at Centrum voor Wiskunde en Informatica, Amsterdam.

Abstract. Some integer factorization algorithms require several vectors in the null space of a sparse $m \times n$ matrix over the field GF(2). We modify the Lanczos algorithm to produce a sequence of orthogonal subspaces of $GF(2)^n$, each having dimension almost N, where N is the computer word size, by applying the given matrix and its transpose to N binary vectors at once. The resulting algorithm takes about $n/(N - 0.76)$ iterations. It was applied to matrices larger than $10^6 \times 10^6$ during the factorizations of 105–digit and 119–digit numbers via the general number field sieve.

1 Introduction

Some integer factorization algorithms require several nonzero vectors $\mathbf{x} \in GF(2)^n$ such that $\mathbf{Bx = 0}$, where $\mathbf{B}$ is a given $m \times n$ matrix over the field GF(2), usually very sparse and with $m < n$. These include the (obsolete) continued fraction method [9, p. 381], quadratic sieve (QS) [13, 14], and number field sieve [3, 11]. For example, when factoring an integer M, the QS method finds congruences

$$a_j^2 \equiv \prod_{i=1}^{m} p_i^{b_{ij}} \pmod{M} \qquad (1 \le j \le n) \ .$$

Here the p_i are primes (or -1) and the b_{ij} are exponents, mostly zero. QS then tries to find $S \subseteq \{1, 2, \cdots, n\}$ such that both sides of $\prod_{j \in S} a_j^2 \equiv \prod_{j \in S} \prod_{i=1}^{m} p_i^{b_{ij}}$ (mod M) are perfect squares. The left product is automatically a square, but the right product is a square only if all exponents are even, i.e., if $\sum_{j \in S} b_{ij} \equiv 0$ (mod 2) for $1 \le i \le m$. This is equivalent to $\mathbf{Bx} \equiv \mathbf{0}$ (mod 2), where $\mathbf{B} = (b_{ij})$, $\mathbf{x} = (x_j)$, and where $x_j = 1$ if $j \in S$ and $x_j = 0$ if $j \notin S$.

Traditionally one has solved $\mathbf{Bx = 0}$ over GF(2) by a variation of Gaussian elimination [9, p. 425], requiring about mn bits. When $\mathbf{B}$ is sparse, one can first apply structured Gaussian elimination [10, §5], replacing $\mathbf{B}$ by a dense matrix with about one third as many rows and columns. A gigabyte does not hold a dense $10^5 \times 10^5$ matrix, whereas we want to solve systems with n around 10^6.

LaMacchia and Odlyzko [10] implement variants to the Lanczos and conjugate gradient methods, as previously suggested by Odlyzko et al. [7, 12]. These methods repeatedly apply a symmetric $n \times n$ matrix to a vector. They store only a few temporary vectors and the original matrix, thus relieving the storage problem if the matrix is sparse. The methods were developed for use on real

L.C. Guillou and J.-J. Quisquater (Eds.): Advances in Cryptology - EUROCRYPT '95, LNCS 921, pp. 106–120, 1995.

matrices, but work over other fields unless one encounters a vector orthogonal to itself. To avoid self–orthogonal vectors, they work in an extension field of GF(2) rather than GF(2) itself.

Wiedemann [15] proposes another iterative algorithm. His algorithm applies a (not necessarily symmetric) $n \times n$ matrix $\mathbf{B}$ to a vector approximately $2n$ times, and constructs the minimal polynomial of $\mathbf{B}$. Using this minimal polynomial, one can find vectors in the null space of $\mathbf{B}$, if $\mathbf{B}$ is singular. The method likewise requires storage only for the matrix $\mathbf{B}$ and for a few temporary vectors.

The Lanczos, conjugate gradient, and Wiedemann algorithms all apply the given matrix (or its transpose) to $\mathcal{O}(n)$ vectors. On a binary computer with N bits per word, one can apply a matrix to N independent vectors over GF(2) at once, using the machine's bitwise operators. We would like to reduce the iteration count from $\mathcal{O}(n)$ to $\mathcal{O}(n/N)$, by accomplishing N times as much work per iteration. Even if we do N times as many operations per iteration after applying the matrix, the total cost of applying the matrix will drop N–fold.

Our variation of Lanczos achieves this objective by decomposing $GF(2)^n$ into several subspaces of dimension almost N which are pairwise orthogonal with respect to the symmetric $n \times n$ matrix $\mathbf{A} = \mathbf{B}^T\mathbf{B}$. The resulting algorithm takes about $n/(N - 0.76)$ iterations. Each iteration applies the matrices $\mathbf{B}$ and $\mathbf{B}^T$ to an $n \times N$ matrix and does a few supplementary operations (i.e., inner products of two $n \times N$ matrices, multiplication of an $n \times N$ matrix by an $N \times N$ matrix, multiplication and inversion of $N \times N$ matrices).

Don Coppersmith published a block Wiedemann algorithm [6] which needs about $3n/N$ applications of $\mathbf{B}$. The present work was inspired by a comment [6, p. 334] that Coppersmith had previously found a block Lanczos algorithm, but before this author had seen [5]. When $N \geq 16$, the present algorithm and [5] each need about 2/3 as many sparse matrix operations as [6], even if $\mathbf{B}$ is not symmetric. This algorithm constructs the orthogonal vectors differently than [5] and needs about 40% as many supplementary operations as [5].

Except for Gaussian elimination, these algorithms are probabilistic. They make random choices, and may fail for some of these choices. I have tried the proposed method on about 50 matrices, and have not experienced failure.

The methods herein work over other finite fields if one can do independent field operations in parallel, analogous to the bitwise operators for GF(2).

2   Notations

Throughout this paper, $\mathbf{A}$ denotes a symmetric $n \times n$ matrix over a field K. Two vectors $\mathbf{v}, \mathbf{w} \in K^n$ are said to be $\mathbf{A}$–orthogonal if $\mathbf{v}^T\mathbf{A}\mathbf{w} = 0$. If $\mathcal{V}$ and $\mathcal{W}$ are subspaces (or subsets) of K^n, then we define the block operations

$$\mathcal{V} + \mathcal{W} = \{\mathbf{v} + \mathbf{w} : \mathbf{v} \in \mathcal{V} \text{ and } \mathbf{w} \in \mathcal{W}\} ,$$
$$\mathbf{A}\mathcal{V} = \{\mathbf{A}\mathbf{v} : \mathbf{v} \in \mathcal{V}\} ,$$
$$\mathcal{V}^T\mathcal{W} = \{\mathbf{v}^T\mathbf{w} : \mathbf{v} \in \mathcal{V} \text{ and } \mathbf{w} \in \mathcal{W}\} .$$

Two subspaces $\mathcal{V}$ and $\mathcal{W}$ of K^n are said to be $\mathbf{A}$–orthogonal if $\mathbf{v}^T\mathbf{A}\mathbf{w} = 0$ for all $\mathbf{v} \in \mathcal{V}$ and $\mathbf{w} \in \mathcal{W}$; this is equivalent to $\mathcal{V}^T\mathbf{A}\mathcal{W} = \{\mathbf{0}\}$.

If $\mathbf{V}$ is an $n_1 \times n_2$ matrix, then $\langle\mathbf{V}\rangle$ denotes the subspace of K^{n_1} generated by the column vectors of $\mathbf{V}$.

If $\mathcal{W}$ is a subspace of K^n, then $\mathcal{O}(\mathcal{W})$ represents a vector in $\mathcal{W}$ or a matrix with column vectors in $\mathcal{W}$. It satisfies $\mathcal{O}(\mathcal{V}) + \mathcal{O}(\mathcal{W}) = \mathcal{O}(\mathcal{V} + \mathcal{W})$. If $\mathbf{M}$ is a matrix of suitable size, then we can replace $\mathcal{O}(\mathcal{W})\mathbf{M}$ by $\mathcal{O}(\mathcal{W})$, but cannot similarly simplify $\mathbf{M}\mathcal{O}(\mathcal{W})$.

We denote the number of bits per computer word by N.

The $k \times k$ identity matrix is denoted by $\mathbf{I}_k$.

3   Standard Lanczos

Suppose $\mathbf{A}$ is a symmetric positive definite $n \times n$ matrix over the field $K = \mathbb{R}$. If $\mathbf{b} \in \mathbb{R}^n$, then the standard Lanczos algorithm solves $\mathbf{A}\mathbf{x} = \mathbf{b}$ by iterating

$$\mathbf{w}_0 = \mathbf{b} \ ,$$

$$\mathbf{w}_i = \mathbf{A}\mathbf{w}_{i-1} - \sum_{j=0}^{i-1} c_{ij}\mathbf{w}_j \quad (i > 0), \quad \text{where} \quad c_{ij} = \frac{\mathbf{w}_j^T\mathbf{A}^2\mathbf{w}_{i-1}}{\mathbf{w}_j^T\mathbf{A}\mathbf{w}_j} \ , \tag{1}$$

until $\mathbf{w}_i = \mathbf{0}$. Using induction on $\max(i, j)$ (and the symmetry of $\mathbf{A}$), we verify

$$\mathbf{w}_j^T\mathbf{A}\mathbf{w}_i = 0 \quad (i \neq j) \ . \tag{2}$$

If $i > n$, then the vectors $\mathbf{w}_0, \mathbf{w}_1, \cdots, \mathbf{w}_i$ are linearly dependent. Suppose $\sum_{j=0}^{i} a_j\mathbf{w}_j = \mathbf{0}$ where $a_i \neq 0$. Pre–multiply by $\mathbf{w}_i^T\mathbf{A}$ to find $a_i\mathbf{w}_i^T\mathbf{A}\mathbf{w}_i = 0$. By positive definiteness, $\mathbf{w}_i = \mathbf{0}$. Let m denote the first value of i such that $\mathbf{w}_i = \mathbf{0}$.

Define

$$\mathbf{x} = \sum_{j=0}^{m-1} \frac{\mathbf{w}_j^T\mathbf{b}}{\mathbf{w}_j^T\mathbf{A}\mathbf{w}_j}\mathbf{w}_j \ . \tag{3}$$

Then $\mathbf{A}\mathbf{x} - \mathbf{b} \in \langle\mathbf{A}\mathbf{w}_0, \ \mathbf{A}\mathbf{w}_1, \cdots, \mathbf{A}\mathbf{w}_{m-1}, \ \mathbf{b}\rangle \subseteq \langle\mathbf{w}_0, \ \mathbf{w}_1, \ldots, \mathbf{w}_{m-1}\rangle$ by (1). By construction, $\mathbf{w}_j^T\mathbf{A}\mathbf{x} = \mathbf{w}_j^T\mathbf{b}$ for $0 \leq j \leq m-1$. Hence $(\mathbf{A}\mathbf{x} - \mathbf{b})^T(\mathbf{A}\mathbf{x} - \mathbf{b}) = 0$ and $\mathbf{A}\mathbf{x} = \mathbf{b}$.

Formula (1) appears to require adding suitable multiples of all earlier $\mathbf{w}_j$ when computing $\mathbf{w}_i$. However, the terms vanish when $j < i - 2$, since

$$\mathbf{w}_j^T\mathbf{A}^2\mathbf{w}_{i-1} = (\mathbf{A}\mathbf{w}_j)^T\mathbf{A}\mathbf{w}_{i-1}$$
$$= \left(\mathbf{w}_{j+1} + \sum_{k=0}^{j} c_{j+1,k}\mathbf{w}_k\right)^T \mathbf{A}\mathbf{w}_{i-1} = 0 \quad (j < i-2) \tag{4}$$

by (1) and (2). Hence (1) simplifies to

$$\mathbf{w}_i = \mathbf{A}\mathbf{w}_{i-1} - c_{i,i-1}\mathbf{w}_{i-1} - c_{i,i-2}\mathbf{w}_{i-2} \quad (i \geq 2) \ . \tag{5}$$

The Lanczos algorithm requires at most n iterations. Each iteration of (5) applies $\mathbf{A}$ to one vector $\mathbf{w}_{i-1}$. The computations

$$c_{i,i-1} = \frac{(\mathbf{Aw}_{i-1})^{\mathrm{T}}(\mathbf{Aw}_{i-1})}{\mathbf{w}_{i-1}^{\mathrm{T}}(\mathbf{Aw}_{i-1})} \quad \text{and} \quad c_{i,i-2} = \frac{(\mathbf{Aw}_{i-2})^{\mathrm{T}}(\mathbf{Aw}_{i-1})}{\mathbf{w}_{i-2}^{\mathrm{T}}(\mathbf{Aw}_{i-2})}$$

can be done with three inner products if we assume (by induction) that $\mathbf{w}_{i-1}$, $\mathbf{w}_{i-2}$, $\mathbf{Aw}_{i-1}$, $\mathbf{Aw}_{i-2}$, and $\mathbf{w}_{i-2}^{\mathrm{T}}\mathbf{Aw}_{i-2}$ are known. Another inner product is used for $\mathbf{w}_i^{\mathrm{T}}\mathbf{b}$ while updating the partial sum of $\mathbf{x}$ in (3). If $\mathbf{A}$ averages d nonzero entries per row or column, then the cost per iteration is $\mathcal{O}(dn)$ to multiply by $\mathbf{A}$ and $\mathcal{O}(n)$ for the other vector arithmetic. The total time is $\mathcal{O}(dn^2) + \mathcal{O}(n^2)$.

The storage requirement is nominal: $\mathcal{O}(1)$ temporary vectors of length n, plus the matrix $\mathbf{A}$ itself.

4 Lanczos over Other Algebraic Domains

The Lanczos iterations (1), (3), and (5) use only rational arithmetic operations (no square roots or transcendental functions). If there is no round–off error, then the final vector $\mathbf{x}$ is an exact solution, not an approximation. As Odlyzko et al. [7] [10, §3] [12] observe, this makes Lanczos usable in other algebraic domains, although the method was discovered by numerical analysts.

Let $\mathbf{A}$ be a symmetric $n \times n$ matrix over a field K. Assume that $\mathbf{w}_i \neq \mathbf{0}$ for $0 \leq i < m$ and that $\mathbf{w}_m = \mathbf{0}$. When $K = \mathbb{R}$ and $\mathbf{A}$ is positive definite, the Lanczos vectors $\{\mathbf{w}_i\}_{i=0}^{m-1}$ satisfy

$$\begin{aligned}
\mathbf{w}_i^{\mathrm{T}}\mathbf{Aw}_i &\neq 0 && (0 \leq i < m) , \\
\mathbf{w}_j^{\mathrm{T}}\mathbf{Aw}_i &= 0 && (i \neq j) , \\
\mathbf{A}\mathcal{W} &\subseteq \mathcal{W}, && \text{where} \quad \mathcal{W} = \langle \mathbf{w}_0, \mathbf{w}_1, \cdots, \mathbf{w}_{m-1} \rangle .
\end{aligned} \tag{6}$$

If $\mathbf{b} \in \mathcal{W}$ and $\mathbf{x}$ is defined by (3), then we claim (6) implies $\mathbf{Ax} = \mathbf{b}$, without further assumptions about the field K (the proof in §3 assumes $K = \mathbb{R}$). As before, $\mathcal{W}^{\mathrm{T}}(\mathbf{Ax} - \mathbf{b}) = \{\mathbf{0}\}$ by construction of $\mathbf{x}$. Also

$$\mathcal{W}^{\mathrm{T}}\mathbf{A}(\mathbf{Ax} - \mathbf{b}) = (\mathbf{A}\mathcal{W})^{\mathrm{T}}(\mathbf{Ax} - \mathbf{b}) \subseteq \mathcal{W}^{\mathrm{T}}(\mathbf{Ax} - \mathbf{b}) = \{\mathbf{0}\} .$$

We know that $\mathbf{Ax} - \mathbf{b} \in \mathcal{W}$, say $\mathbf{Ax} - \mathbf{b} = \sum_{i=0}^{m-1} c_i\mathbf{w}_i$. Pre–multiply by $\mathbf{w}_i^{\mathrm{T}}\mathbf{A}$ to conclude $c_i = 0$ since $\mathbf{w}_i^{\mathrm{T}}\mathbf{Aw}_i$ is assumed nonzero. Since i is arbitrary, $\mathbf{Ax} = \mathbf{b}$.

When $K \neq \mathbb{R}$, if $\mathbf{w}_i$ is computed by (1) or (5), then the requirement $\mathbf{w}_i^{\mathrm{T}}\mathbf{Aw}_i \neq 0$ if $\mathbf{w}_i \neq \mathbf{0}$ in (6) may fail. If $|K| \gg n$, then we may be able to tolerate this risk [7, 10], esp. if we can rerun our problem with slightly different data whenever it fails.

We want to apply Lanczos to the field GF(2), over which about half of all vectors are $\mathbf{A}$–orthogonal to themselves, and need to vary our methods. The field GF(2) has its advantages, since one can apply the matrix $\mathbf{A}$ to N different vectors in $\mathrm{GF}(2)^n$ at once, using bitwise operators. We generalize (6) to allow a sequence of subspaces in place of the vectors $\{\mathbf{w}_i\}$, and adapt the Lanczos iteration (1) to the new framework.

5 Sequences of Orthogonal Subspaces

Let $\mathbf{A}$ be a symmetric matrix over a field K. Block Lanczos algorithms [8, Chapter 7] modify (6) to produce a sequence of subspaces $\{\mathcal{W}_i\}_{i=0}^{m-1}$ of K^n which are pairwise $\mathbf{A}$–orthogonal. The condition $\mathbf{w}_i^{\mathrm{T}}\mathbf{A}\mathbf{w}_i \neq 0$ in (6) is replaced by a requirement that no nonzero vector in $\mathcal{W}_i$ be $\mathbf{A}$–orthogonal to all of $\mathcal{W}_i$.

Definition 1. A subspace $\mathcal{W} \subseteq K^n$ is said to be $\mathbf{A}$-*invertible* if it has a basis $\mathbf{W}$ of column vectors such that $\mathbf{W}^{\mathrm{T}}\mathbf{A}\mathbf{W}$ is invertible.

The property of being $\mathbf{A}$–invertible is independent of the choice of basis, since any two bases for $\mathcal{W}$ are related by an invertible transformation.

If $\mathcal{W}$ is $\mathbf{A}$–invertible, then any $\mathbf{u} \in K^n$ can be uniquely written as $\mathbf{v} + \mathbf{w}$ where $\mathbf{w} \in \mathcal{W}$ and $\mathcal{W}^{\mathrm{T}}\mathbf{A}\mathbf{v} = \{0\}$. Indeed, if the columns of $\mathbf{W}$ are a basis for $\mathcal{W}$, then $\mathbf{w} = \mathbf{W}\left(\mathbf{W}^{\mathrm{T}}\mathbf{A}\mathbf{W}\right)^{-1}\mathbf{W}^{\mathrm{T}}\mathbf{A}\mathbf{u}$.

The generalization of (6) to subspaces is

$$
\begin{aligned}
&\mathcal{W}_i && \text{is } \mathbf{A}\text{–invertible}, \\
&\mathcal{W}_j^{\mathrm{T}}\mathbf{A}\mathcal{W}_i = \{0\} && (i \neq j), \\
&\mathbf{A}\mathcal{W} \subseteq \mathcal{W}, && \text{where} \quad \mathcal{W} = \mathcal{W}_0 + \mathcal{W}_1 + \cdots + \mathcal{W}_{m-1}.
\end{aligned}
\tag{7}
$$

Assume (7). Given $\mathbf{b} \in \mathcal{W}$, we can construct an $\mathbf{x} \in \mathcal{W}$ such that $\mathbf{A}\mathbf{x} = \mathbf{b}$. Let $\mathbf{x} = \sum_{j=0}^{m-1} \mathbf{w}_j$, where $\mathbf{w}_j \in \mathcal{W}_j$ is chosen so that $\mathbf{A}\mathbf{w}_j - \mathbf{b}$ is orthogonal to all of $\mathcal{W}_j$. If the columns of $\mathbf{W}_j$ form a basis for $\mathcal{W}_j$, then

$$
\mathbf{x} = \sum_{j=0}^{m-1} \mathbf{W}_j(\mathbf{W}_j^{\mathrm{T}}\mathbf{A}\mathbf{W}_j)^{-1}\mathbf{W}_j^{\mathrm{T}}\mathbf{b}
\tag{8}
$$

generalizes (3).

Fix $N > 0$. At step i, we will have an $n \times N$ matrix $\mathbf{V}_i$ which is $\mathbf{A}$–orthogonal to all earlier $\mathbf{W}_j$. The initial $\mathbf{V}_0$ is arbitrary. We select $\mathbf{W}_i$ using as many columns of $\mathbf{V}_i$ as we can, subject to the requirement that $\mathbf{W}_i$ be $\mathbf{A}$–invertible. More precisely, we try to replace the Lanczos iterations (1) by

$$
\begin{aligned}
&\mathbf{W}_i = \mathbf{V}_i\mathbf{S}_i, \\
&\mathbf{V}_{i+1} = \mathbf{A}\mathbf{W}_i\mathbf{S}_i^{'\mathrm{T}} + \mathbf{V}_i - \sum_{j=0}^{i} \mathbf{W}_j\mathbf{C}_{i+1,j} && (i \geq 0), \\
&\mathcal{W}_i = \langle \mathbf{W}_i \rangle.
\end{aligned}
\tag{9}
$$

Stop iterating if $\mathbf{V}_i^{\mathrm{T}}\mathbf{A}\mathbf{V}_i = 0$, say for $i = m$.

Here $\mathbf{S}_i$ is an $N \times N_i$ projection matrix chosen so that $\mathbf{W}_i^{\mathrm{T}}\mathbf{A}\mathbf{W}_i$ is invertible while making $N_i \leq N$ as large as possible. The matrix $\mathbf{S}_i$ should be zero except for exactly one 1 per column and at most one 1 per row. These ensure that $\mathbf{S}_i^{\mathrm{T}}\mathbf{S}_i = \mathbf{I}_{N_i}$ and that $\mathbf{S}_i\mathbf{S}_i^{\mathrm{T}}$ is a submatrix of $\mathbf{I}_N$ reflecting the vectors selected from $\mathbf{V}_i$. For example, if $N = 3$, then $\mathbf{S}_i = \begin{pmatrix} 0 & 1 & 0 \\ 0 & 0 & 1 \end{pmatrix}^{\mathrm{T}}$ selects the second and third columns of $\mathbf{V}_i$ for inclusion in $\mathbf{W}_i$.

Formula (9) for $\mathbf{V}_{i+1}$ tries to generalize (1) while ensuring $\mathbf{W}_j^{\mathrm{T}} \mathbf{A} \mathbf{V}_{i+1} = \{0\}$ for $j \leq i$ if the earlier $\mathbf{W}_j$ exhibit the desired $\mathbf{A}$–orthogonality. We use

$$\mathbf{C}_{i+1,j} = (\mathbf{W}_j^{\mathrm{T}} \mathbf{A} \mathbf{W}_j)^{-1} \mathbf{W}_j^{\mathrm{T}} \mathbf{A} (\mathbf{A} \mathbf{W}_i \mathbf{S}_i^{\mathrm{T}} + \mathbf{V}_i) . \tag{10}$$

The terms $\mathbf{V}_i - \mathbf{W}_i \mathbf{C}_{i+1,i}$ in (9) select any columns of $\mathbf{V}_i$ not used in $\mathbf{W}_i$; those columns are known to be $\mathbf{A}$–orthogonal to $\mathbf{W}_0$ through $\mathbf{W}_{i-1}$, and the choice of $\mathbf{C}_{i+1,i}$ adjusts them so they are $\mathbf{A}$-orthogonal to $\mathbf{W}_i$ as well. Without the $\mathbf{V}_i$ term, $\mathrm{rank}(\mathbf{V}_{i+1})$ would be bounded by $\mathrm{rank}(\mathbf{A} \mathbf{W}_i \mathbf{S}_i^{\mathrm{T}}) \leq \mathrm{rank}(\mathbf{V}_i)$, and would soon drop to zero.

Theorem 2. *Equations (9) and (10) imply (7) if $\mathbf{V}_m = 0$. Furthermore,*

$$\mathbf{W}_j^{\mathrm{T}} \mathbf{A} \mathbf{V}_i = 0 \qquad (0 \leq j < i \leq m) . \tag{11}$$

PROOF. The selection of $\mathbf{S}_i$ ensures $\mathcal{W}_i = \langle \mathbf{W}_i \rangle$ is $\mathbf{A}$–invertible. The equation $\mathbf{W}_j^{\mathrm{T}} \mathbf{A} \mathbf{V}_i = 0$ implies $\mathbf{W}_j^{\mathrm{T}} \mathbf{A} \mathbf{W}_i = 0$ since $\mathbf{W}_i = \mathbf{V}_i \mathbf{S}_i$. It also implies $\mathbf{W}_i^{\mathrm{T}} \mathbf{A} \mathbf{W}_j = 0$ since $\mathbf{A}$ is symmetric.

We prove (11) by induction on i. Let $0 \leq k < m$ and assume (11) holds for $0 \leq j < i \leq k$. This assumption is vacuously true when $k = 0$. If $0 \leq j \leq k$, then

$$\begin{aligned}
\mathbf{W}_j^{\mathrm{T}} \mathbf{A} \mathbf{V}_{k+1} &= \mathbf{W}_j^{\mathrm{T}} \mathbf{A} (\mathbf{A} \mathbf{W}_k \mathbf{S}_k^{\mathrm{T}} + \mathbf{V}_k) - \sum_{i=0}^{k} \mathbf{W}_j^{\mathrm{T}} \mathbf{A} \mathbf{W}_i \mathbf{C}_{k+1,i} \\
&= \mathbf{W}_j^{\mathrm{T}} \mathbf{A} (\mathbf{A} \mathbf{W}_k \mathbf{S}_k^{\mathrm{T}} + \mathbf{V}_k) - \mathbf{W}_j^{\mathrm{T}} \mathbf{A} \mathbf{W}_j \mathbf{C}_{k+1,j} = 0
\end{aligned}$$

by induction and the choice (10) of $\mathbf{C}_{k+1,j}$.

A corollary to (11) is $\mathcal{W}_j^{\mathrm{T}} \mathbf{A} \mathcal{W}_i = \{0\}$ if $i \neq j$.

Post-multiply the defining equation (9) for $\mathbf{V}_{i+1}$ by $\mathbf{S}_i$:

$$\begin{aligned}
\mathbf{V}_{i+1} \mathbf{S}_i &= \mathbf{A} \mathbf{W}_i \mathbf{S}_i^{\mathrm{T}} \mathbf{S}_i + \mathbf{V}_i \mathbf{S}_i - \sum_{j=0}^{i} \mathbf{W}_j \mathbf{C}_{i+1,j} \mathbf{S}_i \\
&= \mathbf{A} \mathbf{W}_i + \mathbf{W}_i - \sum_{j=0}^{i} \mathbf{W}_j \mathbf{C}_{i+1,j} \mathbf{S}_i .
\end{aligned}$$

This and (9) give

$$\begin{aligned}
\mathbf{A} \mathbf{W}_i &= \mathbf{V}_{i+1} \mathbf{S}_i - \mathbf{W}_i + \sum_{j=0}^{i} \mathbf{W}_j \mathbf{C}_{i+1,j} \mathbf{S}_i = \mathbf{V}_{i+1} \mathbf{S}_i + \mathcal{O}(\mathcal{W}) , \\
\mathbf{V}_i &= \mathbf{V}_{i+1} - \mathbf{A} \mathbf{W}_i \mathbf{S}_i^{\mathrm{T}} + \sum_{j=0}^{i} \mathbf{W}_j \mathbf{C}_{i+1,j} = \mathbf{V}_{i+1} - \mathbf{A} \mathbf{W}_i \mathbf{S}_i^{'\mathrm{T}} + \mathcal{O}(\mathcal{W}) .
\end{aligned} \tag{12}$$

By hypothesis, $\mathbf{V}_m = 0 = \mathcal{O}(\mathcal{W})$. By backwards induction and (12), $\mathbf{AW}_i = \mathcal{O}(\mathcal{W})$ and $\mathbf{V}_i = \mathcal{O}(\mathcal{W})$ for $0 \leq i \leq m - 1$. Hence $\mathbf{A}\mathcal{W} \subseteq \mathcal{W}$. $\qquad \square$

The subspaces $\mathcal{W}_i$ generated by (9) have dimension at most N. This is immediate from (9) since $\mathcal{W}_i = \langle \mathbf{V}_i \mathbf{S}_i \rangle$ and $\mathbf{V}_i$ is an $n \times N$ matrix.

6 Simplifying the Block Lanczos Recurrence

In standard Lanczos, we simplified (1) to (5). The computation of $\mathbf{w}_i$ from $\mathbf{Aw}_{i-1}$ requires adjustments only by scalar multiples of $\mathbf{w}_{i-1}$ and $\mathbf{w}_{i-2}$, not by $\mathbf{w}_j$ for $j \leq i - 3$.

We would like to similarly optimize the computation of $\mathbf{V}_{i+1}$ in (9), using the invariant (11). We have some freedom in the choice of $\mathbf{S}_i$ since $\langle \mathbf{V}_i \rangle$ may have multiple bases.

If $j < i$, then the term $\mathbf{W}_j^T \mathbf{AV}_i$ in (10) vanishes by (11). We attempt to simplify $\mathbf{W}_j^T \mathbf{A}^2 \mathbf{W}_i$, using (9), (11), and the methods of (4):

$$
\begin{aligned}
\mathbf{W}_j^T \mathbf{A}^2 \mathbf{W}_i &= \left(\mathbf{S}_j^T \mathbf{S}_j\right)\left(\mathbf{W}_j^T \mathbf{A}^2 \mathbf{W}_i\right) = \mathbf{S}_j^T (\mathbf{AW}_j \mathbf{S}_j^T)^T \mathbf{AW}_i \\
&= \mathbf{S}_j^T \left(\mathbf{V}_{j+1} - \mathbf{V}_j + \mathcal{O}(\mathcal{W}_0 + \cdots + \mathcal{W}_j)\right)^T \mathbf{AW}_i \qquad (j < i) \quad (13) \\
&= \mathbf{S}_j^T \mathbf{V}_{j+1}^T \mathbf{AW}_i - \mathbf{W}_j^T \mathbf{AW}_i = \mathbf{S}_j^T \mathbf{V}_{j+1}^T \mathbf{AW}_i \ .
\end{aligned}
$$

If $\mathbf{S}_{j+1} = \mathbf{I}_N$ (so that $\mathbf{V}_{j+1} = \mathbf{W}_{j+1}$) and if $j < i - 1$, then (13) vanishes since $\mathbf{W}_{j+1}^T \mathbf{AW}_i = 0$.

In other cases equation (13) does not similarly simplify. We may be unable to force $\mathbf{C}_{i+1,j} = 0$ for $j = i - 2$, but do want to force $\mathbf{C}_{i+1,j} = 0$ for $j \leq i - 3$. Then the recurrence (9) will simplify to

$$
\mathbf{V}_{i+1} = \mathbf{AW}_i \mathbf{S}_i^T + \mathbf{V}_i - \mathbf{W}_i \mathbf{C}_{i+1,i} - \mathbf{W}_{i-1} \mathbf{C}_{i+1,i-1} - \mathbf{W}_{i-2} \mathbf{C}_{i+1,i-2} \quad (i \geq 2).
\tag{14}
$$

Although (14) has more terms than (5), the time per iteration and the temporary storage requirements will remain acceptable using (14). Equation (14) remains valid for $i = 0$ and $i = 1$ if we define $\mathbf{V}_j = 0$ and $\mathbf{W}_j = 0$ for $j < 0$.

To achieve (14), we require that (13) vanish whenever $j \leq i - 3$. That is, we require $\mathbf{V}_{j+1}$ to be $\mathbf{A}$-orthogonal to $\mathbf{W}_{j+3}$ through $\mathbf{W}_m$. We achieve this by requiring that all vectors in $\mathbf{V}_{j+1}$ be used either in $\mathbf{W}_{j+1}$ or in $\mathbf{W}_{j+2}$. More precisely, we require

$$
\langle \mathbf{V}_{j+1} \rangle \subseteq \mathcal{W}_0 + \mathcal{W}_1 + \cdots + \mathcal{W}_{j+2} \qquad (j \geq -1) \ .
\tag{15}
$$

Assuming (15), we try to simplify the matrix equation (14). Denote

$$
\mathbf{W}_i^{\mathrm{inv}} = \mathbf{S}_i \left(\mathbf{W}_i^T \mathbf{AW}_i\right)^{-1} \mathbf{S}_i^T = \mathbf{S}_i \left(\mathbf{S}_i^T \mathbf{V}_i^T \mathbf{AV}_i \mathbf{S}_i\right)^{-1} \mathbf{S}_i^T \ .
\tag{16}
$$

Each $\mathbf{W}_i^{\mathrm{inv}}$ is a symmetric $N \times N$ matrix. Eliminate all references to $\mathbf{W}_i = \mathbf{V}_i \mathbf{S}_i$:

$$
\begin{aligned}
\mathbf{V}_{i+1} &= \mathbf{A}\mathbf{V}_i\mathbf{S}_i\mathbf{S}_i^T + \mathbf{V}_i - \mathbf{V}_i\mathbf{S}_i\mathbf{C}_{i+1,i} - \mathbf{V}_{i-1}\mathbf{S}_{i-1}\mathbf{C}_{i+1,i-1} - \mathbf{V}_{i-2}\mathbf{S}_{i-2}\mathbf{C}_{i+1,i-2} \\
&= \mathbf{A}\mathbf{V}_i\mathbf{S}_i\mathbf{S}_i^T + \mathbf{V}_i - \mathbf{V}_i\mathbf{W}_i^{\mathrm{inv}}\mathbf{V}_i^T\mathbf{A}(\mathbf{A}\mathbf{V}_i\mathbf{S}_i\mathbf{S}_i^T + \mathbf{V}_i) \\
&\quad - \mathbf{V}_{i-1}\mathbf{W}_{i-1}^{\mathrm{inv}}\mathbf{V}_{i-1}^T\mathbf{A}^2\mathbf{V}_i\mathbf{S}_i\mathbf{S}_i^T - \mathbf{V}_{i-2}\mathbf{W}_{i-2}^{\mathrm{inv}}\mathbf{V}_{i-2}^T\mathbf{A}^2\mathbf{V}_i\mathbf{S}_i\mathbf{S}_i^T \ .
\end{aligned}
$$

$$(17)$$

Equation (17) appears to require four inner products: $\mathbf{V}_i^T\mathbf{A}\mathbf{V}_i$, $\mathbf{V}_i^T\mathbf{A}^2\mathbf{V}_i$, $\mathbf{V}_{i-1}^T\mathbf{A}^2\mathbf{V}_i$, and $\mathbf{V}_{i-2}^T\mathbf{A}^2\mathbf{V}_i$. We can express the latter two inner products in terms of the first two, using (10), (11), (12), and (14):

$$
\begin{aligned}
\mathbf{S}_{i-1}^T\mathbf{V}_{i-1}^T\mathbf{A}^2\mathbf{V}_i &= (\mathbf{A}\mathbf{W}_{i-1})^T\mathbf{A}\mathbf{V}_i = (\mathbf{V}_i\mathbf{S}_{i-1} + \mathcal{O}(\mathcal{W}_0 + \cdots + \mathcal{W}_{i-1}))^T\mathbf{A}\mathbf{V}_i \\
&= \mathbf{S}_{i-1}^T\mathbf{V}_i^T\mathbf{A}\mathbf{V}_i \ ,
\end{aligned}
$$

$$
\begin{aligned}
\mathbf{S}_{i-2}^T\mathbf{V}_{i-2}^T\mathbf{A}^2\mathbf{V}_i &= (\mathbf{A}\mathbf{W}_{i-2})^T\mathbf{A}\mathbf{V}_i = (\mathbf{V}_{i-1}\mathbf{S}_{i-2} + \mathcal{O}(\mathcal{W}_0 + \cdots + \mathcal{W}_{i-2}))^T\mathbf{A}\mathbf{V}_i \\
&= \mathbf{S}_{i-2}^T\mathbf{V}_{i-1}^T\mathbf{A}\mathbf{V}_i \\
&= \mathbf{S}_{i-2}^T\mathbf{V}_{i-1}^T\mathbf{A}\Big(\mathbf{A}\mathbf{W}_{i-1}\mathbf{S}_{i-1}^T + \mathbf{V}_{i-1} \\
&\quad - \mathbf{W}_{i-1}\mathbf{C}_{i,i-1} + \mathcal{O}(\mathcal{W}_{i-2} + \mathcal{W}_{i-3})\Big) \\
&= \mathbf{S}_{i-2}^T\mathbf{V}_{i-1}^T\mathbf{A}\Big(\mathbf{I}_n - \mathbf{V}_{i-1}\mathbf{W}_{i-1}^{\mathrm{inv}}\mathbf{V}_{i-1}^T\mathbf{A}\Big)\Big(\mathbf{A}\mathbf{W}_{i-1}\mathbf{S}_{i-1}^T + \mathbf{V}_{i-1}\Big) \\
&= \mathbf{S}_{i-2}^T\Big(\mathbf{I}_N - \mathbf{V}_{i-1}^T\mathbf{A}\mathbf{V}_{i-1}\mathbf{W}_{i-1}^{\mathrm{inv}}\Big) \\
&\quad \Big(\mathbf{V}_{i-1}^T\mathbf{A}^2\mathbf{V}_{i-1}\mathbf{S}_{i-1}\mathbf{S}_{i-1}^T + \mathbf{V}_{i-1}^T\mathbf{A}\mathbf{V}_{i-1}\Big) \ .
\end{aligned}
$$

Hence (17) simplifies to

$$
\mathbf{V}_{i+1} = \mathbf{A}\mathbf{V}_i\mathbf{S}_i\mathbf{S}_i^T + \mathbf{V}_i\mathbf{D}_{i+1} + \mathbf{V}_{i-1}\mathbf{E}_{i+1} + \mathbf{V}_{i-2}\mathbf{F}_{i+1} \tag{18}
$$

for $i \geq 0$, where

$$
\begin{aligned}
\mathbf{D}_{i+1} &= \mathbf{I}_N - \mathbf{W}_i^{\mathrm{inv}}\Big(\mathbf{V}_i^T\mathbf{A}^2\mathbf{V}_i\mathbf{S}_i\mathbf{S}_i^T + \mathbf{V}_i^T\mathbf{A}\mathbf{V}_i\Big) \ , \\
\mathbf{E}_{i+1} &= -\mathbf{W}_{i-1}^{\mathrm{inv}}\mathbf{V}_i^T\mathbf{A}\mathbf{V}_i\mathbf{S}_i\mathbf{S}_i^T \ , \\
\mathbf{F}_{i+1} &= -\mathbf{W}_{i-2}^{\mathrm{inv}}\Big(\mathbf{I}_N - \mathbf{V}_{i-1}^T\mathbf{A}\mathbf{V}_{i-1}\mathbf{W}_{i-1}^{\mathrm{inv}}\Big) \\
&\quad \Big(\mathbf{V}_{i-1}^T\mathbf{A}^2\mathbf{V}_{i-1}\mathbf{S}_{i-1}\mathbf{S}_{i-1}^T + \mathbf{V}_{i-1}^T\mathbf{A}\mathbf{V}_{i-1}\Big)\mathbf{S}_i\mathbf{S}_i^T \ .
\end{aligned} \tag{19}
$$

We define $\mathbf{W}_j^{\mathrm{inv}}$ and $\mathbf{V}_j$ to be $\mathbf{0}$ and $\mathbf{S}_j$ to be $\mathbf{I}_N$ for $j < 0$.

7 Finding Dependencies over GF(2)

Let $\mathbf{B}$ be an $n_1 \times n_2$ matrix over the field $K = \mathrm{GF}(2)$, where $n_1 < n_2$. Then there exist at least $n_2 - n_1$ linearly independent vectors $\mathbf{x} \in K^{n_2}$ such that $\mathbf{B}\mathbf{x} = \mathbf{0}$. Some integer factorization algorithms [9, pp. 380ff.][11] require finding several (perhaps ten) such $\mathbf{x}$. In practice, the matrix $\mathbf{B}$ is large but very sparse.

The Lanczos algorithm requires that the matrix be symmetric. We let $n = n_2$ and $\mathbf{A} = \mathbf{B}^T\mathbf{B}$. This $\mathbf{A}$ is symmetric, and any solution of $\mathbf{Bx} = \mathbf{0}$ will satisfy $\mathbf{Ax} = \mathbf{0}$ (although the converse need not be true if the rank of $\mathbf{A}$ is less than n_1).

Let N denote the computer word size, typically 32 or 64. Select a random $n \times N$ matrix $\mathbf{Y}$ over $GF(2)$, compute $\mathbf{AY}$, and attempt to find an $n \times N$ matrix $\mathbf{X}$ such that $\mathbf{AX} = \mathbf{AY}$. If we succeed, then the column vectors of $\mathbf{X} - \mathbf{Y}$ will be random vectors in the null space of $\mathbf{A}$. If the rank of $\mathbf{A}$ is at least $\text{rank}(\mathbf{B}) - N + 1$, then one can combine columns of $\mathbf{X} - \mathbf{Y}$ to find vectors in the null space of $\mathbf{B}$.

After selecting $\mathbf{Y}$, we initialize $\mathbf{V}_0 = \mathbf{AY}$ and proceed through the Lanczos iterations (18) until some $\mathbf{V}_i^T\mathbf{AV}_i = \mathbf{0}$, say for $i = m$. Compute

$$\mathbf{X} = \sum_{i=0}^{m-1} \mathbf{W}_i \left(\mathbf{W}_i^T\mathbf{AW}_i\right)^{-1} \mathbf{W}_i^T\mathbf{V}_0 = \sum_{i=0}^{m-1} \mathbf{V}_i\mathbf{W}_i^{\text{inv}}\mathbf{V}_i^T\mathbf{V}_0 \ . \tag{20}$$

Denote $\mathcal{W} = \mathcal{W}_0 + \cdots + \mathcal{W}_{m-1}$ and $\mathcal{W}_m = \langle\mathbf{V}_m\rangle$. Then $\mathcal{W}_m$ is $\mathbf{A}$–orthogonal to itself and to $\mathcal{W}$. By construction, $\mathbf{AX} - \mathbf{V}_0 \in \mathcal{W} + \mathcal{W}_m$. If $\mathbf{V}_m = \mathbf{0}$, then $\mathbf{AX} = \mathbf{V}_0 = \mathbf{AY}$.

Often the algorithm terminates with $\mathbf{V}_m^T\mathbf{AV}_m = \mathbf{0}$ but $\mathbf{V}_m \neq \mathbf{0}$. We argue heuristically that $\mathbf{A}\mathcal{W}_m \subseteq \mathcal{W}_m$. The matrix $\mathbf{V}_m$ is $\mathbf{A}$–orthogonal not only to itself but to $\mathbf{W}_j$ for $j < m$ by (11). In practice, this $\mathbf{V}_m$ has small rank (perhaps two). All $\mathbf{V}_j$ and $\mathbf{W}_j$ are contained in the Krylov subspace

$$\mathcal{V} = \langle\mathbf{V}_0\rangle + \langle\mathbf{AV}_0\rangle + \langle\mathbf{A}^2\mathbf{V}_0\rangle + \cdots \ . \tag{21}$$

If N is large (say $N \geq 16$), then the final $\mathbf{V}_m$ typically has precisely those vectors in $\mathcal{V}$ which are $\mathbf{A}$–orthogonal to all of $\mathcal{V}$, including themselves. See §8 for comments about the size of this subspace if $\mathbf{A}$ is random (which $\mathbf{B}^T\mathbf{B}$ definitely is not). If this assumption about $\mathcal{V}$ is correct, then $\mathcal{V}$ is the direct sum of $\mathcal{W}$ and of $\langle\mathbf{V}_m\rangle = \mathcal{W}_m$. Hence $\mathbf{A}\mathcal{W}_m \subseteq \mathbf{A}\mathcal{V} \subseteq \mathcal{V} = \mathcal{W} + \mathcal{W}_m$. Suppose $\mathbf{w}_0 + \mathbf{w}_1 + \cdots + \mathbf{w}_m \in \mathbf{A}\mathcal{W}_m$, where each $\mathbf{w}_j \in \mathcal{W}_j$. We claim that $\mathbf{w}_j = \mathbf{0}$ for $0 \leq j < m$. We check that

$$\mathbf{w}_j^T\mathbf{A}\mathcal{W}_j = (\mathbf{w}_0 + \mathbf{w}_1 + \cdots + \mathbf{w}_m)^T\mathbf{A}\mathcal{W}_j \subseteq (\mathbf{A}\mathcal{W}_m)^T\mathbf{A}\mathcal{W}_j$$
$$= \mathcal{W}_m^T\mathbf{A}(\mathbf{A}\mathcal{W}_j) \subseteq \mathcal{W}_m^T\mathbf{A}(\mathcal{W} + \mathcal{W}_m) = \{0\} \ .$$

Since $\mathbf{w}_j \in \mathcal{W}_j$ and $\mathcal{W}_j$ is $\mathbf{A}$–invertible for $j < m$, this shows that $\mathbf{w}_j = \mathbf{0}$.

If this heuristic argument about $\mathcal{V}$ is correct, then the images $\mathbf{A}(\mathbf{X} - \mathbf{Y})$ and $\mathbf{AV}_m$ are both in $\mathcal{W}_m$. The total rank of $\mathbf{X} - \mathbf{Y}$ and $\mathbf{V}_m$ is at most $2N$. Using Gaussian elimination, one can take linear combinations of $\mathbf{X} - \mathbf{Y}$ and $\mathbf{V}_m$ to find vectors in the null space of $\mathbf{A} = \mathbf{B}^T\mathbf{B}$. The same construction can be used to find vectors in the null space of $\mathbf{B}$. More precisely, let $\mathbf{Z}$ denote the $n \times 2N$ matrix formed by concatenating the columns of $\mathbf{X} - \mathbf{Y}$ and $\mathbf{V}_m$. Compute $\mathbf{BZ}$, and find a matrix $\mathbf{U}$ (at most $2N \times 2N$) whose columns span the null space of $\mathbf{BZ}$. Then output a basis for $\mathbf{ZU}$. In practice, one does not compute $\mathbf{U}$ explicitly, but applies the same column operations to $\mathbf{Z}$ and $\mathbf{BZ}$.

8 Selecting S_i and W_i

Recurrence (9) does not specify how to select S_i and $W_i = V_iS_i$, except that

- $W_i^T A W_i$ must be invertible;
- rank(W_i) should be as large as possible;
- Any column of V_{i-1} which was not used in W_{i-1} must be used now (15).

Let Q be a symmetric $N \times N$ matrix over a field K. Let $r = \mathrm{rank}(Q)$. We claim that if we select any r linearly independent columns of Q, then the symmetric $r \times r$ submatrix of Q with the same row indices is invertible. After renumbering the rows and columns, we may write $Q = \begin{bmatrix} Q_{11} & Q_{12} \\ Q_{21} & Q_{22} \end{bmatrix}$. Here Q_{11} is the symmetric $r \times r$ matrix which we claim is invertible, Q_{22} is symmetric, and $Q_{21} = Q_{12}^T$. The assumption that the first r columns generate all of Q means that there exists an $r \times (N-r)$ matrix T such that $\begin{bmatrix} Q_{12} \\ Q_{22} \end{bmatrix} = \begin{bmatrix} Q_{11} \\ Q_{21} \end{bmatrix} T$. This translates into $Q_{12} = Q_{11}T$ and $Q_{22} = Q_{21}T = Q_{12}^T T = T^T Q_{11} T$. Hence

$$Q = \begin{bmatrix} Q_{11} & Q_{11}T \\ T^T Q_{11} & T^T Q_{11} T \end{bmatrix} = \begin{bmatrix} I_r & 0 \\ T^T & I_{N-r} \end{bmatrix} \begin{bmatrix} Q_{11} & 0 \\ 0 & 0 \end{bmatrix} \begin{bmatrix} I_r & T \\ 0 & I_{N-r} \end{bmatrix} ,$$

implying $\mathrm{rank}(Q_{11}) = \mathrm{rank}(Q) = r$.

Consequently all maximal A–invertible subspaces of $\langle V_i \rangle$ have dimension $\mathrm{rank}(V_i^T A V_i)$. The selection of W_i can first include anything required by (15), namely $V_i(I_N - S_{i-1}S_{i-1}^T)$. If the corresponding columns of $V_i^T A V_i$ are linearly dependent but the columns in V_i are independent, then the algorithm fails. Otherwise choose a spanning set of columns for $V_i^T A V_i$ while trying to include the required columns, and choose S_i accordingly. Figure 1 summarizes my procedure. Afterwards I check whether all nonzero columns of V_{i+1} were chosen in S_i and/or S_{i-1}.

The pseudocode asserts that some $M[c_k, c_j]$ or $M[c_k, c_j + N]$ is nonzero, where $k \geq j$. At all times, if M_1 denotes the left half of M and M_2 denotes its right half, then $M_1 = M_2 T$ since only row operations are performed. We attempt to get I_N on the left of M, but occasionally zero an entire row. Let $\overline{S} = \{c_1, \cdots, c_{j-1}\} \setminus S$ denote the rows of M which have been zeroed. We consider three sets of vectors in $GF(2)^n$:

1. Dependency vectors v_k for $k \in \overline{S}$. These satisfy $T^T v_k = T v_k = 0$. The k-th element of v_k is 1; the index of any other nonzero element of v_k is in S. Row k of M and column k of M_2 are zero.
2. Rows c_j to c_N of M_2. These rows are linearly independent, since the initial $M_2 = I_N$ had full rank.
3. Columns k of T, for $k \in S$ (same as row vectors, since T is symmetric). Column k of $M_2 T$ matches that in I_N.

Cmt. Inputs: $\mathbf{T} = \mathbf{V}_i^{\mathrm{T}} \mathbf{A} \mathbf{V}$ and $\mathbf{S}_{i-1}$, where $\mathbf{A}$ (and hence $\mathbf{T}$) is symmetric.
Cmt. Outputs: Set S for diagonal of $\mathbf{S}_i \mathbf{S}_i^{\mathrm{T}}$, and $\mathbf{W}_i^{\mathrm{inv}} = \mathbf{S}_i \left(\mathbf{S}_i^{\mathrm{T}} \mathbf{T} \mathbf{S}_i \right)^{-1} \mathbf{S}_i^{\mathrm{T}}$.
Construct an $N \times 2N$ block matrix $\mathbf{M}$, with $\mathbf{T}$ on the left and $\mathbf{I}_N$ on the right.
Cmt. Algorithm performs row operations on $\mathbf{M}$. It may zero an entire row.
Number columns of $\mathbf{T}$ as $c_1, c_2, \cdots, c_N$, with columns in $\mathbf{S}_{i-1}$ coming last.
Initialize $S = \emptyset$.   **Cmt.** S has the columns selected from $\mathbf{V}_i$ for $\mathbf{W}_i$.
do $j = 1$ **to** N
  do $k = j$ **to** N **while** $(\mathbf{M}[c_j, c_j] = 0)$
    if $(\mathbf{M}[c_k, c_j] \neq 0)$ Exchange rows c_j and c_k of $\mathbf{M}$.
  end while
  if $(\mathbf{M}[c_j, c_j] \neq 0)$ **then**   **Cmt.** Use column c_j of $\mathbf{V}_i$ in $\mathbf{W}_i$.
    $S = S \cup \{c_j\}$
    Divide row c_j of $\mathbf{M}$ by $\mathbf{M}[c_j, c_j]$.   **Cmt.** A no–op over GF(2).
    Add multiples of row c_j to other rows of $\mathbf{M}$, to zero rest of column c_j.
  else   **Cmt.** No pivot element found in column c_j.
    Cmt. Column c_j of $\mathbf{T}$ is linear combination of earlier ones. Skip it.
    do $k = j$ **to** N **while** $(\mathbf{M}[c_j, c_j + N] = 0)$
      if $(\mathbf{M}[c_k, c_j + N] \neq 0)$ Exchange rows c_j and c_k of $\mathbf{M}$.
    end do
    assert $(\mathbf{M}[c_j, c_j + N] \neq 0)$
    Add multiples of row c_j to other rows, to zero rest of column $c_j + N$.
    Zero row c_j of $\mathbf{M}$.   **Cmt.** Will be zero for rest of algorithm.
  end if
  Cmt. Column c_j will remain unchanged for the duration of the algorithm.
end do
Copy right half of $\mathbf{M}$ into $\mathbf{W}_i^{\mathrm{inv}}$.

Fig. 1. Pseudocode for selecting $\mathbf{S}_i$ and $\mathbf{W}_i$

Sets 1 and 2 have a total of $|\overline{S}| + (N - j + 1) = N - |S|$ linearly independent vectors. The $|S|$ independent vectors in Set 3 are orthogonal to all vectors in Sets 1 and 2, and therefore span the orthogonal complement.

If $(\mathbf{M}_2 \mathbf{T})[c_k, c_j] = 0$ for all $k \geq j$, then column c_j of $\mathbf{T}$ is orthogonal to everything in Sets 1 and 2. This column must be a linear combination of vectors in Set 3. Let the dependency vector for $\mathbf{T}$ be $\mathbf{v} = \mathbf{v}_{c_j}$. By symmetry, this $\mathbf{v}$ is orthogonal to all of Set 3 and must be a linear combination of vectors in Sets 1 and 2. Since element c_j of $\mathbf{v}$ is nonzero, some vector in Set 1 or Set 2 must be nonzero there. This translates into $\mathbf{M}_2[c_k, c_j] \neq 0$ for some $k \geq j$, as required.

8.1 Rank of W_i

Let $f_N(X)$ be the generating function for the corank of a random $N \times N$ matrix over the field $GF(p)$. That is, the coefficient of X^m in $f_N(X)$ is the probability that the matrix has rank exactly $N - m$. If $q = 1/p$, then

$$
\begin{aligned}
f_0(X) &= 1 \ , \\
f_1(X) &= 1 - q + qX \ , \\
f_N(X) &= (1 - q + q^N X)f_{N-1}(X) + (q - q^N)f_{N-2}(X) \qquad (N \geq 2) \ .
\end{aligned}
\tag{22}
$$

Equation (22) is derived by checking which elements in the first row of the random matrix are zero, using the methods of [2]. It implies the recurrence $f_N(pX) = f_N(X) + (1 - q^N)X f_{N-1}(X)$ for $N \geq 1$. As $N \to \infty$, $f_N \to f$ where $f(pX) = (1 + X)f(X)$. The solution is

$$
f(X) = c_q \left(1 + \frac{X}{p - 1} + \frac{X^2}{(p - 1)(p^2 - 1)} + \frac{X^3}{(p - 1)(p^2 - 1)(p^3 - 1)} + \cdots \right) \ .
$$

Here $c_q = (1 - q)(1 - q^3)(1 - q^5) \cdots$ to force $f(1) = 1$.

The expected rank $\mathrm{Er}_N = N - f_N'(1)$ satisfies

$$
\begin{aligned}
\mathrm{Er}_0 &= 0 \ , \\
\mathrm{Er}_1 &= 1 - q \ , \\
\mathrm{Er}_N &= (1 - q)(1 + \mathrm{Er}_{N-1}) + q^N \mathrm{Er}_{N-1} + (q - q^N)(2 + \mathrm{Er}_{N-2}) \\
&= (1 + q - 2q^N) + (1 - q + q^N)\mathrm{Er}_{N-1} + (q - q^N)\mathrm{Er}_{N-2} \qquad (N \geq 2) \ .
\end{aligned}
\tag{23}
$$

When $p = 2$ and $q = 1/2$, equation (23) implies $\mathrm{Er}_N \geq N - 1 + 2^{-N}$ for $N \geq 0$. For large N, approximate numerical values are $c_2 \approx 0.419422418$ and $\mathrm{Er}_N \approx N - 0.764499780$.

We conjecture that the average number of iterations needed is about n/Er_N for large N and n, subject to $N \ll n \ll 2^{N^2/2}$. The experimental data in Table 1 of §10 support this conjecture.

9 Cost of Block Lanczos

Each iteration computes $\mathbf{A}\mathbf{V}_i$, $\mathbf{V}_i^T \mathbf{A}\mathbf{V}_i$, and $\mathbf{V}_i^T \mathbf{A}^2 \mathbf{V}_i = (\mathbf{A}\mathbf{V}_i)^T \mathbf{A}\mathbf{V}_i$ from $\mathbf{V}_i$. After choosing $\mathbf{S}_i$, it computes $\mathbf{W}^{\mathrm{inv}}$ in (16). It updates the partial sums of $\mathbf{X} - \mathbf{Y}$ using (20), and computes $\mathbf{V}_{i+1}$ using (18) and (19).

Equation (20) appears to require one inner product per $\mathbf{V}_i^T \mathbf{V}_0$ and one multiplication of the $n \times N$ matrix $\mathbf{V}_j$ by a $N \times N$ matrix. Henk Boender [1] observes that most of these inner product computations can be exchanged for some $N \times N$ matrix computations. Equation (15) implies $\langle \mathbf{V}_0 \rangle \subseteq \mathcal{W}_0 + \mathcal{W}_1$; if $i \geq 2$, then $\mathbf{V}_0^T \mathbf{A}\mathbf{V}_i = 0$ by (11). By (18),

$$
\begin{aligned}
\mathbf{V}_{i+1}^T \mathbf{V}_0 &= \left(\mathbf{A}\mathbf{V}_i \mathbf{S}_i \mathbf{S}_i^T + \mathbf{V}_i \mathbf{D}_{i+1} + \mathbf{V}_{i-1} \mathbf{E}_{i+1} + \mathbf{V}_{i-2} \mathbf{F}_{i+1} \right)^T \mathbf{V}_0 \\
&= \mathbf{D}_{i+1}^T \mathbf{V}_i^T \mathbf{V}_0 + \mathbf{E}_{i+1}^T \mathbf{V}_{i-1}^T \mathbf{V}_0 + \mathbf{F}_{i+1}^T \mathbf{V}_{i-2}^T \mathbf{V}_0 \ .
\end{aligned}
$$

The inner products $\mathbf{V}_i^T\mathbf{V}_0$, $\mathbf{V}_{i-1}^T\mathbf{V}_0$, and $\mathbf{V}_{i-2}^T\mathbf{V}_0$ are known by induction.

With this improvement, the cost of (16), (18), (19), and (20) is (for $i > 2$):

- One application of $\mathbf{A} = \mathbf{B}^T\mathbf{B}$ to $\mathbf{V}_i$ to compute $\mathbf{A}\mathbf{V}_i$.
- Two inner products of pairs of $n \times N$ matrices to compute the $N \times N$ matrices $\mathbf{V}_i^T\mathbf{A}\mathbf{V}_i$ and $\mathbf{V}_i^T\mathbf{A}^2\mathbf{V}_i$. (We assume that $\mathbf{V}_{i-1}^T\mathbf{A}\mathbf{V}_{i-1}$ and $\mathbf{V}_{i-1}^T\mathbf{A}^2\mathbf{V}_{i-1}$ are known by induction.)
- Selection of $\mathbf{S}_i$ subject to (15) and the computation of $\mathbf{W}_i^{\mathrm{inv}}$, as in §8.
- A few $N \times N$ matrix operations to compute $\mathbf{D}_{i+1}$, $\mathbf{E}_{i+1}$, and $\mathbf{F}_{i+1}$ in (19).
- Four multiplications of $n \times N$ matrices by $N \times N$ matrices and four additions of $n \times N$ matrices when computing $\mathbf{V}_{i+1}$ and the new partial sum of $\mathbf{X} - \mathbf{Y}$. (The post–multiplication by $\mathbf{S}_i\mathbf{S}_i^T$ in (18) is easy since $\mathbf{S}_i\mathbf{S}_i^T$ is diagonal with zeros and ones. When $K = \mathrm{GF}(2)$, it can be done via bitwise ANDs.)

The conjectured iteration count is $n/\mathrm{Er}_N = \mathcal{O}(n/N)$ if the initial $\mathbf{V}_0$ is random.

If the matrix $\mathbf{B}$ averages d nonzero entries per column, then each multiplication by $\mathbf{A}$ takes time $\mathcal{O}(dn)$. An $\mathcal{O}(Nn)$ algorithm for inner products over $\mathrm{GF}(2)$ circularly shifts the first argument by k bits, and uses n ANDs and $n - 1$ XORs to construct N bits of the inner product; this step is repeated for $0 \le k \le N - 1$. The multiplications of an $n \times N$ matrix by an $N \times N$ matrix can similarly be done with $\mathcal{O}(Nn)$ operations on N–bit words.

The net time is $\mathcal{O}(dn) + \mathcal{O}(Nn)$ per iteration and $\mathcal{O}(dn^2/N) + \mathcal{O}(n^2)$ for the algorithm. Gaussian elimination takes time $\mathcal{O}(n^3)$. Block Lanczos is asymptotically superior (in time and space) to Gaussian elimination if $d \ll n$ and remains competitive with Gaussian elimination if $d = \mathcal{O}(n)$.

Coppersmith [6, pp. 342–343] shows how to compute the inner products and the products of $n \times N$ by $N \times N$ matrices more efficiently.

When $\mathbf{S}_{i-1} = \mathbf{I}_N$ (so that $\mathbf{W}_{i-1} = \mathbf{V}_{i-1}$), the formula for $\mathbf{F}_{i+1}$ simplifies to zero and the term $\mathbf{V}_{i-2}\mathbf{F}_{i+1}$ can be omitted from (18).

Storage requirements are low. Other than the matrix $\mathbf{A}$ itself, the algorithm needs only $\mathbf{V}_{i+1}$, $\mathbf{V}_i$, $\mathbf{V}_{i-1}$, $\mathbf{V}_{i-2}$, $\mathbf{A}\mathbf{V}_i$, the partial sums of $\mathbf{X} - \mathbf{Y}$, and some $N \times N$ matrices. The same storage can be used for $\mathbf{A}\mathbf{V}_i$ and $\mathbf{V}_{i+1}$. The implementation may need storage for the intermediate vector $\mathbf{B}\mathbf{V}_i$ during the computation of $\mathbf{A}\mathbf{V}_i = \mathbf{B}^T(\mathbf{B}\mathbf{V}_i)$, but this can be avoided if $\mathbf{B}$ is stored by rows. At the end, the algorithm needs storage for $\mathbf{V}_m$, $\mathbf{B}\mathbf{V}_m$, $\mathbf{X} - \mathbf{Y}$, and $\mathbf{B}(\mathbf{X} - \mathbf{Y})$.

10 Experimental Results

In June, 1994, we applied the algorithm to an $828,077 \times 833,017$ matrix with $26,886,496$ nonzero entries as part of the record factorization of the 162–digit number $(12^{151} - 1)/11$ via the special number field sieve. The program was run on one processor of a Cray C90 at the Academic Computing Center Amsterdam (SARA), with $N = 64$. The algorithm terminated after 3.4 hours with $m = 13,098$ and $\dim(\mathcal{W}) = 828,075$.

Two days later we applied the algorithm to a $1,284,719 \times 1,294,861$ matrix with 38,928,220 nonzero entries as part of the factorization of a 105–digit cofactor of $3^{367} - 1$ via the general number field sieve (GNFS). The program terminated after 7.3 hours with $m = 20,319$ and $\dim(\mathcal{W}) = 1,284,712$.

The program sizes were about 330 Mb and 480 Mb, respectively, compared to the machine's 2147 Mb. For Gaussian elimination, on a dense matrix with one third as many rows and columns, the sizes would be 9 Gb and 22 Gb.

Table 1 has the observed dimensions of $\mathcal{W}_0$ through $\mathcal{W}_{m-1}$. The "probability" column is based on the coefficients of $f_{64}(X)$. The low values (6 and 18) for $\dim(\mathcal{W}_i)$ occurred at the end, specifically when $i = m - 1$ and the Krylov subspace (21) had been exhausted.

Table 1. Dimensions of $\mathcal{W}_i$ for two big systems

$\dim(\mathcal{W}_i)$	Predicted probability	$828,077 \times 833,017$ Frequency	Percent	$1,284,719 \times 1,294,861$ Frequency	Percent
6	.00000	0	0.000	1	0.005
18	.00000	1	0.008	0	0.000
59	.00004	1	0.008	0	0.000
60	.00133	15	0.115	32	0.157
61	.01997	290	2.214	431	2.121
62	.13981	1854	14.155	2946	14.499
63	.41942	5508	42.052	8333	41.011
64	.41942	5429	41.449	8576	42.207
Total		13,098		20,319	
$\sum \dim(\mathcal{W}_i)$		828,075		1,284,712	

Scott Contini and Arjen Lenstra implemented the algorithm on Bellcore's MasPar 16K-1 with 16K processors [4]. The MasPar took 2.5 CPU days to find 151 column dependencies of a $1,472,607 \times 1,475,898$ matrix with 79,661,432 nonzero entries, after ignoring its 102 densest rows. These dependencies were later combined to get 49 true dependencies and to factor the 119–digit cofactor of the partition number $p(13171)$ via GNFS. Memory requirements were 24 Kb per node, or 390 Mb total. They estimate that Gaussian elimination would have required two CPU weeks and 15 Gb, after reducing the problem to a dense $362,397 \times 362,597$ system.

Don Coppersmith [5, §12] reports a need to preprocess the data to eliminate singletons, which are rows containing a single nonzero element. One normally does such preprocessing, since the corresponding column cannot be in a dependency. We have not experienced his difficulty when we omit this step, but lack a theoretical explanation for why the problem does not occur in our variation.

11   Acknowledgements

This work was supported by Stieltjes Institute for Mathematics, Leiden and by Centrum voor Wiskunde en Informatica, Amsterdam. The CPU time on the Cray C90 was provided by the Dutch National Computing Facilities Foundation NCF, with financial support from the Netherlands Organization for Scientific Research NWO. Thanks to Scott Contini of Bellcore for his constructive comments on an earlier version of this manuscript.

References

1. Henk Boender, Private communication, 1994.
2. Richard P. Brent and Brendan D. McKay, *On determinants of random symmetric matrices over* $\mathbb{Z}_m$, Ars Combinatoria **26A** (1988), 57–64.
3. J.P. Buhler, H.W. Lenstra, Jr., and Carl Pomerance, *Factoring integers with the number field sieve*, The Development of the Number Field Sieve (Berlin) (A.K. Lenstra and H.W. Lenstra, Jr., eds.), Lecture Notes in Mathematics, vol. 1554, Springer–Verlag, Berlin, 1993, pp. 50–94.
4. Scott Contini and Arjen K. Lenstra, *Implementation of blocked Lanczos and Wiedemann algorithms*, In preparation, 1995.
5. Don Coppersmith, *Solving linear equations over GF(2): Block Lanczos algorithm*, Linear Algebra and its Applications **192** (1993), 33–60.
6. ———, *Solving homogeneous linear equations over GF(2) via block Wiedemann algorithm*, Mathematics of Computation **62** (1994), no. 205, 333–350.
7. Don Coppersmith, Andrew M. Odlyzko, and Richard Schroeppel, *Discrete logarithms in GF(p)*, Algorithmica **1** (1986), 1–15.
8. Jane K. Cullum and Ralph A. Willoughby, *Lanczos algorithms for large symmetric eigenvalue computations. Vol. I Theory*, Birkhäuser, Boston, 1985.
9. Donald E. Knuth, *Seminumerical algorithms*, The Art of Computer Programming, vol. 2, Addison–Wesley, Reading, MA, 2nd ed., 1981.
10. B.A. LaMacchia and A.M. Odlyzko, *Solving large sparse systems over finite fields*, Advances in Cryptology, CRYPTO '90 (A.J. Menezes and S.A. Vanstone, eds.), Lecture Notes in Computer Science, vol. 537, Springer–Verlag, pp. 109–133.
11. A.K. Lenstra, H.W. Lenstra, Jr., M.S. Manasse, and J.M. Pollard, *The factorization of the ninth Fermat number*, Mathematics of Computation **61** (1993), no. 203, 319–349.
12. A.M. Odlyzko, *Discrete logarithms in finite fields and their cryptographic significance*, Advances in Cryptology: Proceedings of EUROCRYPT 84 (New York) (T. Beth, N. Cot, and I. Ingemarsson, eds.), Lecture Notes in Computer Science, vol. 209, Springer–Verlag, pp. 224–314.
13. Carl Pomerance, *The quadratic sieve factoring algorithm*, Advances in Cryptology, Proceedings of EUROCRYPT 84 (New York) (T. Beth, N. Cot, and I. Ingemarsson, eds.), Lecture Notes in Computer Science, vol. 209, Springer–Verlag, pp. 169–182.
14. Robert D. Silverman, *The multiple polynomial quadratic sieve*, Mathematics of Computation **48** (1987), no. 177, 329–339.
15. Douglas H. Wiedemann, *Solving sparse linear equations over finite fields*, IEEE Trans. Inform. Theory **32** (1986), no. 1, 54–62.

How to Break Another
"Provably Secure" Payment System

Birgit Pfitzmann, Matthias Schunter   Michael Waidner[*]

Universität Hildesheim
Institut für Informatik
Marienburger Platz 22
D-31141 Hildesheim, Germany

Institut für Rechnerentwurf und Fehler-
toleranz, Universität Karlsruhe
Zirkel 2
D-76128 Karlsruhe, Germany

☎ ++49-5121-883-{738, 788}
Fax ++49-5121-883-732
{pfitzb, schunter}@informatik.uni-
hildesheim.de

☎ (Zurich) ++41-1-724-8220
Fax ++41-1-710-3608
wmi@zurich.ibm.com

*   on leave to IBM Zurich Research Laboratory, Rüschlikon.

Abstract: At Eurocrypt '94, Stefano D'Amiano and Giovanni Di Crescenzo presented a protocol for untraceable electronic cash based on non-interactive zero-knowledge proofs of knowledge with preprocessing. It was supposed to be provably secure given this and a few other general cryptographic tools.

We show that this protocol nevertheless does not provide any untraceability and has some further weaknesses. We also break another "provably secure" system proposed by Di Crescenzo at CIAC 94.

This is the second case of problems with "provably secure" payment systems. Moreover, yet another system with this name tacitly solves a much weaker problem than the seminal paper by Chaum, Fiat, and Naor and most other "practical" papers in this field (de Santis and Persiano, STACS 92). We therefore identify some principal problems with definitions and proofs of such schemes, and sketch better ways to handle them.

1 Introduction

Untraceable electronic cash seems to be a field where schemes called "provably secure" are easier to break than those constructed ad-hoc — this is the second such case after [PfWa_92] broke and repaired a system from [Damg_90].[1]

In the cryptologic community, electronic cash means digital payment systems, nowadays offline, that do not fully rely on the tamper-resistance of devices, and that provide some privacy to their users. It is sometimes said that electronic cash is just like paper cash, only more secure and transferable over networks. We do *not* say so, and have experienced that it is a very confusing thing to say — people associate quite a lot of different properties with cash, and electronic cash fulfils some of them, but others not, and has some new unexpected properties.[2] Moreover, electronic cash systems differ a lot among themselves.

---

[1]   To avoid misunderstandings from the start: Any possible sideswipes are certainly not directed at proving security in cryptography. Quite the contrary, we would like to advocate that at least all the systems called provably secure are indeed proved.

L.C. Guillou and J.-J. Quisquater (Eds.): Advances in Cryptology - EUROCRYPT '95, LNCS 921, pp. 121-132, 1995.
© Springer-Verlag Berlin Heidelberg 1995

For instance, before [ChFN_90], the term electronic cash was even used for online payment systems, since no offline ones (at least in the model where one does not fully rely on the tamper-resistance of devices) were known and it was widely believed that the known systems were as close as one could get to simulating paper cash.

Hence it should always be said *what* electronic cash one is talking about. The current style of expressing this is, in our opinion, one reason for the problems with "provably secure" electronic cash systems. Hence, in Section 2, we just give a brief sketch of the definition against which the systems we break are *claimed* to be secure. We already mention that this is a very weak definition. In Section 3, we describe the system from [AmCr_94]. In Section 4, we show that it does not even fulfil the weak definition. Actually, we think that any similar use of non-interactive zero-knowledge proofs with preprocessing will have such a problem. In Section 5, we describe and break the system from [Cres_94]. This time, we think the claimed restriction of the round complexity cannot be achieved at all, except under a very unreasonable additional assumption.

In Section 6, we give some recommendations. Mostly, these are hints as to what can be done to avoid "provably secure" systems being broken in the future, and to make clearer what properties individual systems are even *trying* to achieve. In addition, we explain some more properties that electronic cash can and should have in practice and that are usually omitted in theoretical papers like the ones considered here or [SaPe_92, FrYu_93].

Some History

It is not necessary to know much about the history of electronic cash for the purpose of this paper. For completeness, however, note that electronic payment systems that *do* fully rely on tamper-resistance usually follow the description in [EvGY_83], whereas those that *only* rely on cryptology for security started with [Chau_83, Chau_85] and similar papers; [Chau_89] is the most comprehensive one. The latter were all on-line systems, i.e., they need on-line verification of payments. The first off-line electronic cash system was presented in [ChFN_90]. At the same time, [Damg_90] presented the first electronic cash system claimed to be provably secure under a well-known cryptologic assumption. This was still an on-line system, whereas the corrected version in [PfWa_92] also has an extension to off-line payments, combining ideas of [Damg_90, ChFN_90]. The newest efficient, not provably secure systems are [Bran_94, Ferg_94]. Further provably secure systems were presented in [FrYu_93, SaPe_92]. The latter solves a weaker problem than the other papers mentioned and relies on non-interactive zero-knowledge proofs in the shared random-string model. The system in [AmCr_94] can be seen as an adaptation of that scheme to the other model for non-interactive zero-knowledge proofs, the model with preprocessing, and to multiple transfers of the "same coin" and divisibility of "coins" into smaller pieces.

---

[2] Just two examples: You cannot simply undo a payment by giving a "coin" back — you can solve some disputes about whether a payment was made by asking the payer to pay the same "coin" to the payee again.

2 The Weak Definition

2.1 Definition Sketch

The electronic payment systems in [AmCr_94, Cres_94], like most others in cryptology since [ChFN_90], are so-called coin systems. Their most important parts are three protocols:

- **Withdrawal**, where a user obtains information called "electronic coin" from a bank in exchange for real money (usually deducted from the user's bank account). In all existing constructions, an electronic coin somehow contains a string called "coin number" and a signature by the bank on the coin number with a special key, such that any signature with this key is interpreted as "the signed string is a coin of value so-and-so". (Thus there are different keys for different denominations.)

- **Payment**, where one user passes an "electronic coin" to another user.

- **Deposit**, where a user exchanges an "electronic coin" for real money again at a bank.

In addition, there is a protocol for establishing a bank account and a start-up protocol for the bank. In real life, several further protocols are needed, see Section 6.

The requirements made in [AmCr_94] can be stated as follows. (This description is more informal, but more correct than the original one. Some general problems with requirements at this level of abstraction are mentioned in Section 6.)

- **Unforgeability**: With k withdrawals, it should be infeasible for the users to compute $k+1$ different coins.

- **Untraceability**: After observing k withdrawals and k deposits, the bank should not be able to find out which deposit corresponds to which withdrawal, and thus who paid to whom. (Hence the "electronic coin" must change its outlook between the withdrawal and the deposit — this is where interesting cryptology, such as privacy homomorphisms, comes in.)

- **No double-spending**: If a user spends the same coin twice, she will be identified as soon as both payees deposit their copies at the bank.

- **No framing**: Nobody can wrongly be blamed as a double-spender.

Extensions to the basic scheme in [AmCr_94] are supposed to offer the following two properties:

- **Transferability**: "Coins" received in a payment cannot only be deposited at the bank, but also forwarded to other users.

- **Divisibility**: "Coins" can be divided into smaller pieces that can be spent individually.

2.2 Weakness of the Definition of Untraceability

The degree of untraceability required in [AmCr_94] and sketched above is very low: There is no untraceability against *payees*, since the adversary is assumed to have observed withdrawals and deposits only. Usually, the adversary against untraceability is assumed to have observed payments, too; see [ChFN_90, FrYu_93]. This deviation from the usual definition cannot be just a typo in [AmCr_94], since the protocol in [AmCr_94] for identifying double-spenders *depends* on the ability of each payee to identify the corresponding payer.

The lower level of untraceability is unacceptable for several reasons:

1. Protocols providing unconditional untraceability in the usual sense, i.e., against all coalitions of bank *and* payees, exist already, e.g., [ChFN_90, FrYu_93, Bran_94, Ferg_94].

2. In practice, untraceability by the payee is often just as important as untraceability by the bank. For instance, people will not want a public transport company to be able to record all their bus rides any more than a bank (In particular, since this company would know the times and places, whereas the bank alone would not.)

3. The same low level of untraceability can be achieved much more easily than described in [AmCr_94]. We sketch this in Section 4.5, after the original protocol.

In the following sections, however, we show that the systems do not even provide this weak level of untraceability.

3 The System Presented at Eurocrypt 94

Initialization

For initialization, the bank in [AmCr_94] publishes some system parameters and public keys, especially a public key of an arbitrary signature scheme and a string commitment scheme.

Each user publishes a public key of a signature scheme, too. Additionally, each user performs the preprocessing phase of a non-interactive zero-knowledge proof (NIZKP) system of knowledge with preprocessing, with the user as the prover and the bank as the verifier.

It is important (although not stated in [AmCr_94]) that the bank's result, β, of this preprocessing must be published, too: If a payee receives a coin, he needs this information for verifying that the coin is valid. More precisely, he needs β as an additional input to the protocol called V_2.

Withdrawal

If a payer wants to withdraw a coin, she chooses a coin identifier, c, randomly, computes a commitment, *com*, of c, and sends it to the bank. The bank sends a signature on *com* back to the payer.

Based on the information used to compute *com*, the signature received from the bank, and the information generated in the preprocessing phase of the NIZKP system, the payer computes a non-interactive zero-knowledge proof, *cp*, that will convince a payee that the payer knows:

- a commitment on c and
- a valid signature by the bank on this commitment.

Payment

The payer forwards c and cp to the payee. The payee verifies the proof cp, using the result, β, of the bank during the preprocessing phase with the payer.[3]

Additionally, the payer signs the pair (ID_{Payee}, c) and forwards this to the payee. This signature is indeed needed by the payee: If c is spent twice, the payee is asked to identify the party that paid c to him.

The payee can reuse the coin or deposit it. Deposit is simply a payment to the bank. (Note that all payments after the first one do not involve *computing* a proof of knowledge, but passing the received one on, since only the first payer has the secret inputs needed to compute such a proof.)

Double-Spending Detection

If the bank receives the same coin in two deposits (i.e., the strings c of the two coins are equal, the strings cp may or may not be), say from users A and B, it determines all previous owners of c, based on the signatures forwarded in payments. In [AmCr_94], this is done by broadcasting a request to all users; we do it more practically below. At the end, either

1. an owner X is detected who does not cooperate, i.e., does not reveal the source of c,
2. an owner X is detected who has paid the coin twice, or
3. two withdrawals of the same c are detected.

In the first two cases, X is assumed to be an attacker. This is clearly not necessarily true in Case 1: X could simply be unreachable or have lost the necessary information due to a technical defect. I.e., this approach works in a technical sense, but would often yield incorrect results in practice.

Case 2 poses technical problems, see Section 4.3.

Case 3 is not considered in [AmCr_94], but it is no problem: In each of the withdrawals, the appropriate sum was deducted from an account, hence there is no fraud at all.

It should be mentioned that the schemes for double-spending detection based on secret sharing [FrYu_93, Bran_94, Ferg_94] can be applied to *transferable* schemes, too [Antw_90, ChPe_93], i.e., feasible solutions to this problem exist already.

4 Weaknesses of this System

4.1 The Main Untraceability Flaw

Unfortunately, the protocol in [AmCr_94] does not provide any untraceability, not even the weak version required there:

---

[3] The protocol in [AmCr_94] is ambiguous about whether c must be passed or only cp, but it must: There is no guarantee that one cannot construct two different NIZKPs on the same c, hence c is needed for the uniqueness of coins. Thus, in [AmCr_94 p.156], T should be the statement that a commitment and signature on the given coin identifier c are known, and c should be an input to V_2, too.

During the deposit of a coin (c, cp), the bank must verify that cp is a valid NIZKP. This verification depends on the preprocessing phase of the user X who originally withdrew the coin; i.e., just like the payee in a payment, the bank must know which string β to use in the verification protocol V_2. It therefore knows the identity of the payer during the preprocessing with whom this particular β was generated, i.e., it has traced this payer.

This is a general problem with NIZKP with preprocessing: At least as long as preprocessing is a 2-party protocol, any reference to it, i.e., any proof, identifies the two parties. Thus at least the real untraceability against payee and bank seems impossible to achieve by straightforward use of this primitive. If preprocessing were performed anonymously before opening an account, all payments of one payer could still be linked, which would lead to identification after a while. The only remedy would be to execute anonymous preprocessing once for each coin to be withdrawn, where all preprocessings have to be unlinkable to the withdrawals. This would be very inconvenient for the users, and the immediate distribution of all the resulting strings β to all payees would contradict the offline properties of the cash system.

4.2 An Untraceability Flaw in Double-Spending Detection

Even without the flaw due to individual preprocessing, the user's privacy is not really protected: Until the double-spender is detected, the bank cannot *prove* that a coin was double-spent, i.e., the users have to believe this. Thus, a dishonest bank could always claim that c was deposited twice in order to get the payees to disclose the payers.

The general rule with untraceable protocols should be that a proof of wrong-doing must be shown before any sensitive information is asked to be disclosed, if at all.

4.3 Finding the Wrong Double-Spender

First, it is wrong to call any user a double-spender who has spent the same coin twice: She may have received the coin twice, e.g., from an attacker at different times. In this case, the honest user is likely to be found before the attacker and punished as the double-spender.

A first refinement, which avoids this problem and the broadcast channel, works as follows: The bank recursively asks each owner X of c for the identity of the user who paid c to X, starting with $X = A, B$. At the end, Case 1 and 3 are as above, and Case 2 is modified to

2. an owner X is detected who has paid the coin twice and cannot show that she also received it twice.

However, there remains a problem, since the signatures are only on the coin and the identity of the payee. Consider the graph of payments with one coin in Figure 1.

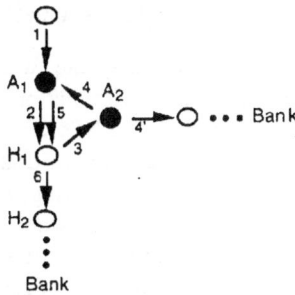

Fig. 1: Payment graph with one coin for an attack on the double-spending detection.

A_1 and A_2 are attackers, the others are honest users. A_2 is the double-spender. However, when the bank traces the coin back, A_2 and H_2 both show they got the coin from H_1. H_1 has two completely identical signatures from A_1, so she cannot prove she got the coin twice. Thus H_1 is punished, and A_1 never has to reveal that he also got the coin from A_2, which leaves A_2 without risk of detection.

To prevent this attack, passing one coin twice to the same payee must yield different signatures, and the payee must be able to verify this. This can be done interactively with a random challenge or a transaction number from the payee. Non-interactively, one can use time-stamps if synchronized clocks are available. Otherwise, the payer uses transaction numbers and the payee must compare a newly received coin with all previously received ones. This cannot introduce a new untraceability problem, since this scheme has no untraceability in this respect anyway.

4.4 No Divisibility Together with Transferability

The divisibility scheme [AmCr_94, Sect. 5] may be correct or not — at least for transferable coins it does not make much sense: Divisibility is based on a predefined tree structure in each coin, and all important complexity parameters (communication during withdrawal and payment and storage space for a coin) are linear in the number of nodes in this tree. Thus, this scheme has no real advantage over using coins of the smallest denomination, such as centimes, only.

To see this complexity, note that a completely separate signature by the bank on a value d_l at each node, l, of the tree must be transferred or stored, respectively, in all these situations. (If the coin is not transferable, only the signature on the root of the transferred part of the tree has to be transferred.)

Moreover, the divisibility scheme *requires* bit-by-bit commitments, whereas much more efficient string commitment schemes exist [Naor_91]. Hence, independent of developments in the computational complexity of the general cryptographic tools, it will remain inefficient. .

4.5 Achieving Weak Untraceability More Easily

As announced in Section 2.2, we show that the weak untraceability required in [AmCr_94] can be achieved more easily than with that protocol.

Since the payee knows the payer anyway, no NIZKP is needed for proving the validity of a coin to a payee. Instead, the payer can forward all information obtained during withdrawal (i.e., the chosen coin identifier, c, the string, r, used to commit to c, and the bank's signature on the commitment) to the payee, who can easily and noninteractively verify it. During deposit, the payee proves knowledge of a signature on a commitment to the coin identifier, using any appropriate zero-knowledge proof scheme (not necessarily noninteractive now!). This would also remove the privacy flaw exhibited above.

5 The System Presented at CIAC 94

There is no space to describe the system from [Cres_94] completely. However, the problems can be understood without many details. We first describe a general problem with the requirements and assumptions, and then show that even those requirements are not fulfilled.

5.1 Extreme Non-Interactiveness and the Consequences

The main purpose of [Cres_94] is to make all transactions completely non-interactive — even a withdrawal consists of just one message, from the user to the bank (a broadcast message to everybody, so that they can later recognize the coin). Now, however, the bank cannot announce if it rejects the withdrawal since no money is left on the user's bank account. Hence either the bank has to accept all withdrawal wishes, or payees will accept the coin but not be able to deposit it. Although this does not contradict the requirements made, it must clearly be seen as an error.

Here we more or less agree with [SaPe_92] who claimed that their 2-message withdrawal was optimal. The only alternative seems to be that everybody mirrors the bank account of everybody else, and all transactions are carried out via a broadcast channel, so that everybody can decide locally if a withdrawal is possible or not. This is not too far from the protocol in [Cres_94], where broadcast channels are also used extensively, but completely unrealistic in practice.

Thus for any further consideration of that system, one has to give up this notion of non-interactiveness and assume that the bank broadcasts if it accepts or rejects a withdrawal.

5.2 No Untraceability

Very obviously, there is no untraceability at all in the scheme, since each coin is identified by a string $(g_1, ..., g_n)$ that was broadcast during withdrawal and is transferred in clear in each transaction with this coin.

5.3 Cryptographic Security Problems

For spending a coin, a kind of signature scheme is used, where a tuple of numbers (w_1, ..., w_n) is signed modulo a Blum integer x_u by computing the square roots r_i of exactly those values w_i that are quadratic residues. It is said that such a signature can easily be checked by a recipient. However, deciding quadratic residuosity modulo Blum integers is assumed to be hard. As an extreme consequence, someone can forge a signature by claiming that none of the w_i's was a quadratic residue.

Moreover, as is well-known from Rabin's scheme, such a scheme can be broken if a chosen-message attack is possible. This is not the case for the original payer in [Cres_94]. However, the payee also has to use this scheme when he deposits the coin. Now w_i is constructed as $r_i d_i g_i$ for certain fixed values r_i and d_i, but with g_i chosen last and by the payer. The payer is restricted in this choice, but this is only verified by the bank. Thus a collusion of a bank and one payer can perform the chosen-message attack and compute the secret key of a payee.

Furthermore, the system starts with the bank choosing several random strings, which should be obtained from the source for shared random strings instead.

6 Recommendations

We do not try to repair the schemes considered above — in both cases we argued that major changes would be needed. We do not doubt that general cryptologic tools are powerful enough to construct many kinds of electronic payment systems. The question that interests us more is how one can avoid that further payment systems called "provably secure" with several, partly obvious, errors are published at refereed conferences. Considering this and experiences that many people both inside and outside cryptology have no idea what exactly all these papers about cryptologic electronic payment systems are trying to achieve, we think that the part of cryptology that deals with large protocols has rather a poor standard of definitions for an exact (and security-related) science. (Of course, the proof sketches in [AmCr_94, Cres_94] were far too short, too.) We see three types of problems with the current definitions of payment systems:

1. There is no correct definition yet.[4]

2. The definitions are incomplete, i.e., properties that would be needed in practice and that can be achieved, are not mentioned and thus often forgotten in constructions. For instance, if one does not trust the bank with respect to framing, one should also require that the bank cannot falsely claim that all the money was already withdrawn from an account, and that a payee can prove that he deposited money if it later fails to turn up on his statement of account. Moreover, it may be required that the bank can only deposit money to the particular payee that the payer intended to pay it to, to say nothing about receipts for payments.

3. Far fewer names for properties are in use than different properties, and thus things are often considered equal that are not. This was seen above in the use of "untraceability" for a far weaker property than in other papers without ever saying so. Actually, the same weak definition is already used in [SaPe_92]. Similarly, it is hard to find in

[OkOh_91] and not at all in [OkOh_92] that untraceability there is weak in a different sense: all payments by one customer can be linked, whereas usually, they are all independent.

We do not claim to have a complete definition at present (actually, if we claimed to have one that fitted on two pages you should be very suspicious), but we can give some hints as to what such a definition should look like.

a) Short statements of cryptographic properties (formal or informal) should always come with an explicit *trust model*, i.e., *for* whom a property is guaranteed, and which *other* participants have to be trusted to guarantee this. For instance, untraceability is usually *for* payers (but some schemes by Chaum offer untraceability for payees instead), and nobody else is trusted, but [AmCr_94] requires trust in all payees.

b) A top-level specification is needed where the *properties* required can be expressed independently of the trust model or even the details of the protocols to be proved, but still in a way that can later be combined with a trust model in a formally precise way.[5]

   • For complete protocols, this can mean a specification by an "ideal" protocol performed by a centralized trusted host, together with a definition what it means for a "real", decentralized protocol with untrusted participants to simulate the ideal protocol. This would extend the similar and still unfinished definitions for multi-party function evaluation (cf. [Yao_82, Beav_91, MiRo_91] for the oldest and newest versions).

   • For individual properties, at least integrity properties, it can mean a specification in any usual formal specification language, if one defines two things: (1) Once and for all, what it means for such a specification to be fulfilled in a cryptologic sense, that is, under a trust model (i.e., with certain participants deviating from their protocols and colluding), and often with error probabilities or in a complexity theo-

---

[4] To justify this briefly, we just look at what is both in public and our own opinion the best one, in [FrYu_93], although we know that we are now being pedantic and that it is good that such a definition sketch exists at all.

First, it is only semi-formal (no probability spaces defined), and most active attacks are not mentioned.

Secondly, it is really not quite correct: The property called "unforgeable" here and "unexpandable" in [FrYu_93] is unfulfillable in any similar form: It says that from the views of customers of n withdrawal protocols, it should be infeasible to compute $n + 1$ distinct coins that will lead to successful purchase protocols. Both the algorithms that compute and spend the "coins" must be adversaries here; hence the term "coin" cannot be defined as "correct coin", but any tape content that will lead to a successful purchase. Now, for any secure system, one can construct a stupid adversary that extends coins by random bits at the end, but still spends them according to their original part. Thus it can formally spend many "distinct" coins. To express what was really meant, this property and that of double-spending detection must be defined in one piece, roughly saying that more successful purchases than withdrawals lead to identification.

Thirdly, some standard requirements are missing, e.g., "no framing" from above, or that the bank should not be forced to carry out a withdrawal if the user has no real money. (The latter can trivially be fulfilled with the protocols in [FrYu_93], but we saw that the protocol in [Cres_94] does *not* fulfil it.)

[5] Just to prevent misunderstanding: We do not mean "logics of authentication", which people outside cryptology use as top-level specifications for lack of such things from inside cryptology, although it seems clear that monotonic logics cannot possibly fulfil what most people expect of them, and although the formal semantics of those logics makes assumptions about cryptologic primitives that the definitions of these primitives do not justify.

retic sense only. (2) What aspects of the protocols considered are visible to the outside world ("interface behaviour"), i.e., are mentioned in the specification. E.g., the problem with the notion of "coin" when held by an adversary (Footnote 4) disappears if one specifies in terms of inputs that start transactions, e.g., "withdraw, *amount*", and outputs denoting the result of such transactions (such as "deposit accepted") alone (cf. [Pfit_93] for such a definition of the (much simpler) properties of signature schemes).

c) Properties that deal with provability can only be defined by reference to a verification protocol. For instance:

• "No framing" needs a protocol **prove_framing** between a bank, a third party, and possibly the accused payer. The requirement of the bank is roughly that it can convince any honest third party that someone was a double-spender if double-spending occurred; the requirement of each payer P is that no honest third party will be convinced of P double-spending if P follows her protocols.[6]

• The fact that the bank cannot deduct more real money from a bank account than the electronic money the payer requested and received must be defined with a protocol **audit_withdrawals** between the bank, a third party, and the payer. In constructions, the bank may show signed withdrawal requests to the third party (which must be produced and stored in the withdrawals), and if the payer claims she did not receive the corresponding coins, it must basically be possible to send them again.

Acknowledgment

We thank *Stig Mjølsnes* for a discussion at Eurocrypt 94 where we already agreed that a cash system requiring occasional broadcast and cooperation of all honest users was not what one would expect, *Joachim Biskup* for helpful comments on this paper, and *Mihir Bellare* and *Jens Krüger* for interesting discussions about requirements on payment systems. Furthermore, we would like to thank the anonymous reviewer who pointed out that some attacks on the system (or, according to Giovanni di Crescenzo, the weakness of the definition mentioned in Section 2.2) have been described in a letter by Jacques Traore. Unfortunately, we have not been able to contact him or to obtain a copy of his letter.

References

AmCr_94   Stefano D'Amiano, Giovanni Di Crescenzo: Methodology for Digital-Money Based on General Cryptographic Tools; Pre-proceeding of Eurocrypt '94, May 9-12, 1994, University of Perugia, Italy 151-162.

Antw_90   Hans van Antwerpen: Electronic Cash; Centre for Mathematics and Computer Science (CWI), Amsterdam, October 11, 1990.

Beav_91   Donald Beaver: Secure Multiparty Protocols and Zero Knowledge Proof Systems Tolerating a Faulty Minority; Journal of Cryptology 4/2 (1991) 75-122.

Bran_94   Stefan Brands: Electronic Cash Systems Based on The Representation Problem In Groups Of Prime Order; Crypto '93, LNCS 773, Springer-Verlag, Berlin 1994, 302-318.

---

[6] Moreover, one may have the funny requirement of a partly dishonest payer that she is not falsely convicted of more double-spending than she actually carried out.

Chau_83   David Chaum: Blind Signature System; Crypto '83, Plenum Press, New York 1984, 153.

Chau_85   David Chaum: Security without Identification: Transaction Systems to make Big Brother Obsolete; Communications of the ACM 28/10 (1985) 1030-1044.

Chau_89   David Chaum: Privacy Protected Payments – Unconditional Payer and/or Payee Untraceability; SMART CARD 2000: The Future of IC Cards, Proceedings of the IFIP WG 11.6 International Conference; Laxenburg (Austria), 19.-20. 10. 1987, North-Holland, Amsterdam 1989, 69-93.

ChFN_90   David Chaum, Amos Fiat, Moni Naor: Untraceable Electronic Cash; Crypto '88, LNCS 403, Springer-Verlag, Berlin 1990, 319-327.

ChPe_93   David Chaum, Torben P. Pedersen: Transferred Cash Grows in Size; Eurocrypt '92, LNCS 658, Springer-Verlag, Berlin 1993, 390-407.

Cres_94   Giovanni Di Crescenzo: A Non-Interactive Electronic Cash System; Proc. Italian Conference on Algorithms and Complexity, CIAC 94, LNCS 778, Springer-Verlag, Heidelberg 1994, 109-124.

Damg_90   Ivan Bjerre Damgård: Payment Systems and Credential Mechanisms with Provable Security Against Abuse by Individuals; Crypto '88, LNCS 403, Springer-Verlag, Berlin 1990, 328-335.

EvGY_83   Shimon Even, Oded Goldreich, Yacov Yacobi: Electronic wallet; Crypto '83, Plenum Press, New York 1984, 383-386.

Ferg_94   Niels Ferguson: Single Term Off-Line Coins; Eurocrypt '93, LNCS 765, Springer-Verlag, Berlin 1994, 318-328.

FrYu_93   Matthew Franklin, Moti Yung: Secure and Efficient Off-Line Digital Money; 20th International Colloquium on Automata, Languages and Programming (ICALP), LNCS 700, Springer-Verlag, Heidelberg 1993, 265-276.

MiRo_91   Silvio Micali, Phillip Rogaway: Secure Computation (Chapters 1-3); Laboratory for Computer Science, MIT, Cambridge, MA 02139, USA; distributed at Crypto '91.

Naor_91   Moni Naor: Bit Commitment Using Pseudorandomness; Journal of Cryptology 4/2 (1991) 151-158.

OkOh_91   Tatsuaki Okamoto, Kazuo Ohta: How to Utilize the Randomness of Zero-Knowledge Proofs; Crypto '90, LNCS 537, Springer-Verlag, Berlin 1991, 456-475.

OkOh_92   Tatsuaki Okamoto, Kazuo Ohta: Universal Electronic Cash; Crypto '91, LNCS 576, Springer Verlag, Berlin 1992, 324-337.

Pfit_93   Birgit Pfitzmann: Sorting Out Signature Schemes; 1st ACM Conference on Computer and Communications Security, 3.-5.11.1993, Fairfax, acm press 1993, 74-85.

PfWa_92   Birgit Pfitzmann, Michael Waidner: How to Break and Repair a "Provably Secure" Untraceable Payment System; Crypto '91, LNCS 576, Springer Verlag, Berlin 1992, 338-350.

SaPe_92   Alfredo de Santis, Guiseppe Persiano: Communication Efficient Zero-Knowledge Proofs of Knowledge (With Applications to Electronic Cash); STACS 92, 9th Annual Symposium on Theoretical Aspects of Computer Science, LNCS 577, Springer-Verlag, Heidelberg 1992, 449-460.

Yao_82   Andrew C. Yao: Protocols for Secure Computations; 23rd Symposium on Foundations of Computer Science (FOCS) 1982, IEEE Computer Society, 1982, 160-164.

Quantum Oblivious Mutual Identification

Claude Crépeau[1][2] and Louis Salvail[2]

[1] LIENS, (CNRS URA1327)
École Normale Supérieure,
45 rue d'Ulm, 75230 Paris CÉDEX 05, FRANCE.
e-mail: crepeau@dmi.ens.fr.
[2] Département d'Informatique et R.O.,
Université de Montréal,
C.P. 6128, succursale centre-ville,
Montréal (Québec), Canada H3C 3J7.
e-mail: salvail@iro.umontreal.ca.

Abstract. We consider a situation where two parties, *Alice* and *Bob*, share a common secret string and would like to mutually check their knowledge of that string. We describe a simple and efficient protocol based on the exchange of quantum information to check mutual knowledge of a common string in such a way that honest parties will always succeed in convincing each other, while a dishonest party interacting with an honest party will have vanishingly small probability of convincing him. Moreover, a dishonest party gains only a very small amount of information about the secret string from running the protocol: whoever enters the protocol with no knowledge of the secret string would have to enter this protocol an exponential number of times in order to gain non-negligible information about the string.

Our scheme offers an efficient identification technique with a security that depends on no computational assumption, only on the correctness of quantum mechanics. We believe such a system should be used in smart-cards to avoid frauds from typing PIN codes to dishonest teller machines.

1 Introduction

Weren't you worried the last time you typed your PIN (Personal Identification Number) to an unknown teller machine that it could be a fake and that its sole purpose could be to steal your PIN? According to a recent headline of the NY Times [26] maybe you should worry:

"ONE LESS THING TO BELIEVE IN: FRAUD AT FAKE CASH MACHINE"

The problem with current identification systems is that the customer must trust the equipment to which he types his PIN. It is completely trivial to modify a teller machine to memorize the PIN numbers that people type to it. PINs are meant to be checked, not given. (Consult [1] for an extensive study of frauds at teller machines.)

Of course, it is always possible to completely solve the problem of identification and authentication of messages by classical methods [9] that require

L.C. Guillou and J.-J. Quisquater (Eds.): Advances in Cryptology - EUROCRYPT '95, LNCS 921, pp. 133-146, 1995.
© Springer-Verlag Berlin Heidelberg 1995

exchanging passwords which length are proportional to the number of uses. Unfortunately, this is completely impractical: we want to rely on the existence of a *short* secret to check identity.

A similar approach has been suggested in a computational model through the construction of pseudo-random generators [7] or pseudo-random families of functions [17] which requires only short secrets seeds. These solutions make sense only in a context where we put computational restrictions on the participants. For powerful parties it is trivial to fake identities.

In the computational model, more sophisticated tools were developed for this purpose: Zero-Knowledge Proofs of Identity [15] introduced in order to provide means by which an honest party may convince another party of his identity in a way that cannot be replayed successfully to another party. This is true even if the verifying party tries his best to extract valuable information out of the proving party. Moreover, a dishonest party attempting to prove an invalid identity will be detected by the verifying party except with vanishingly small probability.

Non-Computational Protocols

The major drawback with these proofs and other computational techniques is that deep down their security must rely on some computational assumption: the proof of knowledge can be checked if the identifying string is the solution to some hard public problem. If one can solve this problem, he can fake identities. This is the case even if we build the protocol from *perfect* cryptographic tools such as ideal Bit Commitment or ideal Oblivious Transfer.

In the current paper, we consider a situation where two parties, *Alice* and *Bob*, share a common secret string and would like to mutually check their knowledge of that string without disclosing it. This problem has been extensively studied by Fagin, Naor and Winkler [14] who provide a large number of scenarios where the problem may be considered. From the cryptographic point of view only one of their solutions may be considered secure: a solution based on the existence of a one-out-of-two Oblivious Transfer [13] which uses $\Omega(n^2)$ Transfers to do the job for an n-bit secret string.

We describe a simple and efficient protocol based on the exchange of quantum information to check mutual knowledge of a common string in such a way that honest parties will always succeed in convincing each other (except with vanishingly small probability), while a dishonest party interacting with an honest party will have only vanishingly small probability of convincing him. Moreover, a dishonest party gains only a very small amount of information about the secret string from running the protocol: whoever enters the protocol with no knowledge of the secret string would have to enter this protocol an exponential number of times in order to gain non-negligible information about the string.

Our scheme, based on coding theory, depends on no computational assumption, has a total running time of $O(n^{2.376})$ and a total of $O(n^2)$ photons are transmitted (if implemented with a one-out-of-two Oblivious Transfer, only $O(n)$ such Transfers are necessary). We also present a scheme which uses only $O(n)$ photons but that cannot tolerate the transmission errors of a real quantum channel. We suggest the reader consults [8] for more details of quantum cryptography.

We believe such a system should be used in smartcards to avoid frauds from typing PIN codes to dishonest teller machines. A PIN could still be used to activate the functions of the card but it should be typed directly to the card (a device you might as well trust since your bank gives it to you, and they have your money anyway!). The card would identify itself with tellers only through our mechanism: no PIN ever exchanged.

The practical difficulty of our scheme is to embed the necessary technology for the Quantum Oblivious Transfer on a card. Since none of the technology is very fancy we believe such cards could be mass produced (see Section 4.6).

2 Preliminaries

2.1 Notations and Tools

For $b \in \{0,1\}$ we define the selection function $(x,y)_{[b]} = \begin{cases} x \text{ if } b = 0 \\ y \text{ if } b = 1 \end{cases}$ where x and y are scalars. In general, if x and y are vectors with n components and $b \in \{0,1\}^n$ then $(x,y)_{[b]}$ is the concatenation of $(x_i, y_i)_{[b_i]}$ for all $i \in \{1, 2, \ldots, n\}$. If we use a single x instead of a pair then $(x)_{[b]}$ is the concatenation of the x_i's for $b_i = 1$. For instance $(1001)_{[1010]} = 10$. We denote by $\Delta(s, \widehat{s})$ the Hamming distance between s and $\widehat{s}$.

Now let us define what we mean by a secure identification protocol in our context. Suppose $\phi_{(n)} \in_R \{0,1\}^n$ is the secret information shared by $\mathcal{A}lice$ and $\mathcal{B}ob$. In the following we write $\mathcal{B}ob^*$ (resp. $\mathcal{A}lice^*$) to designate somebody trying to impersonate $\mathcal{B}ob$ (resp. $\mathcal{A}lice$).

Definition 1. $\mathcal{B}ob^*$ (or $\mathcal{A}lice^*$) has almost no information about ϕ indexed by a security parameter n, $\phi = \{\phi_{(n)} \in_R \{0,1\}^n\}_{n>0}$, if $\mathcal{B}ob^*$'s (or $\mathcal{A}lice^*$'s) information about ϕ can be modeled by a set $\Phi = \{\Phi_{(n)} \subseteq \{0,1\}^n\}_{n>0}$ such that for some $\alpha > 0$ and all sufficiently large n we have:

$$\frac{\#\Phi_{(n)}}{\#\{0,1\}^n} \geq 1 - 2^{-\alpha n} \text{ and } P\left(\phi_{(n)} \notin \Phi_{(n)}\right) < 2^{-\alpha n}$$

and for each $\phi' \in \Phi_{(n)}$ we have:

$$P\left(\phi_{(n)} = \phi' | \phi_{(n)} \in \Phi_{(n)}\right) = \frac{1}{\#\Phi_{(n)}}.$$

The Shannon information given by Φ about ϕ is such that

$$\mathcal{I}(\phi_{(n)} | \Phi_{(n)}) \leq 2^{-\widehat{\alpha} n}, \widehat{\alpha} > 0.$$

Without loss of generality, we use ϕ for an instance $\phi_{(n)}$ whenever the context permits. We denote by ϕ_i the ith bit of ϕ.

An identification scheme hides the secret information ϕ if when a cheating party runs the protocol with no information about ϕ he ends up with almost no information about ϕ.

Definition 2. An identification scheme **hides the secret information** ϕ shared by *Alice* and *Bob*, if for some $\alpha' > 0$, when *Bob** (or *Alice**), who starts with no information about ϕ, cheats the protocol $poly(n)$ times, he has probability greater than $1 - 2^{-\alpha'n}$ to end up with almost no information about ϕ.

The random variable ϕ^* denotes the choice made by *Alice** or *Bob** to run the protocol given that she (or he) has almost no information about ϕ.

If a malicious party $\mathcal{P}^*$ has almost no information about ϕ then he cannot guess any bit ϕ_i with non-negligible bias.

Property 1 *If $\mathcal{P}^*$ has almost no information about $\phi = \{\phi_{(n)} \in_R \{0,1\}^n\}_{n>0}$ there exists $\beta > 0$ such that for all $\phi^* \in \{0,1\}^n$ and each position i independently, $|P(\phi_i^* = \phi_i) - \frac{1}{2}| \le 2^{-\beta n}$.*

If $\mathcal{P}^*$ has almost no information about ϕ and executes a protocol leaking almost no information about ϕ then $\mathcal{P}^*$ has still almost no information.

Property 2 *If Φ_0, Φ_1 are two sets that give almost no information about ϕ then $\Phi = \Phi_0 \cap \Phi_1$ gives almost no information about ϕ.*

These two properties are straightforward applications of the above definitions.

2.2 Simple Quantum Transmission

In this paper we consider the most simple idealization of a quantum transmission. There is only four different ways to transmit photons corresponding to the four polarization angles $0°, 45°, 90°, 135°$ that we denote $|\leftrightarrow\rangle, |\nearrow\rangle, |\updownarrow\rangle, |\nwarrow\rangle$ respectively. If *Alice* wants to send $b \in \{0,1\}$, she used the following encoding rules:

1. $b = 0$ is randomly encoded by $|\leftrightarrow\rangle$ or $|\nearrow\rangle$.
2. $b = 1$ is randomly encoded by $|\updownarrow\rangle$ or $|\nwarrow\rangle$.

At the receiving end *Bob* chooses how he measures the incoming photon either by reading it in rectilinear basis ($|\leftrightarrow\rangle, |\updownarrow\rangle$), denoted $\oplus$, or in the diagonal basis ($|\nearrow\rangle, |\nwarrow\rangle$), denoted $\times$. We suppose this is the only choice he can make. If the photon π encodes a bit b in the rectilinear (resp. diagonal) basis and the receiver measures it in the rectilinear (resp. diagonal) basis then he gets the bit b (except if an error occurred during transmission). If π is measured in the diagonal (resp. in the rectilinear) basis then a random bit is received. For a basis $\phi \in \{\oplus, \times\}$ and a bit b we write $\phi_{[b]}$ for the transmission of the encoded bit b in a photon polarized in the basis ϕ. For more details about how quantum transmission works in general consult [8].

3   A basic quantum identification

Suppose *Alice* and *Bob* who have the secret string ϕ^A and ϕ^B respectively want to test whether $\phi^A = \phi^B$ without revealing their values. In order to achieve this, they are willing to use quantum and public channels. The transmission of $c \in_R \{0,1\}$ polarized in basis $\phi \in_R \{\boxplus, \times\}$ hides all information about ϕ to anybody who has almost no a priori information on the values of c and ϕ. This suggests the use of the secret ϕ^A to encode securely the transmission basis.

Suppose *Alice* and *Bob* share $\phi = \phi^A = \phi^B \in_R \{\boxplus, \times\}^n$. In order for *Alice* to prove to *Bob* that she knows ϕ, she could transmit a random string c taken from a sparse but large subset of $\{0,1\}^n$ polarized in basis ϕ. Therefore, it suffices for *Alice* to choose a random codeword c from a code C_n which she sends to *Bob*. He then measures it in ϕ to obtain the decoded string $\hat{c}$. If the quantum channel is noiseless then $\hat{c} = c$. (In the more realistic case, the quantum channel would be modeled as a binary symmetric channel with parameter ϵ.) If there is a large number of codewords in C_n it could be the case that measuring c in basis ϕ^* hides almost all information about ϕ. Conversely if *Alice* does not know ϕ it could be very unlikely that she succeeds to send $\hat{c}$ close to a codeword c, as long as codewords are not too close to each other. Protocol $ident(\phi^A, \phi^B)$, shown below, implements this idea given ϵ the error rates of the quantum channel.

Protocol 3.1 ($ident(\phi^A, \phi^B)$)

> *1: Alice and Bob agree on a binary linear (n, k_n, d_n)-code $C_n \in \mathcal{C}$ by specifying a generating matrix G for C_n.*
>
> *2: Alice chooses a random word $x \in_R \{0,1\}^{k_n}$ and takes $c = xG$.*
>
> *3: $\overset{n}{\underset{i=1}{DO}}$*
>
> — *Alice sends to Bob a photon polarized in $\phi^A_{i[c_i]}$.*
>
> — *Bob measures the incoming photon in the basis ϕ^B_i and obtains $\hat{c}_i$.*
>
> *4: Bob accepts if when decoding $\hat{c}$, he obtains $c' \in C_n$ such that $\Delta(c', \hat{c}) \leq (\epsilon + \epsilon_0)n$ for $\epsilon_0 > 0$.*

Suppose *Alice* and *Bob* share the same private sequence $\phi = \phi^A = \phi^B$. This implies that, for any $\epsilon_0 > 0$ and except with vanishingly small probability, *Bob* will decode $\hat{c}$ as c which is at Hamming distance less than $(\epsilon + \epsilon_0)n$ from $\hat{c}$ given n sufficiently large.

Now consider a malicious *Alice* (denoted by *Alice**) is trying to impersonate the real *Alice*. We assume that *Alice** knows almost nothing about ϕ^B at the beginning. Therefore she will have roughly half the positions different from *Bob* and thus sends random bits in half the positions. It is easy to show that if the minimum distance d_n of C_n is such that $d_n > 2n(\epsilon + \epsilon_0)$, *Alice** will be detected with probability greater than $1 - 2^{-\alpha n}$ for $\alpha > 0$. By this attempt, she will not learn more than a vanishingly small amount of Shannon information about ϕ^B.

3.1 $ident(\phi, \phi^*)$ with a dishonest $\mathcal{B}ob$

Let $\phi \in_R \{\boldsymbol{+}, \boldsymbol{\times}\}^n$ be $\mathcal{A}$lice's secret and let c be the transmitted random codeword taken from C_n. $\mathcal{B}ob^*$ chooses a set of bases ϕ^* and measures each photon i in the basis ϕ_i^* in order to obtain $\hat{c}$. Roughly speaking, one half of the bases of ϕ^* will match with the bases of ϕ. Thus approximatively half of the bits he will receive are the bits of the codeword sent by $\mathcal{A}$lice. The other bits (the positions i for which $\phi_i \neq \phi_i^*$) are not correlated with the bits transmitted. For $\mathcal{B}ob^*$ to be unable to determine a substantial amount of information about ϕ, the code must be chosen so that any half of the bits of the codewords are purely random. Hence, if the proportion of the bits $\mathcal{B}ob^*$ sees about a codeword is random and the rest of the bits he received are not correlated (thus random) the thing he gets is a purely random string. Given that the same would happend for all but a few ϕ^* almost no further information about ϕ can be determined.

A similar concept in a different setting was studied by [4],[24] ((n, j, k)-functions) and [10] (t-resilient functions). The next definition is taken from [4].

Definition 3 [4]. For any integers n, j, k such that $n \geq j + k$, $j > 0$ and $k > 0$, a function $f : GF(Q)^n \rightarrow GF(Q)^j$ is (n, j, k) if, no matter how one fixes any k of its inputs, each of the Q^j outputs can be produced in exactly Q^{n-j-k} different ways by varying the remaining $n - k$ symbols.

If each symbol of $f(x)$ is obtained by computing a weighted sum on a subset of the digits in x then $f(x)$ is said to be xor-(n, j, k). In [4] the function f is from n-bit strings to j-bit strings, here we consider an arbitrary field. The following theorem showing how to construct (n, j, k)-functions, was originally proved for functions over binary strings. It is straightforward to generalize it to functions over arbitrary fields.

Theorem 4 [4]. *For a set of values (n, j, k), there exists an xor-(n, j, k)-function from $GF(Q)^n$ to $GF(Q)^j$ if and only if there exists an $[n, j, k + 1]$ linear codes C_n over $GF(Q)$.*

If G is the generating matrix of a $[n, j, k + 1]$-code C_n then the function $f(x) = Gx^T$ (x^T is x transposed) is $xor - (n, j, k)$. Saying that f is (n, j, k) implies that $f^{-1}(x)$ is a set which have uniform projection on any k coordinates [10]:

Definition 5 [10]. A set $S \subseteq GF(Q)^n$ has a uniform projection on any j components if for all $\omega \in \{0, 1\}^n$ of weight j and all $a \in GF(Q)^k$, the set $S_{\omega, a} = \{x \in S : (\bot, x)_{[\omega]} = a\}$ is such that $\#S_{\omega, a} = \frac{\#S}{Q^k}$.

If $f(x) = Gx^T$ is (n, j, k) then $f^{-1}(x) = Hx^T$, where H is the parity check matrix for C_n. The matrix H is also the generating matrix for the dual $C_n^\perp$ of C_n. The next theorem makes the connection between uniform projections and some conditions on the dual of the codes C_n.

Theorem 6. *If there exists a family of codes $C = \{C_n\}_{n>0}$ such that for n sufficiently large the dual $C_n^{\perp}$ of C_n has minimum distance d'_n with $\frac{d'_n}{n} \geq \zeta > \frac{1}{2}$, protocol $ident(\phi^A, \phi^*)$ hides almost all information about ϕ^A to Bob^* except with probability exponentially small in n. This holds given an idealized quantum transmission.*

proof sketch: By property 1, except with vanishingly small probability $\Delta(\phi^*, \phi^A) \geq (1 - \zeta)n$. Thus, the number of positions for which Bob^* sees the bits of c is less than ζn. Let H be the generating matrix for $C_n^{\perp}$ of minimum distance ζn. The function Hx^T has uniform projection on any ζn components. Thus, any ζn bits of a codeword c (those for which $Hc^T = 0$) are random when $c \in_R C_n$. Since the other $(1 - \zeta)n$ positions are random, the string $\hat{c}$ he received is purely random. Therefore, the set $\Phi_1 = \{\phi | \Delta(\phi, \phi^*) \geq (1 - \zeta)n\}$ models the knowledge leaked to Bob^* by the actual execution of the protocol. It is easy to show that Φ_1 gives almost no information about ϕ. Let Φ_0 be the model for Bob^*'s knowledge about ϕ^A before the actual execution. The set $\Phi = \Phi_0 \cap \Phi_1$ models Bob^*'s knowledge after the current execution given he had almost no information when entering the protocol. By property 2 the set Φ gives almost no information about ϕ^A. For Bob^* executing $poly(n)$ times the protocol, the set

$$\Phi = \bigcap_{i=1}^{poly(n)} \Phi_i$$

where Φ_i models the information leaked about ϕ for the ith execution gives almost no information about ϕ^A. We conclude that $ident(\phi^A, \phi^*)$ hides almost all information about ϕ^A to Bob^*. $\square$

3.2 Code Properties

Over the last few sections we have suggested some conditions on our codes. Let us now summarize the properties that families of codes C must satisfy to guarantee the security of protocol $ident(\phi^A, \phi^B)$ while preserving efficiency of the scheme.

1. Given our mode of transmission via the quantum channel, we want C to be a family of binary codes.
2. Each $C_n \in C$ must be efficiently constructible and efficiently decodable.
3. Each $C_n \in C$ must have minimal distance d_n such that $\frac{d_n}{n} \geq 2(\epsilon + \epsilon_0)$ for $\epsilon_0 > 0$.
4. The dual $C_n^{\perp}$ of $C_n \in C$ must have minimal distance d'_n such that $\frac{d'_n}{n} \geq \zeta$ for $\zeta > \frac{1}{2}$.

The set of conditions above is bad news. First of all, these conditions cannot be satisfied because of Plotkin's bound on codes [20] when $\zeta > \frac{1}{2}$. Fortunately, modifications to the protocol (for instance by using more than two transmission bases) open the possibility of relaxing this condition to $\zeta > 0$. But even then, no known family of codes satisfies these four conditions at once. It is nevertheless easy to find codes meeting any three of them. For instance, concatenated codes [16] achieve conditions 1, 2 and 3, random binary linear codes meet conditions 1,3 and 4, while Reed-Solomon codes meet 2,3 and 4.

It is common in coding theory to take care of arbitrary long messages via block codes. These codes are of no help in our setting because their duals have small minimum distance. This is easy to see since it is sufficient to observe a constant number of bits to tell if a word is a candidate codeword or not.

On top of these problems due to coding theory, more fundamental problems arise from our protocol: we have made a very strong assumption that $Alice^*$ and Bob^* send and receive photons in only two possible bases. In reality we would have to deal with the fact that they can use very different quantum states and quantum measurements. It is indeed completely unknown to us if this protocol is safe under these general conditions.

The main problem in quantum cryptography is to provide proofs for the security of cryptographic primitives assuming the most general quantum measurements an opponent could make. Nevertheless, the full security of the quantum bit commitment primitive has been obtained in [5] and quantum oblivious transfer has been shown secure against a large set of measurements[21]. Basing our identification scheme on such primitives gives more freedom on the codes while, at the same price, providing security against any quantum measurements. Oblivious transfer has already been used by [14] to solve the problem of identification. In the next section, we present a different solution based on quantum oblivious transfer and theorem 6.

4 The Final Protocol

No existing family of codes meets the four conditions above. One way around this problem is to drop condition 1. To do so we need a means of transferring elements of a larger field $GF(Q)$ at once. This is exactly the purpose of a $\binom{2}{1}$-OST, a *One-Out-of-Two Oblivious String Transfer* [11]. We can thus modify our protocol to use this primitive instead of the quantum transmission of section 3. A nice side effect of this modification is that transmission errors also go away.

Doing this modification is not very costly since a $\binom{2}{1}$-OST can be implemented using a constant number of $\binom{2}{1}$-OT$_2$ [11], which in turn can be obtained from the quantum transmission [6]. The solution we describe next works over $GF(4)$ (and thus we use a $\binom{2}{1}$-OT$_4$).

The Appendix provides a modified protocol (from [6]) for Quantum Oblivious Transfer ($\binom{2}{1}$-OT$_4$) sufficient for this application. That protocol also relies on the existence of a bit commitment. To avoid computational assumptions again at that level we recommend using the Quantum Bit Commitment Scheme of [5]. The protocol of the next subsection combined with the one from the Appendix constitute the complete Quantum Oblivious Mutual Identification.

4.1 Protocol

Suppose $\mathcal{A}lice$ and $\mathcal{B}ob$ share $\phi = \phi^A = \phi^B \in_R \{0,1\}^n$. In order for $\mathcal{A}lice$ to prove to $\mathcal{B}ob$ that she knows ϕ, she transmits a random string c taken from a sparse but large set C, in such a way that if they agree on the same ϕ he receives c and verifies that it belongs to C, but otherwise receives a rather random string which is unlikely to be in C.

More precisely, let M be a random $n \times k$ matrix over $GF(4)$. (In banking applications, M is chosen by the bank and may be made public.) Let C_n be the $[n, k_n, d_n]$ linear code over $GF(4)$ generated by M. Let $C_n^\perp$ be the $[n, n - k_n, d_n']$ linear code over $GF(4)$ dual to C_n.

Protocol 4.1 ($identOT(\phi^A)(\phi^B)$)

    1: *Alice picks $r, s \in_R \{00, 01, 10, 11\}^n$.*

    2: $\overset{n}{\underset{i=1}{DO}}$ *Alice runs $\binom{2}{1}$-$OT_4(r_i, s_i)(\phi_i^B)$ with Bob who receives v_i.*

    3: *Bob picks $x, y \in_R \{00, 01, 10, 11\}^n$ and announces it to Alice.*

    4: *Alice picks $c \in_R C_n$, computes and sends $u \leftarrow c \oplus (r \oplus x, s \oplus y)_{[\phi^A]}$.*

    5: *Bob accepts only if $u \oplus v \oplus (x, y)_{[\phi^B]} \in C_n$.*

In the above protocol, in contrast to protocol 3.1, the randomization of step 3 is necessary because the $\binom{2}{1}$-OT_4 no longer provides the fact that a random element is obtained when $\phi_i^* \neq \phi_i^B$, and we do not want $\mathcal{A}lice$ to take advantage of that fact. From the previous sections it is now easy to see why the protocol is correct and secure.

4.2 $identOT(\phi, \phi)$ with honest parties

If $\mathcal{A}lice$ and $\mathcal{B}ob$ share the same private sequence $\phi = \phi^A = \phi^B$ then

$$u \oplus v \oplus (x, y)_{[\phi]} = c \oplus (r \oplus x, s \oplus y)_{[\phi]} \oplus (r, s)_{[\phi]} \oplus (x, y)_{[\phi]} = c.$$

Therefore $\mathcal{B}ob$ accepts.

4.3 $identOT(\phi^*, \phi^B)$ with a dishonest $\mathcal{A}lice$

Suppose a malicious $\mathcal{A}lice$ (denoted by $\mathcal{A}lice^*$) tries to impersonate the real $\mathcal{A}lice$. We assume that $\mathcal{A}lice^*$ knows almost nothing about ϕ^B at the beginning. This section specifies code parameters that allow $\mathcal{B}ob$ to reject $\mathcal{A}lice^*$ with probability exponentially close to one.

Now we show how to choose the parameter k of code in order to eliminate the chances that $\mathcal{A}lice^*$ identify as $\mathcal{A}lice$ successfully.

Theorem 7. *If $d_n \in \Omega(n)$ $\mathcal{B}ob$ will reject Alice* except with probability exponentially small in n.*

proof sketch: $\mathcal{B}ob$'s final calculation is $u \oplus v \oplus (x, y)_{[\phi^B]}$ which by definition of v is $u \oplus (r \oplus x, s \oplus y)_{[\phi^B]}$. $\mathcal{A}lice^*$, who knows r, s, x, y may try to choose a u cleverly to make this a codeword for as many ϕ^B as possible. We show this is not possible.

First notice that for any fixed r, s, when x, y are uniformly chosen at random $\Delta(r \oplus x, s \oplus y) \approx 3n/4$. By the assumption that $\mathcal{A}lice^*$ has almost no information about ϕ^B we know that the number of equally likely candidates for ϕ^B is roughly 2^n. Fix a u and a ϕ^B and assume the corresponding word has syndrome S. We know that all the ϕ' that are different from ϕ^B in the positions where $r \oplus x$ and $s \oplus y$ are the same will not change the result of the calculation. Therefore roughly $2^{n/4}$ values of ϕ^B will yield the same result with syndrome S.

On the other hand, any ϕ' that differ from ϕ^B in positions where $r \oplus x$ and $s \oplus y$ are different will yield a different result. As long as the number of differences is no more than $d_n/2$ the resulting words cannot have the same syndrome because this would imply that the code contains a word of weight less than d_n and no other word of syndrome S can be closer than the one we started from. Therefore $2^{n/4} \sum_{i=1}^{d_n/2} \binom{3n/4}{i}$ possibilities for ϕ^B will yield results of another syndrome and each of these is associated to a single word of syndrome S.

In conclusion, for any fixed u and S, there is always at least $\sum_{i=1}^{d_n/2} \binom{3n/4}{i}$ times more possibilities for ϕ^B that do not yield a word of syndrome S than those that yield a word of syndrome S. If $d_n \geq 2\delta n$ for some constant δ, this value is roughly $2^{\frac{3H(\delta)}{4}n}$. Therefore the probability that $\mathcal{A}lice^*$ gets $\mathcal{B}ob$ to accept is no more than $2^{-\alpha n}$ with $\alpha = 3H(\delta)/4$. □

We need to know more than just the fact that $\mathcal{A}lice^*$ will fail most of the time: we show that in the case of failure she cannot learn much about ϕ_B.

Theorem 8. *If $d_n \in \Omega(n)$ $\mathcal{A}lice^*$ learns almost nothing about ϕ_B, except with probability exponentially small in n.*

proof sketch: By the same counting argument as above, there cannot be more than $2^{(1-3h(\epsilon)/4)n}$ possibilities of ϕ_B that would yield codewords. When rejected, $\mathcal{A}lice^*$'s only gain in knowledge is that the real ϕ_B was not one of those. Thus she can eliminate only that many strings. □

The consequence of these theorems is that with probability $1 - 2^{-\alpha n}$ $\mathcal{A}lice$ will be rejected and in that case she may discard only $2^{(1-3h(\epsilon)/4)n}$ strings as candidates for ϕ^B. She will thus still have almost no information about ϕ^B even after discarding $poly(n) \times 2^{(1-3h(\epsilon)/4)n}$ strings.

4.4 $identOT(\phi^A, \phi^*)$ with a dishonest $\mathcal{B}ob$

Now we analyze the situation from the point of view of an honest $\mathcal{A}lice$ facing a malicious $\mathcal{B}ob^*$.

Theorem 9. *Except with exponentially small probability in n, Protocol $identOT(\phi)$ hides all information about ϕ^A to $\mathcal{B}ob^*$ if the code is a $[n, k_n, d_n]$-code with a dual $[n, n - k_n, d'_n]$-code such that $d'_n \geq (\frac{1}{2} + \gamma)n$ for $0 < \gamma < \frac{1}{2}$.*

proof sketch: For each position i such that $\phi_i^* = \phi_i^A$, Bob will get the codeword position c_i of c. The $\#\{i|\phi_i^A = \phi_i^*\}$ is at most $(\frac{1}{2} + \gamma)n$ except with probability smaller than $2e^{-\gamma^2 n}$. The remaining positions (j) of $(x, y)_{[\phi^B]} \oplus u \oplus v$ for $\phi_j^* \neq \phi_j^A$ are not correlated with c_j as long as r and s were originally chosen at random. Since the dual has minimum distance $d' > (\frac{1}{2} + \gamma)n$ we conclude that theorem 6 applies and that Protocol *IdentOT* hides the information ϕ^A to Bob^*. □

Example: Codes with such properties exist over $GF(4)$. For instance, a random $n \times 0.91n$ 4-ary matrix is likely to define a $[n, 0.91n, 0.02n]$ code with a $[n, 0.09n, 0.52n]$ dual code. Assymptoticaly, the probability that such a matrix do not define a code with these parameters is exponentially small in n.

4.5 Complexity

Protocol 4.1 runs in time $O(n^2)$ (to choose a codeword) and uses $O(n)$ $\binom{2}{1}$-OT$_4$, where n is the security parameter. When combined with the sub-protocol from the Appendix the total running time of the final protocol becomes $O(n^3)$ and a total of $O(n^2)$ photons are transmitted. The running time decreases to $O(n^{2.376})$ if more efficient codes such as the Superconcentrator Codes of Spielman [25] are used in the protocol of the Appendix in both the Oblivious Transfer and the Bit Commitment (time $O(n^2)$) and if all the commitments are done at once in order to save on the time necessary to compute the hash function ($O(n)$ products of a vector by a matrix with a total of $O(n^3)$ operations may be replaced by a matrix product which takes only time $O(n^{2.376})$).

We must point out that despite the fact that our Protocol 4.1 is more efficient than that of Fagin, Naor and Winkler [14] in terms of $\binom{2}{1}$-OT$_2$, when used together with the Quantum Oblivious Transfer their protocol can be made as efficient as ours in terms of photons. This is because their protocol requires n transfers of n-bit strings (which implies $\Omega(n^2)$ $\binom{2}{1}$-OT$_2$) while our protocol requires only n transfers of 2-bit strings (which can be done with $O(n)$ $\binom{2}{1}$-OT$_2$). The fact is that the Quantum Oblivious Transfer can be used to transfer 1, 2 or up to $O(n)$ bits at no extra cost. In order to have a real gain in terms of photons transmitted we need a Quantum O-T that requires only a constant number of photons to transfer a constant number of bits (see open problem 3).

4.6 Implementation Remarks

The protocol from the appendix uses quantum transmission both for Oblivious Transfer and Bit Commitment. At first glance, it seems like the quantum transmission of data must go in both directions, since the Oblivious Transfer goes from *Alice* to *Bob* and the Bit Commitment goes the other way. As pointed out in [12], there is no need for photons traveling both ways. These two protocols may be implemented with the photons going in a single direction. It does not matter who send the photons to who, the same result can be achieved from them. (A similar idea was suggested by Hans-Joachim Knobloch [19].)

Because of the above remark, in a smartcard scenario it suffices to implement on the card the technology for sending polarized photons: a weak light source with a multiple polarizer system. As for the ATM it would have to use the more elaborate technology for making polarization measurements on the incoming photons. Since the distance between the sender and receiver could be of a few millimeters the actual error rate of the quantum transmission would be extremely low (error rates of 0.5% have been observed on hundreds of meters [22]).

5 Conclusion and Open Questions

We have presented a protocol for mutual identification based on the existence of an Oblivious Transfer and have shown improvements to the Quantum Oblivious Transfer in order to combine them in an efficient Quantum Mutual Identification Protocol. Here is a few open problems:

1. It would be interesting to show that the protocol of section 3 is secure even if the participants use arbitrarily complicated physics.
2. Find binary codes satisfying conditions 2,3, and 4.
3. Find a reduction of $\binom{2}{1}$-OST to the quantum transmission that requires only a constant number of photons per word of constant length.
4. Implement the necessary technology on a smartcard!

Acknowledgments

We would like to thank Thomas Beth, Gilles Brassard, Artur Ekert, Richard Jozsa, Denis Langlois, and Bill Wootters for their interest in this research.

References

1. Anderson, R.J., "Why Cryptosystems Fail", in *Proceedings of the 1993 ACM Conference in Computer and Communications Security* pp 215 - 227
2. Ash, R., *Information Theory*, John Wiley & Sons, 1965.
3. Bennett, C.H., G. Brassard, *Quantum Cryptography: Public key distribution and coin tossing*, Proc. of IEEE International Conference on Computers, Systems, and Signal Processing, Banglore, India, December 1984, pp. 175–179.
4. Bennett, C.H., G. Brassard, J.-M. Robert, *Privacy Amplification by Public Discussion*, SIAM Journal on Computing, Vol. 17, No.2, 1988, pp. 210–229.
5. Brassard, G., C. Crépeau, R. Jozsa, D. Langlois, *A quantum bit commitment scheme provably unbreakable by both parties*, Proceeding of the 34th annual IEEE Symposium on Foundations of Computer Science, November 1993,pp. 362–371.
6. Bennett, C.H., G. Brassard, C. Crépeau, M.-H. Skubiszewska, *Practical Quantum Oblivious Transfer*, In proccedings of CRYPTO'91, Lecture Notes in Computer Science, vol 576, Springer Verlag, Berlin, 1992, pp 351–366.

7. Brassard, G., *On computationally secure authentication tags requiring short secret shared keys*, Advances in Cryptology: Proceedings of CRYPTO 82, Plenum Press, 1983, pp.79–86.

8. Brassard, G., *Cryptology column — Quantum cryptography: A bibliography*, Sigact News, vol. 24, no. 3, 1993, pp.16–20.

9. Carter, J.L., M. N. Wegman, *New Hash Functions and Their Use in Authentication and Set Equality*, Journal of Computer and System Sciences, Vol. 22, 1981, pp. 265–279.

10. Chor, B., O. Goldreich, J. Hastad, J. Freidman, S. Rudich, R. Smolensky, *The bit extraction problem or t-resilient functions*, Proc. 26th IEEE Symposium on Foundation of Computer Science, Portland, Oregon, 1985, pp.396–407.

11. Crépeau, C. and M. Sántha. *Efficient reductions among oblivious transfer protocols based on new self-intersecting codes*. In Sequences II, Methods in Communications, Security, and Computer Science, pp. 360–368. Springer-Verlag, 1991.

12. Crépeau, C., *Quantum Oblivious Transfer*, Journal of Modern Optics, Dec. 1994.

13. Even, S., Goldreich, O. and Lempel, A., "A randomized protocol for signing contracts", *Communications of the ACM*, vol. 28, 1985, pp.637–647.

14. Fagin, R., M. Naor and P. Winkler, *Comparing Common Secret Information without Leaking it*, submitted for publication, Communications of the ACM, 1994.

15. Fiat, A and A. Shamir. *How to prove yourself: practical solutions to identification and signature problems*. In A. M. Odlyzko, editor, Proceedings CRYPTO 86, pages 186–194. Springer, 1987. Lecture Notes in Computer Science No. 263.

16. Forney, G. D., *Concatenated Codes*, The M.I.T. Press, 1966.

17. Goldreich, O., S. Goldwasser, and S. Micali. *How to construct random functions*. In Proceedings of the 25th IEEE Symposium on Foundations of Computer Science, pp. 464–479, Singer Island, 1984. IEEE.

18. Kilian, J., *Founding cryptography on oblivious transfer*. In Proc. 20th ACM Symposium on Theory of Computing, pp. 20–31, Chicago, 1988. ACM.

19. Knobloch, H.-J.,*personal communication* through T. Beth.

20. Mac Williams, F.J. and N.J.A. Sloane, *The Theory of Error-Correcting Codes*, North-Holland, 1977.

21. Mayers, D., L. Salvail, *Quantum Oblivious Transfer is Secure Against Individual Measurements*, In the Proceedings of PHYSCOMP 94, Dallas, 1994, pp. 69–77.

22. Muller, A., Breguet, J. and Gisin, N., *Experimental demonstration of quantum cryptography using polarized photons in optical fiber over more than 1 km* In Europhysics Letters, vol. 23, no. 6, 20 August 1993, pp..383–388.

23. Rabin, M. O., *How to exchange secrets by oblivious transfer*, Technical Memo TR-81, Aiken Computation Laboratory, Harvard University, 1981.

24. Robert, J.-M., *Détection et correction d'erreur en cryptographie*, Master thesis, Département d'informatique et de Recherche Opérationnelle, Université de Montréal, Montréal, Québec, Canada, 1985.

25. Spielman, D., *Linear-time Codable and Decodable Error-Correcting Codes*. In Proc. 27th ACM Symposium on Theory of Computing, 1995. ACM.

26. *One Less Thing to Believe In: Fraud at Fake Cash Machine*, New York Times, 13 May 1993, pp. A1 & B9.

Appendix

We refer the reader to [6, 5, 12] for more details on this protocol. Here ϵn is a limit on the number of errors that can be tolerated from real noise. The actual error rate ϵ' should be less than ϵ in order to reject an honest $\mathcal{A}lice$ accidentally only with probability exponentially small in n.

Protocol 5.1 ($\binom{2}{1}$–$OT_4(q_0, q_1)(c)$)

1: $\overset{n}{\underset{i=1}{DO}}$
  – $\mathcal{A}lice$ picks $r_i \in_R \{0.1\}$ and $\beta_i \in_R \{\text{⊕}, \times\}$,
  – $\mathcal{B}ob$ picks $\beta_i' \in_R \{\text{⊕}, \times\}$,
  – $\mathcal{A}lice$ sends to $\mathcal{B}ob$ a photon π_i with polarization $\beta_{i[r_i]}$,
  – $\mathcal{B}ob$ measures photon π_i in basis β_i' and obtains a bit r_i'.

2: $\mathcal{B}ob$ runs $commit(r_1'r_2'...r_n'\beta_1'\beta_2'...\beta_n')$ with $\mathcal{A}lice$.

3: $\mathcal{A}lice$ picks a random bit t and announces it to $\mathcal{B}ob$.

4: if $t = 0$ then
  – $\mathcal{B}ob$ runs $unveil(r_1'r_2'...r_n'\beta_1'\beta_2'...\beta_n')$,
  – $\mathcal{A}lice$ checks that $\#\{i \mid \beta_i = \beta_i' \text{ and } r_i \neq r_i'\} < \epsilon n$,
  – $\mathcal{A}lice$ and $\mathcal{B}ob$ restart the protocol.

5: $\mathcal{A}lice$ announces her choices $\beta_1\beta_2...\beta_n$ to $\mathcal{B}ob$.

6: $\mathcal{B}ob$ randomly selects two subsets $I_0, I_1 \subset \{1, ..., n\}$ subject to $|I_0| = |I_1| = n/2$, $I_0 \bigcap I_1 = \emptyset$ and $\forall i \in I_0, \beta_i = \beta_i'$ or $\forall i \in I_1, \beta_i \neq \beta_i'$, and announces $I_c, I_{\bar{c}}$ to $\mathcal{A}lice$.

7: $\mathcal{A}lice$
  – receives J_0, J_1. and defines $w_0 \leftarrow r_{j_1^0}r_{j_2^0}...r_{j_{n/2}^0}$ and $w_1 \leftarrow r_{j_1^1}r_{j_2^1}...r_{j_{n/2}^1}$
    with $j_l^b \in J_b$ and $j_l^b < j_{l+1}^b$ for $b \in \{0,1\}$ and $1 \leq l < n/2$,
  – computes their syndromes $S_0 \leftarrow Syn(w_0)$ and $S_1 \leftarrow Syn(w_1)$
    with respect to an agreed upon linear code C (consult [6] for details),
  – picks a random privacy amplification function $h : \{0,1\}^{n/2} \to \{0,1\}^2$,
  – computes $\hat{q}_0 \leftarrow q_0 \oplus h(w_0)$ and $\hat{q}_1 \leftarrow q_1 \oplus h(w_1)$,
  – sends $S_0, S_1, h, \hat{q}_0, \hat{q}_1$ to $\mathcal{B}ob$.

8: $\mathcal{B}ob$
  – receives $S_0, S_1, h, \hat{q}_0, \hat{q}_1$,
  – computes a word $\hat{w}$ of syndrome S_c using the decoding algorithm of C from word $w' = r_{i_1}'r_{i_2}'...r_{i_{n/2}}'$ with $i_l \in I_0$ and $i_l < i_{l+1}$ for $1 \leq l < n/2$,
  – computes and returns $q_c \leftarrow \hat{q}_c \oplus h(\hat{w})$.

The privacy amplification function used in Step 7 can be a the concatenation of the XOR of two random subsets of the bits of its input. In the Bit Commitment protocol of Step 2, a single privacy amplification function can be used for all of them.

Securing Traceability of Ciphertexts — Towards a Secure Software Key Escrow System*

(Extended Abstract)

Yvo Desmedt**

Department of Electrical Engineering and Computer Science, University of
Wisconsin–Milwaukee, WI 53201-0784, U.S.A., e-mail: desmedt@cs.uwm.edu

Abstract. The Law Enforcement Agency Field (LEAF), which in Clipper is appended to the ciphertext, allows the Law Enforcement Agency
to *trace* the sender and receiver. To prevent users of Clipper to delete
the LEAF, the Clipper decryption box will not decrypt if the correct
LEAF is not present. Such a solution requires the implementation to be
tamperproof.
In this paper we propose an alternative approach to achieve traceability. Our solution is based on the computational complexity of some well
known problems in number theory. So, our scheme does not require a
tamperproof implementation, nor a secret algorithm. Its applications extend beyond key escrow.

1 Introduction

Clipper [7] (see also [3, 25]) allows the Law Enforcement Agency to wiretap after
having received the proper authorization. Key Escrow Agencies keep shares of
the user's secret key which they reveal to the Law Enforcement Agency once the
proper authorization has been obtained. To identify the sender and receiver, a
field, called the LEAF is attached to the ciphertext. If the LEAF can be removed,
then the Law Enforcement Agency cannot request an appropriate court order
(since one cannot identify sender or receiver). If the transmission method used
is broadcast oriented, such as in cellular telephone, it may be hard to trace the
sender and it may even be harder to trace the receiver. In such application it is
essential that the functionality of the LEAF is secure.

Let us explain how one guarantees in Clipper that the LEAF is present,
i.e., has not been deleted. When a ciphertext is received the correctness of the

---

* DISCLAIMER: This paper is not intended at all as an endorsement of the Clipper
  idea or the idea of Key Escrow. The intend of this paper is scientific, *i.e.*, to propose
  a solution to the open problem whether it is possible to make a software based key
  escrow system without the need of tamperproofness.

** A part of this work has been supported by NSF Grant NCR-9106327. The author is
  solely responsible for the content of this paper. A part of this research was done while
  the author was visiting the Università di Salerno, Italy. This visit was supported in
  part by CNR AI n.94.00011.

L.C. Guillou and J.-J. Quisquater (Eds.): Advances in Cryptology - EUROCRYPT '95, LNCS 921, pp. 147-157, 1995.
© Springer-Verlag Berlin Heidelberg 1995

LEAF is checked. If incorrect, the chip refuses[3] to decrypt. So *tamperproofness* is *essential* to the protection method used in *Clipper*.

In this paper we discuss how an alternative to Clipper can be adapted to secure the LEAF. The main goal of the paper is to demonstrate that it may *not* be necessary to rely on tamperproofness to achieve Key Escrow. Our solution is based on computational number theory.

We propose an ElGamal based scheme in which given the ciphertext (and nothing more) it is possible for the Law Enforcement Agency, knowing a trapdoor, to trace the receiver. The trapdoor will not help the Law Enforcement Agency to decrypt ciphertexts. We will argue that any attempt by the sender to hide the identity of the receiver by altering the ciphertext will imply that the receiver can no longer decrypt it properly. We also note that for public key schemes, as RSA, it is very easy for the sender to hide the identity of the receiver when sending ciphertext (see Section 3.1).

To understand the importance of this paper we remind the reader that NIST (National Institute of Standards and Technology, US) has set up a key escrow software working group [27, 28]. The only software solutions presented so far [33] allows any hacker to modify the program so that the ciphertext becomes untraceable. In our approach any software implementation of it seems secure (under reasonable assumptions) against such hacker's attempts.

We note that extensive research has gone towards achieving anonymity and untraceability, *e.g.*, [4, 5, 6], but very little to guarantee traceability against senders and receivers who conspire.

In [9], Threshold Decryption was proposed as an alternative to Clipper. Threshold Decryption allows shareholders to decrypt ciphertext (or a session key) without the need to reveal their shares to a third party. Several Threshold Decryption schemes have been presented. One [10] is based on the ElGamal cryptosystem [14], others [11, 16] are based on RSA [30] and a "proven secure" one is presented in [8]. It is not the purpose of this paper to discuss these ideas in more details. The scheme we present is compatible with the the ElGamal based system Threshold Decryption method, but it is not excluded that this idea would work for other cryptosystems as well.

In Section 3.1 we argue that in existing schemes it is easy to bypass traceability of the sender/receiver of the ciphertext. In Section 3 we discuss the main ideas that we will use to achieve traceability and then discuss how this idea can be set up in an ElGamal setting. Before proposing the actual scheme (see Section 5) we first discuss in Section 4 the privacy of the plaintext. In Section 6 we explain why the traceability of the scheme seems secure.

For the reader who is not familiar with basic algebraic notations and with basic cryptosystems we use, we first overview these.

---

[3] The security of this check is not very high as was recently demonstrated by Matt Blaze [26]. He demonstrated that, by doing some trial-and-error on the LEAF, Clipper could be used with an incorrect LEAF, implying that the Law Enforcement Agency is unable to eavesdrop. Future implementations of Clipper may have a higher security against such attacks.

2 Background

First we remind the reader how the ElGamal [14] public key encryption scheme works. In the ElGamal scheme, we have a prime p and an element g of large enough order, $e.g.$, a primitive element. Each user publishes as public key $y_j = g^{s_j} \bmod p$, where s_j is the user's secret key (chosen uniformly random in Z_{p-1}). To encrypt the message M $(0 \leq M < p)$ the sender chooses a uniformly random r in Z_{p-1}, which we denote as $r \in_R Z_{p-1}$, (or $r \in_R Z_{\mathrm{ord}\, g}$ if ord g is known) and sends the receiver $(R, C) = (g^r, M \cdot y_j^r) \bmod p$, where y_j is the receiver's public key. Decryption is straightforward. Indeed, the receiver knowing s_j computes $M = C \cdot R^{-s_j} \bmod p$. To simplify notations, we often drop, in the rest of the paper, the mod p.

We now remind the reader of some basic algebraic [22] notations we will use in this text. Let G be a finite group G and $g \in G$. The subgroup of elements $\{g, g^2, \ldots, g^{d-1}, g^d = 1\}$ is called cyclic and is denoted as $\langle g \rangle$. The smallest d for which $g^d = 1$ is called the order of g and we denote it as: ord(g). The cardinality of a finite group G is called the order of G.

3 Towards the scheme

3.1 Failure of existing schemes

For many cryptosystems the ciphertext itself does not reveal the identity of the sender nor the identity of the receiver. In public key systems the public key is receiver dependent, but that does not mean that an eavesdropper can find the identity of the receiver by observing one ciphertext. In the RSA system the ciphertext is a number between 0 and $n-1$, but if only one ciphertext block is sent this does not necessarily allow the eavesdropper to find n, which would identify the receiver. Worse, the sender can hide n by using similar techniques as were developed in [16]. Indeed, to hide n, the sender adds random multiples of n to the ciphertext, $i.e.$, $r \cdot n$, where $0 \leq r < \lfloor 2^{2\lceil \log_2 n \rceil}/n \rfloor$. From [16] follows that if plaintext is uniformly distributed (which can roughly be approached using a source coder [18]) then n is well hidden[4]. Traditional schemes based on discrete logarithm have similar problems, as is easy to show.

Above demonstrates that the idea of using a public key scheme is not sufficient to guarantee that one is able to trace the receiver from the ciphertext only.

---

[4] In fact, the following families of distributions $U(n)$ and $V(n)$ are statistically indistinguishable [21], where $U(n)$ corresponds with choosing uniformly a number between 0 and $2^{2|n|} - 1$ and $V(n)$ corresponds with choosing uniformly a number between 0 and $\left(n \cdot \lfloor 2^{2|n|}/n \rfloor\right) - 1$, as is easy to prove. From this follows that taking polynomially many samples from $U(n)$ and $V(n)$ is indistinguishable.

3.2 Main ideas

To trace the receiver, the main idea is to put redundancy in the ciphertext, which identifies the receiver. The redundancy originates from the public key. The redundancy needs to be added such that if the sender can remove the redundancy, then the receiver can no longer decrypt the ciphertext.

Let us first discuss how this could be realized in general. We continue with our public key scenario. Assume that all ciphertexts sent to whoever all belong to some public set, e.g., Z_p, where p is a prime which has been standardized. To achieve traceability all ciphertexts sent to the receiver, R, should belong to a subset S_R, which is unique for each receiver. To avoid any false identifications, one might require that the subsets are disjoint. So, if given the ciphertext, the Law Enforcement Agency can find (in polynomial time) the subset S_R, then it is able to identify the receiver.

Let us first observe that the requirement that subsets S_R and $S_{R'}$ ($R \neq R'$) be disjoint, is too strong. It is sufficient that the probability of a false identification is very small. One could wonder how one could achieve these subsets. As illustration, we explain this now for the ElGamal scheme.

3.3 ElGamal based subsets

Observe that in the ElGamal scheme $R = g^r \bmod p$ belongs to the group generated by g. To make the ciphertext receiver dependent, R could belong to a receiver dependent subgroup of Z_p^0, where p is a standard prime. To achieve this, let g_j depend on the user j such that $\langle g_j \rangle$ (the cyclic group generated by g_j) is different from $\langle g_k \rangle$ when $j \neq k$.

We now discuss how such $\langle g_j \rangle$ could be obtained. In our first scheme, let m be such that the number of users one can expect is maximum $2^m - 1$. The Law Enforcement Agency chooses different large primes q_i and a prime p, such that $q_i \mid p - 1$. The Law Enforcement Agency also chooses a $g \in Z_p$ of order $Q = \prod_{i=1}^m q_i$. The Law Enforcement Agency keeps all q_i secret.

When a user j wants to register a public key, the Law Enforcement Agency chooses a unique binary user identifier $\mathbf{e}_j = (e_1, \ldots, e_m)$, $\mathbf{e}_j \neq \mathbf{0}$, and stores $(j, \mathbf{e}_j)$ in its data base. It then constructs

$$g_j = g^{\frac{Q}{\prod_{i=1}^m q_i^{e_i}}} \pmod{p}$$

and gives g_j to the user. The user then proceeds as in ElGamal, it is, chooses a secret $s_j \in_R Z_{p-1}^*$ and constructs $y_j = g_j^{s_j} \bmod p$. The Law Enforcement can easily verify that $y_j \in \langle g_j \rangle$ and that s_j is relatively prime to $p - 1$, by computing $\text{ord}(y_j)$. Indeed, $\text{ord}(y_j) = \text{ord}(g_j)$ iff $y_j = g_j^{s_j}$ and s_j is relatively prime to $p - 1$ [22, p. 45] and [2, p. 89]. Section 3.4 reminds the reader how the Law Enforcement Agency can verify this.

The public key of the user is (g_j, y_j), p is standard. In our first scheme the encryption and decryption will be similar as in ElGamal, except that $r \in_R Z_{p-1}^*$, i.e., relatively prime to $p - 1$.

Observe that there is absolutely no need for the Law Enforcement Agency to reveal e_j to user j. It might be better for the Law Enforcement Agency to keep its data base (i.e., (j, e_j)) secret, which we will motivate in the final paper. Also, it might be better that for all users j the Hamming weight of e_j is identical, which requires one to make m a little larger (see final paper).

3.4 Tracing a receiver

Let us explain how the Law Enforcement Agency can find the receiver if R is properly constructed. To check this, one first checks whether $R^Q \equiv 1$ mod p, if not then $R \notin \langle g \rangle$. Let us assume it is. Observe that the order of g_j is $\prod_{i=1}^{m} q_i^{e_i}$ and that in a finite cyclic group, in this case Z_p^*, the order of a subgroup identifies uniquely the subgroup [22, p. 45]. So to find j it is sufficient for the Law Enforcement Agency to find the order of R modulo p. This can easily be computed. Indeed if $R^{Q/q_1} \equiv 1$ mod p then $e_1 = 0$, else $e_1 = 1$. To find e_i, one continues similarly computing R^{Q/q_i} mod p. Once the Law Enforcement Agency has found e, it is able to find j using its data base.

In Section 6.2 we will argue (not prove) that the traceability of ciphertext in the scheme is secure against senders who try to fool the Law Enforcement Agency.

3.5 Towards the scheme

The final scheme (Section 5) is almost identical to the one we have discussed in Section 3.3. However, privacy issues, which we discuss in the next section, will force us to slightly modify the scheme.

4 Privacy

Several security aspects must be addressed, but we postpone most of these until we have discussed our scheme in full detail. Here we discuss whether using some g_j which is not a primitive element reduces the security of ElGamal. We also discuss whether the Law Enforcement Agency can decrypt the ciphertext without the help or knowledge of the receiver or the Key Escrow Agencies. We assume in this section that the reader is familiar with basic group theory [22].

First of all we observe that it is not necessary that the plaintext M be in $\langle g_j \rangle$. This implies that, given C, one might be able to compute a part of M. If Q is public, then there exists $M \in Z_p$ such that $M^Q \neq 1$ mod p, but $y_j^Q = 1$ mod p for all public keys y_j, so raising $C^Q \equiv M^Q$ mod p. To analyze the signficance of this we consider the factorization of $p - 1 = q_1^{a_1} \cdots q_m^{a_m} \cdot q_{m+1}^{a_{m+1}} \cdots q_l^{a_l}$. If for all i $(1 \leq i \leq l)$ $a_i = 1$ and $l = m + 1$ and $q_{m+1} = 2$, then the standard ElGamal scheme leaks similar knowledge. (Indeed from the Legendre symbol $(y_j \mid p)$ and $(R \mid p)$ one can compute $(y_j^r \mid p)$ and then $(M \mid p) = (C \mid p) \cdot (y_j^r \mid p)$.)

Since the Law Enforcement Agency knows the factorization it is however able to find much more (without the help of the Key Escrow Agencies) about M than other eavesdroppers can. Indeed, *for example*, since it knows ord(g_j) it can compute $C^{\mathrm{ord}(g_j)} \boxminus M^{\mathrm{ord}(g_j)}$ mod p. Whether such projections really tell something useful about the plaintext is doubtful. However, one could argue that the Law Enforcement Agency (as long as it has not received proper authorization) should not be able to know more about the plaintext than outsiders. We now address this.

5 The scheme

The user's public key is as in Section 3.3, *i.e.*, similar as in the ElGamal scheme with the main exception that g_j is user dependent. Users use this variant of ElGamal *only* to exchange a common session key (so a conventional cryptosystem is used for the actual encryption). It is, the sender chooses a uniformly random $M \in_R Z_p$ and to send it to the receiver, the sender computes $(R, C) = (g^r, M \cdot y_j^r)$ mod p, where $r \in_R Z_{p-1}^*$. Once M is obtained, sender and receiver hash M to obtain the session key.

The goal of this hash function is to stop the Law Enforcement Agency to learn something extra about the session key. So the hash function should destroy the algebraic properties.

To avoid false identifications in a voluntary[5] escrow system, non-registered receivers should not be allowed to use a $\langle g_j \rangle$ already used by a registered user. This can be achieved when all non-registered users use the *same* g'. This g' is nothing else than a special[6] g_j which would never be assigned to a registered user. If the Law Enforcement Agency identifies an R (of a ciphertext) in $\langle g' \rangle$, then it knows that the ciphertext is sent to a non-registered user. Since a public key system requires an authority (authorities) to manage the public key directory to guarantee authenticity [29], we assume that this authority enforces non-registered user to use g' as the first part of their public key.

In a non-voluntary key escrow system the authority that manages the public key directory will enforce the users to register. This authority could be similar as the telephone monopoly (see also [25]), we therefore call it the monopoly. In this scenario, non-registered users do not have a public key[7], but such users might choose an g_j already in use! To find these violations the monopoly could trace the receiver randomly using non-cryptographic technology and check whether the claimed identity corresponds with the real one (such checks would not affect the privacy of the conversation itself, as we have discussed).

---

[5] Clipper has been called a "voluntary" key escrow system.

[6] It is important that ord(g') has no small factor. If it would, then it would allow a "dishonest" sender to expand slightly the bandwidth of the subliminal channel explained in Section 6.2, by having $g_i = g'$. This follows from the fact that ElGamal leaks a little, as explained in Section 4.

[7] They have to use the "public key" system as a conventional cryptosystem or have to rely on non-official authorities that guarantee authenticity of non-registered "public keys". One could compare this with users setting up their own telephone company to hide from the rest of the world.

5.1 Enhancements

To any interested individual, a special office of the Law Enforcement Agency could prove e.g., in zero-knowledge [20], that each prime factor of the order of g is large. If Q and g is public each user can for himself verify that $g_j \in \langle g \rangle$ by checking whether $g_j^Q = 1 \bmod p$. Indeed in a finite cyclic group, in this case Z_p^*, there is at maximum one subgroup of a given order [22, p. 45]. The user should also verify whether $g_j \neq 1 \bmod p$.

To reduce the trust in the Law Enforcement Agency, the concept of mental games (secure distributed computation) [34, 19] can be used allowing several parties to be involved when g and q_i are chosen.

6 Security

We have already addressed the privacy aspect of our scheme. We now wonder whether a collaborating sender and receiver can bypass the traceability. In other words, can a sender send a ciphertext to a receiver using the secret key of the receiver in such a way that the receiver can decrypt it, but the Law Enforcement Agency is mislead, it is identifies the wrong receiver. Before discussing whether we can hide the identity of the receiver, we discuss the model in which we might work.

6.1 Model

We are only concerned about tracing registered receivers (it is receivers whose g_j is constructed by the Law Enforcement Agency). We further restrict ourselves to receivers and senders who never communicated secretly before. Indeed if they ever did, then they could have exchanged a common secret key to be used for any future communication. We also should restrict ourselves to senders and receivers who have no common knowledge besides[8] the registered public keys to be used in the escrow system.

We have to address the question whether a sender can find a method to hide the identity of the receiver. This problem is similar to the subliminal problem introduced by Simmons [32] (for a study in the context of key escrow, consult [23]). In Simmons case two known senders were using an authentication system to hide a secret message, by using a *non-specified* protocol. In our case the users might use a non-specified algorithm to send secret data.

It seems that the model of subliminal-freeness [12] covers this. However, it is immediately clear that if the senders and receivers share a secret key, that

---

[8] This also excludes public keys used for signatures! Indeed, it has been observed that if a sender knows the receiver's public key for a signature scheme, e.g., DSS [13], then the sender can request the receiver to sign a non-escrowed public key. The sender, after having checked the authenticity of this key, would use it to encrypt data untraceably.

they can send ciphertext which is untraceable (from the point of view of somebody who only has access to the ciphertext and not the physical means of communication). So we must exclude such scenarios. This means in the context of subliminal-freeness that the sender and receiver should not have common knowledge, it is have no knowledge tapes.

6.2 How secure

We were unable to prove that it is impossible to hide the identity of the receiver when sending a ciphertext. Although this might disappoint a theoretician, one should not forget that nevertheless Rivest, Shamir and Adleman *only* argued (and never proved) that their scheme is as secure as factoring, no better attack to invert the RSA function has been found so far.

First let us observe that it seems that the system can be used to send a narrowband subliminal message. Indeed suppose a sender only sends M that contain redundancy. When a sender wants to send a subliminal message to receiver j he uses the public key (g_i, y_i) of *another*[9] receiver i. Clearly j decrypts this ciphertext wrongly and the redundancy will not appear. The mere fact that the receiver cannot properly decrypt is the subliminal information. It is clear that the bandwidth is extremely low. Clearly the Law Enforcement Agency will trace the wrong receiver. One could correctly observe that in our scheme M should be uniform, so above attacks assumes that receiver and sender had some common knowledge, namely the redundancy pattern to check for. However, such common knowledge violates the model.

Let us now restrict our analysis to the case where the receiver (alone[10]) can compute the correct M from the ciphertext and the eavesdropper cannot. Let us assume that the receiver uses ElGamal to decrypt. In this scenario we argue (no proof) that the problem of fooling the Law Enforcement Agency is hard if factoring and discrete log are hard. To succeed, a sender will on input M and (g_j, y_j) compute some (R, C) (not necessarily the specified one) and from (R, C) the receiver can compute the correct M. Then if the receiver (alone) must be able to compute M, one must have that $g_j^{r s_j} = C \cdot M^{-1} \bmod p$. So, given g_j and $y_j = g_j^{s_j} \bmod p$, the sender must compute $R = g_j^r$ and $g_j^{r s_j} \bmod p$. Finding $g_j^{r s_j}$ when given (g_j, R, y_j) is the Diffie-Hellman problem, which is believed to be as hard as finding r (which has been proven for some parameters [24]). So it seems that the sender must know r. To fool the Law Enforcement Agency the sender must manage to compute an R such that $\mathrm{ord}(R) \neq \mathrm{ord}(g_j)$ and $\mathrm{ord}(R) = \mathrm{ord}(g_i)$ for some receiver i. If $R = g_j^r$ this means that $\gcd(r, Q) \neq 1$. So, if the sender must know r then the sender can find a factor of Q. Similar reasonings can be followed for trivial variants of ElGamal.

---

[9] Another, more successful, variant of the idea of using another person's key to send subliminal data has recently been described [15] (see also [17]). An active party, for example, the phone company, can, however, easily prevent the last attack [15].

[10] In the final paper, we extend the analysis allowing for multiple conspiring receivers.

Finally it seems that knowledge of other g_i will not help a dishonest sender, since finding the order of g_i is believed as hard as factoring $p - 1$ [1] and even if g is public finding the discrete log of g_i is believed to be hard.

7 Conclusion

In this paper we proposed an encryption scheme to solve the open problem, posed by NIST, whether a software based approach to Key Escrow is possible.

For our scheme, it seems that if the encryption is done in software the LEAF, being an integral part of the encrypted session key, cannot be removed without destroying ciphertext (the encrypted session key) so badly that it can no longer be decrypted. Another advantage of this scheme is that there is no need to use a tamperproof chip and/or a secret algorithm. The scheme has its full power, enhancing the privacy protection of the user, if Threshold Decryption [10] is used to provide the Law Enforcement Agencies with the decrypted session keys.

The ideas proposed in this paper have broader applications than their use in the context of Clipper only. Indeed these can be used in any setting in which the receiver has to be traceable. We discussed this in the context of Clipper due to its importance at the time this paper was written.

Let us briefly discuss some practical aspects. Let us assume that roughly 10^9 users would use the system then $m = 30$. If each q_i is 320 bits, then Q has 9600 bits, which means that p is only 10 times larger than usually used (we assume that nowadays p of 1024 bits is typical). So our scheme is (less than) 100 times slower than standard ElGamal.

Finally our scheme allows for several enhancements as the use of a verifiable secret sharing scheme [25].

Our paper introduces the following problems. Since the security of our scheme is only heuristic, the open problem is whether one can design a public key encryption scheme in which traceability is proven secure. Also can one make an RSA type variant. Finally can one improve the speed of the scheme.

Acknowledgement

The author thanks Yair Frankel, Sandia National Laboratories, for his comment [15].

References

1. Adleman, L. M., McCurley, K. S.: Open problems in number theoretic complexity. In Discrete Algorithms and Complexity, Proceedings of the Japan-US Joint Seminar (Perspective in Computing series, Vol. 15) (June 4–6, Kyoto, Japan 1986) D. Johnson, T. Nishizeki, A. Nozaki, and H. Wilf, Eds. Academic Press Inc., Orlando, Florida pp. 263–286.
2. Berlekamp, E. R.: Algebraic Coding Theory. Aegen Park Press 1984.

3. Beth, T.: Zur Sicherheit der Informationstechnik. Informatik-Spektrum **13** (1990) 204–215.

4. Chaum, D.: Untraceable electronic mail, return addresses, and digital pseudonyms. Commun. ACM **24** (1981) 84–88.

5. Chaum, D.: Blind signatures for untraceable payments. In Advances in Cryptology. Proc. Crypto'82 (Santa Barbara, 1983) D. Chaum, R. Rivest, and A. T. Sherman, Eds. Plenum Press N. Y. pp. 199–203.

6. Chaum, D.: The dining cryptographers problem: unconditional sender and recipient untraceability. Journal of Cryptology 1 (1988) 65–75.

7. A proposed federal information processing standard for an escrowed encryption standard (EES). Federal Register July 30, 1993.

8. De Santis, A., Desmedt, Y., Frankel, Y., Yung, M.: How to share a function securely. In Proceedings of the twenty-sixth annual ACM Symp. Theory of Computing (STOC) (May 23–25, 1994) pp. 522–533.

9. Desmedt, Y., Frankel, Y., Yung, M.: A scientific statement on the Clipper chip technology and alternatives September 1993. Comment sent to NIST.

10. Desmedt, Y., Frankel, Y.: Threshold cryptosystems. In Advances in Cryptology — Crypto '89, Proceedings (Lecture Notes in Computer Science 435) (1990) G. Brassard, Ed. Springer-Verlag pp. 307–315.

11. Desmedt, Y., Frankel, Y.: Shared generation of authenticators and signatures. In Advances in Cryptology — Crypto '91, Proceedings (Lecture Notes in Computer Science 576) (1992) J. Feigenbaum, Ed. Springer-Verlag pp. 457–469.

12. Desmedt, Y. G.: Subliminal-free cryptosystems. Submitted to the Journal of Cryptology April 1989, revised version submitted May 3, 1994.

13. A proposed federal information processing standard for digital signature standard (DSS). Federal Register August 1991.

14. ElGamal, T.: A public key cryptosystem and a signature scheme based on discrete logarithms. IEEE Trans. Inform. Theory **31** (1985) 469–472.

15. Frankel, Y.: February 1995. Personal communication.

16. Frankel, Y., Desmedt, Y.: Parallel reliable threshold multisignature. Tech. Report TR-92-04-02 Dept. of EE & CS, Univ. of Wisconsin–Milwaukee April 1992.

17. Frankel, Y., Yung, M.: Escrowed encryption systems visited: Threats, attacks, analysis and designs. Manuscript, November 1994.

18. Gallager, R. G.: Information Theory and Reliable Communications. John Wiley and Sons New York 1968.

19. Goldreich, O., Micali, S., Wigderson, A.: How to play any mental game. In Proceedings of the Nineteenth annual ACM Symp. Theory of Computing, STOC (May 25–27, 1987) pp. 218–229.

20. Goldreich, O., Micali, S., Wigderson, A.: Proofs that yield nothing but their validity or all languages in NP have zero-knowledge proof systems. Journal of the ACM **38** (1991) 691–729.

21. Goldwasser, S., Micali, S., Rackoff, C.: The knowledge complexity of interactive proof systems. Siam J. Comput. **18** (1989) 186–208.

22. Jacobson, N.: Basic Algebra I. W. H. Freeman and Company New York 1985.

23. Kilian, J., Leighton, T.: Failsafe key escrow. Tech. rep. Massachusetts Institute of Technology Technical Report MIT/LCS/TR-636 Cambridge, Massachusetts August 1994.

24. Maurer, U. M.: Towards the equivalence of breaking the diffie-hellman protocol and computing discrete logarithms. In Advances in Cryptology — Crypto '94,

Proceedings (Lecture Notes in Computer Science 839) (1994) Y. G. Desmedt, Ed. Springer-Verlag pp. 271–281.

25. Micali, S.: Fair public-key cryptosystems. In Advances in Cryptology — Crypto '92, Proceedings (Lecture Notes in Computer Science 740) (1993) E. F. Brickell, Ed. Springer-Verlag pp. 113–138.

26. Flaw discovered in federal plan for wiretapping. The New York Times June 2, 1994.

27. Opportunity to join a cooperative research and development consortium to develop secure software encryption with integrated cryptographic key escrowing techniques August 24, 1993. NIST.

28. NIST responses to questions from the senate subcommittee on technology and the law, May 3, 1994.

29. Popek, G. J., Kline, C. S.: Encryption and secure computer networks. ACM Computing Surveys 11 (1979) 335–356.

30. Rivest, R. L., Shamir, A., Adleman, L.: A method for obtaining digital signatures and public key cryptosystems. Commun. ACM 21 (1978) 294–299.

31. Schnorr, C. P.: Efficient identification and signatures for smart cards. In Advances in Cryptology — Crypto '89, Proceedings (Lecture Notes in Computer Science 435) (1990) G. Brassard, Ed. Springer-Verlag pp. 239–252.

32. Simmons, G. J.: The prisoners' problem and the subliminal channel. In Advances in Cryptology. Proc. of Crypto 83 (1984) D. Chaum, Ed. Plenum Press N.Y. pp. 51–67.

33. Walker, S. T., Balenson, D. M.: A software key escrow approach, June 10, 1994. Trusted Information Systems, Inc.

34. Yao, A. C.: How to generate and exchange secrets. In 27th Annual Symp. on Foundations of Computer Science (FOCS) (1986) IEEE Computer Society Press pp. 162–167.

Secure Multiround Authentication Protocols

Christian Gehrmann

Dept. of Information Theory, Lund University,
Box 118, S-221 00, Lund, Sweden,
Tel: +46-46-104353,
Fax: +46-46-104714,
Email: chris@dit.lth.se

Abstract. Gemmell and Naor proposed a new protocol for unconditionally secure authentication of long messages. However Gehrmann showed that the proof of the security of the protocol was incorrect. Here we generalize the multiround protocol model. We prove the security of a 3-round protocol and give for this case a new easy implementable construction which has a key size close to the fundamental lower bound for even extremely long messages. Furthermore, we give a proof of a secure multiround protocol for an arbitrary number of rounds.

1 Introduction

Gemmell and Naor [1] proposed an unconditionally secure authentication scheme (or A-code) without secrecy in which, by following a protocol, codewords are passed back and forth. This scheme makes it possible, albeit at the expense of increasing data exchange, to authenticate very large messages while keeping the key size small. It was shown by Johansson, Kabatanskii and Smeets in [2] that for single round authentication the key size is bounded by the logarithm of the message size.

Denote by $\log^{(k)}(x)$ a k-times logarithm $\log(\log(...log(x)))$. Gemmell and Naor showed that their k-round protocol for a message size of n bits demands a key size

$$H(K) \approx \log^{(k)}(n) + 2\log(\frac{1}{P_s}),\qquad(1)$$

where P_s is the probability of a successful substitution attack for the scheme.

However, the security analysis made by Gemmell and Naor only took into account a certain substitution attack. In [3] Gehrmann, by considering the impersonation attack, showed that protocols where the number of rounds is even are of no interest. Furthermore, he introduced a special six step substitution attack for which the probability calculation of P_s made by Gemmell and Naor did not hold. In this paper, we push the analysis further. We propose new protocols and prove their security.

In Section 2 we give a classification of possible attacks on multiround protocols. In the next section we propose a new 3-round protocol and also give a specific construction based on Reed-Solomon codes. Using the tools of Section 2 we give a proof of its security. In Section 4 we show how to make a secure protocol for an arbitrary number of rounds.

L.C. Guillou and J.-J. Quisquater (Eds.): Advances in Cryptology - EUROCRYPT '95, LNCS 921, pp. 158-167, 1995
© Springer-Verlag Berlin Heidelberg 1995

2 Attacks on multiround protocols

We will consider a system for message authentication in which a transmitter A wants to send a message to a receiver B by exchanging codewords. The transmission channel is supposed to be controlled by an opponent O, who can send own (new) codewords over the channel to A and B or substitute codewords sent by A or B with own new ones. A and B are assumed to share secret information, i.e., the key, unknown to the opponent. We denote by k the total number of codewords sent over the channel to authenticate the original message. Since the original message is sent by A we will denote this message by m^A. The corresponding message observed by B may have been changed by the opponent and will be denoted by m^B. The codewords used by the protocol in the authentication process we will denote by subindex. For example m_i denotes the i-th exchanged codeword. A $k = 3$ round authentication protocol is shown in the figure below.

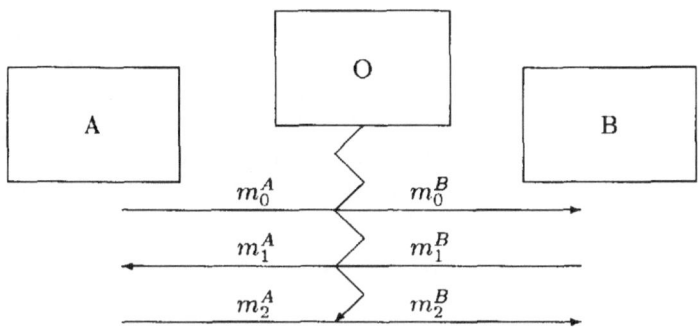

Figure 1: Multiround authentication for $k = 3$.

We will use the following notation:

$\underline{m}^A = m_0^A, (m_1^A), ..., m_{k-1}^A$: a by A sent and (received) codeword sequence.

$\underline{m}^B = (m_0^B), m_1^B, ..., (m_{k-1}^B)$: a by B sent and (received) codeword sequence.

$\underline{M}$: the set of possible codeword sequences.

K: the secret key.

P_I: probability of a successful impersonation attack.

P_S: probability of a successful substitution attack.

For the Gemmell and Naor multiround protocol it was shown [3] that the last round should be omitted if k is even. This implies that in contrast with the Gemmell and Naor model it always have to be the receiver that detects the intruder. Hence, in our generalized multiround authentication model we will in the sequel assume k to be odd.

The attack described by Gehrmann follows an asynchronous model, which was the model used by Gemmell and Naor. The attacks we describe in this paper will also be in accordance with an asynchronous protocol model. To avoid meaningless complications we make the assumption that B(A) immediately after receiving an even(odd) codeword responds with his own odd(even) codeword, i.e., we don't take into account the time delay for A and B. Depending on the secret key, not all sequences $\underline{M}$ are acceptable sequences. Denote by $\underline{M}(K)$ the subset of $\underline{M}$ that are allowable sequences under the specific key K. Depending on the odd number received codeword, his own chosen even number codeword and the key K, A will create and observe a sequence $\underline{m}^A \in \underline{M}(K)$. In a similar way B, by receiving even numbered codewords and responding odd numbered codewords, will observe a sequence $\underline{m}^B$. He accepts the sequence as a message from A if and only if $\underline{m}^B \in \underline{M}(K)$. We will deal with two attack scenarios depending on the capabilities given to the opponent.

2.1 Ordinary substitution attacks

This case corresponds to the ordinary model for single round authentication [4], [5], [6]. Here we assume that O has no capability to choose the message m^A to be authenticated. This original message m^A is for example in the Gemmell and Naor protocol equivalent with the first codeword m_0^A sent by A. All even numbered codewords observed by A and all odd numbered codewords observed by B are chosen by A and B respectively and hence the opponent O has only control over the codewords in the set

$$\mathcal{O} = \{m_0^B, m_1^A, m_2^B, ..., m_{k-2}^A, m_{k-1}^B\}.$$

O succeeds in a substitution attack if and only if he, during observation of the even codewords of $\underline{m}^A$ and odd codewords of $\underline{m}^B$, creats a sequence of codewords from the set $\mathcal{O}$, such that the corresponding sequence observed by B $\underline{m}^B \in \underline{M}(K), m^B \neq m^A$.

However as in the attack in [3], the order of the codewords substituted by the opponent could be chosen in an asynchronous way. Changing the order may increase P_s and thus all possible order of substitution must be taken into account when calculating P_s. Denote by t one particular substitution order (or type of attack) and by $\underline{m}(t), m_i(t) \in \mathcal{O}$, the corresponding sequence of codewords created by O (O-sequence). Furthermore, denote by $T_s(k)$ the set of valid differently ordered substitution sequences for a k-round protocol and by $|T_s(k)|$ the cardinality of this set.

We illustrate the notation with the following example.

Example 1. There are $|T_s(3)| = 3$ different types of attacks for the $k = 3$ round protocol. We have listed all the types in the table below together with the complete sequence of codewords observed on the channel(the codewords that O uses for a substitution attack are marked with arrows).

O-sequence	Whole sequence
$\underline{m}(1) = m_0^B, m_2^B, m_1^A$	$m_0^A, \rightarrow m_0^B, m_1^B, \rightarrow m_2^B, \leftarrow m_1^A, m_2^A$
$\underline{m}(2) = m_0^B, m_1^A, m_2^B$	$m_0^A, \rightarrow m_0^B, m_1^B, \leftarrow m_1^A, m_2^A, \rightarrow m_2^B$
$\underline{m}(3) = m_1^A, m_0^B, m_2^B$	$m_0^A, \leftarrow m_1^A, m_2^A, \rightarrow m_0^B, m_1^B, \rightarrow m_2^B$

Here $\underline{m}(3)$ is of the attack type analyzed in [3].

We have the following theorem on the number of different ordered O-sequences.

Theorem 1.

$$|T_s(k)| = \binom{k}{\frac{(k-1)}{2}}. \tag{2}$$

Proof. The subsequences $m_0^B, m_2^B, \dots$ and $m_1^A, m_3^A, \dots$ in $\underline{m}(t)$ have to be ordered. There are $\frac{(k-1)}{2}$ codewords in the A-subsequence, these should be chosen out of k differents positions in the O-sequence. Hence there are $\binom{k}{\frac{(k-1)}{2}}$ different O-sequences of length k.

Using the introduced definitions P_s may be written as

$$P_s = \max_{t \in T_s(k)} \max_{\underline{m}(t), m^B \neq m^A} \Pr\{\underline{m}^B \in \underline{M}(K)\}. \tag{3}$$

2.2 Chosen message substitution attacks

We will in the sequel deal with an attack model where the original message might freely be chosen by the opponent. This includes that O has the capability to choose *both* the first original message m^A *and* m^B, the one to substitute it with. When the opponent is also given the additional possibility to freely choose the original message m^A that should be authenticated, the situation looks different. The first codeword m_0^A sent by A may consists of just m^A or m^A and a second part created by A. In the chosen-message substitution scenario we assume that O may freely choose the part m^A of m_0^A and O thus has control over the set

$$\mathcal{O}' = \{m_0'^A, m_0^B, m_1^A, m_2^B, \dots, m_{k-2}^A, m_{k-1}^B\},$$

where the $'$ marks that O maybe not might control m_0^A completely. Similar to the ordinary substitution attack case we denote by $T_c(k)$ the different ordered O-sequences of length $k + 1$. We then have the following Corollary.

Corollary 2.

$$|T_c(k)| = \binom{k+1}{\frac{(k+1)}{2}}. \tag{4}$$

Proof. This case corresponds to that for Theorem 1 with a O-sequence of length $k + 1$ and an A-subsequence of length $\frac{(k+1)}{2}$.

Similar to the ordinary substitution attack case we denote one particular ordered sequence by $\underline{m}(t)$. Denote by P'_s the probability of successful substitution attack for the chosen message scenario. P'_s may be written as

$$P'_s = \max_{t \in T_c(k)} \max_{\underline{m}(t), m^B \neq m^A} \Pr\{\underline{m}^B \in \underline{M}(K)\}. \tag{5}$$

Adding the chosen-message attack to the model increases the demands on the authentication protocol. A protocol that is secure in a model including the chosen-message attack is obviously also secure for the ordinary substitution attack model. We will from now on only consider protocol that are secure in the stronger chosen-message attack model sense. Change the demands on the protocols and only letting them be secure for the ordinary substitution attack might reduce the complexity.

Remark. In the single round authentication case there are different definitions of P_s in use. To ensure secure single authentication when the chosen-message attack is added to the model it is necessary to use the max max definition of P_s as for example made in [2], [7].

3 Secure k = 3 round protocols

We start with a special treatment for $k = 3$. We propose a protocol which is a modified version of that of Gemmell and Naor [1].

Let Q be a power of a prime and denote by C a code over $GF(Q)$ with length n. Let $p > \frac{1}{n}$ and the minimum distance d of the code C satisfy

$$d \geq n - np.$$

Denote by C^A an A-code for which the probability of a successful substitution attack is less or equal to p_s and the probability of a successful impersonation attack is less or equal to $p_I < p_s$. Let $C_i(m) \in GF(Q)$ be the code symbol at i-th coordinate of the codeword corresponding to message m. Denote by $a \circ b$ a concatenation between two words a and b. We suggest the following protocol for authentiation of the message m^A:

(i) A chooses a random number $j, 1 \leq j \leq l$, where l is chosen according to the desired security. A sends the codeword $m_0^A = j \circ m^A$

(ii) B receives codeword m_0^B and chooses a random number $i, 1 \leq i \leq n$. B sends codeword $m_1^B = i$.

(iii) A receives codeword m_1^A and uses the code C^A to transmit
$m_2^A = C^A(m_1^A, C_{m^A}(m_0^A)) = C^A(m_1^A, C_{m^A}(j \circ m^A))$.

(iv) B receives codeword m_2^B and calculates $C^A(m_1^B, C_{m^B}(m_0^B))$ and accepts the codeword sequence as authentic if and only if
$m_2^B = C^A(m_1^B, C_{m_1^B}(m_0^B))$.

Theorem 3. *Let* $a = \max_{m,i,c} |\{j : C_i(j \circ m) = c\}|$. *For the* $k = 3$ *round protocol above*

$$P'_s = \max(\frac{a}{l} + (1 - \frac{a}{l})p_s, p + (1 - p)p_s) \qquad (6)$$

the probability for a successful substitution attack when we also take into account the chosen-message attack.

Proof. Denote by $P_t = \max_{\underline{m}(t), m^B \neq m^A} \Pr\{\underline{m}^B \in \underline{M}(K)\}$. According to (4) there are $|T_c(3)| = \binom{4}{2} = 6$ possible attacks, when we also take into account the chosen-message attack, corresponding to the sequences

(1) $\quad \rightarrow m_0'^A, \rightarrow m_0^B, m_1^B, \leftarrow m_1^A, m_2^A, \rightarrow m_2^B$.
(2) $\quad \rightarrow m_0'^A, \leftarrow m_1^A, m_2^A, \rightarrow m_0^B, m_1^B, \rightarrow m_2^B$.
(3) $\quad \rightarrow m_0'^A, \rightarrow m_0^B, m_1^B, \rightarrow m_2^B, \leftarrow m_1^A, m_2^A$.
(4) $\quad \rightarrow m_0^B, m_1^B, \rightarrow m_0'^A, \rightarrow m_2^B, \leftarrow m_1^A, m_2^A$.
(5) $\quad \rightarrow m_0^B, m_1^B, \rightarrow m_0'^A, \leftarrow m_1^A, m_2^A, \rightarrow m_2^B$.
(6) $\quad \rightarrow m_0^B, m_1^B, \rightarrow m_2^B, \rightarrow m_0'^A, \leftarrow m_1^A, m_2^A$.

(3),(4),(6) Here O sends the last codeword m_2^B before receiving m_2^A and hence he gets no information about the secret key and the probability of a successful attack $P_3 = P_4 = P_6 = p_I < p_s$.

(1) O chooses a $m_0^B \neq m_0'^A$ and receives m_1^B. He succeeds with his attack by just letting $m_1^A = m_1^B$ and $m_2^B = m_2^A$ if $C_{m_1^B}(m_0^B) = C_{m_1^B}(m_0'^A)$. Otherwise $m_1^B, C_{m_1^B}(m_0^B) \neq m_1^B, C_{m_1^B}(m_0'^A)$ independent of the choice m_1^A and O has to find $C^A(m_1^B, C_{m_1^B}(m_0^B))$ given $C^A(m_1^A, C_{m_1^A}(m_0'^A))$. By the definition of the A-code the probability for this $\leq p_s$. B chooses m_1^B uniformly over $\{1, n\}$ and this together with the definition of the code C gives $\Pr\{C_{m_1^B}(m_0^B) = C_{m_1^B}(m_0'^A)\} = p$ and hence the overall probability

$$P_1 = p + (1 - p)p_s.$$

(2) O receives m_1^B after choosing m_1^A and hence

$$\Pr\{m_1^B, C_{m_1^B}(m_0^B) = m_1^A, C_{m_1^A}(m_0'^A)\} \leq \frac{1}{n} < p.$$

If $m_1^B \neq m_1^A$ O must find $C^A(m_1^B, C_{m_1^B}(m_0^B))$ given $C^A(m_1^A, C_{m_1^1}(m_0'^A))$ and by the definition of the A-code the probability for that is less or equal to p_s and hence the overall probability satisfies

$$P_2 < p + (1 - p)p_s.$$

(5) If O after receiving m_1^B finds an m^A such that $C_{m_1^B}(m_0^B) = C_{m_1^B}(m_0'^A)$ he will succeed by choosing $m_1^A = m_1^B$ and $m_2^B = m_2^A$. Recall that $a = \max_{m,i,c} |\{j : C_i(j \circ m) = c\}|$ from the statement of the theorem. From

this definition and the fact that A choses j uniformly and at random over $\{1, l\}$ follows

$$\max_{m^A} \Pr\{C_{m_1^B}(m_0^B) = C_{m_1^B}(m_0'^A)|C_{m_1^B}(m_0^B)\} =$$

$$\max_{m^A} \Pr\{C_{m_1^B}(m_0^B) = C_{m_1^B}(j \circ m^A)|C_{m_1^B}(m_0^B)\} = \frac{a}{l}.$$

Otherwise $m_1^A, C_{m_1^B}(m_0^B) \neq m_1^A, C_{m_1^B}(m_0'^A)$ independent of the choice m_1^A and O has to find $C^A(m_1^B, C_{m_1^B}(m_0^B))$ given $C^A(m_1^A, C_{m_1^A}(m_0'^A))$. By the definition of the A-code the probability for this event $\leq p_s$. Hence the overall probability

$$P_5 = \frac{a}{l} + (1 - \frac{a}{l})p_s.$$

Now using (5) gives the desired result.

We will continue with giving an efficient construction by using Reed-Solomon codes (RS-codes).

Construction: For simplicity let $m = m^A$. Let $Q = 2^r, r = v2^{v-t-1}, l = 2^t$ and let C be an RS-code over $GF(Q)$ with $k = 2^s, r - s = t$. Hence

$$n = Q = 2^r$$
$$d = n - k = 2^r - 2^s.$$

Thus $p = (n - d)/n = k/n = 2^s/2^r = 2^{r-t}/2^r = 2^{-t}$. Further let $j \circ m$ be regarded as the k-tuple $(j \circ m_0, m_1, \cdots, m_{k-1})$ over $GF(Q)$, where j is the t first bits and m_0 the $r - t$ next bits of the element $j \circ m_0 \in GF(Q)$. Additionally let the code symbol of index β be obtained by evaluating the polynomial $C_\beta(j \circ m) = j \circ m_0 + m_1\beta + \cdots + m_{k-1}\beta^{k-1}$ as in the description of RS-codes [8]. Let the code C^A be the A-code obtained from a RS-code over $GF(2^v), k = 2^{v-t}$, as suggested in [2], i.e., $p_I = 2^{-v}, p_s = 2^{v-t}/2^v = 2^{-t}$. Thus we have a construction which needs t random bits at the transmission side, r random bits at the receiver side and with a key size of $2v$ bits. Furthermore, the construction allows a message size:

$$\log|M| = r(2^s - 1) + r - t = r2^{r-t} - t = v2^{v-t-1}2^{v2^{v-t-1}-t} - t. \tag{7}$$

Theorem 4. *For the construction above*

$$P_s' < 2^{1-t}. \tag{8}$$

Proof. Denote as in Theorem 3 by $a = max_{m,i,c}|\{j : C_i^1(j \circ m) = c\}|. \forall m, \beta, c$ the number of solutions of the following equation with respect to j

$$C_\beta(j \circ m) = j \circ m_0 + m_1\beta + \cdots + m_{k-2}\beta^{k-1} = c$$

is less or equal to 1 and hence $a = 1$. Further according to (6)

$$P_s = \max(\frac{a}{l} + (1 - \frac{a}{l})p_s, p + p_s - pp_s) =$$
$$= \max(\frac{1}{2^t} + (1 - \frac{1}{2^t})p_s, p + p_s - pp_s) =$$
$$= \max(2^{-t} + 2^{-t} - 2^{-2t}, 2^{-t} + 2^{-t} - 2^{-2t}) < 2^{1-t}.$$

The message size, key size and authenticator length for different parameters of the construction is listed in Table 1 below together with the concatenated RS-codes single authentication construction parameters of [7].

t	length of key	$\approx$ message length	
		new	[7]
20	42	22	120
20	44	2^{29}	2^8
20	46	2^{78}	2^{11}
20	48	2^{179}	2^{13}
40	82	82	210
40	84	2^{50}	2^{10}
40	86	2^{139}	2^{12}
40	88	2^{320}	2^{14}

Table 1. Key and message size in bits for the construction with $P_s < 2^{1-t}$.

It follows from the table above that the 3-round authentication system realizes very long message authentication by using short keys and that the key size even for extremely long messages is close to the Gilbert, MacWilliams and Sloane [9] famous square root bound for single authentication, i.e., $2\log(\frac{1}{P_s})$. The data expansion of the protocol is just $v2^{v-t} + t$ bits for a protocol with a key size of $2v$ bits and with $P_s' < 2^{1-t}$, and is hence almost negligible.

4   k $\geq$ 5 round protocol

As we have shown in the previous section, the 3-round protocol gives a secure authentication system with very short keys and few random bits for most practical situations. However to make the treatment complete we now also prove the security of a multiround protocol for an arbitrary number of rounds. Consider the following protocol:

Let p be a security parameter and C^r a code over $GF(Q_r), Q_r \geq \frac{2^{k-r}}{p}$ with length n_r and with minimum distance d satisfying

$$d \leq n_r - n_r\frac{p}{2^{k-1-r}}$$

and let C^A be a Cartesian A-code with a probability for a successful substitution and impersonation attack less than p. Furthermore, let $\forall r \leq k - 2, 1 \leq i_r \leq n_r$ and $\forall r \leq k - 4, 1 \leq j_r \leq \frac{2^{k-1}}{p}$, where the i's and j's chosen uniformly and random by either A or B when using the protocol. Let m_A and m_B be defined as

$$m_A = (C_{i_{k-2}^A}^{k-2}(\cdots(C_{i_2^A}^2(C_{i_1^A}^1(m^A), i_1^A), i_2^A, j_0^A)\cdots), i_{k-2}^A, j_{k-4}^A), \qquad (9)$$

$$m_B = (C_{i_{k-2}^B}^{k-2}(\cdots(C_{i_2^B}^2(C_{i_1^B}^1(m^B), i_1^B), i_2^B, j_0^B)\cdots), i_{k-2}^B, j_{k-4}^B). \qquad (10)$$

A $k \geq 5$ secure protocol is described below.

(i) $r = 0$, A chooses a random number j_0^A, and sends the codeword $m_0^A = (j_0^A, m^A)$.

(ii) $r = r + 1$, B receives codeword m_{r-1}^A and chooses two random numbers i_r^B, j_r^B. B sends codeword $m_r^B = (i_r^B, j_r^B)$.

(iii) If $r = k - 2$ then step v).

(iv) $r = r + 1$, A receives codeword m_{r-1}^A and chooses two random numbers i_r^A, j_r^A. A sends codeword $m_r^A = (i_r^A, j_r^A)$, back to step ii).

(v)   A receives codeword m_{k-2}^A and uses the A-code to transmit codeword $m_{k-1}^A = C^A(m_A)$, where m_A is given by (9).

(vi) B receives codeword m_{k-1}^B, calculates $C^A(m_B)$, where m_B is given by (10) and accepts the codeword sequence as authentic if and only if $m_{k-1}^B = C^A(m_B)$.

Theorem 5. *For the protocol above*

$$P_s' < 2(1 - \frac{1}{2^k})p. \tag{11}$$

Proof. Among all $T_c(k)$ attacks we will consider only two types; i) either we follow the order of the protocol, i.e., the sequence

$$m_0^A, m_0^B, m_1^B, m_1^A, ..., m_{k-1}^A, m_{k-1}^B,$$

or ii) any of all the other types of attacks.

i) This case corresponds to that analyzed in [1] with adding the j's to the protocol. However the j's are not affecting the choice of indices and the proof still holds, giving the probability of successful attack less than

$$2(1 - \frac{1}{2^{k-1}})p.$$

ii) As in the proof of Theorem 3 if O sends the last codeword m_{k-1}^B before receiving m_{k-1}^A he succeeds with probability at most p. Thus assume m_{k-1}^B is the last codeword in the attack sequence and that after the codeword $l \geq 2, m_l^A$ (or $l \geq 3, m_l^B$) the order of the protocol is followed. An arbitrary attack sequence not equal to that of i) may then be described as

$$\cdots, m_{l-2}^A, m_{l-1}^A, m_l^A, m_l^B, m_{l+1}^B, \cdots, m_{k-1}^A, m_{k-1}^B$$

or

$$\cdots, m_{l-2}^B, m_{l-1}^B, m_l^B, m_l^A, m_{l+1}^A, \cdots, m_{k-1}^A, m_{k-1}^B,$$

where the dots $\cdots$ marks any allowable combination of the remaining part of the codewords. O succeeds by just forwarding the codewords between A and B if

$$(C_{i_l^A}^l(\cdots), i_l^A, j_{l-2}^A) = (C_{i_l^B}^l(\cdots), i_l^B, j_{l-2}^B). \tag{12}$$

O sends the codeword m_{l-2}^B (or m_{l-2}^A) before receiving m_{l-2}^A (or m_{l-2}^B) and he has thus no knowledge of j_{l-2}^A (or j_{l-2}^B) when choosing j_{l-2}^B (or j_{l-2}^A). Hence the probability of equality in (12) at most $\frac{p}{2^{k-1}}$. If $j_{l-2}^A \neq j_{l-2}^B$ it follows from i) that he succeeds with probability at most $2(1 - \frac{1}{2^{k-1}})p$ and thus the overall probability of successful attack

$$\frac{p}{2^{k-1}} + (1 - \frac{p}{2^{k-1}})(2(1 - \frac{1}{2^{k-1}})p) < 2(1 - \frac{1}{2^k})p.$$

5 Conclusion

We have generalized the model of the Gemmell and Naor multiround protocol. A proof a secure 3-round protocol was given together with a very efficient construction for this protocol. The construction demands only a key size of 88 bits for a protocol which authenticate a message of size up to 2^{320} bits with probability of successful attack less than 2^{-39} and with a very small size of the data expansion. The 3-round protocol gives an unconditionally secure authentication system for most practical message sizes. Finally we have given a proof of a secure protocol for an arbitrary number of rounds.

References

1. P. Gemmell, M. Naor,"Codes for interactive authentication", *Proceedings of CRYPTO '93*, 1993, pp. 355-367.
2. T. Johansson, G. Kabatanskii, B. Smeets, "On the relation between A-codes and codes correcting independent errors", *Proceedings of Eurocrypt '93*, 1993, pp. 1-11.
3. C. Gehrmann, "Cryptanalysis of the Gemmell and Naor Multiround Authentication Protocol", *Proceedings of CRYPTO '94*, 1994, pp. 121-128.
4. G.J. Simmons, "A survey of Information Authentication", in *Contemporary Cryptology, The science of information integrity*, ed. G.J. Simmons, IEEE Press, New York, 1992.
5. J.L. Carter, M.N. Wegman, "New hash functions and their use in authentication and set equality", *J. Computer and System Sci.*, Vol 22, 1981, pp. 265-279.
6. D.R. Stinson, "Universal hashing and authentication codes", Design, Codes and Cryptography, vol. 4, no. 4, 1994. pp. 369-380.
7. J. Bierbrauer, T. Johansson, G. Kabatanskii, B. Smeets, "On Families of Hash Functions via Geometric Codes and Concatenation", *Proceedings of CRYPTO '93*, 1993, pp. 331-342.
8. I.S. Reed, G. Solomon, "Polynomial Codes over certain Finite Fields", J. Soc. Ind. Appl. Math., vol. 8, June 1960, pp. 300-304.
9. E. Gilbert, F.J. MacWilliams, N. Sloane, "Codes Which Detect Deception". Bell System Technical Journal. Vol. 53. No. 3. March 1974, pp. 405-424.

Verifiable Secret Sharing as Secure Computation

Rosario Gennaro* and Silvio Micali

Laboratory for Computer Science
Massachusetts Institute of Technology

Abstract. Verifiable Secret Sharing is a fundamental primitive for secure cryptographic design. We present a stronger notion of verifiable secret sharing and exhibit a protocol implementing it. We show that our new notion is preferable to the old ones whenever verifiable secret sharing is used as a tool within larger protocols, rather than being a goal in itself. Indeed our definition, and so our protocol satisfying it, provably guarantees reducibilty. Applications of this new notion in the field of secure multiparty computation are also provided.

1 Introduction

Secret Sharing and Verifiable Secret Sharing (VSS for short) are fundamental notions and tools for secure cryptographic design. Despite the centrality and the maturity of this concept (almost 10 years passed from its original introduction), we shall advocate that a stronger and better definition of a VSS is needed in order to achieve the very desirable property of reducibility for secure protocols. We shall then provide the first provably correct implementation of this stronger definition.

REDUCIBILITY is an essential tool for secure protocols design because it allows such protocols to be built in a *modular* way. In our case this means that one can first design a protocol P for a given task assuming that an "abstract" and "perfect" VSS protocol exists. Then one designs a correct (according to a given definition!) VSS protocol Q. Finally one substitutes Q in place of the abstract sub-protocol in P. However this way of proceeding yields a secure protocol P if and only if the definition of VSS satisfied by Q enjoys the reducibility property. Unfortunately no prior notion of a VSS provably guarantees reducibility. Thus it is a goal of this paper to provide such a definition and a VSS protocol that satisfies it.

THE INTUITIVE NOTION OF A VSS. As first introduced by Chor, Goldwasser, Micali and Awerbuch in [4], a VSS protocol consists of a two-stage protocol. Informally, there are n players, t of which may be *bad* and deviate from their prescribed instructions. One of the players, the *dealer*, possesses à value s as a secret input. In the first stage, the dealer commits to a unique value v (no matter what the bad players may do); moreover, $v = s$ whenever the dealer is honest. In the second stage, the already committed value v will be recovered by all good players (no matter what the bad players might do).

---

* Contact author. Email address: `rosario@theory.lcs.mit.edu`. Research supported by NSF grant no.9121466-CCR

PRIOR WORK. Several definitions and protocols for VSS have been proposed in the past ten years (E.g., [4, 1, 3, 7, 10].) We contend, however, that these notions and these protocols are of *very limited use*. In fact, their security concerns "begin when the dealer's secret is committed, and end when it is recovered." That is these notions do not concern themselves with what happens when VSS is used as a subprotocol. Because in many applications running a *single* VSS protocol is exactly what is wanted, these prior definitions and protocols are totally adequate in those scenarios. However, they are not adequate in more general scenarios since it is by now a well-known phenomenon that protocols that are secure by themselves, cease to be secure when used as a sub-protocols. In these cases the security of the entire protocol must be proven "from scratch" (for instance, this is the case in [7] where they use VSS as a tool to reach Byzantine agreement) rather than in a more natural and elegant "modular way." A notion of security that guarantees reducibility has been presented by Micali and Rogaway [9] for the problem of *function evaluation*, but not for general multiparty protocols.

OUR WORK. We extend reducibility-guaranteeing notions of security to verifiable secret sharing protocols and concretely exhibit VSS protocols that provably satisfy these notions. More precisely, in this paper we achieve the following goals:

1. We propose a new definition of VSS casted as a special istance of secure function evaluation.
2. We compare our new notion with the previously proposed ones, and show that it is *strictly and inherently stronger*.
3. We modify earlier VSS protocols of [1, 10] and show that our new protocol is secure according to our notion.
4. Finally we present some applications that use VSS as a subroutine and need our stronger notion of VSS to be secure. These include shared authentications and signatures and a very elegant proof of the security of the completeness theorem on multiparty secure computation by Ben-Or, Goldwasser and Wigderson. [1]

2 Prior work

In order to focus on the difficulties that are proper of VSS, we shall deal with a simple computational model, both when reviewing prior work and when presenting our new one.

COMPUTATIONAL MODEL. We consider n players communicating via a very convenient synchronous network. Namely, to avoid the use of Byzantine Agreement protocols we allow players to *broadcast* messages, and, in order to avoid the use of cryptography, we assume that each pair of players is connected by a *private* communication channel (i.e., no adversary can interfere with or have any access to messages between good players).

We model the corrupted processors as being coordinated by an *adversary* $\mathcal{A}$. This adversary will be *dynamic* (i.e., decides during the execution of the protocol which processors corrupt); *all-powerful:* (i.e., can perform arbitrarily long computations); and *completely-informed* (i.e., when corrupting a player she finds out all his computational history: private input, previous messages sent and received, coin tosses, etc.). Further, the adversary is also allowed *rushing* (i.e., in a given round of communication, bad players receive messages before the good ones and, based on those messages, the adversary can decide whom to corrupt next). We say that such an adversary is a t-adversary ($0 \leq t \leq n$) if t is an upper bound on the number of processors she can corrupt (t is also referred to as the *fault-tolerance* of the protocol.) This computational model is precisely discussed in [7] and [9].

PRIOR DEFINITIONS OF VSS. To exactly capture the informal idea of a VSS, has proven to be an hard task in itself. The definition reviewed below is that of [7], which relies on the notion of a *fixed event*:

Definition: We say that an event $\mathcal{X}$ is *fixed* at a given round in an execution E of a protocol, if $\mathcal{X}$ occurs in any execution E' of the protocol coinciding with E up to the given round.

Definition 1. Let P be a pair of protocols where the second is always executed after the first one, $P =$(Share-Verify, Recover). In protocol Share-Verify, the identity of the *dealer* is a common input to all players, and the secret is a private input to the dealer; the output of player P_i is a value $verification_i \in \{yes, no\}$. In protocol Recover, the input of each player P_i is his computational history at the end of the previous execution of Share-Verify; the output of each P_i is a string σ_i. We say P is a VSS protocol with fault-tolerance t if the following 3 properties are satisfied:

1. *Acceptance of good secrets:* In all executions of Share-Verify with a t-adversary $\mathcal{A}$ in which the dealer is good, $verification_i = yes$ for all good players P_i.
2. *Verifiability:* If less than t players output $verification = no$ at the end of Share-Verify then at this time a value σ has been fixed and at the end of Recover all good players will output the same value σ and moreover if the dealer is good $\sigma =$ the secret. [2]
3. *Unpredictability* In a random execution of Share-Verify with a good dealer and the secret chosen randomly in a set of cardinality m any t-adversary $\mathcal{A}$ won't be able to predict the secret better than at random i.e. if $\mathcal{A}$ outputs a number a at the end of Share-Verify then $Prob[a = s] = \frac{1}{m}$

SECURE COMPUTATION. Let us summarize the definition of secure function evaluation of [9]. Informally the problem is the following: n players $P_1, \ldots, P_n$, holding, respectively, private inputs $x_1, \ldots, x_n$, want to evaluate a vector-valued function f on their individual secret inputs without revealing them (more than already implied by f's output). That is, they want to compute $(y_1, \ldots, y_n) = f(x_1, \ldots, x_n)$ such that each player P_i will learn exactly y_i.

This goal is easily achievable if there is an external and trusted party, who privately receives all individual inputs and then computes and privately hands out all individual outputs. Of course, even in this *ideal* scenario, the adversary can create some problems. She can corrupt a player P_i before he gives his input x_i to the external party and change it with some other number $\tilde{x}_i$. And she can still corrupt players after the function has been evaluated and learn their outputs. These problems should, however, be regarded as *inevitable*. Indeed, following [8], [9] call a protocol for evaluating f secure if it

---

[2] Notice that if we simply ask in the Verifiability condition that "all the good players output the same number σ at the end of the Recover phase" it would not be sufficient for our purposes. In fact, we would still allow the adversary to decide during Recover what value σ the good players will output. Thus Share-Verify would not model a secret commitment as required.

approximates the above ideal scenario "as closely as possible." The nature of this approximation is informally summarized below.

Definition *(Initial configuration, traffic, input and output):* Let us define the following quantities within the context of a protocol P.

The *initial configuration* for P is a vector ic, whose ith component, $ic_i = (x_i, r_i)$ consists of the private input and the random tape of player P_i.

The *traffic* of player P_i in protocol P at round q, t_i^q, is the set of messages sent and received by P_i up to that round.

A *local input function* $\mathbf{I} = (I_1, \ldots, I_n)$ for P is an n-tuple of functions such that there exists a specific round r such that, by applying I_i to the traffic t_i^r, we get the input player P_i is "contributing to the computation." $\mathbf{I}(\mathbf{ic})$ will denote the vector of those values when P is run on initial configuration ic.

A *local output function* $\mathbf{O} = (O_1, \ldots, O_n)$ for P is an n-tuple of functions such that by applying O_i to the final traffic t_i^{final} of player P_i we get his output.

Definition *(Adversary view):* The *adversary view*, $VIEW_{Network}^{\mathcal{A}}$, during P is the probability distribution over the set of computational histories (traffic and coin tosses) of the bad players.

Definition *(Simulator and ideal evaluation oracle):* A simulator Sim is an algorithm that "plays the role of the good players". The adversary interacts with the simulator as if she was interacting with the network. The simulator tries to create a view for the adversary that is indistinguishable from the real one. He does this without knowing the input of the players, but it is given access to a special oracle called the *ideal evaluation oracle.* For a protocol P with local input function $\mathbf{I}$ evaluatable at round r, the rules of the interaction between Sim and the oracle are the following:

- if $\mathcal{A}$ corrupts player P_i before round r Sim gets from the oracle the input x_i and gives it to $\mathcal{A}$.
- at round r Sim gets the output y_i' for all the players corrupted so far, where $(y_1', \ldots, y_n') = f(x_1', \ldots, x_n')$ where $x_i' = x_i$ if P_i is still good, otherwise $x_i' = I_i(t_i^r)$
- if $\mathcal{A}$ corrupts a player P_i after round r, Sim gets from the oracle the pair (x_i, y_i') and gives it to $\mathcal{A}$.

Definition 2. *Secure Function Evaluation:* Let f be a vector-valued function, P a protocol, Sim a simulator, and $\mathbf{I}$ and $\mathbf{O}$ local input and output functions. We say that P securely evaluates the function f if

- *Correctness:* If ic is the initial configuration of the network, then
  1. $x_i = I_i(\mathbf{ic})$ for all good players P_i
  2. with high probability, $\mathbf{O}(\mathbf{t}^{final}) = f(\mathbf{I}(\mathbf{ic}))$
  (I.e. no matter what the adversary does, the function is evaluated during the protocol on some definite inputs defined by the local input functions over the traffic of the players. These inputs coincide with the original inputs for the good players)
- *Privacy:* For all initial configurations ic, if $VIEW_{Sim}^{\mathcal{A}}$ is the adversary view of the simulated execution of the protocol, we have that

$$VIEW_{Network}^{\mathcal{A}} \approx VIEW_{Sim}^{\mathcal{A}}$$

(I.e., the two views are statistically indistinguishable.)

There are many reasons for which this definition captures correctly the notion of a secure computation. In particular, the following one: the definition in [9] allows one to prove formally many desirable properties of secure protocols, the most interesting for us being *reducibility*:

Theorem 3 [9]. *Let f and g be two functions. Suppose there is a protocol P that securely evaluates f in the model of computation in which it can perform ideal evaluations of g. Suppose also that there is a protocol Q that securely computes g. Denote with P^Q the protocol in which the code for Q is substituted in P in the places where P ideally computes g. Then P^Q is secure.*

Interested readers are referred to the original paper [9] for a proof of this statement and a complete and a formal description of their definition.

3 Our definition of VSS

In this section we provide a new definition of VSS that guarantees reducibility. The key idea for achieving this property is to cast VSS in terms of secure function evaluation. Accordingly, we shall define two special functions SHAR and REC, and demand that both of them be securely evaluated in the sense of [9].

We assume a network of n players $P_1, \ldots, P_{n-1}$ and P_n, where $P_n = D$ the dealer. Let Σ be a set. Consider the vector space Σ^n and the following metric on it: given two vectors $\mathbf{a}, \mathbf{b}$ in Σ^n, let us define the distance between them as the number of components in which they differ; that is, $d(\mathbf{a}, \mathbf{b}) = |\{1 \le i \le n, a_i \ne b_i\}|$ We define the t-disc of $\mathbf{a}$ as the set of points at distance $\le t$ from $\mathbf{a}$ i.e. $disc_t(\mathbf{a}) = \{\mathbf{b} \in \Sigma^n : d(\mathbf{a}, \mathbf{b}) \le t\}$

We will define again VSS as a pair of protocols, called **Share-Verify** and **Recover**, that compute, respectively, two functions, SHAR and REC, satisfying the following properties. SHAR is the function we use to share the secret among the players. It is defined on the entire space for the $n-1$ players (their private input does not matter in this phase) and on two finite special sets R and S for the dealer. S is the space of possible secrets while R is a set of random strings. We will ask even after seeing any l shares $(l \le t)$ all secrets are equally likely to generate those shares. We call this property t-*uniformity* (see 2 below). Similarly REC is the function we use to reconstruct the secret. We will run it on the output of the previous phase. What we want is that we will be able to do so even if up to t components of the output of the sharing process are arbitrarily changed. We call this property t-*robustness* of the function REC (see 3 below).

Definition: Two functions SHAR and REC are a *sharing-reconstructing pair* with parameter t if they have the following properties:

1. (*Domain.*)
$$\text{SHAR} : \Sigma^{n-1} \times (R \times S) \to \Sigma^n$$
$$\text{REC} : \Sigma^n \to \Sigma^n$$

2. (*t-uniformity.*) $\forall l \le t$ there exists an integer n_l such that $\forall s_{i_1}, \ldots, s_{i_l} \in \Sigma$ and $\forall v_1, \ldots, v_{n-1} \in \Sigma$, $\forall s \in S$, and $\forall \mathbf{x} \in \Sigma^n$ such that $\forall j \in [1, l]$ $x_{i_j} = s_{i_j}$, there exist exactly n_l values $r_1, \ldots, r_{n_l} \in R$ such that for $i = 1, \ldots, n_l$,

$$\text{SHAR}(v_1, \ldots, v_{n-1}, r_i \circ s) = \mathbf{x}$$

3. (*t-robustness.*) $\forall\, v_1, \ldots, v_{n-1} \in \Sigma,\ \forall\, s \in S,\ \forall\, r \in R,$
   if $\mathbf{x} \in disc_t(\text{SHAR}(v_1, \ldots v_{n-1}, r \circ s))$, then

$$\text{REC}(\mathbf{x}) = (s, s, \ldots, s)$$

A VSS protocol will be composed by two protocols that *securely* evaluate these two functions; the second being evaluated over the output of the first.

Definition 4. A VSS protocol of fault-tolerance t is a pair of protocols (Share-Verify, Recover) such that

 - Share-Verify securely evaluates the function $\mathbf{y} = \text{SHAR}(x_1, \ldots, x_{n-1}, r \circ s)$,
 - Recover securely evaluates the function $\text{REC}(\mathbf{y})$, and
 - SHAR and REC are a sharing-reconstructing pair with parameter t.

Remarks: Though the above definition may appear "tailored on some specific VSS protocols," in the final paper we shall argue that it does not loose any generality. Also, as we shall see below, by demanding that both components (and particularly the second one) of a share-reconstructing pair be securely evaluated, we are putting an unusually strong requirement on a VSS protocol. But it is exactly this requirement that will guarantee the desired reducibility property.

4 Comparison with previous definitions of VSS

Let us compare now Definition 4 and Definition 1, our token example of prior VSS definitions. To begin with, there is a minor syntactical difference between the two definitions: according to Definition 1 when good players find out the dealer is bad they just stop playing and output *verification* = *no*. In our new definition instead the computation goes on, no matter what. This discrepancy can be eliminated by having protocols in the first definition agree on a default value when the dealer is clearly bad and protocols in the second definition always output *verification* = *yes* at the end of Share-Verify (since we are dealing with a secure function evaluation, we are guaranteed that all good players will output a common value). With these minor changes we can prove the following (the proof can be found in the appendix):

Theorem 5. *If P is a VSS protocol of fault-tolerance t satisfying Definition 4, then P is also a protocol of fault-tolerance t satisfying Definition 1.*

Now the natural question to ask is: Are Definitions 3 and 1 equivalent? That is, if a given VSS protocol P' satisfies Definition 1, does it also satisfy Definition 3? The answer to this important question, provided by the following Theorem 3, is NO. And it better be that way if we want to preserve reducibility of VSS protocols. Indeed as we will see later the formal specifications of Definition 1 are not sufficient to guarantee composition of VSS protocols inside larger protocols.

Theorem 6. *Definition 4 is strictly stronger than Definition 1, that is, there are VSS protocols satisfying Definition 1, but not Definition 4.*

The proof of this theorem (see the appendix for a detailed proof) is based on the fact that usually VSS protocols (consider for example the one in [1]) reconstruct the secret by having each player distributing his own share to all other players. This does satisfy

Definition 1 since there we do not put any requirement on the secrecy of the shares. But this does not satisfy Definition 4 since doing so we do not compute *securely* the function REC. The problem is that the players reveal too much information about their own input share during the protocol. In other words we want that, when we compute REC($\mathbf{y}$) over $\mathbf{y} = $ SHAR($v_1, \ldots, v_{n-1}, r \circ s$), no knowledge about $\mathbf{y}$ should leak except the secret s (including no extra information about the secret s itself). The rationale for asking this is again the fact that we want our VSS protocols to be secure not just by themselves but when used inside subroutines of more complex protocols. Leaking knowledge about the shares (or extra knowledge about the secret) may create problems to the security of the overall protocol. Consider the following example.

Consider a VSS protocol P, satisfying Definition 1, in which the secret is a 3-colorable graph. During the **Recover** protocol the graph is reconstructed together with a 3-coloring of it kindly provided by the dealer. Notice that Definition 1 is not violated, but notice also that an adversary gains from the execution of such a protocol some knowledge about the secret she could not obtain by herself. This in turns means that there exists no simulator for this protocol and so that Definition 4 cannot be satisfied. And the serious problem with P is that, if used inside a larger protocol in which it is crucial that the knowledge of that particular 3-coloring stays hidden, P, though "secure" as a VSS protocol on its own, jeopardizes the security of the larger protocol.

This problem is solved by our definition substituting property 3 (unpredictability) with a stronger one based on zero-knowledge and simulatability, which is exactly what we do by asking for a secure computation of the functions SHAR and REC. In particular the secure computation of the function REC is the most important difference between the two definitions. In particular we want to stress the importance of maintaining the secrecy of the shares of each individual good player. In the appendix we will show that protecting the secrecy of the shares is necessary for some applications in which VSS is used as a subroutine inside some specific protocols.

Probably one of the reasons this point may appear somewhat moot is that in Shamir's secret sharing scheme [11] the shares consist of the value of a polynomial of degree t with free term s. For a t-adversary who corrupts exactly t players, knowing the secret is equivalent to knowing the shares of all players. In fact, knowing the t shares of the corrupted players and the secret at the end of **Recover**, she has $t + 1$ points of the t-degree polynomial, and by evaluating the so inferred polynomial at the names of all good players, she easily computes all shares. However, we object that what happens to be true for the VSS protocols based on Shamir's scheme, may not be true for all VSS protocols [3]. And one should not "wire in" a general definition what happens to be true in a specific case. Moreover, even in Shamir-based VSS protocols, if the polynomial has degree strictly larger than the number of corrupted players, then it is no longer true that for the adversary knowledge of the secret is equivalent to knowledge of all shares. Indeed, as we will show later this is the crucial point in our application protocols described in the appendix. *It is thus needed that the knowledge gainable by an adversary at the end of a secure VSS protocol exactly coincides with the original secret whenever the dealer is honest.*

---

[3] For example consider Blakley secret sharing scheme [2] in which the secret is a point in a $t + 1$-dimensional space and shares are random hyperplanes passing through that point.

5 A VSS protocol that satisfies our definition

In this section we will exhibit a VSS protocol satisfying our definition and of fault-tolerance $\frac{n}{3} - 1$, by modifying an older protocol of Ben-Or, Goldwasser, and Wigderson [1]. The modification actually occurs only in the **Recover** part, and uses techniques also developed by [1], but within their "computational protocol" rather than in their VSS protocol. We have also found a protocol of fault-tolerance $\frac{n}{2} - 1$ based on Tal Rabin's protocol [10], but we will not describe it here for space limitations.

Suppose we are dealing with a t-adversary $\mathcal{A}$. Let $n = 3t + 4$ and $P_1, \ldots, P_{n-1}$, $P_n = D$ be the set of players, D being the dealer. We will make all our computations modulo a large prime $p > n$. It is known from the error-correcting codes theory that if we evaluate a polynomial f of degree $t + 1$ over the $n - 1$ different points i for $i = 1, \ldots, n - 1$ then given the sequence $s_i = f(i)$ we can reconstruct the coefficients of the polynomial in polynomial time even if up to t elements in the sequence are arbitrarily changed. This is the well known Berlekamp-Welch variant of the Reed-Solomon error-correcting code. For details readers can refer to a standard text like [12]. Let K be a security parameter. With K/n we mean $\lceil \frac{K}{n} \rceil$.

The protocol appears in the boxes. The **Share-Verify** part is identical to the one in [1]. The **Recover** protocol is modified with respect to the one in [1] in order to make it a secure computation of the function REC. The basic idea is that each player will distribute his share to all the other players, but covering it up appropriately with some randomness so that no information about the share is revealed but the secret reconstruction process is not compromised.

Theorem 7. *The protocol $P = (\text{Share} - \text{Verify}, \text{Recover})$ is a VSS protocol according to Definition 4 with fault-tolerance $\frac{n}{3} - 1$*

Remark: A completely error-free version of this protocol can be obtained as in [7] by running a different zero-knowledge proof that the shares lie on a single polynomial. The proof uses a bivariate polynomial and it is out of the scope of this paper. Details can be found in [1, 7]. Notice that this is the first time a formal proof of the security of the protocol of [1] appeared.

Remark: Notice how, assuming the dealer is honest, at the end of the **Recover** phase the adversary, even knowing the secret s and t shares of the corrupted players, knows nothing about the shares of the honest players (since the polynomial is of degree $t + 1$ and to interpolate it $t + 2$ points are needed).

Protocol Share-Verify from [1]

1. The dealer chooses a random polynomial $f_0(x)$ of degree $t + 1$ with the only condition that $f_0(0) = s$ his secret. Then he sends to player P_i the share $s_i = f_0(i)$. Moreover he chooses $2K$ random polynomials $f_1, \ldots, f_{2K}$ of degree $t + 1$ as well and sends to P_i the values $f_j(i)$ for each $j = 1, \ldots, 2K$.
2. Each player P_i broadcasts K/n random bits $\alpha_{(i-1)K/n+j}$ for $j = 1, \ldots, K/n$
3. The dealer broadcasts the polynomials $g_j = f_j + \alpha_j f_0$ for all $j = 1, \ldots, K$
4. Player P_i checks if the values he holds satisfy the polynomials broadcast by the dealer. If he finds an error he broadcasts a complaint. If more than t players complain then the dealer is faulty and all players assume the default zero value to be the dealer's secret.
5. If less than t players complained the dealer broadcasts the values he sent in the first round to the players who complained.
6. Each player P_i broadcasts K/n random bits $\beta_{(i-1)K/n+j}$ for $j = 1, \ldots, K/n$
7. The dealer broadcasts the polynomials $h_j = f_{K+j} + \beta_j f_0$ for all $j = 1, \ldots, K$
8. Player P_i checks if the values he holds and the values broadcast by the dealer in round 5 satisfy the polynomials broadcast by the dealer. If he finds an error he broadcasts a complaint. If more than t players complain then the dealer is faulty and all players assume the default zero value to be the dealer's secret.

Protocol Recover (modified)

1. Each player P_i chooses a random polynomial h_i of degree $t + 1$ such that $h_i(0) = s_i$ his own input share. He sends to player P_j the value $h_i(j)$
2. Each player P_i chooses random polynomials $p_i(x), q_{i,1}(x), \ldots, q_{i,2K}(x)$ of degree $t + 1$ and with free term 0. He sends to player P_j the values $p_i(j), q_{i,1}(j), \ldots, q_{i,2K}(j)$
3. Each player P_i broadcasts K random bits $\gamma_{l,(i-1)K/n+m}$ for $l = 1, \ldots, n$ and $m = 1, \ldots, K/n$
4. Each player P_i broadcasts the following polynomials $r_j = q_{i,j} + \gamma_{i,j} p_i$ for each $j = 1, \ldots, K$
5. Each player P_i checks that the information player P_l sent him in round 1 is consistent with what player P_l broadcast in round 3. If there is a mistake or P_l broadcast a polynomial with non-zero free term P_i broadcasts bad_l. If there are more than t players broadcasting bad_l, player P_l is disqualified and all the other players assume 0 to be P_l's share. Otherwise P_l broadcasts the information he sent in round 1 to the players who broadcast bad_l
6. Each player P_i broadcasts K random bits $\delta_{l,(i-1)K/n+m}$ for $l = 1, \ldots, n$ and $m = 1, \ldots, K/n$
7. Each player P_i broadcasts the following polynomials $r_j = q_{i,K+j} + \delta_{i,j} p_i$ for each $j = 1, \ldots, K$
8. Each player P_i checks that the information player P_l sent him in round 1 and broadcast in round 5 is consistent with the polynomials player P_l broadcast in round 7. If there is a mistake or P_l broadcast a polynomial with non-zero free term P_i broadcasts bad'_l. If more than t players broadcast bad'_l then P_l is bad and all players assume his share to be 0.
9. Each player P_i distributes to all other players the following value $s_i + p_1(i) + p_2(i) + \ldots + p_n(i)$ then interpolates the polynomial $F(x) = f_0(x) + p_1(x) + p_2(x) + \ldots + p_n(x)$ using the error correcting algorithm of Berlekamp and Welch. The secret will then be $s = F(0) = f(0)$.

6 Conclusion

In the past cryptographic schemes and protocols used to be considered secure until not broken. Due to the increasing use and importance of cryptography, this approach is no more acceptable. To call a protocol *secure* we need a proof of its security. This means that we need definitions and methods to be able to prove security.

Following this philosophy we have presented a new and stronger definition for one of the most important cryptographic protocols: Verifiable Secret Sharing. We argued that this definition is the correct one especially when VSS is to be used as a sub-protocol inside larger protocols (which is probably the most common case for VSS). We also presented a protocol which provably satisfies our new definition. Finally some applications of this new protocol (and of our new definition of VSS) are described in the appendix.

References

1. Michael Ben-Or, Shafi Goldwasser, and Avi Wigderson. Completeness theorems for non-cryptographic fault-tolerant distributed computation. In *20th ACM Symposium on Theory of Computing*, pages 1–10, 1988.
2. G.R. Blakley. Safeguarding cryptographic keys. In *National Computer Conference*, pages 313–317, 1979.
3. David Chaum, Claude Crepeau, and Ivan Damgard. Multiparty unconditionally secure protocols. In *20th ACM Symposium on Theory of Computing*, pages 11–19, 1988.
4. Benny Chor, Shafi Goldwasser, Silvio Micali, and Baruch Awerbuch. Verifiable secret sharing and achieving simultaneity in the presence of faults. In *26th IEEE Symposium on Foundations of Computer Science*, pages 383–395, 1985.
5. Yvo Desmedt and Yair Frankel. Shared generation of authentication and signatures. In *CRYPTO '91*, Lecture Notes in Computer Science, pages 457–469. Springer-Verlag, 1991.
6. Yvo Desmedt, Yair Frankel, and Moti Yung. Multi-receiver/multi-sender network security: efficient authenticated multicast/feedback. In *INFOCOM*, pages 2045–2054, 1992.
7. Paul Feldman and Silvio Micali. An optimal probabilistic protocol for synchronous byzantine agreement. In *20th ACM Symposium on Theory of Computing*, 1988.
8. Oded Goldreich, Silvio Micali, and Avi Wigderson. How to play any mental game. In *19th ACM Symposium on Theory of Computing*, pages 218–229, 1987.
9. Silvio Micali and Philip Rogaway. Secure computation. In *CRYPTO '91*, Lecture Notes in Computer Science. Springer-Verlag, 1991. Current version available from the authors.
10. Tal Rabin and Michael Ben-Or. Verifiable secret sharing and multiparty protocols with honest majority. In *21st ACM Symposium on Theory of Computing*, 1989.
11. Adi Shamir. How to share a secret. *Communications of the ACM*, 22(11):612–613, 1979.
12. W.Peterson and E.Weldon. *Error Correcting Codes*. MIT Press, second edition, 1972.

Appendix

This appendix is dedicated to the proofs of some of the statements in the the paper.

Proof of Theorem 5

Let P be a VSS protocol of fault-tolerance t satisfying Definition 4. Let ic be the initial configuration vector of the network.

P satisfies the Verifiability property of Definition 1. Indeed at the end of the Share-Verify phase, let $\sigma = I_n(ic_n)$ i.e. the value contributed by the dealer to the computation. This value is *fixed* at the end of Share-Verify since it is a function of the traffic of the dealer. Moreover because of the t-robustness of the function REC we have the same value σ will be output by all good players at the end of the Recover part. Indeed because of the t-robustness property it does not matter that t bad players may change their input before computing the function REC. Finally, if the dealer is good then $\sigma = I_n(ic_n)$ is equal to the secret s.

P also satisfies the Unpredictability property of Definition 1. First notice that the t-uniformity property implies that the output of the function SHAR is composed of t-wise independent uniformly distributed random variables and so it is impossible (in an information-theoretic sense) to predict the secret better than at random for any algorithm that has knowledge of only $l \leq t$ components of such output. However a t-adversary has not only that knowledge but she also has a view of the entire protocol (i.e. traffic of bad players and messages broadcast by good players). But here is where secure computation comes to our rescue. Because of the security of the evaluation of the function SHAR the adversary can create the entire view by herself using the simulator, and so basically the other information is irrelevant. More precisely let's assume for the sake of contradiction that protocol P does not satisfy the Unpredictability condition. This means that there is an adversary $\mathcal{A}$ that does not corrupt the dealer and such that $Prob[\mathcal{A}$ guesses $s] > \frac{1}{m}$ where $m = |S|$ is the cardinality of the space of possible secrets. We will exhibit an algorithm $\mathcal{G}$ (for guesser) that guesses the secret with the same probability but having access only to t components of the output of the function SHAR. Since Share-Verify is a protocol that securely evaluates the function SHAR it must have a simulator Sim. $\mathcal{G}$ runs Sim with the adversary $\mathcal{A}$ and uses $\mathcal{A}$'s guess for the secret as his guess. Since the simulated view of the protocol is indistinguishable from the real one, $\mathcal{A}$ will guess the correct secret with probability bigger than $\frac{1}{m}$. Notice that by running the simulator $\mathcal{G}$ will only know at most t components of the output of the function SHAR, the ones which will be answered by the oracle in response to request of corruption by $\mathcal{A}$. So we have a contradiction since we found an algorithm that predicts the secret better than at random with knowledge of only t components of the output of SHAR, which contradicts the t-uniformity property. □

Proof of Theorem 6

The proof of this theorem is based on the fact that usually VSS protocols (consider for example the one in [1]) reconstruct the secret by having each player distributing his own share to all other players. This does satisfy Definition 1 since there we do not put any requirement on the secrecy of the shares. But, as the following lemma shows, this does not satisfy Definition 4 since doing so we do not compute *securely* the function REC.

Lemma 8. *The protocol P in which at round 1 every player broadcasts or distributes his own share is not a secure computation of the function REC.*

Proof. P is a 1-round protocol. Consider the following t-adversary $\mathcal{A}_1$. $\mathcal{A}_1$ corrupts t players (say $P_1, \ldots, P_t$) right at the beginning of the protocol. So in the simulated execution the simulator Sim will receive from the oracle their inputs (shares) $x_1, \ldots, x_t$. Now Sim has to simulate the broadcast messages of the good players. No matter what *any* Sim does, there are only two possibilities:

- either the shares broadcast by Sim do not interpolate a secret
- or they do interpolate a secret s', but since Sim has no knowledge at this point of what the "true" secret s is, only with probability $\frac{1}{|S|}$ $s = s'$

In both cases however the simulated execution is distinguishable from the real one to $\mathcal{A}_1$ and so protocol P is not secure according to definition 2.

The above proof is sufficient to show that protocol P is not secure. However let us describe an alternative proof of this statement, based on a different adversary. This will allow us to exemplify a different problem with protocol P. This will help the reader understand better the modification we will do to the Recover protocol in order to achieve security for VSS protocols.

Proof. (alternative proof of Lemma 8)
Consider the following t-adversary $\mathcal{A}_2$. $\mathcal{A}_2$ corrupts $t - 1$ players (say $P_1, \ldots, P_{t-1}$) right at the beginning of the protocol. So in the simulated execution the simulator Sim will receive from the oracle their inputs (shares) $x_1, \ldots, x_{t-1}$. Now Sim has to simulate the broadcast messages of the good players. Sim does so and broadcasts $x'_t, \ldots, x'_n$. As before no matter what *any* Sim does, there are only two possibilities:

- either the shares broadcast by Sim do not interpolate a secret (in this case, as above, the simulation is already a bad one)
- or they do interpolate a secret s',

At this point $\mathcal{A}_2$ corrupts her last player (say P_t) and so gets from the oracle the true pair (x_t, s) and only with probability $\leq \frac{1}{|S|}$ we have that $x'_t = x_t$ and $s = s'$ and so the simulated execution will not succeed in convincing the adversary that she is talking with the real network, hence P is not secure according to definition 2.

Proof of Theorem 7

Let SHAR be the following function:

$$\text{SHAR}(v_1, v_2, \ldots, v_{n-1}, r \circ s) = (s_1, s_2, \ldots, s_{n-1}, \epsilon)$$

with $s_i = f_0(i)$ where $f_0(x) = s + a_1 x + \ldots + a_{t+1} x^{t+1}$ and $r = a_1 \circ \ldots \circ a_d$ (i.e. the polynomial is created using the coin tosses r of the dealer). Then we can state that

Lemma 9. *Protocol* Share-Verify *securely evaluates the function* SHAR *according to Definition 2.*

Proof. If we look back at Definition 2 we see that we have to check that both conditions, correctness and privacy, are satisfied in the *blended* way they interact in the simulation process through the common input and output functions.

So first of all let's define these functions for our protocol. Remember that with t_i we define the traffic of player P_i. Clearly for all players P_i $i \leq n$ the input function always returns the empty string, $I_i(t_i) = \epsilon$, since the players do not contribute any input during the computation of the function SHAR. For the dealer, $D = P_{n+1}$, the input function is a little bit more complicated. Let us denote with m_i the message the dealer broadcast to player P_i in round 5 if P_i complained in round 4, or the message the dealer sent to player P_i in round 1 if P_i did not complain. Then $I_D(t_D) = f(0)$ where $f = BW(m_1, \ldots, m_n)$ is the $t + 1$-degree polynomial resulting from the Berlekamp-Welch interpolation of the m_i's. The output function is simpler: $O_i(t_i) = m_i$ (where $m_D = \epsilon$). Now we can check both conditions.

Correctness: First we have to prove that for all good players P_i, $I_i(t_i)$ is equal to the correct input. In our specific case this has to be checked only for the dealer. If the dealer is good, $m_i = f(i)$ where f is a $t + 1$-degree polynomial with free term s the secret. So

$I_D(t_D) = s$ if the dealer is good. The second correctness condition is that with high probability $O(t) = \text{SHAR}(I(t))$. In our case this means that with high probability the values m_t held by good players must be on a single polynomial of degree $t + 1$. This is true with probability $\geq 2^{-\frac{2K}{3}}$ since at least $\frac{2K}{3}$ bits are chosen truly randomly by good players in rounds (2) and (6). Each bit represents a "question" that a bad dealer who distributed bad shares will be able to answer correctly in the following round only with probability $\frac{1}{2}$ (i.e. if he predicted the bit correctly when he distributed the shares). Hence the bound on the probability of error.

Privacy: We have to exhibit a simulator for the protocol. We distinguish 2 cases:

Case A: The dealer is corrupted before round 1. Then the simulator will just follow the instructions of the players, with the only exception that it will turn them over to the adversary in case of corruption. Since the players do not contribute any input to the computation this will reduce the simulated execution to one of VSS with a bad dealer. So the simulation will be indistinguishable to the eyes of the adversary.

Case B: The dealer is not corrupted before round 1. Then the simulator in round 1 will just create a random "fake" secret s' and will share it to the players according to the protocol instructions with a polynomial f'. If the dealer is not corrupted at all during the protocol then everything will run smoothly since to the eyes of the adversary the execution will look like an ordinary VSS with a good dealer (again this is true because the players do not contribute any input to the computation). If the dealer is corrupted after round 1 however the adversary and the simulator will get from the oracle the true input s of the dealer. At that point the simulator turns over the control of the dealer to the adversary, but changes the polynomial used to share the secret to a new polynomial f'' such that $f''(0) = s$ and $f''(i) = f'(i)$ for all players P_i that were corrupted so far by the adversary. The simulator changes accordingly the random polynomials f_t used for the zero-knowledge proof to make them consistent with whatever has been broadcast so far. The simulator can always do this since the adversary has at most t points of a $t + 1$-degree polynomial. For the rest of the simulation the simulator will use the polynomial f'' for the computation of the good players still under his control. We claim that this execution is indistinguishable to the adversary from a real one. This is so because the only thing different from a true execution is the fact that the shares the adversary gets before corrupting the dealer are created using a different polynomial than the real one, but thanks to the properties of polynomials this is not a problem for the simulator once the dealer is corrupted.

Let REC be the function

$$\text{REC}(s_1, \ldots, s_{n-1}, \epsilon) = (s, \ldots, s, s)$$

where s is the result of the Berlekamp-Welch "interpolation" of the s_i.

Lemma 10. *Protocol* Recover *securely evaluates the function* REC *according to Definition 2.*

Remark: Before embarking in the formal proof of this lemma let us give some intuition of why this is true. For example notice how the adversaries $\mathcal{A}_1$ and $\mathcal{A}_2$ that we described in the proof of Lemma 8 are helpless with the new protocol Recover-A. We added round 1 so that the simulator can learn the true secret *before* the shares have been given out publicly and this takes care of an adversarial attack like $\mathcal{A}_1$. Moreover we added the "masking" polynomials p_i so that the players reveal shares of a random polynomial F whose only property is that $F(0) = s$, so while reconstructing the secret no information is revealed about the true input shares; this solves the problem raised by adversary $\mathcal{A}_2$.

Proof. As in the proof of Lemma 9 we start by defining the input and output functions of our protocol. The input function I_i of player P_i is defined as follow: let $m_{i,j}$ be the message P_i sends to player P_j at round 1; $I_i(t_i) = h_i(0)$ where $h_i = BW(m_{i,1}, \ldots, m_{i,n})$ is the $t + 1$-degree polynomial resulting from the Berlekamp-Welch interpolation of the $m_{i,j}$'s (if there is no such polynomial then assume $I_i(t_i) = 0$). The output function is the following: let M_i be the message broadcast by player P_i at round 9; $O_i(t_i) = F(0) = s$ where $F = BW(M_1, \ldots, M_n)$ is the $t + 1$-degree polynomial resulting from the Berlekamp-Welch interpolation of the M_i's.

Correctness: It is clear that for all good players $I_i(t_i) = s_i$ the correct input share. Then we have to check that at least with high probability $O(t) = \text{REC}(I(t))$. In our case this means to prove that

$$\text{REC}(M_1, \ldots, M_n, \epsilon) = \text{REC}(s_1, \ldots, s_n, \epsilon)$$

Now this equation is not satisfied if one of the following things happens:

- either a bad player P_l succeeds in sharing random "garbage" instead of the values $p_l(j)$ in round 2 (in this case the M_i's will not interpolate a polynomial)
- or P_l does distribute $p_l(j)$ in round 2 but manages to use a polynomial with free term different than zero (in this case the M_i's will reconstruct a different secret)

Since the sharing process is exactly identical to the one of the protocol Share-Verify, we already know that P_l succeeds in any of the two cases only with probability $2^{-\frac{2K}{3}}$. So since there are at most $\frac{n}{3}$ bad players, the probability that the protocol computes an incorrect output is at most $\frac{n}{3}2^{-\frac{2K}{3}}$ which for K large enough is exponentially small.

Privacy: We have to exhibit a simulator and then prove that the simulation is indistinguishable from the true network execution. Consider the following simulator Sim_R:

1 At round 1, Sim_R simulates player P_i by choosing a random polynomial h_i' of degree $t + 1$ and sending $h_i'(j)$ to P_j. At this point the simulator is allowed to receive from the oracle the output of the function, so Sim_R will learn the true secret s. If some player P_l is corrupted by the adversary $\mathcal{A}$ at the end of this round (or in the following rounds), then both Sim_R and $\mathcal{A}$ learn the true share s_l and Sim_R has to change the polynomial h_i' accordingly so that $h_i'(0) = s_l$ but without changing its value on points already known to the adversary. Sim_R can always do this because the adversary has at most t points of a $t + 1$ degree polynomial.

2-8 During rounds 2 to 8 the simulator just follows plainly the instructions of the players. Since what players do in these rounds is completely random and not related to their inputs, Sim_R will always be able to create an indistinguishable view.

9 Finally at round 9, Sim_R chooses a polynomial g of degree $t + 1$ such that $g(0) = s$ and then for each player P_i Sim_R broadcasts $g(i) + p_1(i) + \ldots + p_n(i)$ where p_j is the polynomial distributed by player P_j during rounds 2-8 of the simulation. The Reed-Solomon interpolation of these values will give as result s. If a player P_l is corrupted at the end of this round, then both Sim_R and $\mathcal{A}$ will learn from the oracle the true input share s_l. If $s_l \neq g(l)$ then Sim_R just changes the value of p_l at the point l so to make the entire sum consistent with what broadcast.

The simulation is indistinguishable from a real execution to the eyes of the adversary. In fact as we already said, in round 2-8 all messages are random and unrelated to the input so the simulator can easily play the role of the good players. In round 1 the adversary sees at most only t shares of the real input of a good player. Because of the property of Shamir secret sharing scheme, these shares are completely random and so can be simulated even with no knowledge of the real input (as in the case of the simulator). In round 9 the real share is broadcast "hidden" by some random "garbage", this will allow the simulator to broadcast the message of a good player with the right distribution even without knowing the real input.

Applications

Unfortunately there is no space in this extended abstract for a detailed description of the applications. However let us briefly sketch some of them.

A PROOF FOR THE BGW PROTOCOL: In the final paper we will describe what is probably the nicest feature of our new definition and new protocols for VSS. We will present a very simple and "modular" proof for the theorem of Ben-Or, Goldwasser and Wigderson that any function can be computed securely with fault-tolerance of $\frac{n}{3}$. This result first appeared in a STOC abstract without a formal proof [1]

They claim that it is enough to prove that it is possible to compute securely addition and multiplication, since any function F can indeed be reduced to an arithmetic circuit whose nodes are indeed just addition and multiplication. By repeatingly using these 2 protocols we can then compute securely the entire function. Notice however that this line of reasoning gives for granted the reducibility property for secure protocols. However at that time a satisfactory definition of security had not been presented and the reducibilty theorem was generally considered true but never proven.

However we stand on the privileged position of having had these tasks completed for us by [9]. This will allow us to present a simple and modular proof for the theorem of [1]. Notice that in order to use the reducibility theorem we *need* to use our new notion and protocols for VSS. This does not mean that the theorem of [1] is wrong as it is, but that in order to have our simpler and modular proof their VSS protocol is not appropriate.

SHARED AUTHENTICATIONS AND SIGNATURES: A first idea would be to use these shares for *authentication* purposes. More precisely after being shared among the players, the secret is given to some *authenticators*. At a certain point if a subset of the players (of opportune size) wants to prove their identity to or sign a message for an authenticator they just recover the secret together with her and if the reconstruction succeeds then the authenticator knows she is dealing with the right people. However since the shares are never revealed the authenticator (or any of the players) will never be able to prove herself as someone else.

The idea of shared authentication and signatures appears in [5, 6]; we will briefly describe their scheme here. All operations (as usual) are modulo a big public prime p. A trusted key distribution center generates two random polynomials $Q_0(x)$ and $Q_1(x)$ of degree t and shares them among the players, i.e. player P_i receives $Q_0(i)$ and $Q_1(i)$. The authenticators receive $Q_0(0)$ and $Q_1(0)$. Later suppose players $P_{i_1}, \ldots, P_{i_s}$ ($s \geq t$) want to sign a message M for the authenticator. P_{i_j} broadcasts the message $Q_0(i_j) + MQ_1(i_j)$. The authenticator interpolates the broadcast shares into a polynomial F and checks that $F(0) = Q_0(0) + MQ_1(0)$. This protocol has two problems:

(1) If a player wants to jam the signature process all he has to do is to broadcast garbage (they solve it assuming players to be honest).

(2) It is a one-time scheme since after signing two different messages M_1 and M_2 the secret key of player P_i is easily computable.

Using our VSS Recover protocol to reconstruct the secret $Q_0(0) + MQ_1(0)$ we will solve both problems at once. Indeed bad shares contributed by players will be detected and eliminated. Moreover the share broadcast by player P_i betrays nothing of the true secret $(Q_0(i), Q_1(i))$ and so even after numerous signatures no information leaks about it; so we can do without the trusted combiner.

Finally notice that since usually the key management center is trusted, we do not need the verification process during the Share-Verify part making that part much simpler.

Efficient Secret Sharing Without a Mutually Trusted Authority

Extended Abstract

Wen-Ai Jackson, Keith M. Martin* and Christine M. O'Keefe*

Department of Pure Mathematics, The University of Adelaide, Adelaide SA 5005,
Australia

Abstract. Traditional secret sharing schemes involve the use of a mutually trusted authority to assist in the generation and distribution of shares that will allow a secret to be protected among a set of participants. In contrast, this paper addresses the problem of establishing a secret sharing scheme for a given access structure *without* the use of a mutually trusted authority. A general protocol is discussed and several implementations of this protocol are presented. The efficiency of these implementations is considered. The protocol is then refined and constructions are presented for mutually trusted authority free threshold schemes.

1 Introduction

A *secret sharing scheme* is a method by which a *secret* can be protected among a group of *participants*. Each participant holds a private *share* of the secret. Only certain sets of participants (*authorised sets*) are desired to be able to reconstruct the secret from their respective pooled shares. The collection of these subsets is the *access structure* of the secret sharing scheme. For the purposes of most of this paper, sets of participants that are not in the access structure (*unauthorised sets*) will not be able to determine any more information about the secret than is known publicly. Such schemes are often referred to as being *perfect*.

It is natural to make the assumption that if a set A of participants contains a subset that belongs to an access structure then A is itself in that access structure. We call access structures with this property *monotone*. A secret sharing scheme on n participants in which precisely all subsets of size at least k ($1 \leq k \leq n$) are in the access structure is known as a (k, n)-*threshold scheme* (we say that the access structure is (k, n)-*threshold*). Threshold schemes were the first types of secret sharing scheme proposed ([2, 10]).

We make a subtle distinction between two types of secret that can be protected by a secret sharing scheme. A secret is said to be *explicit* if it takes a fixed value that is predetermined by factors outside the secret sharing scheme design. In other words, the scheme is designed to protect a particular predetermined

---

* This work was supported by the Australian Research Council

L.C. Guillou and J.-J. Quisquater (Eds.): Advances in Cryptology - EUROCRYPT '95, LNCS 921, pp. 183-193, 1995.
© Springer-Verlag Berlin Heidelberg 1995

number within a given domain. This might be, say, a bank account number, the number of a security box or an enabling code. A secret is said to be *implicit* if it does not take a predetermined value. In this case the secret sharing scheme must protect a secret, but the value of the secret can be *any* number within a specified domain. A secret sharing scheme is more likely to have an implicit secret either in a situation where there is no obvious number associated with the secret, such as when the scheme is to be used to demonstrate concurrence in an access control protocol, or in a situation where the implicit secret value is *subsequently* adopted as, say, a cryptographic key or the number of a secure vault. In these situations the implicit secret, and/or the shares that generate it, must be part of the initialisation of the device that verifies the secret. For instance, an application was described in [6] where the shares of an implicit secret were manually incorporated into the initialisation process of the locking mechanism of a vault door.

Traditional models for secret sharing schemes rely on the existence of a *Mutually Trusted Authority* (*MTA*) to set up the scheme in the first place. This authority must be trusted by all the participants and can either be human (perhaps an organisation) or be a device. If the secret is explicit then the MTA is trusted with the knowledge of the explicit secret and with the generation and distribution of suitable shares that relate to the secret in question. In the case of an implicit secret, the MTA is further responsible for the generation of the implicit secret that is to be shared among the participants of the scheme.

We study here secret sharing schemes that do *not* require the existence of an MTA during their set-up protocols. We will thus refer to such schemes as being *MTA-free*. In an MTA-free scheme the participants generate their own shares. The MTA-free schemes that we consider all have implicit secrets. Unless there is a singleton participant set in the access structure of a secret sharing scheme, it does not seem very likely that a protocol can be devised which allows a group of participants to generate shares to protect an explicit secret. If there is a singleton participant set in the access structure then, since that participant effectively knows the secret directly from their share, that participant could (in theory) play the role of an MTA and generate shares of the (explicit) secret for the other participants. Indeed, a traditional secret sharing scheme can be thought of as a secret sharing scheme of this type where the MTA is an extra participant, in the access structure as a singleton set.

We note first that there does exist one family of monotone access structures which can be easily realised by MTA-free secret sharing schemes. A (*unanimous*) (n, n)-threshold scheme can be constructed without an MTA, as follows. Let w be a fixed positive integer.

- Each participant chooses a (random) share from $\mathbf{Z}_w$;
- The (implicit) secret is the sum of the participants' shares modulo w.

The first paper to consider constructions of more general MTA-free schemes was by Meadows [9]. In this novel paper a (k, n)-threshold scheme was proposed which allows the first k participants to generate their own (random) shares. Unfortunately a 'black box' is then required to generate the shares of the remaining

$n - k$ participants. This black box must be trusted with the knowledge of all the shares and with the value of the (implicit) secret. Thus by our definition the black box is playing the role of an MTA. The only possible advantage of this protocol is that the value of the implicit secret is directly determined from the shares chosen by the first k participants. However this does not appear to be much different from a scheme set up by a (device-based) MTA that selects the implicit secret using a random number generator.

In 1991 Ingemarsson and Simmons [6] reconsidered the design of MTA-free schemes for general monotone access structures and suggested an elegant protocol. The basic idea of [6] is that the n participants first generate shares of an (MTA-free) unanimous (n, n)-threshold scheme. The implicit secret of this unanimous scheme becomes the secret of the final scheme. Each participant then acts as their own MTA and sets up a private secret sharing scheme to protect their share of the unanimous scheme among a number of the other participants. Thus a participant's share in the unanimous scheme becomes the explicit secret of their private secret sharing scheme. It is quite possible that after carrying out this protocol, no participant will actually know the access structure of the induced secret sharing scheme. In [6] it is suggested that this procedure has the potential to realise an MTA-free scheme for any monotone access structure. We will later prove this suggestion to be correct.

The main aim of this paper is to start with a monotone access structure Γ, and determine which initial MTA-free scheme and which private secret sharing schemes should be used in order to realise an MTA-free scheme for Γ. There is not necessarily a unique way of doing this and so we are particularly interested in trying to find efficient and economical methods. These are based on trying to minimise the number of shares that have to be generated, mutually communicated and stored by the participants in the scheme. In doing so we show that *not all* of the participants need to generate shares in the first instance. It is often the case that in an efficient realisation of an MTA-free scheme for Γ, some participants need only store shares that have been generated by other participants.

In Section 2 we discuss the concept of access structure domination, which is fundamental to the rest of the paper. Section 3 is about MTA-free schemes in general and includes a construction protocol that will work for any monotone access structure. In Section 4 we concentrate on MTA-free threshold schemes and show that some variations of the standard protocol can be used to improve scheme efficiency. We have been forced to omit most proofs in this extended abstract, but it is hoped that by way of examples we can illustrate the main ideas behind the constructions presented. Complete proofs will be provided in the full paper.

2 Access Structure Domination

Let Γ be a monotone access structure defined on a participant set $\mathcal{P}$. The monotonicity of Γ ensures that we can find a collection Γ^- of *minimal* authorised sets in Γ and a set Γ^+ of maximal unauthorised sets. Note that a participant need not belong to any minimal set in Γ. If every participant does belong to a minimal set then we say that Γ is *connected*. We recall from [1] that Γ can be considered as a logical expression with the participants being boolean variables. Let $\Gamma^- = \{C_1, \ldots, C_r\}$, let $+$ denote logical OR and let juxtaposition denote logical AND. Then the disjunctive normal form of the *logical equivalent* of Γ is $\Gamma = C_1 + \cdots + C_r$. It follows that a subset A of participants is in the access structure Γ if and only if the logical equivalent of Γ is *true* when the variables in A are all true. For example, let $\mathcal{P} = \{a, b, c, d\}$ and $\Gamma^- = \{\{a, b, c\}, \{c, d\}\}$. Then we write $\Gamma = abc + cd$, or equivalently $\Gamma = (ab + d)c$.

We now recall from [8] a useful family of monotone access structures that can be derived from Γ. Let $A \subseteq \mathcal{P}$. We define the *contraction* $\Gamma \cdot A$ of Γ at A to be the monotone access structure on $\mathcal{P}$ given by

$$B \in \Gamma \cdot A \iff B \cup A \in \Gamma.$$

Conceptually, $\Gamma \cdot A$ is the access structure that results if the shares belonging to the participants in A are publicly revealed. For example, if $\Gamma = abc + cd$ then $\Gamma \cdot c = ab + d$ and $\Gamma \cdot d = c$.

Now let Γ_0 be a monotone access structure defined on $\mathcal{P} = \{p_1, \ldots, p_n\}$. Associate with each $p_i \in \mathcal{P}$ a monotone access structure Γ_i defined on $\mathcal{P}$. Let $\Gamma = (\Gamma_0; \Gamma_1, \ldots, \Gamma_n)$ be the monotone access structure defined on $\mathcal{P}$ that is formed by replacing p_i by Γ_i in the logical equivalent of Γ_0.

Example 1. Let $\mathcal{P} = \{a, b, c, d\}$. Let $\Gamma_0 = abcd$, $\Gamma_a = c$, $\Gamma_b = c + d$, $\Gamma_c = d$ and $\Gamma_d = d$. Then $\Gamma = (\Gamma_0; \Gamma_a, \Gamma_b, \Gamma_c, \Gamma_d) = c(c + d)dd = cd$. Similarly, if $\Gamma_0 = abcd$, $\Gamma_a = a + c$, $\Gamma_b = b + d$, $\Gamma_c = $'true' (in other words, $(\Gamma_c)^- = \{\emptyset\}$) and $\Gamma_d = $'true' then $\Gamma = (\Gamma_0; \Gamma_a, \Gamma_b, \Gamma_c, \Gamma_d) = (a + c)(b + d) = ab + ad + bc + cd$.

For $A \subseteq \mathcal{P}$, let $X(A) = \{p_i \mid A \in \Gamma_i\}$. We can also describe $\Gamma = (\Gamma_0; \Gamma_1, \ldots, \Gamma_n)$ in the following way.

Lemma 1. *Let $\Gamma, \Gamma_0, \Gamma_1, \ldots, \Gamma_n$ be monotone access structures defined on $\mathcal{P} = \{p_1, \ldots, p_n\}$. Then the following two statements are equivalent:*

1. *$\Gamma = (\Gamma_0; \Gamma_1, \ldots, \Gamma_n)$;*
2. *For every $A \subseteq \mathcal{P}$ we have $A \in \Gamma$ if and only if $X(A) \in \Gamma_0$.*

Now let Γ_0 and Γ be distinct monotone access structures defined on $\mathcal{P} = \{p_1, \ldots, p_n\}$. Using terminology suggested in [11], we say that Γ_0 *dominates* Γ if there exist monotone access structures $\Gamma_1, \ldots, \Gamma_n$ such that

1. $\{p_i\} \in \Gamma_i$ (for each i, $1 \leq i \leq n$);
2. $\Gamma = (\Gamma_0; \Gamma_1, \ldots, \Gamma_n)$.

Thus, from Example 1, we see that $\Gamma_0 = abcd$ dominates $\Gamma = ab + ad + bc + cd$. We say that Γ_0 *directly* dominates Γ if there does *not* exist a monotone access structure Γ' (distinct from Γ_0 and Γ) such that Γ_0 dominates Γ' and Γ' dominates Γ. We now classify all the monotone access structures that are (directly) dominated by a given monotone access structure.

Theorem 2. *Let Γ_0 and Γ be monotone access structures defined on $\mathcal{P}$. Then Γ_0 dominates Γ if and only if $\Gamma_0 \subseteq \Gamma$.*

The next result is an interpretation of the main theorem in [11].

Result 3. *Let Γ_0 and Γ be monotone access structures defined on $\mathcal{P}$. Then Γ_0 directly dominates Γ if and only if there exists a (unique) maximal unauthorised subset B of Γ_0 such that $\Gamma = \Gamma_0 \cup \{B\}$.*

3   Mutually Trusted Authority free Schemes

We first give a basic model for secret sharing (see, for example, [13]). We will use the *entropy* function in our definition (see, for example, [5] for an introduction to entropy and its properties). Let $\mathcal{P} = \{p_1, \ldots, p_n\}$ be a participant set and let s be a secret. Let participant p_i receive a share from a set $[p_i]$ and let the secret come from a set $[s]$. A *secret sharing scheme* for Γ is a probability distribution ρ defined on a set of *distribution rules* $\Omega \subseteq [p_1] \times \cdots \times [p_n] \times [s]$ such that for $A \subseteq \mathcal{P}$,

1. if $A \in \Gamma$ then $H(s|A) = 0$;
2. if $A \notin \Gamma$ then $H(s|A) > 0$.

If it is the case that for each $A \notin \Gamma$ we have $H(s|A) = H(s)$ then the secret sharing scheme is said to be *perfect*. We call $H(p_i)$ the *size* of the share associated with p_i, and $H(s)$ the *size* of the secret. It can be seen (for example [13]) that in any perfect secret sharing scheme, if $p_i \in A$ for some minimal authorised set A then $H(p_i) \geq H(s)$. If $H(p_i) = H(s)$ for all such p_i then we say that the perfect secret sharing scheme and its access structure are *ideal*. We note ([2, 10]) that ideal (k, n)-threshold schemes can be found for all $1 \leq k \leq n$.

In a traditional secret sharing scheme, an MTA selects a distribution rule π from Ω with probability $\rho(\pi)$ and then distributes the appropriate shares to the participants of the scheme. In an MTA-free scheme the participants indirectly select a (random) distribution rule through the generation of their own (random) shares.

Consider the following extension of the protocol in [6] for setting up an MTA-free scheme. Firstly, let a subset $\mathcal{P}_0$ of the participants in $\mathcal{P}$ generate shares of a perfect scheme M_0 for some access structure Γ_0. Let x_i denote the share generated by p_i ($p_i \in \mathcal{P}_0$). Then let each $p_i \in \mathcal{P}_0$ construct a private perfect secret sharing scheme M_i for Γ_i on $\mathcal{P}$ to protect the explicit secret x_i. Thus x_i can be obtained by p_i or by any authorised set in Γ_i. In the degenerate case where $\Gamma_i^- = \{\emptyset\}$, p_i publicly reveals (*broadcasts*) their share. Otherwise p_i communicates the shares of M_i to the participants included in M_i. This process creates a new perfect secret sharing scheme M for access structure $\Gamma = (\Gamma_0; \Gamma_1, \ldots, \Gamma_n)$, where for each $p_i \notin \mathcal{P}_0$ we take Γ_i to be such that $\Gamma_i^- = \{\emptyset\}$ and thus for each i ($1 \leq i \leq n$), $\{p_i\} \in \Gamma_i$. For the structure of the distribution rules of M in terms of those of M_0 and M_i, we refer to [8, 14], where constructions of this type were fully described.

Simmons [11] asked which access structures Γ could be realised from Γ_0 in this manner. In other words, *which access structures Γ are dominated by Γ_0?* Theorem 2 concisely answers this question by showing that these are precisely the access structures Γ such that $\Gamma \supseteq \Gamma_0$. We approach the problem from another direction in this paper. Namely, given an access structure Γ, exactly which access structures $\Gamma_0, \Gamma_1, \ldots, \Gamma_n$ such that $\Gamma = (\Gamma_0; \Gamma_1, \ldots, \Gamma_n)$ should be chosen in order to (efficiently) generate an MTA-free scheme for Γ?

3.1 The Base Access Structure

The first issue to be considered in the design of an MTA-free scheme for Γ is which initial access structure Γ_0 should be chosen. We refer to Γ_0 as the *base* access structure. We assume that each $p \in \mathcal{P}_0$ independently generates a random share of M_0 from set $[p]$.

Theorem 4. *Let Γ_0 be the access structure of a secret sharing scheme M_0 defined on $\mathcal{P}$ such that for each $p \in \mathcal{P}$ the share held by p in M_0 is independently and randomly chosen from the set $[p]$. Then Γ_0 has a unique minimal authorised set $\mathcal{P}_0$.*

If M_0 is a perfect secret sharing scheme then it follows from Theorem 4 that M_0 is a unanimous threshold scheme defined on $\mathcal{P}_0$, and thus that the participants in $\mathcal{P} \setminus \mathcal{P}_0$ need not generate shares in the initial stage of the protocol. Thus from Theorems 2 and 4 we have the following criteria for selection of the base access structure Γ_0.

- $\Gamma_0 \subseteq \Gamma$;
- Γ_0 is unanimous threshold on $\mathcal{P}_0$ ($P_0 \subseteq \mathcal{P}$).

3.2 Measures of Efficiency

There are three parameters that we might want to minimise for reasons of economy and efficiency in an MTA-free scheme. These are the total size $g(\mathcal{P})$ of shares *generated* by the participants, the total size $c(\mathcal{P})$ of shares *communicated* by participants, and the total size $s(\mathcal{P})$ of shares *stored* by participants in the scheme. For a participant $p \in \mathcal{P}$ let $g(p)$ be the sum of the sizes of shares generated by p, $c(p)$ be the sum of the sizes of shares communicated by p and $s(p)$ be the sum of the sizes of shares stored by p.

In the light of Section 3.1, we assume that M_0 is chosen to be ideal and hence the shares of M_0 (and thus the secrets of M_i, for $p_i \in \mathcal{P}_0$) all have the same size h. For the purposes of the discussion immediately following we will take h as the size of a 'unit' share.

Let $p \in \mathcal{P}$. If $p \notin \mathcal{P}_0$ then $g(p) = c(p) = 0$. Otherwise, if $p \in \mathcal{P}_0$ then p generates one share x_p of the initial scheme and then a number (possibly zero) of shares of a private scheme to protect x_p. These extra shares are communicated to some of the other participants. Thus if $p \in \mathcal{P}_0$ then $g(p) = 1 + c(p)$. Hence in total,

$$g(\mathcal{P}) = c(\mathcal{P}) + |\mathcal{P}_0|. \tag{1}$$

A participant $p \in \mathcal{P}_0$ either keeps the share x_p secure or broadcasts it and hence does not need to store it. Let $\mathcal{P}_0'$ be the subset of $\mathcal{P}_0$ who store their shares. All shares of private schemes that are communicated to $p \in \mathcal{P}$ are stored securely. Hence we have in total,

$$s(\mathcal{P}) = c(\mathcal{P}) + |\mathcal{P}_0'|. \tag{2}$$

Then from (1) and (2) we have,

$$g(\mathcal{P}) = s(\mathcal{P}) + |\mathcal{P}_0 \setminus \mathcal{P}_0'| \; ; \qquad c(\mathcal{P}) = s(\mathcal{P}) - |\mathcal{P}_0'|. \tag{3}$$

Thus from (3) we see that $s(\mathcal{P})$ alone is an effective measure of efficiency since $g(\mathcal{P})$ and $c(\mathcal{P})$ are directly proportional to $s(\mathcal{P})$. In the event that two different schemes have the same total storage $s(\mathcal{P})$ then (3) suggests that a scheme which has a large value of $|\mathcal{P}_0'|$ relative to $|\mathcal{P}_0|$ is preferable.

Minimising storage has been the most studied measure of efficiency for traditional secret sharing schemes (for example [3, 4, 8, 14]). Efficiency rates can be calculated from the *contribution vector* (or *convec*) of the scheme. This is the vector $(c_1, \ldots, c_n) = (1/H(s))(H(p_1), \ldots, H(p_n))$. For perfect secret sharing schemes the most common efficiency measures are the *information rate*, which is the minimum $1/c_i$ ($1 \leq i \leq n$), and the *average information rate*, which is $n/(c_1 + \cdots + c_n)$. For MTA-free schemes we see that the total storage $s(\mathcal{P})$ is $c_1 + \cdots + c_n$. For simplicity, in this paper we only consider minimising the total storage, which is equivalent to maximising the average information rate.

3.3   The Private Access Structures

The problem of constructing an efficient MTA-free scheme for Γ is thus the problem of selecting a base access structure Γ_0 (subject to the constraints of Section 3.1) and a collection of private access structures Γ_p ($p \in \mathcal{P}_0$) in such a way that $s(\mathcal{P})$ is minimised. As a standard for comparison we use what we call the *Basic* construction. This is in effect the most 'obvious' way of constructing an MTA-free scheme for Γ. The Basic construction is used in the proof of Theorem 2.

The Basic Construction

Γ_0	(n, n)-threshold on $\mathcal{P}$
Γ_p	$p + \Gamma$ (for each $p \in \mathcal{P}$)
M_0	ideal unanimous threshold scheme on $\mathcal{P}$
M_p	perfect scheme for Γ_p (for each $p \in \mathcal{P}$)

Example 2. Let $\mathcal{P} = \mathcal{P}_0 = \{a, b, c, d\}$ and $\Gamma = ab + ac + bcd$. *Applying the Basic construction gives* $\Gamma_0 = abcd$, $\Gamma_a = a + bcd$, $\Gamma_b = b + ac$, $\Gamma_c = c + ab$ *and* $\Gamma_d = d + ab + ac$. *Since* $\Gamma_a, \Gamma_b, \Gamma_c, \Gamma_d$ *are all ideal (see [12]) we can find ideal* M_a, M_b, M_c, M_d *and thus a scheme* M *for* Γ *with convec* $(c_a, c_b, c_c, c_d) = (4, 4, 4, 2)$ *(for example, participant a generates one unit share, and receives one unit share from each of b, c, d).*

We show that, by applying contractions, the Basic construction can be quite considerably improved upon. We call this modified construction method the *Contraction* construction.

The Contraction Construction

Γ_0	(a, a)-threshold on $\mathcal{P}_0$, for some $\mathcal{P}_0 = \{p_1, \ldots, p_a\} \in \Gamma$
Γ_1	$p_1 + \Gamma$
Γ_2	$p_2 + \Gamma \cdot p_1$
$\vdots$	$\vdots$
Γ_a	$p_a + \Gamma \cdot p_1 p_2 \ldots p_{a-1}$
M_0	ideal unanimous threshold scheme on A
M_i	perfect scheme for Γ_i (for each i, $1 \le i \le a$)

Example 3. Let $\mathcal{P}$ and Γ be as in Example 2. Applying the Contraction construction with $\mathcal{P}_0 = \{a, b\}$ and $\Gamma_0 = ab$ gives $\Gamma_a = a + bcd$, $\Gamma_b = b + c$. Since Γ_a, Γ_b are ideal (see [12]) we can find ideal M_a, M_b and thus a scheme M_1 for Γ with convec $(1, 2, 2, 1)$. Alternatively, applying the Contraction construction with $\mathcal{P}_0 = \{b, c, d\}$ and $\Gamma_0 = bcd$ gives $\Gamma_b = b + ac$, $\Gamma_c = c + a$, $\Gamma_d = d + a$. Since $\Gamma_b, \Gamma_c, \Gamma_d$ are ideal (see [12]) we can find ideal M_b, M_c, M_d and thus a scheme M_2 for Γ with convec $(3, 1, 2, 1)$. Both M_1 and M_2 are considerably more efficient in terms of total storage (and information rate) than the scheme M constructed in Example 2. Scheme M_1 is slightly more efficient than M_2.

4 Mutually Trusted Authority free Threshold Schemes

We now consider the special case of realising an MTA-free (k, n)-threshold scheme. We assume that $k < n$ since the case $k = n$ was covered in Section 1. Let $\mathcal{P} = \{p_1, \ldots, p_n\}$, let $1 \le k < n$ and let Γ be (k, n)-threshold on $\mathcal{P}$. The most efficient MTA-free scheme for Γ we have seen thus far is by the Contraction construction applied to a minimal authorised set of Γ. In this case,

- $\Gamma_0 = (k, k)$-threshold on $\mathcal{P}_0 = \{p_1, \ldots, p_k\}$;
- $\Gamma_i = p_i + \big((k-i+1, n-i)$-threshold on $\{p_{i+1}, \ldots, p_n\}\big)$ (for each i, $1 \le i \le k$).

Using ideal threshold schemes $M_0, M_1, \ldots, M_a$ we can calculate the convec for the resulting scheme M. Each p_i $(1 \le i \le k)$ stores their share of M_0 and receives one share from each of $p_1, \ldots, p_{i-1}$. Each p_i $(k+1 \leq i \leq n)$ receives one share from each of $p_1, \ldots, p_k$. Thus $c_i = i$ $(1 \leq i \leq k)$ and $c_i = k$ $(k+1 \leq i \leq n)$. So

$$s(\mathcal{P}) = \frac{1}{2}k(k+1) + k(n-k) = nk - \frac{1}{2}k(k-1). \tag{4}$$

We show that the Contraction construction gives an optimal construction for threshold schemes under certain assumptions.

Theorem 5. *Let M be an MTA-free (k, n)-threshold scheme that is constructed from an ideal unanimous threshold scheme M_0 on a k-subset A of $\mathcal{P}$, and, for each $p \in A$, a perfect scheme M_p for some Γ_p. Then $s(\mathcal{P}) \geq nk - (1/2)k(k-1)$.*

We now show that if the protocol for establishing an MTA-free scheme is generalised to permit the use of secret sharing schemes that are not perfect then we can always improve on the total storage given by the Contraction construction. Let $0 \leq c \leq k$. A (c, k, n)-*ramp scheme* on an n-set $\mathcal{P}$ is a secret sharing scheme such that for $A \subseteq \mathcal{P}$,

1. if $|A| \geq k$ then $H(s|A) = 0$;
2. if $|A| \leq c$ then $H(s|A) = H(s)$.

Ramp schemes such that $H(p) = H(s)/(k-c)$ $(p \in \mathcal{P})$ can be constructed from ideal (k, n)-threshold schemes ([7]).

We now present a construction which works for all $k \leq \frac{1}{2}(n+1)$. This construction relies on implementing the private access structures by using ramp schemes as opposed to perfect threshold schemes. For this reason we call it the *Private Ramp* construction.

The Private Ramp Construction

Γ_0	(n, n)-threshold on $\mathcal{P}$
Γ_p	$p + ((k, n-1)$-threshold on $\mathcal{P} \setminus \{p\})$
M_0	ideal (n, n)-threshold scheme on $\mathcal{P}$
M_p	modified $(0, k, n-1)$-ramp scheme on $\mathcal{P} \setminus \{p\}$

We say that M_p is 'modified' because $p \in \Gamma_p$ and so every distribution rule of M_p must also distribute a copy of the secret of M_p to participant p.

Example 4. Let Γ be $(2, 3)$-threshold defined on $\mathcal{P} = \{a, b, c\}$. Using the Contraction construction with $\mathcal{P}_0 = \{a, b\}$ and $\Gamma_0 = ab$, gives $\Gamma_a = a + bc$, $\Gamma_b = b + c$ and a scheme M_1 for Γ with convec $(c_a, c_b, c_c) = (1, 2, 2)$. Using the Private Ramp construction gives $\Gamma_a = a + bc$, $\Gamma_b = b + ac$, $\Gamma_c = c + ab$. We then use the $(0, 2, 2)$-ramp schemes M_a, M_b, M_c to construct a scheme M_2 for Γ with convec $(2, 2, 2)$. Thus M_2 has a total storage of 6 which is slightly more than the total storage 5 of M_1.

Thus for the $(2,3)$ case, the Private Ramp construction did not perform as well as the Contraction construction. In the Private Ramp construction each participant $p \in \mathcal{P}$ generates a share of unit size and then receives $n-1$ other shares, each of size $1/k$, from the other participants. So for the Private Ramp construction,

$$s(\mathcal{P}) = n(1 + \frac{n-1}{k}) = \frac{n(k+n-1)}{k}. \tag{5}$$

Thus we can see from (4) and (5) that, generally, the Contraction construction has a lower total storage than the Private Ramp construction when k is small with respect to n, but the Private Ramp construction is an improvement on the Contraction construction when k is close to $\frac{1}{2}(n+1)$.

We show now that if a ramp scheme is used to implement the base access structure instead of the private access structures then we can do even better. The *Base Ramp* construction has the added advantage that it works for all $1 \le k \le n - 1$.

The Base Ramp Construction

Γ_0	(n,n)-threshold on $\mathcal{P}$
Γ_p	$p + ((k, n-1)\text{-threshold on } \mathcal{P} \setminus \{p\})$
M_0	$(k-1, n, n)$-ramp scheme on $\mathcal{P}$
M_p	ideal scheme for Γ_p $(p \in \mathcal{P})$

Example 5. As in Example 4, let Γ be $(2,3)$-threshold defined on $\mathcal{P} = \{a, b, c\}$. Using the Base Ramp construction gives $\Gamma_a = a + bc$, $\Gamma_b = b + ac$, $\Gamma_c = c + ab$ and a scheme M_3 for Γ with convec $(3/2, 3/2, 3/2)$. Thus M_3 has a total storage of $9/2$ which is an improvement on the total storage 5 of M_1 using the Contraction construction, and on the total storage 6 of M_2 using the Private Ramp construction.

Note that the Base Ramp construction is essentially the same as the Basic construction for a (k, n)-threshold scheme, except with a different M_0. Each participant $p \in \mathcal{P}$ generates a share of size $1/(n-k+1)$ and then receives $n-1$ other shares, each of size $1/(n-k+1)$, from the other participants. So for the Base Ramp construction,

$$s(\mathcal{P}) = n(\frac{n}{n-k+1}) = \frac{n^2}{n-k+1}. \tag{6}$$

Thus from (4) and (6), and from (5) and (6), we see that the Base Ramp scheme is an improvement on both the Contraction construction and the Private Ramp construction (when applicable) for (k, n)-threshold schemes. We note that the Base Ramp Construction for threshold schemes can be generalised to other monotone access structures.

We conclude by presenting a table containing the values for the total storage of various MTA-free (k, n)-threshold schemes under the constructions discussed in this paper (* denotes that the construction is not possible for these parameters).

(k, n)	Total Storage			
	Basic	Contraction	Private Ramp	Base Ramp
$(2, 3)$	9	5	6	9/2
$(2, 4)$	16	7	10	16/3
$(2, 5)$	25	9	15	25/4
$(3, 4)$	16	9	*	8
$(3, 5)$	25	12	35/3	25/3
$(4, 5)$	25	14	*	25/2
$(5, 10)$	100	40	28	100/6
$(10, 20)$	400	155	58	400/11

Table 1. Various total storage values for MTA-free (k, n)-threshold schemes.

References

1. J. Benaloh and J. Leichter. Generalized secret sharing and monotone functions. *Adv. in Cryptology - CRYPTO'88, Lecture Notes in Comput. Sci*, 403:27–35, 1990.
2. G. R. Blakley. Safeguarding cryptographic keys. *Proceedings of AFIPS 1979 National Computer Conference*, 48:313–317, 1979.
3. E. F. Brickell and D. R. Stinson. Some improved bounds on the information rate of perfect secret sharing schemes. *J. Cryptology*, 5:153–166, 1992.
4. R. M. Capocelli, A. De Santis, L. Gargano, and U. Vaccaro. On the size of shares for secret sharing schemes. *J. Cryptology*, 6:157–167, 1993.
5. R. G. Gallager. *Information theory and reliable communication*. John Wiley and Sons, New York, 1968.
6. I. Ingemarsson and G. J. Simmons. A protocol to set up shared secret schemes without the assistance of a mutually trusted party. *Adv. in Cryptology - EURO-CRYPT'90, Lecture Notes in Comput. Sci.*, 473:266–282, 1991.
7. W.-A. Jackson and K. M. Martin. A combinatorial interpretation of ramp schemes. Submitted, 1994.
8. K. M. Martin. New secret sharing schemes from old. *J. Combin. Math Combin. Comput.*, 14:65–77, 1993.
9. C. Meadows. Some threshold schemes without central key distributors. *Congressus Numerantium*, 46:187–199, 1985.
10. A. Shamir. How to share a secret. *Comm. ACM*, 22(11):612–613, 1979.
11. G. J. Simmons. The consequences of trust in shared secret schemes. *Adv. in Cryptology - EUROCRYPT'93, Lecture Notes in Comput. Sci.*, 765:448–452, 1994.
12. G. J. Simmons, W.-A. Jackson, and K. Martin. The geometry of shared secret schemes. *Bull. Inst. Combin. Appl.*, 1:71–88, 1991.
13. D. R. Stinson. An explication of secret sharing schemes. *Des. Codes Cryptogr.*, 2:357–390, 1992.
14. D. R. Stinson. Decomposition constructions for secret sharing schemes. *IEEE Trans. Inform. Theory*, 40:118–125, 1994.

General Short Computational
Secret Sharing Schemes

Philippe Béguin[1] and Antonella Cresti[2],[*]

[1] Laboratoire d'Informatique [**]
Ecole Normale Supérieure, 75230 Paris Cédex 05, France
e-mail: beguin@truffe.ens.fr

[2] Dipartimento di Scienze dell'Informazione
Università di Roma "La Sapienza", 00198 Roma, Italy
e-mail: antone@dsi.uniroma1.it

Abstract. A secret sharing scheme permits a secret to be shared among participants in such a way that only qualified subsets of participants can recover the secret. If any non qualified subset has absolutely no information about the secret, then the scheme is called perfect. Unfortunately, in this case the size of the shares cannot be less than the size of the secret. Krawczyk [9] showed how to improve this bound in the case of computational threshold schemes by using Rabin's information dispersal algorithms [14], [15].
We show how to extend the information dispersal algorithm for general access structure (we call access structure, the set of all qualified subsets). We give bounds on the amount of information each participant must have. Then we apply this to construct computational schemes for general access structures. The size of shares each participant must have in our schemes is nearly minimal: it is equal to the minimal bound plus a piece of information whose length does not depend on the secret size but just on the security parameter.

1 Introduction

Secret sharing is an important tool in security and cryptography. An important issue in secret sharing theory is the size of the share distributed, since the security of a system degrades as the amount of information that must be kept secret increases. A very strong requirement is that all qualified subsets of participants can reconstruct the secret but all other subsets obtain no information (in an information–theoretic sense) about the secret. These schemes are called *perfect* secret sharing schemes. Unfortunately, in this case the size of the shares cannot

---

[*] Part of this work was done while the author was visiting the Laboratoire d'Informatique of the Ecole Normale Supérieure, France.
[**] Supported by the Centre National de la Recherche Scientifique URA 1327.

L.C. Guillou and J.-J. Quisquater (Eds.): Advances in Cryptology - EUROCRYPT '95, LNCS 921, pp. 194-208, 1995.
© Springer-Verlag Berlin Heidelberg 1995

be less than the size of the secret. However, the proof of this lower bound uses the notion of information–theoretic secrecy.

A natural question is whether one can do better for secret sharing if the notion of secrecy is computational, namely, against resource bounded adversaries: i.e. any qualified subset can reconstruct the secret but any other subset obtains no computational information about the secret. These schemes are called *computational* secret sharing schemes.

Krawczyk [9] proposed a computational m-threshold scheme, where m shares recover the secret but $m - 1$ shares give no (computational) information on the secret, in which shares corresponding to a secret uniformly chosen in a set S are of size $\log |S|/m$ (where $|S|$ denotes the cardinality of the set S) plus a short piece of information whose length does not depend on the secret size but just on the security parameter. In our paper, $|X|$ denotes the cardinality of the set X, whereas in Krawczyk's one, $|x|$ denotes the size of x for $x \in X$.

The scheme of Krawczyk is very simple and combines in a natural way traditional (perfect) secret sharing schemes, encryption, and known information dispersal algorithms. It is provable secure given a secure (private key) encryption function.

A natural and open question is whether the space efficiency can be carried over more general access structures than just threshold schemes: one of the problems was to find an information dispersal algorithm for general access structures.

In this paper, we define information dispersal algorithms for general access structures; we show bounds on the size of pieces each participant must have and we give practical constructions of algorithms that reach these bounds. Then we apply these results to computational secret sharing schemes. We show how to realize computational secret sharing schemes for general access structures that are nearly optimal: the size of each share is equal to the minimal theoretical bound plus a piece of information whose length does not depend on the secret size but just on the security parameter.

2 Perfect Secret Sharing (PSS)

Let $\mathcal{P} = \{P_1, \ldots, P_n\}$ be the set of participants. Denote by $\mathcal{A} \subseteq 2^{\mathcal{P}}$ the family of subsets of participants which we desire to be able to recover the file; $\mathcal{A}$ is called the *access structure*. It is reasonable to require that $\mathcal{A}$ be *monotone*, that is if $A \in \mathcal{A}$ and $A \subseteq A' \subseteq \mathcal{P}$, then $A' \in \mathcal{A}$.

If $\mathcal{A}$ is an access structure on $\mathcal{P}$, then $B \in \mathcal{A}$ is a *minimal* authorized subset if $A \notin \mathcal{A}$ whenever $A \subset B$. The set of minimal authorized subsets of $\mathcal{A}$ is denoted $\mathcal{A}^0$ and is called the *basis* of $\mathcal{A}$. $\mathcal{A}$ is uniquely determined as a function of $\mathcal{A}^0$, as we have $\mathcal{A} = \{B \subseteq \mathcal{P} : \exists A \subseteq B, A \in \mathcal{A}^0\}$. We say that $\mathcal{A}$ is the *closure* of $\mathcal{A}^0$ and write $\mathcal{A} = cl(\mathcal{A}^0)$.

Given an access structure $\mathcal{A}$, on a set $\mathcal{P} = \{P_1, \ldots, P_n\}$ of participants, let be K the space of secrets, and let $\{p_K(k)\}_{k \in K}$ be a probability distribution on K. Let a secret sharing scheme for secrets in K be fixed. For any participant $P_i \in \mathcal{P}$, let us denote by V_i the set of all possible shares given to participant P_i.

Given a set of participants $A = \{P_{i_1}, \ldots, P_{i_r}\} \subseteq \mathcal{P}$, where $i_1 < i_2 < \ldots < i_r$, denote by V_A the set $V_{i_1} \times \cdots \times V_{i_r}$. A secret sharing scheme for secrets in K and a probability distribution $\{p_K(k)\}_{k \in K}$ naturally induce a probability distribution on V_A, for any $A \subseteq \mathcal{P}$. Denote such probability distribution by $\{p_{V_A}(a)\}_{a \in V_A}$. Finally, denote by $H(K)$ the entropy of $\{p_K(k)\}_{k \in K}$ and by $H(V_A)$ the entropy of $\{p_{V_A}(a)\}_{a \in V_A}$, for any $A \in 2^{\mathcal{P}}$.

Following the information-theoretical approach of [6] and [4] we have the following definition.

Definition 1. Let $\mathcal{A}$ be an access structure on a set $\mathcal{P}$ of participants. We say that a secret sharing scheme for secrets in K is *perfect* for the access structure $\mathcal{A}$ on $\mathcal{P}$, if the following two properties hold:

1. *Any qualified subset can reconstruct the secret:*
   Formally, for all $A \in \mathcal{A}$, it holds $H(K|V_A) = 0$.
2. *Any non-qualified subset has absolutely no information on the secret:*
   Formally, for all $A \notin \mathcal{A}$, it holds $H(K|V_A) = H(K)$.

3 Information Dispersal Algorithms (IDA)

We analyze the problem of distributing pieces of a file f among a set of users in such a way that some predefined subsets of users can, pooling together their pieces, reconstruct the entire file f. An information dispersal algorithm differs from a secret sharing scheme as there are no restriction whatsoever about the sets which are not in $\mathcal{A}$. Rabin ([14],[15]) first considered the problem and introduced the *Information Dispersal Algorithms*. His schemes are intended for the distribution of a piece of information among n active processors, in such a way that the recovery of the information is possible in presence of m active processors, where m and n are parameters satisfing $1 \leq m \leq n$. The basic idea of his algorithms is to add to the information some amount of redundancy and then to partition it into n fragments, each transmitted to one of the parties. Reconstruction of f is possible out of m fragments. Information dispersal algorithms have several applications to secure and reliable storage of information in computer networks. Moreover they can be applied to fault-tolerant transmission of information and to communication between processors in parallel computers.

Subsequently, Naor and Roth [12], using integer linear programming techniques, proposed an information dispersal algorithm over arbitrary graphs. In their model, an arbitrary file f is distributed among the nodes of the graph in such a way that each node of the graph, by accessing the memory of its own and of its adjacent nodes, can reconstruct the contents of f. Their scheme can be applied to store files in distributed networks.

In this paper, we define information dispersal algorithms in a similar way than secret sharing schemes, using an information–theoretical approach. Let $\mathcal{P} = \{P_1, \ldots, P_n\}$ be the set of participants; we denote by $\mathcal{A}$ the access structure that is the subsets of participants which we desire to be able to recover the file: $\mathcal{A}$

is monotone. We define in a similar way respect to secret sharing schemes the *minimal authorized* subsets, the *basis* and the *closure* of $\mathcal{A}$.

If F is the set of files, $\{p_F(f)\}_{f \in F}$ a probability distribution on F, and an information dispersal algorithm for files in F is fixed, we define as $V_i, V_A, H(K)$ respectively $G_i, G_A, H(F)$.

Definition 2. Let $\mathcal{A}$ be an access structure on a set $\mathcal{P}$ of participants. We say that an algorithm Σ to distribute a file in F according with the probability distribution $\{p_F(f)\}_{f \in F}$ is an Information Dispersal Algorithm (IDA) if any qualified subset can reconstruct the file. Formally, for all $A \in \mathcal{A}$, it holds

$$H(F|G_A) = 0.$$

The following lemma holds.

Lemma 3. Let $\mathcal{A}$ be an access structure on a set $\mathcal{P}$ of participants. Any information dispersal algorithm for $\mathcal{A}$, for any $A \in \mathcal{A}$, must give to at least a participant $P_j \in A$ a fragment from a domain G_j such that $H(G_j) \geq H(F)/|A|$.

Proof. Let $A \in \mathcal{A}$. Consider the conditional mutual information $I(G_A; F)$. It can be written either as $H(G_A) - H(G_A|F)$ or as $H(F) - H(F|G_A)$. Hence, from (4) of Appendix A and from 1. of definition 2 we have

$$H(G_A) = H(F) + H(G_A|F) \geq H(F). \tag{1}$$

From (5) and (8) of appendix A it follows that $\sum_{P_i \in A} H(G_i) \geq H(G_A)$. So, from (1) one gets that there exists a participant P_j in A such that $H(G_j) \geq H(F)/|A|$.
$\square$

3.1 A Simple Information Dispersal Algorithm

We outline the following simple information dispersal algorithm $\Sigma(m, b)$, where b is the number of file fragments and m is the minimum number of fragments required to reconstruct the file. It is a simple version of Rabin's information dispersal algorithm. The algorithm is based on Reed Solomon erasure codes. The information $f \in F$ to be shared is first partitioned into m equal parts where each part is viewed as an element over a finite field (e.g. $GF(q)$, for a large enough q). These m elements are then viewed as coefficients of a polynomial of degree $m-1$, and the b fragments for distribution are obtained by evaluating this polynomial in b different points (we need $q \geq b$). Clearly the whole information can be reconstructed (by interpolation) from any m fragments.

Assuming the uniform probability distribution over the set of files F, we have $H(F) = \log|F| = m \log q$. Moreover, for all participant $P_i \in \mathcal{P}$ it holds $H(G_i) \leq \log|G_i| = \log q = \log|F|/m$.

Observe that we have the requirement $q \geq b$. This implies $\log|F| = m \log q \geq m \log b$. So when the parameters m and b are big the algorithm works for large files.

3.2   The Size of Pieces

The efficiency of any information dispersal algorithm, is computed regarding the size of pieces given to each participant. So, even in the case of general access structures, we are interested to minimize the size of fragments distributed to participants.

Information Rate

If we are interested in limiting the maximum size of fragments for each participant (i.e., the maximum quantity of information that must be given to any participant), then a worst-case measure of the maximum of $H(G_i)$ over all $P_i \in \mathcal{P}$ naturally arises. Analogously to definition of *information rate* for secret sharing schemes presented in [2], we give the following definition.

Definition 4. We define the *information rate* of an information dispersal algorithm Σ for the access structure $\mathcal{A}$, when the probability distribution on the set of files F is Π_F, as

$$\varrho(\mathcal{A}, \Pi_F, \Sigma) = \frac{H(F)}{\max\{H(G_i) \ : \ 1 \leq i \leq n\}}.$$

The following theorem holds.

Theorem 5. Let $\mathcal{A}$ be an access structure on a set $\mathcal{P}$ of participants. The information rate of any information dispersal algorithm Σ for $\mathcal{A}$ satisfies

$$\varrho(\mathcal{A}, \Pi_F, \Sigma) \leq \varrho_{\max},$$

where $\varrho_{\max} = \min\{|A| : A \in \mathcal{A}^0\}$.

Proof. Let $A \in \mathcal{A}^0$ such that $|A| = \varrho_{\max}$. From Lemma 3, any information dispersal algorithm Σ for F must give to at least a participant $P_j \in A$ a fragment such that $H(G_j) \geq H(F)/\varrho_{\max}$. So, for any information dispersal algorithm Σ, $\max\{H(G_i) \ : \ 1 \leq i \leq n\} \geq H(G_j) \geq H(F)/\varrho_{\max}$. Hence,

$$\varrho(\mathcal{A}, \Pi_F, \Sigma) = \frac{H(F)}{\max\{H(G_i) \ : \ 1 \leq i \leq n\}} \leq \varrho_{\max}.$$

$\blacksquare$

Average Information Rate

In many cases it is preferable to limit the sum of the size of fragments given to all participants. In such a cases the arithmetic mean of the size of fragments for each participant is a more appropriate measure.

Definition 6. We define the *average information rate* of an information dispersal algorithm Σ for an access structure $\mathcal{A}$ when the probability distribution on the set of files F is Π_F, as

$$\tilde{\varrho}(\mathcal{A}, \Pi_F, \Sigma) = \frac{nH(F)}{\sum_{i=1}^n H(G_i)}.$$

Consider the following linear programming problem LP1.

$$
\begin{array}{l}
\text{Minimize} \qquad\qquad M = \sum_{i=1}^n \alpha_i \;\; \text{subject to:} \\[2mm]
\qquad\qquad \alpha_i \geq 0, \quad 1 \leq i \leq n \\[2mm]
\qquad \sum_{P_i \in A} \alpha_i \geq 1, \quad \forall A \in \mathcal{A}^0
\end{array}
$$

Let $M_{\min}^{(1)}$ the solution of the linear programming problem LP1. The following theorem holds.

Theorem 7. Let $\mathcal{A}$ be an access structure on a set $\mathcal{P}$ of participants. The average information rate of any information dispersal algorithm Σ for $\mathcal{A}$ satisfies

$$\tilde{\varrho}(\mathcal{A}, \Pi_F, \Sigma) \leq \frac{n}{M_{\min}^{(1)}}.$$

Proof. Let $A \in \mathcal{A}^0$. From Lemma 3, it follows that $\sum_{P_i \in A} H(G_i) \geq H(G_A) \geq H(F)$. Let $x_i = H(G_i)/H(F)$, for $1 \leq i \leq n$. We have $x_i \geq 0$, and $\sum_{P_i \in A} x_i \geq 1$, $\forall A \in \mathcal{A}^0$. Then $\sum_{i=1}^n x_i \geq M_{\min}^{(1)}$. Hence,

$$\tilde{\varrho}(\mathcal{A}, \Pi_F, \Sigma) = \frac{n}{\sum_{i=1}^n x_i} \leq \frac{n}{M_{\min}^{(1)}}.$$

$\square$

3.3 General Information Dispersal Algorithms

The schemes in this section are obtained supposing the uniform probability distribution on F: hence $H(F) = \log|F|$. Let $\mathcal{A}$ be an access structure on a set $\mathcal{P}$ of participants. For all $P_i \in \mathcal{P}$, let p_i, q_i be some positive integers such that

$$\sum_{P_i \in A} \frac{p_i}{q_i} \geq 1 \qquad \forall A \in \mathcal{A}^0. \tag{2}$$

Let m be the least common multiple of $q_1, \ldots, q_n$, let $x_i = p_i/q_i$, for $i = 1, \ldots, n$, and let $b = \sum_{i=1}^n m \cdot \frac{p_i}{q_i}$. We assume the information dispersal algorithm $\Sigma(m, b)$ described in section 3.1 which works for parameters b (number of file fragments) and m (number of required fragments to reconstruct the file). We explain later how to choose the values x_i, for $i = 1, \ldots, n$, in order to optimize the information rate or the average information rate of the scheme.

Distribution Scheme

- Using $\Sigma(m, b)$ partition the file $f \in F$ into b fragments, $f_1, \ldots, f_b$.
- Assign to each participant P_i, $m \cdot x_i$ distinct fragments $f_1^{(i)}, \ldots, f_{m \cdot x_i}^{(i)}$ (this is always possible since $\sum_{i=1}^{n} m \cdot x_i$ is equal to b, the number of available fragments).

The fragment of each participant P_i consists on $g_i = (f_1^{(i)}, \ldots, f_{m \cdot x_i}^{(i)})$, for $i = 1, \ldots, n$.

Reconstruction Scheme

- Each set of participants A in the access structure collect their fragments.
- Using $\Sigma(m, b)$ reconstruct f out of the collected values.

Proposition 8. The above scheme constitutes an information dispersal algorithm for the access structure $\mathcal{A}$.

Proof. For all $A \in \mathcal{A}^0$, from condition (2) there holds $\sum_{P_i \in A} m \cdot x_i \geq m$. From this fact, and from the properties of the algorithm $\Sigma(m, b)$ derives the feasibility for a set of participants A to reconstruct the file f out of the fragments. ⊟

3.4 How to Optimize the Information Rate

We now show how to choose the values x_i, for $1 \leq i \leq n$ in order to maximize the information rate. We propose two techniques. Both of them are optimal as they reach the lower bound proved in section 3.2. Moreover, the algorithm obtained applying the second technique gives to participants fragments no longer than necessary.

First Technique

We propose the following simple method to choose the values $x_i = p_i/q_i$, for all $i = 1, \ldots, n$. For all participants $P_i \in \mathcal{P}$, let be $p_i = 1$, and $q_i = \min\{|A| : A \in \mathcal{A}^0 \text{ and } P_i \in A\}$. The following theorem holds.

Theorem 9. The above x_i satisfies condition (2). Moreover, the information dispersal algorithm Σ_1 obtained taking these values maximizes the information rate, that is $\varrho(\mathcal{A}, \Sigma_1) = \varrho_{max}$.

Proof. Let be $A \in \mathcal{A}^0$. For all $P_i \in A$, there holds $q_i \leq |A|$. Hence,

$$\sum_{P_i \in A} \frac{p_i}{q_i} \geq \sum_{P_i \in A} \frac{1}{|A|} = |A| \cdot \frac{1}{|A|} = 1.$$

So each x_i satisfies equality (2).

To prove that the scheme Σ_1 reaches the bound on information rate, observe that each participant P_i receives m/q_i fragments, each of size $\log|F|/m$. Moreover, for all $i = 1, \ldots, n$, $q_i = \min\{|A| : A \in \mathcal{A}^0 \text{ and } P_i \in A\} \geq \min\{|A| : A \in \mathcal{A}^0\} = \varrho_{\max}$. So, for $i = 1, \ldots, n$ there holds

$$\log|G_i| = \frac{\log|F|}{m} \cdot \frac{m}{q_i} = \frac{\log|F|}{q_i} \leq \frac{\log|F|}{\varrho_{\max}}.$$

As $H(G_i) \leq \log|G_i|$ and $H(F) = \log(F)$, then

$$H(G_i) \leq \frac{H(F)}{\varrho_{\max}}.$$

Then $\max\{H(G_i) : 1 \leq i \leq n\} \leq H(F)/\varrho_{\max}$, so $\varrho(\mathcal{A}, \Sigma_1) \geq \varrho_{\max}$. From Theorem 5 it follows the equality. $\qquad\square$

Second Technique

This second technique provide an information dispersal algorithm with maximal average information rate among the schemes with maximal information rate: this algorithm reaches the bound on information rate and gives to participants fragments no longer than necessary. It should be found solving the following linear programming problem called LP2.

> Minimize $\qquad M = \sum_{i=1}^{n} \beta_i$ subject to:
>
> $$0 \leq \beta_i \leq 1/\varrho_{\max}, \quad 1 \leq i \leq n$$
>
> $$\sum_{P_i \in A} \beta_i \geq 1, \quad \forall A \in \mathcal{A}^0$$

Let be $M_{\min}^{(2)} = \sum_{i=1}^{n} \beta_i^*$ the solution to the linear programming problem LP2. As each β_i^*, for $i = 1, \ldots, n$ is rational we can express it as a fraction $\beta_i^* = p_i/q_i$. The following theorem holds.

Theorem 10. The above β_i^* satisfies equation (2). Moreover, the information dispersal algorithm Σ_2 obtained taking $x_i = \beta_i^*$, for $i = 1, \ldots, n$, maximizes the information rate, that is $\varrho(\mathcal{A}, \Sigma_2) = \varrho_{\max}$.

Proof. From definition of β_i^*, for all $A \in \mathcal{A}^0$, $\sum_{P_i \in A} \beta_i^* \geq 1$, so equality (2) is satisfied.

To prove that the scheme Σ_2 reaches the bound on information rate, observe that each participant P_i receives $m \cdot \beta_i^*$ fragments, each of size $\log|F|/m$. So, for $i = 1, \ldots, n$ there holds

$$\log|G_i| = \frac{\log|F|}{m} \cdot m \cdot \beta_i^* = \log|F| \cdot \beta_i^* \leq \frac{\log|F|}{\varrho_{\max}}.$$

Then $\varrho(\mathcal{A}, \Sigma_2) \geq \varrho_{\max}$. From Theorem 5 it follows the equality. $\qquad\square$

In this way, the sum of fragments distributed to participants is minimized, while all fragments are less than $\log|F|/\varrho_{\max}$.

3.5 How to Optimize the Average Information Rate

The linear optimization problem LP1 described in section 3.2 will be used to maximize the average information rate.

Let be $M_{min}^{(1)} = \sum_{i=1}^{n} \alpha_i^*$ the solution to the linear programming problem LP1. As each α_i^*, for $i = 1, \ldots, n$ is rational we can express it as a fraction $\alpha_i^* = p_i/q_i$. The following theorem holds.

Theorem 11. The above α_i^* satisfies equation (2). Moreover, the information dispersal algorithm Σ_3 obtained taking $x_i = \alpha_i^*$, for $i = 1, \ldots, n$, maximizes the average information rate, that is $\tilde{\varrho}(\mathcal{A}, \Sigma_3) = \frac{n}{M_{min}^{(1)}}$.

Proof. From definition of α_i^*, for all $A \in \mathcal{A}^0$, $\sum_{P_i \in A} \alpha_i^* \geq 1$, so equality (2) is satisfied.

To prove that the scheme Σ_3 reaches the bound on information rate, observe that each participant P_i receives $m \cdot \alpha_i^*$ fragments, each of size $\log |F|/m$. So, for $i = 1, \ldots, n$ there holds

$$\log |G_i| = \frac{\log |F|}{m} \cdot m \cdot \alpha_i^* \leq H(F) \cdot \alpha_i^*.$$

Then

$$H(G_i) \leq H(F) \cdot \alpha_i^*.$$

So, $\sum_{i=1}^{n} H(G_i) \leq H(F) \cdot \sum_{i=1}^{n} \alpha_i^* = H(F) \cdot M_{min}^{(1)}$. Hence,

$$\tilde{\varrho}(\mathcal{A}, \Sigma_3) = \frac{nH(F)}{\sum_{i=1}^{n} H(G_i)} \geq \frac{n}{M_{min}^{(1)}}.$$

From Theorem 7 it follows the equality. $\qquad\square$

3.6 Comparison of the Techniques

Observe that the solution $M_{min}^{(2)}$ to the optimization problem LP2 is in general bigger than the solution $M_{min}^{(1)}$ to the problem LP1 in which the fragments may be bigger than $\log |F|/\varrho_{max}$. The following example shows that the three previous schemes don't give in general the same result.

Let $\mathcal{P} = \{P_1, \ldots, P_6\}$, and let be $\mathcal{A}^0 = \{\{P_1, P_2\}, \{P_2, P_3, P_4\}, \{P_2, P_5, P_6\}\}$ the basis of the access structure $\mathcal{A}$ on $\mathcal{P}$.

In the next table, we present for the three schemes Σ_1, Σ_2 and Σ_3 the following values in order to outline the differences between the techniques : the value $x_i = \frac{\log |G_i|}{\log |F|}$ for each $i = 1, \ldots, 6$, the value ϱ and the value $\tilde{\varrho}$.

	x_1	x_2	x_3	x_4	x_5	x_6	ϱ	$\tilde{\varrho}$
Σ_1	1/2	1/2	1/3	1/3	1/3	1/3	2	18/7
Σ_2	1/2	1/2	1/2	0	1/2	0	2	3
Σ_3	0	1	0	0	0	0	1	6

3.7 Modified Schemes

As we observe in section 3.1 the simple information dispersal algorithm $\Sigma(m, b)$ has the requirement $\log |F| = m \log q \geq m \log b$. So when the parameters m and b are big our algorithms work only for large files. Moreover, for special access structures m and b could be exponentially big in n.

To solve this problems, we propose now two information dispersal algorithms which work even for small files and which nearly reach the theoretical minimal bounds proved in section 3.2. The first algorithm is intended to maximize the information rate, while the second analizes the average information rate.

Algorithm Σ_4

This algorithm is simply an adaptation of algorithm Σ_1 to handle the case of small files. So, while for the algorithm Σ_1 there were for all $P_i \in \mathcal{P}$, $p_i = 1$ and $q_i = \min\{|A| : A \in \mathcal{A}^0 \text{ and } P_i \in A\}$, now, for the algorithm Σ_4 we take $p'_i = 1$ and $q'_i = \varrho_{\max} = \min\{|A| : A \in \mathcal{A}^0\}$. So, $m = \varrho_{\max}$, and $b = n$. Clearly $q'_i \leq q_i$, hence from theorem 9, for all $A \in \mathcal{A}^0$,

$$\sum_{P_i \in A} \frac{p'_i}{q'_i} \geq \sum_{P_i \in A} \frac{p_i}{q_i} \geq 1.$$

This proves that p'_i and q'_i satisfy inequality (2).
Each participant receives a fragment of size $\log |F| / \varrho_{\max}$. So, $\varrho(\mathcal{A}, \Sigma_4) \geq \varrho_{\max}$, and it follows that with this algorithm we obtain optimal information rate. Moreover, the constraint on the file size is $\log |F| \geq \varrho_{\max} \log n$, and $\varrho_{\max}$ satisfies $\varrho_{\max} \leq n$. So the algorithm Σ_4 is useful even for small files.

Algorithm Σ_5

This algorithm is an adaptation of algorithm Σ_3 which can be used with small files and which gives an average information rate close to optimum.
Suppose we have found the razios α_i^* and $M_{\min}^{(1)}$ solving the optimization problem LP1. Suppose $|F|^{1/n} > n(M_{\min}^{(1)} + 1)$. Let k be the biggest integer such that $|F|^{1/kn} > kn M_{\min}^{(1)} + n$, and let be $b = \sum_{i=1}^{n} \lceil \alpha_i^* \cdot k \cdot n \rceil$. Use in the distribution scheme for Σ_5 an information dispersal algorithm with parameters $(k \cdot n, b)$. Observe that $b \leq kn M_{\min}^{(1)} + n < |F|^{1/kn}$, so $\log |F| > kn \log b$. Then the simple algorithm of section 3.1 applies. We give to each participant P_i, $\lceil \alpha_i^* \cdot k \cdot n \rceil$ distinct fragments, each of size $\frac{\log |F|}{kn}$. So $\log |G_i| = \frac{\log |F|}{kn} \lceil \alpha_i^* \cdot k \cdot n \rceil$.
Now we give a bound on the average information rate of the algorithm.

$$\sum_{i=1}^{n} \log |G_i| \leq \sum_{i=1}^{n} (\frac{\log |F|}{kn}(\alpha_i^* \cdot k \cdot n + 1)) = (M_{\min}^{(1)} + \frac{1}{k}) \log |F|.$$

Hence, from theorem 7

$$\frac{n}{M_{\min}^{(1)} + 1/k} \leq \bar{\varrho}(\mathcal{A}, \Sigma_5) \leq \frac{n}{M_{\min}^{(1)}}.$$

It follows that with this algorithm we obtain an average information rate nearly optimal. Moreover, the constraint on the file size is $|F|^{1/n} > n\left(M_{\min}^{(1)} + 1\right)$, that is $\log|F| \geq n\log\left(n\left(M_{\min}^{(1)} + 1\right)\right)$, and $M_{\min}^{(1)}$ satisfies $M_{\min}^{(1)} \leq n$. So the algorithm Σ_5 is useful even for small files.

4 Computational Secret Sharing Schemes

In a computational secret sharing scheme, any qualified subset can reconstruct the secret but any non qualified subset obtain no computational information on the secret. The scheme presented here is a generalization of Krawczyk's one [9].

In this section we use the information dispersal algorithms described in the previous section to construct *short* secret sharing schemes for general access structures. We can choose $IDA \equiv \Sigma_i$ with $i \in \{1, \ldots, 5\}$, depending if we want optimize the information rate or the average information rate.

Let $\mathcal{P} = \{P_1, \ldots, P_n\}$ be the set of participants and $\mathcal{A}$ be an access structure on $\mathcal{P}$. Let S be the space of secrets we want to share. We assume a secure (length preserving) private key encryption function with space of plaintext S, denoted ENC. Let be K the space of keys and F the space of ciphertexts of ENC. We now assume a perfect secret sharing scheme PSS for the access structure $\mathcal{A}$ and the set on secrets K and an information dispersal algorithm IDA for the same access structure and the set of files F.

For each participant $P_i \in \mathcal{P}$ we denote by G_i the set of possible fragments given to participant P_i with IDA, and by V_i the set of possible shares given to participant P_i with PSS. Moreover, we denote by $W_i = G_i \times V_i$ the set of possible shares given to participant P_i in the scheme SS which we are going to describe. We consider uniform probability distributions both over S and over K.

Distribution Scheme of SS:

- Chose a random encryption key $k \in K$. Encrypt the secret $s \in S$ using the encryption function ENC under the key k, let $f = ENC_k(s)$.
- Using IDA partition the encrypted file f into n fragments $g_1, \ldots, g_n$, and distribute them to the participants in $\mathcal{A}$.
- Using PSS generate n shares for the key k, denoted $v_1, \ldots, v_n$, and distribute them to the participants in $\mathcal{A}$.

The share of each participant P_i, $i = 1, \ldots, n$ consists on $w_i = (g_i, v_i)$.

Reconstruction Scheme of SS:

- Each set of participants A in the access structure collect their shares.
- Using IDA reconstruct f out of the collected values g_i for all $P_i \in A$.
- Using PSS recover the key k out of v_i for all $P_i \in A$.
- Decrypt f using k to recover the secret s.

The next theorem is similar to Krawczyk's one [9].

Theorem 12. The above scheme SS constitutes a computationally secure secret sharing scheme for the access structure $\mathcal{A}$ provided that ENC is a secure encryption function and PSS a perfect secret sharing scheme.

Proof. The feasibility for a set of participants A to reconstruct the encrypted secret f out of the fragments is inherited from the properties of the algorithm IDA. Also the reconstruction of the key k out of v_i for all $P_i \in A$ is guaranteed by the secret sharing scheme PSS. Knowledge of f and k permits deriving s using the decryption function to each set of participants in the access structure $\mathcal{A}$.

As for the secrecy against a coalition of participants B not belonging to the access structure, the intuitive idea is the following. The fragments corresponding to f of all participants in B give no more information on s than f itself. On the other hand, the fragments corresponding to k of all participants in B give no information at all on k. Therefore participants in B cannot learn something about s. $\qquad\square$

The Size of Shares

The length of each share for $i = 1, \ldots, n$, is $\log |W_i| = \log |G_i| + \log |V_i|$. So, the length of the shares depends both on the information dispersal algorithm and on the perfect secret sharing scheme used to construct the scheme. Depends on what information dispersal algorithm we choose, the size of G_i is minimal for the corresponding definition. And the size of V_i does not depend on the secret size but only on the security parameter.

However, observe that for general access structures the size $\log |V_i|$ of shares of perfect secret sharing schemes used in order to share the enciphering key k should be exponentially large respect to the size of the secret key $\log |K|$: an upper bound better than exponential is not known for the length of shares in the general case. Moreover, Csirmaz [5] proved that there are access structures on n elements so that any perfect secret sharing scheme must assign a share which is of size at least $\dfrac{n}{\log n}$ times the size of the secret k.

We have better upper bounds when the access structure is based on graphs. If, for example, the graph on which the access structure is based is complete multipartite, then there exists an ideal perfect secret sharing scheme for $\mathcal{A}$ (see [3]) and the size of the shares becomes $\log |S|/\varrho_{\max} + \log |K|$. Otherwise, using bounds found in [17] we can say that $\log |V_i| \leq \log |K|(\Delta + 1)/2$, where Δ is the maximum degree of the graph. Moreover, better bounds on V_i can be obtained if the graph is acyclic.

5 Conclusions

We have shown how to realize computational secret sharing schemes for general access structures. Our schemes are nearly optimal: the size of each share is equal to the minimal theoretical bound plus a piece of information whose length does not depend on the secret size but just on the security parameter.

We remark that the size of the shares that must be kept secret is an important issue: in many cases such shares must be kept in mind or in tamper–resistant devices, so they must be very small. In our scheme, only V_i must be secret, the piece G_i could be in a hard disk or in a floppy disk. Moreover, the size of the V_i does not depend on the size of the secret file. Hence for very long files our schemes are very useful.

Finally, observe that computational schemes are not weaker than perfect ones in practical viewpoint since most of the time people uses an encryption function to distribute the shares or a pseudo-random generator to produce them. Hence our schemes are very convenient and practical for very long file.

References

1. J. Benaloh, and J. Leichter, *Generalized secret sharing and monotone functions*, in "Advances in Cryptology - CRYPTO '88", S. Goldwasser Ed., " Lecture Notes in Computer Science", Vol. 403, Springer-Verlag, Berlin, 1988, pp. 27–35.

2. C. Blundo, A. De Santis, L. Gargano, and U. Vaccaro, *On the Information Rate of Secret Sharing Schemes*, in "Advances in Cryptology - CRYPTO 92", Ed. E. Brickell, "Lecture Notes in Computer Science", Vol. 740, E. Brickell Ed., Springer-Verlag, pp. 149–169, 1993.

3. E. F. Brickell and D. M. Davenport, *On the Classification of Ideal Secret Sharing Schemes*, J. Cryptology, Vol. 4, No. 2, pp. 123–124, 1991.

4. R. M. Capocelli, A. De Santis, L. Gargano, and U. Vaccaro, *On the Size of Shares for Secret Sharing Schemes*, Journal of Cryptology, Vol. 6, No. 3, pp. 157–169, 1993.

5. L. Csirmaz, *Size of Shares Must Be Large*, in "Advances in Cryptology – Eurocrypt '94", Lecture Notes in Computer Science, A. De Santis Ed., Springer-Verlag.

6. E. D. Karnin, J. W. Greene, and M. E. Hellman, *On Secret Sharing Systems*, IEEE Trans. on Inform. Theory, Vol. IT-29, No. 1, pp. 35–41, Jan. 1983.

7. I. Csiszar and J. Körner, *Information Theory. Coding Theorems for Discrete Memoryless Systems*, Academic Press, 1981.

8. R. G. Gallager, *Information Theory and Reliable Communications*, John Wiley & Sons, New York, NY, 1968.

9. H. Krawczyk, *Secret Sharing Made Short*, in "Advances in Cryptology - CRYPTO '93", D. Stinson Ed., " Lecture Notes in Computer Science", Vol. 773, Springer-Verlag, Berlin, 1994.

10. K. Kurosawa, W. Ogata, S. Tsujii, *Nonperfect secret sharing schemes*, in "Advances in Cryptology - AUSCRYPT '92".

11. K. Kurosawa, W. Ogata, K. Okada, K. Sakano, S. Tsujii, *Nonperfect secret sharing schemes and Matroids*, in "Advances in Cryptology - EUROCRYPT '93".

12. M. Naor, and R. M. Roth, Optimal File Sharing in Distributed Networks, Proceedings of 32nd IEEE Symposium on Foundations of Computer Science, 1991, pp. 515–525.

13. C. H. Papadimitriou, and K. Steiglitz, *Combinatorial Optimization: Algorithms and Complexity*, Prentice Hall, 1982.

14. M. O.Rabin, *Efficient Dispersal of Information for Security, Load Balancing and Fault Tolerance*, Journal of ACM, Vol. 36, No. 2, 1989, pp. 335–348.

15. M. O. Rabin, *The Information Dispersal Algorithm and its Applications*, in "Sequences: Combinatorics, Compression, Security and Transmission", R. M. Capocelli Ed., Springer-Verlag, 1990, pp. 406–419.

16. A. Shamir, *How to Share a Secret* Communications of the ACM, Vol. 22, n. 11, pp. 612–613, Nov. 1979.

17. D. R. Stinson, *Decomposition Constructions for Secret Sharing Schemes*, IEEE Trans. on Inform. Theory, Vol. IT-40, pp. 118–125, 1994.

A   Information theory

In this appendix we review the information theoretic concepts we are going to use. For a complete treatment of the subject the reader is advised to consult [7] and [8].

Given a probability distribution $\{p(x)\}_{x \in X}$ on a set X, we define the *entropy* of X, $H(X)$, as

$$H(X) = -\sum_{x \in X} p(x) \log p(x)^3.$$

The entropy $H(X)$ is a measure of the average information content of the elements in X or, equivalently, a measure of the average uncertainty one has about which element of the set X has been chosen when the choices of the elements from X are made according to the probability distribution $\{p(x)\}_{x \in X}$. The entropy enjoys the following property

$$0 \le H(X) \le \log |X|, \tag{3}$$

where $H(X) = 0$ if and only if there exists $x_0 \in X$ such that $p(x_0) = 1$; $H(X) = \log |X|$ if and only if $p(x) = 1/|X|, \forall x \in X$.

Given two sets X and Y and a joint probability distribution $\{p(x,y)\}_{x \in X, y \in Y}$ on their Cartesian product, the *conditional entropy* $H(X|Y)$, also called the equivocation of X given Y, is defined as

$$H(X|Y) = -\sum_{y \in Y} \sum_{x \in X} p(y)p(x|y) \log p(x|y).$$

The conditional entropy can be written as $H(X|Y) = \sum_{y \in Y} p(y)H(X|Y = y)$ where $H(X|Y = y) = -\sum_{x \in X} p(x|y) \log p(x|y)$ can be interpreted as the average uncertainty one has about which element of X has been chosen when the choices are made according to the probability distribution $\{p(x|y)\}_{x \in X}$, that is, when it is known that the value chosen from the set Y is y. From the definition of conditional entropy it is easy to see that

$$H(X|Y) \ge 0. \tag{4}$$

---

[3] All logarithms in this paper are of base 2

If we have $n + 1$ sets $X_1, \ldots, X_n, Y$ the entropy of $X_1 \ldots X_n$ given Y can be written as

$$H(X_1 \ldots X_n | Y) = H(X_1 | Y) + H(X_2 | X_1 Y) + \cdots + H(X_n | X_1 \ldots X_{n-1} Y) \quad (5)$$

The *mutual information* between X and Y is defined by

$$I(X; Y) = H(X) - H(X | Y) \quad (6)$$

and enjoys the following properties:

$$I(X; Y) = I(Y; X), \quad (7)$$

and

$$I(X; Y) \geq 0,$$

from which one gets

$$H(X) \geq H(X | Y), \quad (8)$$

with equality if and only if X and Y are independent.

Fair Blind Signatures

Markus Stadler[1], Jean-Marc Piveteau[2], Jan Camenisch[1]

[1] Institute for Theoretical Computer Science
ETH Zürich
CH-8092 Zürich, Switzerland
Email: {stadler, camenisch}@inf.ethz.ch

[2] Union Bank of Switzerland
UBILAB
Bahnhofstrasse 45
CH-8021 Zürich, Switzerland
Email: piveteau@ubilab.ubs.ch

Abstract. A blind signature scheme is a protocol for obtaining a signature from a signer such that the signer's view of the protocol cannot be linked to the resulting message-signature pair. Blind signature schemes are used in anonymous digital payment systems. Since the existing proposals of blind signature schemes provide perfect unlinkability, such payment systems could be misused by criminals, e.g. to safely obtain a ransom or to launder money. In this paper, a new type of blind signature schemes called fair blind signature schemes is proposed. Such schemes have the additional property that a trusted entity can deliver information allowing the signer to link his view of the protocol and the message-signature pair. Two types of fair blind signature schemes are distinguished and several realizations are presented.
Keywords. Blind signatures, fair cryptosystems, electronic payment systems, cryptographic protocols.

1 Introduction

The concept of a blind signature scheme was introduced by Chaum [4]. A blind signature scheme is a cryptographic primitive involving two entities: a sender and a signer. It allows the sender to have a given message signed by the signer, without revealing any information about the message or its signature. Blind signature schemes have been used to realize cryptographic protocols providing the anonymity of some of the participants, e.g. voting protocols and secure electronic payment systems (e.g. [1, 3, 6, 7, 8, 11, 14])

Several realizations of blind signature schemes have been proposed [2, 4, 9]. All the existing proposals provide perfect unlinkability, i.e. it is impossible (in an information theoretical sense) except for the sender to link a message-signature pair to the corresponding instance of the signing protocol.

[1] The first and the third author are supported by the Swiss Federal Commission for the Advancement of Scientific Research (KWF) and by the Union Bank of Switzerland.

L.C. Guillou and J.-J. Quisquater (Eds.): Advances in Cryptology - EUROCRYPT '95, LNCS 921, pp. 209-219, 1995.
© Springer-Verlag Berlin Heidelberg 1995

Unfortunately, this anonymity could be misused by criminals. In anonymous electronic payment systems blind signatures prevent linking the withdrawal of money and the payment made by the same customer. The impossibility to relate withdrawals and payments allows perfect black-mailing [16] or money-laundering. It has been argued that this is not a problem if such payment systems are only used for small amounts. We believe that the problem still exists, especially for fully digital payment systems: it could be possible to automatically perform a large number of payments and thereby transfer huge amounts of money anonymously. Therefore, it would be useful if the anonymity could be removed with the help of a trusted entity, when this is required for legal reasons.

In [13] Micali introduces the concept of fair cryptosystems to prevent the misuse of strong cryptographic systems by criminals. We pursue a similar goal for blind signature schemes by proposing a new type of blind signature schemes, called fair blind signature schemes. They have the additional property that, with the help of a trusted entity, it is possible to link a message-signature pair and the corresponding protocol view of the signer. This concept is discussed in Section 2. Several fair blind signature schemes are presented in the last three sections.

2　The Concept of Fair Blind Signatures

The model of a fair blind signature scheme consists of several senders, a signer and a trusted entity, e.g. a judge, and of two protocols (see Fig. 1):

- A signing protocol involving the signer and a sender.
- A link-recovery protocol involving the signer and the judge.

By executing the signing protocol, the sender obtains a valid signature of a message of his choice such that the signer cannot link his view of the protocol to the resulting message-signature pair. By running the link-recovery protocol, the signer obtains information from the judge that enables him to recognize the corresponding protocol view and message-signature pair. There are two types of fair blind signature schemes, depending on the information the judge receives from the signer during the link-recovery protocol:

- *Type I:* Given the signer's view of the protocol, the judge delivers information that enables the signer (or everybody) to efficiently recognize the corresponding message-signature pair (e.g. the judge can extract the message).
- *Type II:* Given the message signature pair, the judge delivers information that enables the signer to efficiently identify the sender of that message or to find the corresponding view of the signing protocol.

Theoretically, a type I fair blind signature scheme can also be used to link a given message-signature pair to a view of the protocol by running the link-recovery protocol with all views as inputs, but this is inefficient. The same holds for type II schemes.

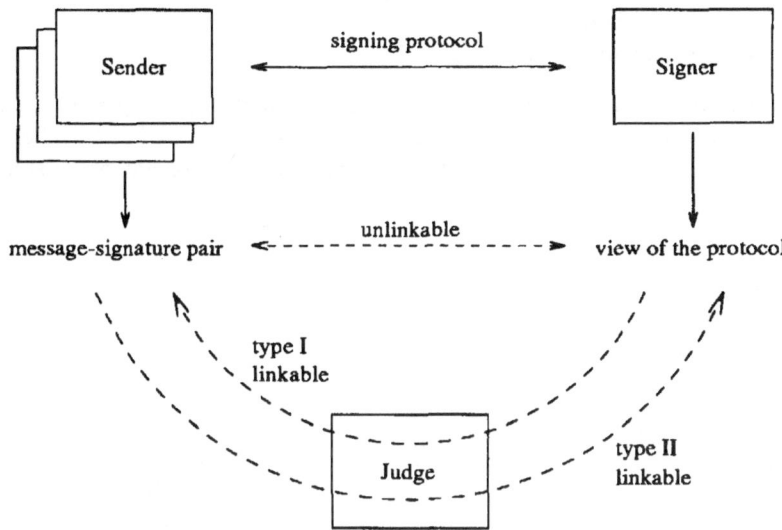

Fig. 1. The model of a fair blind signature scheme

There are different applications for fair blind signatures. One is to provide a tool to prevent money-laundering in anonymous payment systems. In a payment system based on type II fair blind signatures the authorities can determine the origin of dubious money, while in systems based on type I signatures they can find out the destination of suspicious withdrawals.

Another application is the "perfect crime" scenario described in [16]: a customer is blackmailed and forced to anonymously withdraw digital money from his account, acting as an intermediary between the blackmailer and the bank. In a perfectly anonymous payment system, the ransom could not be recognized later, but if a (type I) fair blind signature scheme had been used, the judge, when given the bank's view of the withdrawal protocol, can trace the blackmailed coins. Unfortunately, our realizations of fair blind signatures do not solve the general problem of blackmailing: a cheating sender could try to force the signer to use a different, truly blind signing protocol. The solution of this general blackmailing problem seems to be difficult.

3 Fair Blind Signatures using Cut-and-Choose

We first present a fair blind signature scheme based on Chaum's blind signature scheme and on the well-known cut-and-choose method [4, 6]. The system parameters are as follows:

- (n, e), the signer's public key ($n = pq$ is the product of two large primes and e is an integer relatively prime to $\varphi(n) = (p-1)(q-1)$).
- $E_J(\cdot)$, the enciphering function of a judge's public key cryptosystem.

- $\mathcal{H}$, a one-way hash function.
- k, a security parameter (e.g. $k > 20$).

The sender and the signer first agree on a session identifier ID (each instance of the signing protocol should correspond to a different value of ID). Then, they perform the following protocol (where $\|$ denotes the concatenation of strings).

Sender   **Signer**

for $i = 1, \ldots, 2k$
  randomly choose $r_i \in \mathbb{Z}_n$
  and strings α_i, β_i
  $u_i = E_J(m\|\alpha_i)$
  $v_i = E_J(ID\|\beta_i)$
  $m_i = r_i^e \mathcal{H}(u_i\|v_i) \pmod{n}$

$\xrightarrow{\quad m_i \quad}$

randomly choose a subset
$S \subset \{1, \ldots, 2k\}$ of size k

$\xleftarrow{\quad S \quad}$

for all $i \in S$ $\xrightarrow{\quad r_i,\, u_i,\, \beta_i \quad}$

for every $i \in S$ check $m_i \overset{?}{=}$
$r_i^e \mathcal{H}(u_i\|E_J(ID\|\beta_i)) \pmod{n}$

$b = \left(\prod_{i \notin S} m_i\right)^{1/e} \pmod{n}$

$\xleftarrow{\quad b \quad}$

$s = b/\prod_{i \notin S} r_i \pmod{n}$

The resulting signature consists of s and the set of pairs $T = \{(\alpha_i, v_i) | i \notin S\}$. The signature can be verified by checking that:

$$s^e \overset{?}{=} \prod_{(\alpha,v) \in T} \mathcal{H}(E_J(m\|\alpha)\|v) \pmod{n} \ .$$

At the end of an execution of the signing protocol, the signer is convinced that, with overwhelming probability, each v_i has been formed correctly. Since every v_i depends on ID, it is impossible for a dishonest sender to use information received during different sessions to generate a signature without following the signing protocol. Furthermore, the probability that the sender can obtain a correct signature with forged u_i is negligible.

It is easy to see that this is a fair blind signature scheme of type I and II:

- Given the values u_i, $i \in S$, the judge can disclose the message m (note that it is very unlikely that all of the u_i are forged). Therefore, the scheme is of type I.

- Given the signature (s, T), the judge can easily compute the identification string ID by decrypting the v's in T. Therefore, the scheme is of type II.

The scheme can now be modified in order to be of type I or type II, only:

- Compute all v_i as $v_i = \mathcal{H}(ID||\beta_i)$. Since the judge cannot disclose the session identifier ID anymore, this scheme is of type I, only.
- Compute all u_i as $u_i = \mathcal{H}(m||\alpha_i)$. Since the judge cannot disclose the message m anymore, this scheme is of type II, only.

Unfortunately, this fair blind signature scheme is inefficient: a large amount of data is exchanged during the signing protocol, and the resulting signature is long. More efficient implementations are considered in the next sections.

4 Type I Fair Blind Signatures using Oblivious Transfer

The type I fair blind signature scheme presented in this section is based on a variation of the Fiat-Shamir signature scheme [12] and on the concept of one-out-of-two oblivious transfer [10]. Although the signing protocol is still inefficient, the resulting signature is very short.

4.1 A Variation of the Fiat-Shamir Signature Scheme

Let $n = pq$ be the product of two large primes chosen by the signer such that 3 is relatively prime to $\varphi(n) = (p-1)(q-1)$ and let y be a random value in $\mathbb{Z}_n^*$. The pair (n, y) is the signer's public key. Let further $\mathcal{H}$ denote a one-way hash function and k be a security parameter (e.g. $k > 80$). In contrast to the original Fiat-Shamir signature scheme, this scheme uses third roots instead of square roots. Let us define the sequences

$$y_i = \mathcal{H}(y + i) \pmod{n} , \quad x_i = y_i^{1/3} \pmod{n} , \quad i = 1 \ldots k$$

Note that only the signer, knowing the factorization of n, can compute the sequence x_i. To sign a message m the signer proceeds as follows:

- randomly choose $r \in \mathbb{Z}_n^*$, compute $t = r^3 \pmod{n}$
- compute $c = \mathcal{H}(t||m)$, let c_i denote the i-th bit of c
- compute $s = r \prod_{i=1}^{k} x_i^{c_i} \pmod{n}$
- (s, t) is the signature of the message m and can be verified by checking

$$s^3 \stackrel{?}{=} t \prod_{i=1}^{k} y_i^{c_i} \pmod{n} .$$

4.2 Fair one-out-of-two Oblivious Transfer

One-out-of-two oblivious transfer (OT_2^1, see [10]) is a protocol between a sender and a receiver which allows the receiver to choose one of two messages sent by the sender in a way such that he receives only the chosen message and the sender does not know which message he has chosen (note that we allow the receiver to choose the message in contrast to the original concept introduced in [10]).

Let m_0 and m_1 denote the two messages sent by the sender and let c be the selection bit of the receiver. An execution of an OT_2^1 protocol is then denoted by

Let us now consider a modified implementation which allows a judge, but not the sender, to determine the selection bit. Let us denote an execution of such a "fair"-OT_2^1 by

A fair one-out-of-two oblivious transfer could be realized as follows: Let $n_J = p_J q_J$ be the product of two large primes so that the factorization of n_J is known to the judge only. Let further $g \in QR_{n_J}$ have a large order, and let h be a quadratic non-residue in $\mathbb{Z}_{n_J}^*$ with positive Jacobi symbol. The functions "encr" and "decr" are simple encryption and decryption functions (e.g. DES) used to transfer the messages of the sender.

Receiver		Sender
randomly choose $r \in \mathbb{Z}_{n_J}$		
$t = g^r h^c \pmod{n_J}$	$\xrightarrow{\quad t \quad}$	
		randomly choose $\alpha \in \mathbb{Z}_{n_J}$
		$A = g^\alpha \pmod{n_J}$
		$k_0 = t^\alpha \pmod{n_J}$
		$k_1 = (th^{-1})^\alpha \pmod{n_J}$
		$y_0 = \text{encr}(m_0, k_0)$
		$y_1 = \text{encr}(m_1, k_1)$
	$\xleftarrow{\; A, y_0, y_1 \;}$	
$k_c = A^r \pmod{n_J}$		
$m_c = \text{decr}(y_c, k_c)$		

Because of the quadratic residuosity assumption the sender cannot find out whether the receiver got m_0 or m_1. But the judge can easily compute the selection bit c by checking whether t is a quadratic residue in $\mathbb{Z}_{n_J}^*$ or not. On the other hand, the receiver cannot compute m_{1-c} because he cannot compute k_{1-c} due to the Diffie–Hellman assumption.

4.3 Fair Blind Fiat-Shamir Signatures

With $fair\text{-}OT_2^1$ we can now convert the signature scheme from Section 4.1 into a fair blind signature scheme of type I.

Sender   **Signer**

$$\text{choose } r_1, \ldots, r_k \in \mathbb{Z}_n^*$$
$$t = \prod_{i=1}^k r_i^3 \pmod n$$

$$\xleftarrow{\quad t \quad}$$

randomly choose $\alpha \in \mathbb{Z}_n^*$,
$\tilde{t} = t\alpha^3 \pmod n$

$c = \mathcal{H}(\tilde{t}\|m)$,
c_i is the i-th bit of c

for $i = 1 \ldots k$ do

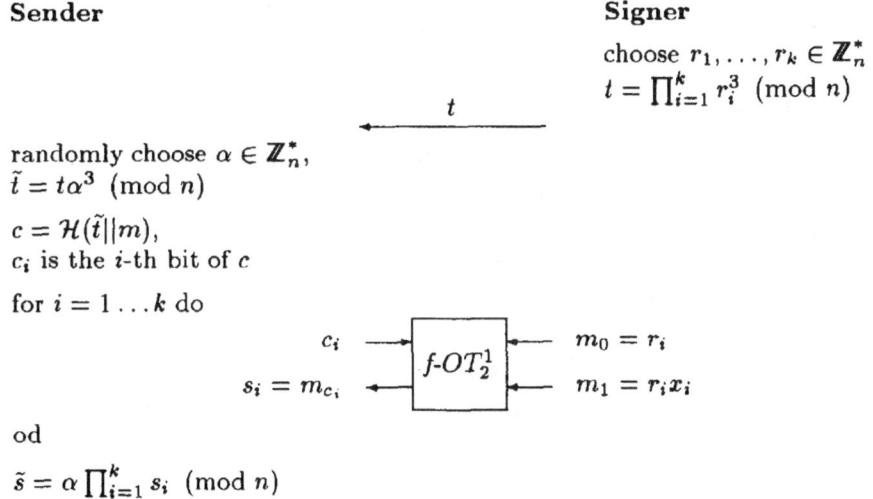

od

$\tilde{s} = \alpha \prod_{i=1}^k s_i \pmod n$

Then the pair $(\tilde{s}, \tilde{t})$ is a valid signature of m (c_i is the i-th bit of $\mathcal{H}(\tilde{t}\|m)$):

$$\tilde{s}^3 = \tilde{t} \cdot \prod_{i=1}^k y_i^{c_i} \pmod n$$

Let us analyze the blindness of this scheme. We assume that the signer cannot determine the selection bits c_i (because of the $fair\text{-}OT_2^1$). So t is the only value the signer could use to recognize the signature later. But for each valid signature $(\hat{s}, \hat{t})$ of a message $\hat{m}$ there is exactly one α with $\hat{t} = t\alpha^3 \pmod n$ and therefore $\hat{s} = \alpha \prod_{i=1}^k r_i x_i^{\hat{c}_i} \pmod n$, where $\hat{c}_i$ is the i-th bit of $\mathcal{H}(\hat{t}\|\hat{m})$. So the resulting signature is independent of the signing protocol and the signature scheme is perfectly blind (from the signer's point of view).

On the other hand, considering the fairness of the scheme, if the signer sends the view of the protocol to the judge, the selection bits c_i can be determined and therefore the challenge c is known. This value could then be put onto a black-list, so that everybody can recognize that message-signature pair later.

5 Fair Blind Signatures with Registration

Our last proposal is again of type I and II, simultaneously. The main idea is that the sender has two pseudonyms registered at the judge. One of the pseudonyms is used during the signing protocol, whereas the other one is part of the signature. Thus the judge, who knows the two corresponding pseudonyms, can link a view of the signing protocol and the corresponding signature.

If the sender uses the same pseudonyms twice, then the signer can link the two corresponding views of the signing protocol, and everyone can easily link the two resulting signatures. So, if different messages are to be unlinkable, the sender has to be registered at the judge for each single message to be signed. This scheme is therefore not suited if perfect anonymity is required, i.e. if different message-signature pairs of the same sender are to be unlinkable.

The system parameters are as follows:

- a group G of prime order q, for which it is hard to compute discrete logarithms, and a publicly known element $g \in G$.
- $y = g^x$, the signer's public key (where x is his secret key).
- $\text{Sig}_J(\cdot)$, the judge's signature scheme, so that everybody can verify messages signed by the judge.
- $\mathcal{H}$, a one-way hash function.

The scheme consists of two protocols, one for registration at the judge, and one for blind signature generation.

The registration protocol

Sender		Judge
	$\xrightarrow{\text{request}}$	randomly choose $A \in G$, $\alpha \in \mathbb{Z}_q$
		$\tilde{A} = A^\alpha$
	$\xleftarrow{A,\ \text{Sig}_J(A\|0)}$	store $(A, \tilde{A})$
	$\xleftarrow{\alpha,\ \text{Sig}_J(\tilde{A}\|1)}$	
$\tilde{A} = A^\alpha$		

The bit appended to the pseudonyms A and $\tilde{A}$ in the signature of the judge prevents a dishonest sender to permute the two pseudonyms.

The signature generation protocol

Sender		Signer

$$A, \mathrm{Sig}_J(A||0)$$
$\longrightarrow$

verifies $\mathrm{Sig}_J(A||0)$

$z = A^x$

z
$\longleftarrow$

$\tilde{z} = z^\alpha$

randomly choose $r \in \mathbf{Z}_q$
$t_1 = g^r,\ t_2 = A^r$

t_1, t_2
$\longleftarrow$

randomly choose $\beta, \gamma \in \mathbf{Z}_q$

$\tilde{t}_1 = t_1^\beta g^\gamma$

$\tilde{t}_2 = t_2^{\alpha\beta} \tilde{A}^\gamma$

$\tilde{c} = \mathcal{H}(m||\tilde{A}||\tilde{z}||\tilde{t}_1||\tilde{t}_2) \pmod q$

$c = \tilde{c}/\beta \pmod q$
$\longrightarrow$

$s = r + cx \pmod q$

s
$\longleftarrow$

$\tilde{s} = \beta s + \gamma \pmod q$

The resulting signature is the 6-tuple

$$(\tilde{A}, \mathrm{Sig}_J(\tilde{A}||1), \tilde{z}, \tilde{t}_1, \tilde{t}_2, \tilde{s})$$

It can be verified by first verifying $\mathrm{Sig}_J(\tilde{A}||1)$ and then by checking whether

$$g^{\tilde{s}} \stackrel{?}{=} \tilde{t}_1 y^{\tilde{c}}, \quad \text{and} \quad \tilde{A}^{\tilde{s}} \stackrel{?}{=} \tilde{t}_2 \tilde{z}^{\tilde{c}}$$

with $\tilde{c} = \mathcal{H}(m||\tilde{A}||\tilde{z}||\tilde{t}_1||\tilde{t}_2) \pmod q$.

This scheme can be viewed as a modification of the Chaum-Pedersen blind signature scheme [9], with the pair $(\tilde{A}, m)$ playing here a role similar to the message in [9]. Therefore, the security of our scheme is strongly related to the security of the Chaum-Pedersen blind signature scheme. Furthermore, the blindness property is easy to verify: as in [9], for any signature $(\tilde{A}, \mathrm{Sig}_J(\tilde{A}||1), \tilde{z}, \tilde{t}_1, \tilde{t}_2, \tilde{s})$ and for any signer's view, there exist α, β, γ such that the signer's view leads to that signature.

6 Conclusions

We have introduced the concept of fair blind signatures, and presented possible realizations. When applied to the design of payment systems protecting privacy, fair blind signatures allow to meet the requirements of all parties: on one hand the customers, who like to have as much privacy protection as possible, on the other hand the authorities (the bank and the judge in our model), who like to prevent criminals from misusing this privacy protection. In the usual case (which means that the judge is not involved in a transaction), the anonymity of the customer's payment is guaranteed. However, in particular situations (e.g. for legal reasons) it is possible to remove this anonymity with the help of the judge.

Fair blind signature offer a satisfactory solution against abuses of the system, like money laundering or blackmailing of customers (as it is the case for a "perfect crime" in the sense of [16]). A solution to the general blackmailing attack seems to be an open problem.

Another subject of investigation is the development of more efficient fair blind signature schemes.

Acknowledgement

We would like to thank U. Maurer and H.P. Frei for their valuable support.

References

1. S. Brands: Untraceable Off-line Cash in Wallets with Observers, *Proceedings of Crypto '93*, LNCS 773, Springer Verlag, pp. 302-318.
2. J. Camenisch, J.-M. Piveteau, M. Stadler: Blind Signatures Based on the Discrete Logarithm Problem, to appear in the proceedings of Eurocrypt '94.
3. J. Camenisch, J.-M. Piveteau, M. Stadler: An Efficient Payment System Protecting Privacy, *Proceedings of ESORICS '94*, Lecture Notes in Computer Science 875, Springer Verlag, pp. 207-215.
4. D. Chaum: Blind Signature Systems, *Proceedings of Crypto '83*, Plenum, p. 153.
5. D. Chaum, E. van Heyst: Group Signatures, *Proceedings of Eurocrypt '91*, Lecture Notes in Computer Science 547, Springer Verlag, pp. 257-265.
6. D. Chaum, A. Fiat, M. Naor: Untraceable Electronic Cash, *Proceedings of Crypto '88*, LNCS 403, Springer Verlag, pp. 319-327.
7. D. Chaum: Privacy Protected Payment, SMART CARD 2000, Elsevier Science Publishers B.V. (North-Holland), 1989, pp. 69-93.
8. D. Chaum, B. den Boer, E.van Heyst, S. Mjølsnes, A. Steenbeek: Efficient Offline Electronic Checks, *Proceedings of Eurocrypt '89*, LNCS 434, Springer Verlag, pp. 294-301.
9. D. Chaum, T. Pedersen: Wallet databases with observers, *Proceedings of Crypto '92*, LNCS 740, Springer Verlag, pp. 89-105.
10. S. Even, O. Goldreich, A. Lempel: A Randomized Protocol for Signing Contracts, *Communications of the ACM*, 28, 1985, pp. 637-647.
11. N. Ferguson: Single Term Off-line Coins, *Proceedings of Eurocrypt '93*, LNCS 765, Springer Verlag, pp. 318-328.
12. A. Fiat, A. Shamir: How to prove yourself: Practical solutions to identification and signature problems, *Proceedings of Crypto '86*, LNCS 263 , Springer Verlag, pp. 186-194.
13. S. Micali: Fair Cryptosystems, Technical Report MIT/LCS/TR-579.b, 1993.

14. T. Okamoto, K. Ohta: Universal Electronic Cash, *Proceedings of Crypto '91*, LNCS 576, Springer Verlag, pp. 324-337.
15. R.L. Rivest, A. Shamir, L. Adleman: A Method for Obtaining Digital Signatures and Public Key Cryptosystems, *Communications of the ACM*, **21**, 1978, pp. 120-126.
16. S. von Solms, D. Naccache: On Blind Signatures and Perfect Crime, *Computer & Security*, **11**, 1992, pp. 581-583.

Ripping Coins for a Fair Exchange

Markus Jakobsson*

Department of Computer Science and Engineering,
University of California, San Diego,
La Jolla, CA 92093

Abstract. A fair exchange of payments for goods and services is a barter where one of the parties cannot obtain the item desired without handing over the item he offered. We introduce the concept of ripping digital coins to solve fairness problems in payment transactions. We demonstrate how to implement coin ripping for a recently proposed payment scheme [9, 8], giving a practical and transparent coin ripping scheme. We then give a general solution that can be used in any payment scheme with a challenge. We also indicate how fairness can be obtained by building a contract into the coin.

1 Introduction

Many payments are practically concurrent with the delivery of the purchase; when you buy milk in a grocery store you cannot leave with the milk without paying, but neither can the store charge you and then not give you your milk. In this situation, there should not be any problems with trust since both parties are physically present. However, if you call a taxi and ask the driver to pick up some goods for you, and then return to deliver them to you, there could be a problem with trust: the taxi driver doesn't want to go and pick the items up without getting paid first (since you will not go with him,) but you don't want to pay him in advance risking that he is not honest. Let us say that the taxi driver's payment would be $100. Paying him half the amount before he leaves, and the rest upon delivery would not solve the problem. However, if you tear the $100 bill in two parts and give the driver one half before he leaves, and the other upon delivery, this *would* solve the problem. Neither you nor the driver will be able to use half a bill, since this will be without value, but when he returns and you give him the second half, he can tape the parts together and use the bill. If you are concerned that the taxi driver might not return just on spite, thus making you lose $100 although *he* will not get it, you can both tear bills and give each other halves in order to keep each other honest. When he returns, you will exchange the second halves, and then perform a standard payment.

Similar issues of trust can arise in many digital cash applications [1, 2, 5, 6, 7, 8, 9, 10, 11, 13], since the two parties in the transaction may be geographically

---

* Research partly done at DigiCash, the Netherlands. Email address: markus@cs.ucsd.edu

L.C. Guillou and J.-J. Quisquater (Eds.): Advances in Cryptology - EUROCRYPT '95, LNCS 921, pp. 220-230, 1995.
© Springer-Verlag Berlin Heidelberg 1995

large distances apart, and also because of the inherent problems of proving that a payment or delivery did or did not take place. In section 2 and 3 we analyze the problem in the context of anonymous electronic cash protocols, and discuss the intuition behind our solution based on coin ripping. In section 4, we show how to rip coins in a newly proposed privacy protecting payment system [9]. In section 5 we show how to rip a coin in any challenge based payment scheme.

A different solution to achieve a fair exchange is to let the challenge be a (randomly) hashed contract. This allows the buyer to prove that a certain deposited coin was designated for a certain purchase by showing the hash preimage to the challenge. However, it gives a solution with the power balance on the Shop's side; if the Shop does not deliver the item paid for, the buyer will have to sacrifice his anonymity in order to make a complaint at the Bank. We will show how this solution works in the Appendix.

Our coin ripping solution gives the buyer the advantage, and lets an honest buyer remain anonymous at all times. If the buyer decides not to fulfil the payment after the ordered item or service is delivered, neither the buyer nor the seller will get the coin, and moreover, the seller will not be able to prove what the buyer has done, but only blacklist him for future transactions (assuming the seller knows who the buyer is, which often is the case.) Often, though, the seller will be the most powerful party of the two and this balance might therefore be preferable. If such pranks are a persistent problem, the solution can be modified to increase the costs of pranks by requiring rip-spent coin payments in both directions.

2 Our Requirements

The payment systems we will apply our coin ripping method to will be assumed to have the following properties:

1 Representations of coins cannot be created without the help of the Bank.
2 The withdrawer will be able to prevent the Bank from correlating the withdrawn coins to the withdrawer when spent and deposited. In existing schemes, withdrawal protocols with this property often use what is called a "blind signature."
3 The receiver of the payment will be able to verify the correctness of a coin without the help of the Bank.
4 Any party or coalition of parties that attempts to spend more coins than they have withdrawn (so called "over-spending,") will be detected by the Bank, and moreover, the Bank will be able to calculate the identity of the over-spender with an overwhelming probability.

We will let the payer *rip-spend* a coin, and add the following requirements while keeping the above properties for both rip-spent and fully spent coins.

5 A rip-spent coin as such is useless for the receiver of the payment, as he will not be able to deposit it, i.e., it is infeasible to create a depositable coin from any number of rip-spent coins.

6 The receiver of the payment will be able to verify that a rip-spent coin is of the right form.

7 If a payer cheats the receiver of the payment by not giving him the second part of the coin later, the payer is prevented from using the coin in any other payments.

8 The receiver of the payment will be able to verify that the second part of the coin fits with the first part, and thereby be convinced that he will be able to deposit it.

A trivial solution is to use a two-spendable coin, i.e., a coin that can be spent twice without revealing the identity of its owner [2, 9] and let the payer rip-spend a coin by performing one of the two spendings, and fully spend it by spend the second. The Bank will only give a person money or credit when *both* of the two corresponding coins are deposited. If the payer decides to cheat the receiver of the ripped coin, he will not be able to use the coin for a payment later, as he will then have spent the coin for a total of three times, and will thus be identified by the Bank as an overspender. This solution does not, however, have the following desirable feature that we will require:

9 A payer may not spend or rip-spend more coins than he has withdrawn, or he will be detected and identified.

(This requirement is a little bit different from requirement 5, as we are referring to fully spent coins in requirement 5, but relate to rip-spent coins here. For a two-spendable coin, the coin owner could instead of correctly spending a coin, spend it twice in a rip-spending fashion without getting caught. We want a solution such that one coin can only be rip-spent once without detection.)

Another setback with this trivial solution is that there will be an added cost of letting users rip-spend coins; instead of one-spendable coins, all users, Shops, and the Bank will have to store and send two-spendable coins, which have larger representations. Two desirable (but not necessary) properties we will require of our new coins will be:

10 Coins that can be ripped will not have a larger representation than normal coins.

11 The Bank can never know whether a coin will be or has been ripped unless the payer defaults on a rip-spent coin; in other words, the ripping will be transparent to the Bank.

3 Intuition

We will here discuss a sketch of a general solution that has all the listed features of the last section. First, we will describe a rather general format for electronic cash systems fitting many proposed schemes [1, 2, 5, 7, 8, 9, 10] to a high degree.

Bank Signatures: The Bank authenticates coins using a signature function pair (S_{Bank}, V_{Bank}), where S_{Bank} is the private function used for producing a signature, and V_{Bank} is the public function used for verification of a signature.

A Coin: A coin is a pair (x, s), where x is a random number known only to the withdrawer and $s = (S_{Bank} \circ g)(x)$ for some function g. The coin scheme must be secure against existential forgery [12], and a signature of a message must not be useful in producing a signature of another legal message. The pair (x, s) must be chosen by both the withdrawer, Alice, and the Bank in a way that achieves *blinding*, i.e., so that the Bank can not know which message was signed. We will in this section disregard many details for the sake of clarity, and avoid the discussion of how schemes are made to be safe against existential forgery and how the blinding will be done.

Spending a Coin: The user spends a coin by engaging in a protocol with the Shop that gives the Shop the transcript $(\overline{x}, c, \overline{s})$, where c is a challenge chosen by the Shop and $\overline{x}$ and $\overline{s}$ are functions of x and s, and of c. The triple $(\overline{x}, c, \overline{s})$ is non-forgeable and its correctness can be verified by the Shop. In many implementations, including this, a one-spendable coin corresponds to a line whose equation in its turn corresponds to the identity of the withdrawer. For each time a coin is spent, the payer has to give a point on this line. Answering two challenges for one coin will reveal the equation of the line, and thereby also the identity of the over-spender. Answering only one challenge per coin will give no Shannon information about the identity of the payer.

Depositing a Coin: The Shop sends the Bank the triple $(\overline{x}, c, \overline{s})$, and the Bank verifies its correctness. The triple is related to the withdrawal session where (x, s) was obtained; thus, the Bank is able to trace coins given two transcripts

Ripping a Coin: The user can rip-spend a coin by applying a one-way function f to either $\overline{x}$, $\overline{s}$, or parts of these. The Shop will at the end of the payment protocol have received the transcript $(\overline{x}', c, \overline{s}')$, where either $\overline{x}'$ or $\overline{s}'$ is the result of applying f to $\overline{x}$ or $\overline{s}$, or parts thereof. We will do this so that the Shop can still verify the correctness of the triple in order to be convinced that it is a valid coin. However, the Bank will not accept the triple $(\overline{x}', c, \overline{s}')$ for credit, as it will not be of the correct form, and so, the Shop cannot deposit it for credit. It is important that the Shop will not be able to produce a valid, depositable transcript $(\overline{x}, c, \overline{s})$ from $(\overline{x}', c, \overline{s}')$. The payer completes the payment (gives the second part of the coin) by sending the Shop information that allows the Shop to calculate the depositable transcript $(\overline{x}, c, \overline{s})$. Should the payer not do this, then the Shop gives the triple $(\overline{x}', c, \overline{s}')$ to the Bank. This will prevent the payer from spending the coin again without being detected, as the identity of the payer will be possible to obtain from two transcripts from the same coin, even though f is applied to the transcript.

It should be apparent that our general outline of the payment protocol using ripping does not necessarily lose any of the listed properties, but we will certainly have to specify and analyze our solution before it is clear that all the properties are indeed retained. We will in the next section give an example of a payment protocols, and show specifically how coins in this can be ripped.

4 Coin-Ripping for a Specific Protocol

We will show how to rip a coin in an abstraction of the protocol by Ferguson [9], but the method is applicable to many other schemes as well. We will make only minor changes to the protocols, and our solution will satisfy all the requirements listed. In our description, Alice will be the payer and Shop the receiver of the payment, and we will use the denotation introduced in the previous section.

Bank Signatures: The Bank authenticates coins using a signature function pair (S_{Bank}, V_{Bank}), where $S_{Bank}(x) = x^{1/e}$ modulo a composite N. Here, $e = e_1 e_2$, where $e_1 \neq \pm 1$ is an odd integer and e_2 is a large prime. (In the original protocol, [9], $e_1 = 1$ was used.)

Original Payment Protocol: The following protocol is executed:

1. Alice sends $\overline{x}$ to the Shop.
2. The Shop constructs a challenge c and sends it to Alice.
3. Alice sends the answer $\overline{s}$ to the Shop, who verifies that $V_{Bank}(\overline{x}) = \Theta(\overline{s}, c)$ for a function Θ.

Rip Coin Payment Protocol: The following protocol is executed to rip-spend the first part of a coin:

1. Alice sends $\overline{x}$ to the Shop.
2. The Shop constructs a challenge c and sends it to Alice.
3. Alice sends the answer $\overline{s}' = \overline{s}^{e_1}$ to the Shop, who verifies that $V_{Bank}(\overline{x}) = (\Theta(\overline{s}, c))^{e_1}$.

Thus, the difference is that instead of giving an eth root as a signature, the payer only gives an e_2^{th} root. This way, the Shop does not get a transcript that he can deposit in return for money, unless it can calculate e_1^{th} roots. Alice spends the second part of the coin by giving $\overline{s}$ to the Shop, who verifies that this is correct. Should Alice not give the Shop this, then the Shop can give the transcript $(\overline{x}, c, \overline{s}')$ to the Bank, who after verifying that it is a correct transcript for a rip-spent coin stores it. Should Alice ever spend the same coin again, then the Bank can find out her identity exactly as for a normal over-spending, and punish her for over-spending. Note that giving the second part of the rip-spent coin does not constitute an over-spending as only one challenge will have been responded to.

Theorem 1: The rip-spending protocol preserves the anonymity of the payer.

Proof: Assume that there is a strategy for the Shop and the Bank to collaborate in the rip-spending protocol to find out any information about the identity of the payer. They can use this strategy to find the same information in the original coin spending protocol as follows: Let $(\overline{x}, c, \overline{s})$ denote a fully spent coin. Let $(\overline{x}, c, \overline{s}')$ be the same coin, but ripped. The Shop can calculate the transcript $(\overline{x}, c, \overline{s}')$ from $(\overline{x}, c, \overline{s})$ as f is public. Therefore, if any information about the identity of a spender of a rip-spent coin can be found, the same information must be possible to obtain for a fully spent coin. Thus, since the original protocol preserves the privacy of the honest user, so must our rip-spending protocol. $\square$

Theorem 2: If the payer after withdrawing k coins successfully can spend or rip-spend $k+1$ coins without detection, then he could do the same for Ferguson's protocol with $S_{Bank}(x) = x^{1/e_2} \, mod \, N$.

Proof: This holds since the rip-spend protocol is identical to Ferguson's original protocol for $S_{Bank}(x) = x^{1/e_2} \, mod \, N$, and any spent coin (using $S_{Bank}(x) = x^{1/(e_1 e_2)}$) can be used to construct a rip-spent coin (for which $S_{Bank}(x) = x^{1/e_2}$.) $\square$

We will now prove that the receiver of a rip-spent coin will not be able to cash it unless he can invert a one-way function:

Theorem 3: The Shop will not be able to construct a cashable coin $(\hat{x}, \hat{c}, \hat{s})$ from a rip-spent coin $(\overline{x}, c, \overline{s}')$ unless he can calculate $e_1{}^{th}$ roots.

Proof: Assume that there is a polynomial-time cheating strategy for the Shop, allowing it to construct a coin $\hat{C}_1 = (\hat{x}, \hat{c}, \hat{s})$ that can be deposited from a rip-spent coin $C_{1/2} = (\overline{x}, c, \overline{s}')$. Call the algorithm for this A.

Let Alice spend a coin $C_1 = (\overline{x}, c, \overline{s})$ in a Shop. The Shop can produce a rip-spent coin $C_{1/2} = (\overline{x}, c, \overline{s}')$ from this by applying f to the $\overline{s}$. Let $C_{1/2}$ be the input to A and call the output $\hat{C}_1$. There are three possibilities:

- $C_1 = \hat{C}_1$:
  This means that given input $C_{1/2} = (\overline{x}, c, \overline{s}')$, where c is chosen according to some strategy $\Phi(\overline{x})$, A produces an output $C_1 = (\overline{x}, c, \overline{s})$, where $\overline{s} = \overline{s}'^{1/e_1}$. In the original protocol, different exponents are used to encode different coin denotations. Thus, being able to produce this kind of transcript $\hat{C}_1$ from $C_{1/2}$ allows us to change the value of a coin in the original protocol.
- $C_1 = (\overline{x}, c_1, \overline{s}_1)$, $\hat{C}_1 = (\overline{x}, c_2, \overline{s}_2)$, $c_1 \neq c_2$:
  This means that given a correct payment transcript for some coin with one challenge answered, we can produce a correct payment transcript for the

same coin but with another challenge answered. Two such transcripts counts as two spendings, and given a one-spendable coin, the Bank could given these two calculate the identity of the spender. This contradicts the fact that the original protocol which we build our protocol on is privacy preserving, as this would give an algorithm for deciding the identity of the spender of a coin in that protocol.

- $C_1 = (\overline{x}_1, c_1, \overline{s}_1)$, $\hat{C}_1 = (\overline{x}_2, c_2, \overline{s}_2)$, $\overline{x}_1 \neq \overline{x}_2$:

  Both C_1 and $\hat{C}_1$ are valid coins. However, since $\overline{x}_1 \neq \overline{x}_2$, they are two coins with *distinct* secret representations, i.e., two different coins. Since they both stem from one withdrawal session, and both will be accepted by the Bank as distinct and valid coins, this will allow a coalition of Shops and Payers to make the Bank accept $k + 1$ deposited coins after only having withdrawn k coins in the original protocol using $S_{Bank}(x) = x^{1/(e_1 e_2)}$. Since coins using $S_{Bank}(x) = x^{1/e_2}$ can be calculated from such coins, this gives us a contradiction if coins cannot be forged in the original protocol. $\square$

We have thus proved that the introduction of coin ripping can be made safely for the protocol by Ferguson.

5   The General Solution

We have in our example assumed the use of RSA-signatures to authenticate the coins. However, we will show how any digital cash scheme where the payment protocol is of a certain general form can be modified to allow ripping. The form is a three move protocol with the payer sending a message $\overline{x}$ to the Shop, who responds with a challenge c of some minimal size followed by the payer sending a response $\overline{s}$ to the challenge c. In this general solution, however, rippability will incur a small storage overhead, as well as a minor communication overhead. The idea is to use a special form of challenge jointly set by the payer and the receiver of the payment.

Let $(f_1, f_2, \oplus)$ be a triple of functions such that it is hard to find a collision (x_1, x_2), (x_1', x_2') such that $f_1(x_1) \oplus f_2(x_2) = f_1(x_1') \oplus f_2(x_2')$. An example of such a triple $(f_1, f_2, \oplus)$ is

$$
\begin{cases}
f_1(x) = g_1{}^x \quad mod \text{ p} \\
f_2(x) = g_2{}^x \quad mod \text{ p} \\
a \oplus b = ab \quad mod \text{ p}
\end{cases}
$$

where p is a prime. This is collision-free if the representation problem is hard; this has been shown being as hard as the discrete logarithm problem [3].

The participants jointly calculate the challenge, c, the following way:

1. Alice chooses a random number r_A and calculates $c_A = f_1(r_A)$. She calculates a commitment *com* to c_A, unconditionally safe for the receiver. Alice sends *com* to the Shop.
2. The Shop sets his share of the challenge, c_S, as $f_2(r_S)$ for some r_S, and sends this c_S to Alice.
3. Alice opens up her commitment.
4. The Shop verifies the correctness of the commitment *com*. The challenge is $c = c_A \oplus c_S$.

The user then rip-spends the coin using this challenge c, spending the coin as he usually would. He spends the other half by revealing r_A to the Shop. In order to deposit the coin, the Shop will have to give the Bank the spent coin *and* the pair (r_A, r_S), thus proving that the second part of the coin has been spent. In order to file a complaint (if Alice did not give the second part of the coin,) the Shop just sends the Bank the spent coin. A coin can be spent all in one part by letting the Shop set both r_A and r_S, and construct the challenge the usual way from these numbers.

Theorem 4: The rip-spending protocol preserves the anonymity of the payer.

This proof is similar to the proof of Theorem 1. Assume that there is a strategy for the Shop and the Bank to collaborate in the rip-spending protocol to find out any information about the identity of the payer. They can use this protocol to help them find the same information in the original coin spending protocol as follows: The Shop sets the challenge as $c = f_1(r_A) \oplus c_S$, where r_A is chosen at random and c_S is chosen according to the cheating strategy for the rip-spent coin. Thus, if the Shop can learn any secret information from a ripped coin, it could learn exactly the same kind of information in the original protocol, so ripping a coin will leak no information. $\square$

Theorem 5: If Alice does not send the second part of the coin then she will be prevented from spending this coin later.

Proof: This holds since two rip-spent coins will correspond to an over-spent coin in the original protocol if the special form of the challenge is not considered, and so the Bank will be able to find the identity of the over-spender with overwhelming probability. To see that the probability of finding a cheating payer's identity remains the same as in the original protocol, note that Alice cannot influence what the final challenge will be, since she cannot find collisions for the commitment scheme, as we are using a commitment scheme unconditionally safe for the receiver. Thus, the probability of finding Alice's identity is not affected by the fact that she sets part of the challenge. $\square$

Theorem 6: The Shop will not be able to construct a cashable coin $(\hat{x}, \hat{c}, \hat{s})$ from a rip-spent coin $(\overline{x}, c, \overline{s}')$ unless it can invert f_1.

Proof: The Shop cannot find an r_A such that $c_A = f_1(r_A)$; since *com* is a commitment of c_A and not r_A, the commitment cannot give any help. Therefore, he cannot find another representation of the challenge used for that coin. The rest of the proof is analogous to the proof of Theorem 3, with the small difference that the Shop in the first case instead will be able to invert f_1. □

Thus, we have proved that rip-spending can be safely introduced in any payment protocol with a challenge of sufficient length.

6 Conclusion

We have shown how coins can be rip-spent in order to achieve fairness in payment protocols. We have given an efficient solution based on an existing payment protocol, and then a general solution for any challenge-based scheme with large enough challenge.

There are many alternative settings and extensions to the protocol we have looked at. For example, we can obtain extensions to multi-spendable coins and divisible coins without any further changes to the protocol; for each rip-spent coin or rip-spent share of coin, the spender will have lost his ability to fully spend this coin or share of coin without being charged with over-spending.

Similarly, observers can be used at the same time as coin ripping. An observer is a tamper-safe piece of hardware that keeps part of the coins to prevent over-spending. It will only answer the correct number of challenges for each coin, e.g., one challenge for a one-spendable coin, two for a two-spendable, etc. Since only one challenge will need to be answered for each one-spendable coin, even if it is ripped and spent in two parts, the observer will not be affected. The ripping algorithm described can be located entirely in the non-observer part of the device.

We note that the efficiency will be basically unchanged with ripping of coins. We will have to communicate one extra message, the "opening up" of the hidden information, and both parties will have to compute one extra function. Also, they will have to store the rip-spent coin along with some tag specifying the transaction involved until the second part will be released.

7 Acknowledgements

Thanks to David Chaum for suggesting the problem and Russell Impagliazzo, Stefan Brands, Niels Ferguson, Giovanni Di Crescenzo and Karin Högstedt for valuable feedback.

References

1. S. Brands, "An Efficient Off-line Electronic Cash System Based On The Representation Problem," CWI Technical Report CS-R9323, April 11, 1993
2. S. Brands, "Untraceable Off-line Cash in Wallet with Observers," Crypto '93, pp. 302-318
3. S. Brands, "An Efficient Off-line Electronic Cash System Based On The Representation Problem", CWI Technical Report CS-R9323, April 11, 1993. Can be obtained by ftp from ftp.cwi.nl, directory pub/CWIreports/AA, filename CS-R9323.ps.Z.
4. D. Chaum, "Achieving Electronic Privacy," Scientific American, August 1992, pp. 96-101
5. D. Chaum, A. Fiat, M. Naor, "Untraceable Electronic Cash," Crypto '88, pp. 319-327
6. D. Chaum, T. Pedersen, "Wallet databases with observers," Crypto '92, pp. 89-105
7. R. Cramer, T. Pedersen, "Improved Privacy In Wallets With Observers," Eurocrypt '93, pp. 329-343
8. N. Ferguson, "Single term off-line coins," Eurocrypt '93, pp. 318-328
9. N. Ferguson, "Extensions of Single-term Coins," Crypto '93, pp. 292-301
10. N. Ferguson, "Single term off-line coins," Technical Report CS-R9318, CWI, Amsterdam, 1993. Anonymous ftp: ftp.cwi.nl:/pub/CWIreports/AA/CS-R9318.ps.Z
11. M. Franklin, M. Yung, "Secure and efficient off-line digital money," Automata Languages and Programming, 20th International Colloquium, ICALP '93, pp. 265-276
12. S. Goldwasser, S. Micali, R. Rivest, "A 'Paradoxical' Solution to the Signature Problem", 25th Annual Symposium on Foundations of Computer Science, 1984, pp. 441-448
13. T. Okamoto, K. Ohta, "Universal Electronic Cash," Crypto '91, pp. 324-337

8 Appendix: Another Approach for Fairness

Another method for achieving fairness is to let the Shop and the payer, Alice, agree on a contract C, the Shop pick a random value r, and use $c = h(C, r)$ as the challenge, where h is a collision-free hash function. When Alice receives the item paid for, she will give a signature $\sigma = S_{Alice}(r)$ as a receipt. We will assume that this signature and the delivery will be simultaneous. (Note that this need not be assumed for ripped coins.) Here, we will require S_{Alice} to be a signature function that is *not* safe against existential forgery, i.e., for which it will be possible to find correct message-signature pairs for random messages. This holds for RSA since we can set σ and then calculate $r = \sigma^{e_{Alice}}$ modulo N_{Alice}.

If the Shop cashes the coin but does not deliver the item ordered, then Alice can go to the Bank and show the Bank (C, r) as a proof that she has paid. The Bank finds the spent coin protocol $(\overline{x}, c, \overline{s})$ for which $c = h(C, r)$ and asks the Shop to show the receipt σ that the ordered item was received. If the receiver of the payment cannot give the Bank σ, then the receiver must not have delivered the item ordered, and the payer wins the case; if the receiver has been given the receipt σ, then he shows this to the Bank and the payer loses the case.

Theorem 7: It is not possible to win a case for a contract that was never agreed on, or where the receipt of delivery was given to the Shop.

Proof: This holds since h is collision-free, meaning that it will not be possible to find two pairs $(r, C), (r', C')$ such that $h(C', r') = h(C, r)$. Thus, it is not possible to find a new contract for a certain challenge, and therefore, Alice cannot falsely claim that the Shop did not follow the contract used for the challenge. $\square$

Theorem 8: The privacy of the payment scheme will not be compromised by the use of signed receipts.

Proof: We will show that The Shop can produce a transcript $(\overline{x}, \overline{s}, C, r, \sigma)$ such that for an arbitrary supposed payer Alice

$$\begin{cases} c = h(C, r) \\ \sigma = S_{Alice}(r), \end{cases}$$

and $(\overline{x}, c, \overline{s})$ is a correct transcript for a spent coin.

Let (x, s) be the representation of a coin *withdrawn by the Shop*. The Shop performs the following steps:

1. Select a random σ.
2. Calculate $r = \sigma^{e_{Alice}}$.
3. Set an arbitrary contract C.
4. Calculate a challenge $c = h(C, r)$.
5. Spend the withdrawn coin using the challenge c.

It is not possible to see that the coin was withdrawn by the Shop and not Alice; therefore, and because of the random invertibility of the signature function S_{Alice}, the Shop can fake transcripts by using his own coins. Thus, he cannot prove that Alice in fact did buy something from him, and the protocol is privacy preserving with the use of contracts if it is privacy preserving without. $\square$

Restrictive Binding of Secret-Key Certificates *

Stefan Brands

CWI, P.O. Box 94079, 1090 GB Amsterdam, The Netherlands
E-mail: brands@cwi.nl

Abstract. Many signature transporting mechanisms require a signer to issue triples, consisting of a secret key, a matching public key, and a certificate of the signer on the public key. Of particular interest are so-called restrictive blind signature issuing protocols, in which the receiver can blind the issued public key and the certificate but not a certain predicate of the secret key.
This paper describes the first generally applicable technique for designing efficient such issuing protocols, based on the recently introduced notion of secret-key certificates. The resulting three-move issuing protocols require the receiver to perform merely a single on-line multiplication, and the property of restrictive blinding can be proved with respect to a plausible intractability assumption. Application of the new issuing protocols results in the most efficient and versatile off-line electronic cash systems known to date, without using the blind signature technique developed by Chaum.

1 Introduction

A restrictive blind signature issuing protocol enables an issuer to issue a triple, consisting of a secret key, a matching public key, and a certificate of the issuer on the public key, in such a way that (1) the public key and the certificate can be blinded by the receiver, whereas (2) a certain non-trivial "blinding-invariant" predicate of the secret key cannot. This notion was introduced in [1] for the case that the certificate is a public-key certificate (that is, a digital signature on the public key), and one particular embodiment was provided. From this embodiment the most efficient privacy-protecting off-line cash system known to that date was constructed, and it was shown how to incorporate several extensions in functionality that had not been realized before. Oddly enough, it is unclear how to generalize the design technique underlying the embodiment in [1] in order to design other embodiments of restrictive blind signature issuing protocols.

Recently [3] the notion of secret-key certificates was introduced. These certificates can be used for secure management of cryptographic keys in much the same way as can public-key certificates. As a consequence it makes sense to issue

---

* Patent pending.

L.C. Guillou and J.-J. Quisquater (Eds.): Advances in Cryptology - EUROCRYPT '95, LNCS 921, pp. 231-247, 1995.
© Springer-Verlag Berlin Heidelberg 1995

secret-key certificates by means of a restrictive blind signature issuing protocol. Is there a methodology for designing such issuing protocols for the new type of certificates? This is the problem that we are concerned with in this paper.

The main result of this paper is a technique for converting secret-key certificate schemes into restrictive blind signature schemes. The new technique can be applied to any signature scheme of the so-called Fiat-Shamir type, if only it can be turned into an ordinary blind signature scheme (as defined by Chaum [9]) by applying a divertability technique due to Okamoto and Ohta [18]. No such generally applicable technique is known for public-key certificates.

The advantages of the resulting issuing protocols over the restrictive blind signature issuing protocol in [1] are threefold: each of the new protocols requires the receiver to perform only a single on-line multiplication (all other computations can be performed off-line), as opposed to several hundred in [1]; issuing protocols can be designed that can be used in conjunction with showing protocols that are as secure as the RSA assumption, instead of the Discrete Log assumption; and it can rigorously be proved that a single receiver cannot blind the blinding-invariant predicate of the secret key, assuming only a plausible intractability assumption—no such proof is known for the scheme in [1].

The new technique has a direct bearing on the design of efficient privacy-protecting mechanisms for signature transport; for details, the reader is referred to [5] and [6]. Since none of the new issuing protocols is a blind signature issuing protocol as defined by Chaum in [9] and his later work, this radically falsifies the popular belief that efficient privacy-protecting off-line electronic cash systems must be based on withdrawal protocols that are blind signature issuing protocols.

This paper is organized as follows. In Sect. 2 background information needed to understand the subsequent exposition is provided. In Sect. 3 the general technique for restrictive blinding of secret-key certificates is explained on the basis of a particular signature scheme. In Sect. 4 the difference between restrictive blinding of secret-key certificates and the blinding technique of Chaum is discussed. Finally, references are provided in Sect. 5 to articles that explain how to apply the new technique to privacy-protecting signature transporting mechanisms.

Due to the imposed page limit, this paper is a shortened version of the original paper [5]. Where radical pruning has taken place, a reference to the full paper has been substituted.

2   Background

A summary is provided in this section of several basic notions; an understanding of these notions is an absolute prerequisite in order to understand the material in Sect. 3. Because in the next section the technique for designing restrictive blind secret-key certificate issuing protocols will be explained for explicitness in terms of the Guillou-Quisquater signature scheme [16], this particular scheme will serve throughout this section to illustrate the basic notions.

2.1 Digital Signatures

In a digital signature scheme, the objects of interest are pairs consisting of a message and a corresponding digital signature of a signer. A signature scheme consists of several items [15]. First, a verification algorithm that determines what exactly constitutes a digital signature on a message. This algorithm usually is deterministic and can hence be represented in terms of an equality relation. Secondly, a key generation algorithm that generates a key pair for the signer. And thirdly, a signature scheme that specifies an *issuing protocol* between the signer and a receiver.

It is the issuing protocol that we are most concerned with in this paper. The primary purpose of any issuing protocol is to provide a means to ensure that the signer can issue message-signature pairs in a one-to-one correspondence with executions of the issuing protocol. There is virtually no limit to the variety of different issuing protocols that may be used by the signer to issue a certain type of digital signature; it is only on the basis of its intended purpose and (presumed) security that one particular issuing protocol will be more suitable than another. "Ordinary" signature issuing protocols are only intended to meet the primary purpose mentioned above, and need normally not be interactive. If other purposes are to be met as well, then interaction may need to be incorporated.

The most practical type of digital signature schemes known to date originates from a technique introduced by Fiat and Shamir [14]. For this reason we will henceforth refer to this type of scheme as a Fiat-Shamir type signature scheme. Signature schemes of the Fiat-Shamir type have in common that they are derived from three-move sound identification protocols that do not leak "useful" information about the secret key of the prover (without being zero-knowledge), by replacing the challenge of the verifier by a one-way hash of the message and the information sent by the prover in the first move; the verification relation of the underlying identification protocol determines the signature verification algorithm, and the modified protocol is the signature issuing protocol. As a by-product, the interaction is no longer needed, since the signer can compute the challenge by itself. Among the known Fiat-Shamir type signature schemes are the Fiat-Shamir scheme itself [14], the Feige-Fiat-Shamir scheme [13], the Schnorr scheme [20], the Guillou-Quisquater scheme [16], the Brickell-McCurley scheme [8], and the Okamoto schemes [17]. Although no reductions from well-known problems to the security of any of these Fiat-Shamir type signature schemes are known, it is generally believed that they are secure.

Since the exposition in Sect. 3 will be in terms of the Guillou-Quisquater scheme, we will summarize this scheme here. The computations in the Guillou-Quisquater signature scheme are performed in a multiplicative group modulo n, denoted by $\mathbb{Z}_n^*$, with n being the product of two distinct large primes. The computations in the exponents are performed modulo a number v. For convenience, but without loss of generality, we will always assume that v is a prime number that is not a proper divisor of the order $\varphi(n)$ of $\mathbb{Z}_n^*$; another suitable choice would have been to take v to be twice such a prime number. Furthermore, in expressions involving multiplications and divisions of numbers in $\mathbb{Z}_n^*$ the "mod n" operator will never be written down explicitly.

Before describing the three constituents of the Guillou-Quisquater signature scheme, we will state the RSA assumption [19], since this is part of what the scheme derives its presumed security from:

Assumption 1. *Let n denote the product of two distinct prime numbers, v a prime number that is co-prime with $\varphi(n)$, and h_0 an element of $\mathbb{Z}_n^*$. No probabilistic polynomial-time algorithm A, on given as input a triple (n, v, h_0) generated according to some appropriate probability distribution, can output $h_0^{1/v}$ with non-negligible probability of success. The success probability is taken over the coin tosses of A and the probability distribution of the input triple.*

An appropriate probability distribution may select n by generating uniformly at random two hard primes of length k, and multiplying them. From now on, the use of the indication "random" will refer to a probability distribution over the specified set that is polynomially indistinguishable from the uniform distribution, and that is independent of any other event. Without loss of generality, we will assume henceforth that n and v are generated independently of h_0 and are always of the correct form (meaning that, for instance, the probability that v is not prime is zero, instead of merely negligible), and that h_0 is generated at random from $\mathbb{Z}_n^*$.

The *key generation algorithm* for the Guillou-Quisquater signature scheme, on given as input a security parameter k, generates a public key $(n, v, h_0, \mathcal{H}(\cdot))$ and a corresponding secret key $x_0 = h_0^{1/v}$, for use by a probabilistic polynomial-time signer $\mathcal{S}_0$. The triple (n, v, h_0) is generated as specified in the RSA assumption, and $\mathcal{H}(\cdot)$ is a polynomial-size description of a hash-function that maps its inputs to $\mathbb{Z}_{2^t}$ for some appropriate t such that $2^t < v$. The hash-function is generated at random from a suitable family of collision-intractable hash functions. This family preferably is correlation-free, as defined by Okamoto [17].

A *digital signature* on a message m is defined to be a pair (r_0, c_0) such that $c_0 = \mathcal{H}(m, r_0^v h_0^{-c_0})$. If the family of hash-function is sufficiently strong (in particular, correlation-free), then it should be infeasible to algebraically combine several message-signature pairs into a new such pair.

In the *signature issuing protocol*, $\mathcal{S}_0$ issues a signature on a message m to a probabilistic polynomial-time receiver $\mathcal{R}_0$ by generating at random a number $w_0 \in \mathbb{Z}_n^*$, and computing $r_0 := x_0^{c_0} w_0$ and $c_0 := \mathcal{H}(m, w_0^v)$. It is generally believed that executions of this issuing protocol do not help a probabilistic polynomial-time attacker to forge signatures, and so that the primary purpose of an issuing protocol is met by this particular protocol.

2.2 Blind Signature Issuing Protocols

Besides meeting the primary purpose of an "ordinary" signature issuing protocol, in a *blind* signature issuing protocol (a notion introduced by Chaum [9]) an additional property must be satisfied. Namely, the receiver in a blind signature issuing protocol must be able to retrieve a pair in such a way that it is uncorrelated to the view of the signer in the issuing protocol. The computations required of the receiver to achieve this property are commonly referred to as "blinding."

Efficient blind signature issuing protocols are known for a variety of digital signatures. Chaum [10] described a blind signature issuing protocol for issuing one particular type of digital signatures, called RSA signatures [19]. Okamoto and Ohta [18] proposed a general technique that applies to a variety of digital signatures of the Fiat-Shamir type; each of these types of signatures can be issued by means of a blind signature issuing protocol.

The technique of Ohta and Okamoto amounts to *not* removing the interaction when applying the technique of Fiat and Shamir for converting a three-move sound identification protocol into a signature issuing protocol; by having the verifying party (receiver) determine the challenge, one may hope that it can blind the issued message-signature pair. Indeed, as shown by Ohta and Okamoto, for many schemes of the Fiat-Shamir type this works if only a certain random self-reducibility property holds.

For the Guillou-Quisquater signature issuing protocol, the modification of Ohta and Okamoto results in the following blind signature issuing protocol:

Step 1. S_0 generates at random a number $w_0 \in \mathbb{Z}_v$, and sends $a_0 := w_0^v$ to $\mathcal{R}_0$.

Step 2. $\mathcal{R}_0$ generates at random a number $t_1 \in \mathbb{Z}_n^*$ and a number $t_2 \in \mathbb{Z}_v$, computes $c_0' := \mathcal{H}(m, t_1^v h_0^{t_2} a_0)$ for a message m of its choice, and sends the challenge $c_0 := c_0' + t_2 \bmod v$ to S_0.

Step 3. S_0 sends the response $r_0 := x_0^{c_0} w_0$ to $\mathcal{R}_0$.

$\mathcal{R}_0$ accepts if and only if $r_0^v h_0^{-c_0} = a_0$. (Note that the protocol described up to this point is identical to the Guillou-Quisquater *identification* protocol, with $\mathcal{R}_0$ generating its challenge from $\mathbb{Z}_v$ and in a particular way.) If $\mathcal{R}_0$ accepts, it computes $r_0' := r_0 t_1 h_0^{c_0' + t_2 \operatorname{div} v}$. As can easily be shown, this is a blind signature issuing protocol, with (r_0', c_0') being a Guillou-Quisquater signature on m.

Although it should be easier to forge signatures when the signer uses the new issuing protocol instead of the previous one, since the message can be chosen depending on the information sent by the signer in the first move and the challenge can be freely chosen, the resulting signature scheme is generally believed to be unforgeable. For the blind Guillou-Quisquater signature issuing protocol this belief can be expressed as follows:

Assumption 2. *For any $l \geq 0$, no probabilistic polynomial-time receiver can determine with non-negligible probability of success $l+1$ distinct pairs, consisting of a message and a corresponding Guillou-Quisquater signature, by performing l executions of the blind Guillou-Quisquater signature issuing protocol with an honest signer.*

2.3 Secret-Key Certificate Schemes

Public-key certificates are a well-known cryptographic tool for secure key management. The idea is to have a chosen party, called the (certificate) issuer, certify the public keys of other parties by digitally signing these public keys with respect to its own public key. As with public-key certificates, with *secret-key certificates* [3] the objects of interest are triples consisting of a secret key, a corresponding public key, and a certificate of an issuer on the public key. However, contrary to public-key certificates, in a secret-key certificate scheme the certificate is *not* a digital signature on the public key; the publicly verifiable relation between a public key and a certificate thereon is such that anyone can generate (in isolation) pairs consisting of a public key and a matching certificate, with a distribution that is indistinguishable from the distribution that applies when the issuing protocol is conducted with the issuer. On the other hand, as with public-key certificates, triples consisting of a secret key, a corresponding public key, and a secret-key certificate on the public key can only be retrieved by performing an issuing protocol with the issuer. Since there is no point in using a public-key certificate scheme if the cryptographic actions that are to be performed with respect to a certified public key can be performed *without* knowing a corresponding secret key, secret-key certificate schemes offer the same functionality as public-key certificates do; see [3] for further details.

Regardless of the type of certificate, we will refer to a pair consisting of a public key and a matching certificate as a *certified public key*, and to a triple consisting of a secret key, a corresponding public key, and a matching certificate as a *certified key pair*. A *certificate scheme* consists of several items. Similar to a digital signature scheme, a verification algorithm is needed that determines what exactly constitutes a certificate on a public key. We also need a key generation algorithm that generates key pairs for the issuer, and a certificate issuing protocol. In addition, we need a key generation algorithm for the receiver of a certified key pair; this algorithm specifies which key pairs are considered valid for certification. Finally, in case of a secret-key certificate scheme there must exist a simulator that simulates certified public keys with the same probability as that by which they are generated by the issuing protocol.

In [3] a particular class of embodiments of secret-key certificate schemes was described, based on signature schemes of the Fiat-Shamir type; each of these embodiments can be proved to be as secure as the Fiat-Shamir type signature scheme from which it has been derived, and triples can be issued without the issuer learning the secret keys. According to this construction, a secret-key certificate scheme can be derived from the Guillou-Quisquater signature, as follows. On input a security parameter k, the *key generation algorithm* generates a public key $(n, v, h, g, \mathcal{H}(\cdot))$ and a corresponding secret key (x, y) for the certificate issuer $\mathcal{S}$. Here, $(n, v, h, \mathcal{H}(\cdot))$ and x are generated as described by the key generation algorithm for the Guillou-Quisquater signature scheme, and g and y are generated according to the same distribution as that by which h and x are generated. In particular, $h = x^v$ and $g = y^v$.

A *secret-key certificate* of S on a public key h_i of a receiver $\mathcal{R}_i$, for some $i \in \mathbb{N}$, is a pair (r, c) such that

$$c = \mathcal{H}(h_i, r^v (h\, h_i)^{-c}).$$

A secret key of $\mathcal{R}_i$ corresponding to its public key h_i is a pair $(s_{0i}, s_{1i}) \in \mathbb{Z}_v \times \mathbb{Z}_n^*$ such that

$$h_i = g^{s_{0i}} s_{1i}^v.$$

(Other choices can be made as well; see [3, 6].)

In [3] an "ordinary" *issuing protocol* is described that enables S to issue a certified key pair in such a way that it cannot learn a secret key corresponding to the public key that $\mathcal{R}_i$ ends up with. In Sect. 3 we will develop a new issuing protocol for this particular secret-key certificate scheme, to demonstrate the new technique for designing restrictive blind signature schemes.

3 Restrictive Blinding of Secret-Key Certificates

We now get to the heart of the matter. In a *restrictive* blind signature issuing protocol, the objects of interest are triples, consisting of a secret key, a matching public key, and a certificate on the public key; contrast this to the "ordinary" blind signature issuing protocols discussed in Subsection 2.2, which are concerned with pairs. One can distinguish between restrictive blind public-key certificate issuing protocols and restrictive blind secret-key certificate issuing protocols, depending on the type of certificates used. Similar to ordinary blind signature issuing protocols, the receiver in a restrictive blind signature issuing protocol should be able to ensure that the public key and the certificate of such a triple are uncorrelated to the view of the signer in the issuing protocol. However, the receiver should *not* be able to modify a certain non-trivial predicate of the secret key while blinding about.

We now proceed to demonstrate that a *restrictive* blind secret-key certificate issuing protocol can be designed for any Fiat-Shamir type signature scheme for which an *ordinary* blind signature issuing protocol can be constructed by applying the blinding technique of Okamoto and Ohta. A formal description of the general construction would be rather unwieldy, and so we will explain the technique on the basis of secret-key certificates derived from Guillou-Quisquater signatures (which we described in Subsection 2.3); this should make it much easier to understand why the construction works. Exactly the same design technique applies to secret-key certificates based on at least any of the following Fiat-Shamir type signature schemes: Fiat-Shamir [14], Brickell-McCurley [8], Feige-Fiat-Shamir [13], Okamoto [17] (several schemes), and Schnorr [20].

3.1 Restrictive Blinding for the Guillou-Quisquater Secret-Key Certificate Scheme

In the restrictive blind issuing protocol, $\mathcal{R}_i$ will receive a certified key pair (s_{0i}, s_{1i}), h_i', (r', c'), with the blinding-invariant predicate of the secret key being equal to $s_{0i} \bmod v$. The apostrophes on the public key and the certificate serve to emphasize that they will be uncorrelated to the view of S. We will assume that S initially provides $\mathcal{R}_i$ with a number s_{0i}, generated according to some appropriate probability distribution over $\mathbb{Z}_v$, and will denote $g^{s_{0i}}$ by h_i. Note that the pair $(s_{0i}, 1)$ is a secret key corresponding to public key h_i; one can think of h_i as the public key that is to be blinded to h'_i. For technical reasons, related to the proof of Proposition 10, we will assume that s_{0i} is generated independently of h. For all practical purposes this is not a restriction.

The issuing protocol is as follows:

Step 1. S generates at random a number $w \in \mathbb{Z}_n^*$, and sends $a := w^v$ to $\mathcal{R}_i$.

Step 2. $\mathcal{R}_i$ generates at random two numbers $s_{1i}, t_1 \in \mathbb{Z}_n^*$, and a number $t_2 \in \mathbb{Z}_v$. $\mathcal{R}_i$ computes $h'_i := h_i s_{1i}^v$, $c' := \mathcal{H}(h'_i, t_1^v (h h_i)^{t_2} a)$, and sends $c := c' + t_2 \bmod v$ to S.

Step 3. S sends $r := (xy^{s_{0i}})^c w$ to $\mathcal{R}_i$.

$\mathcal{R}_i$ accepts if and only if $r^v (h h_i)^{-c} = a$. If this verification holds, $\mathcal{R}_i$ computes $r' := r t_1 (h h_i)^{c' + t_2 \text{ div } v} s_{1i}^{c'}$.

Observe that we have applied the technique of Okamoto and Ohta to the Guillou-Quisquater signature issuing protocol, with S performing the protocol with respect to a combined public key that is the product of its own public key and the "not-yet-blinded" public key of $\mathcal{R}_i$, and $\mathcal{R}_i$ blinding not only c and r but also this combined public key. In an (informal) nutshell, this explains the general technique.

Why should this construction work? For this to become clear, we will prove that our exemplary scheme described above is a restrictive blind secret-key certificate scheme; pay particular attention to Proposition 10, as its proof in essence provides the answer to this question.

Following Feige, Fiat and Shamir [13], we will denote by $\overline{\mathcal{Z}}$ a party $\mathcal{Z}$ that follows the issuing protocol, by $\widehat{\mathcal{Z}}$ a probabilistic polynomial-time party $\mathcal{Z}$ that may deviate from the issuing protocol in an arbitrary way, and by $\widetilde{\mathcal{Z}}$ a party $\mathcal{Z}$ with unlimited computing power that may deviate from the issuing protocol in an arbitrary way.

3.2 Proof of Correctness

The secret-key certificate scheme defined in Subsection 2.3, with the issuing protocol of Subsection 3.1 substituted for the ordinary issuing protocol, defines a new certificate scheme. To prove that our new scheme is indeed a secret-key certificate scheme (correctness), we must prove that $\mathcal{R}_i$ in the issuing protocol receives a certified public key. Moreover, since we have changed the issuing protocol it is not immediately clear whether certified public keys can still be generated with indistinguishable probability distribution, without cooperation of S; we will prove that property next.

Proposition 3. *If $\overline{\mathcal{R}_i}$ accepts, then*

$$(s_{0i}, s_{1i}),\ h'_i,\ (r', c')$$

is a certified key pair.

Proof. See [5]. □

Proposition 4. *The new certificate scheme is a secret-key certificate scheme.*

Proof. See [5]. □

3.3 Proof of the Primary Purpose

We go on to prove that the primary purpose of issuing protocols, namely the one-to-one correspondence between executions of the issuing protocol and certified key pairs, holds for the new certificate scheme. The one-to-one correspondence should hold even if multiple receivers, each of which may perform the issuing protocol for a different number s_{0i}, conspire. A conspiracy should be thought of as being one probabilistic polynomial-time Turing machine, composed of the "participating" receivers; this is easy to formalize, and will be left out here.

We first need two lemmas. For a definition of witness hiding, see Feige and Shamir [12].

Lemma 5. *If the blind Guillou-Quisquater signature issuing protocol is witness hiding, then no conspiracy can compute $g^{1/v}$ with non-negligible probability of success.*

Proof. See [5]. □

Note that a similar result can be proved *unconditionally* for the restrictive blind secret-key certificate schemes that can be derived by our technique from the signature schemes of McCurley-Brickell [8] and Okamoto [17], since these schemes are known to be witness hiding. Furthermore, the result also holds if a conspiracy can "wire-tap" executions of the issuing protocol with honest receivers; in Step 2 of the simulation obviously the same simulation can be used for these honest receivers.

The proof of the following lemma is trivial, and is therefore omitted.

Lemma 6. *If Assumption 2 is true, then the blind Guillou-Quisquater signature issuing protocol is witness hiding.*

We are now prepared to prove the one-to-one correspondence. In the following, a conspiracy is said to be able to *forge a certified key pair* if it can compute with non-negligible probability of success $l + 1$ distinct certified key pairs by performing l executions of the issuing protocol with $\overline{\mathcal{S}}$, for some $l \geq 0$.

Proposition 7. *If Assumption 2 is true, then no conspiracy can forge a certified key pair.*

Proof. See [5]. □

In combination with Proposition 3, which states that an honest receiver receives a certified key pair when it performs an execution of the issuing protocol, this result tells us that there is a one-to-one correspondence between executions of the issuing protocol and certified key pairs. In other words, the primary purpose of issuing protocols is satisfied.

3.4 Proof of the Property of Restrictive Blinding

We next turn to proving that the issuing protocol is a *restrictive blind* signature issuing protocol. We hereto need to prove (1) that the public key and the certificate can be blinded by the receiver, whereas (2) the blinding-invariant predicate of the secret key cannot. We start with the first of these two properties.

Proposition 8. *If $\mathcal{R}_i$ follows the issuing protocol then the issued certified public key is perfectly blinded.*

Proof. See [5]. □

We now turn to proving the second property, for which we need a new assumption. If we modify the Guillou-Quisquater *identification* protocol by having the prover generate h_0 at random in each execution of the protocol, instead of fixing it initially, the proof of soundness for the Guillou-Quisquater identification still applies. In particular, if the prover can compute correct responses with respect to two different challenges of the verifier, then it "knows" $h_0^{1/v}$. Applying the general technique of Fiat and Shamir for converting identification schemes into signature schemes to this modified "identification" protocol (correspondingly taking c equal to $\mathcal{H}(m, h_0, a_0)$), and leaving out the message m, it follows that it should be infeasible to determine in isolation pairs $h_0, (r_0, c_0)$, without knowing $h_0^{1/v}$, such that $c_0 = \mathcal{H}(h_0, r_0^v h_0^{-c_0})$. Of course, this should also hold if we instead consider satisfying $c_0 = \mathcal{H}(h^{-1} h_0, r_0^v h_0^{-c})$, for a predetermined and randomly chosen $h \in \mathbb{Z}_n^*$. (This variation is considered merely for a technical reason, which will become clear in the proof of the next proposition; alternatively, we could have redefined the verification relation for the issuing protocol.) Substituting $h\, h_i$ for h_0, we get the following assumption.

Assumption 9. *There exists a probabilistic polynomial-time Turing machine M (the knowledge extractor) such that for any probabilistic polynomial-time Turing machine A with work tape WT and random tape RT, and any sufficiently large k, if A on input $(n, v, h, \mathcal{H}(\cdot))$ outputs with nonnegligible probability of success a pair $h_i, (r, c)$ such that*

$$c = \mathcal{H}(h_i, r^v (h\, h_i)^{-c}),$$

then $M(A, WT, RT, (n, v, h, \mathcal{H}(\cdot))) = (h\, h_i)^{1/v}$ *with non-negligible probability. The probability is taken over the coin tosses of A and the distribution of the input tuple, and the distribution of the input tuple is as specified for the Guillou-Quisquater signature scheme.*

This assumption is all the more plausible considering that we allow an attacker to forge pairs only *from scratch*, and that the knowledge extractor should succeed with only non-negligible probability.

In the following proposition, we will study the issuing protocol with respect to an "isolated" receiver. We will construct a simulator A that moves to a third step only after any l executions of the issuing protocol have been simulated. To ensure that A always halts in polynomial time in a defined state, we can let A halt if some polynomial amount of time has expired without any requests for executions of the issuing protocol. To keep the proof simple we will not incorporate such a notion of timing, but will instead implicitly assume its presence; this detail can easily be filled in.

Proposition 10. *If the RSA assumption and Assumption 9 are true, then $\widehat{\mathcal{R}_i}$ cannot retrieve with non-negligible probability of success a certified key pair (s_{0i}^*, s_{1i}), h_i, (r, c) for which s_{0i}^* differs modulo v from its blinding-invariant part s_{0i}.*

Proof. Suppose that $\widehat{\mathcal{R}_i}$ can misuse l executions of the issuing protocol to extract with non-negligible probability of success a certified key pair such that $s_{0i}^* \neq s_{0i} \bmod v$. We will construct a polynomial-time algorithm A for breaking the RSA assumption.

Algorithm A, on given as input a triple (n, v, h_0) generated as specified in the RSA assumption, performs the following steps:

Step 1. (Simulate the initial key generation.) Set $g := h_0$. Generate at random an element $x \in \mathbb{Z}_n^*$ and an element $s_{0i} \in \mathbb{Z}_v$ (according to the probability distribution by which $\overline{S}$ generates the blinding-invariant parts), and compute $h := x^v g^{-s_{0i}}$. Generate $\mathcal{H}(\cdot)$ in the same way as described in the key generation for the Guillou-Quisquater signature scheme. The simulated public key of $\overline{S}$ is $(n, v, h, g, \mathcal{H}(\cdot))$.

Step 2. Simulate for $\widehat{\mathcal{R}_i}$ the actions that $\overline{S}$ would perform, as follows:
- (Generation of the blinding-invariant part of the secret key.) Use s_{0i} as the invariant part of the secret key for $\widehat{\mathcal{R}_i}$.
- (The issuing protocol.)
  Step 1. Generate at random an element w of $\mathbb{Z}_n^*$. Compute $a := w^v$, and send a to $\widehat{\mathcal{R}_i}$.
  Step 2. Receive c from $\widehat{\mathcal{R}_i}$.
  Step 3. Compute $r := x^c w$, and send r to $\widehat{\mathcal{R}_i}$.
  Continue this simulation until l executions of the issuing protocol with $\widehat{\mathcal{R}_i}$ have been performed.

Step 3. Check if $\widehat{\mathcal{R}}_i$ has on its tapes a certified key pair (s_{0i}^*, s_{1i}), h_i', (r', c') such that $s_{0i}^* \neq s_{0i} \bmod v$. If not, then halt.

Step 4. Run the knowledge extractor M on all tapes of A and $\widehat{\mathcal{R}}_i$, and the tuple $(n, v, h, \mathcal{H}(\cdot))$. Compute $e, f \in \mathbb{N}$ such that $(s_{0i} - s_{0i}^*)e = 1 + f v$, by applying the extended Euclidean algorithm (such a pair exists since $s_{0i} \neq s_{0i}^* \bmod v$). Denoting the output of M by d, compute $(d/x\,s_{1i})^e g^{-f}$, and output the outcome.

Note that we have made use of the knowledge extractor M of Assumption 9 in Step 4 of this simulation, and have viewed $\widehat{\mathcal{R}}_i$ and A as one Turing machine (this can easily be formalized, and is omitted here).

By definition of the key generation of A in Step 1, the public key in Step 1 is simulated with the same probability distribution as that by which $\overline{\mathcal{S}}$ generates its public key. The response that is computed by A in the simulated issuing protocol is the same as the response that $\overline{\mathcal{S}}$ would compute:

$$
\begin{aligned}
r^v &= (x^c w)^v \\
&= (x^v)^c w^v \\
&= (h g^{s_{0i}})^c a \\
&= (h\, h_i)^c a.
\end{aligned}
$$

From this it easily follows that the view of $\widehat{\mathcal{R}}_i$ that is provided by A in the simulated issuing protocol has the same distribution as that provided by $\overline{\mathcal{S}}$ in the issuing protocol, regardless of the probability distribution by which $\widehat{\mathcal{R}}_i$ generates its challenges and despite of the tricky way in which A generates h. Hence, Step 4 is reached by supposition with non-negligible probability.

By Assumption 9, the output d of M in Step 4 is equal to $(h\,h_i')^{1/v}$ with non-negligible probability, and in that case we have

$$
\begin{aligned}
((d/x\,s_{1i})^e)^v &= ((d/x\,s_{1i})^v)^e \\
&= (d^v / x^v s_{1i}^v)^e \\
&= (h\,h_i' / x^v s_{1i}^v)^c \\
&= (x^v g^{-s_{0i}} g^{s_{0i}^*} s_{1i}^v / x^v s_{1i}^v)^e \\
&= (g^{s_{0i}^* - s_{0i}})^e \\
&= g^{(s_{0i}^* - s_{0i})e} \\
&= g^{1+f v} \\
&= g\,(g^f)^v.
\end{aligned}
$$

Since $g = h_0$, it follows that $(d/x\,s_{1i})^e g^{-f} = h_0^{1/v}$ and so the output of A in Step 4 is equal to $h_0^{1/v}$ with non-negligible probability. This contradicts the RSA assumption. Hence, if Assumption 9 and the RSA assumption are true then it cannot be the case that $s_{0i}^* \neq s_{0i} \bmod v$ with non-negligible probability of success. $\qquad\square$

Observe that the proof of this strong result makes use of the fact that the simulator can determine the public key of $\overline{S}$ in terms of s_{0i}, *without this changing the distribution of the views for* $\mathcal{R}_i$. As a quick comparison with the proof of Proposition 4 will reveal, this proof technique owes its existence to the simulatability of certified public keys that is inherent to secret-key certificates.

3.5 Sequential Executions of the Issuing Protocol

Obviously, in applications of practical interest there are *multiple* receivers, as has been emphasized throughout by the use of a subscript i. We can distinguish between multiple executions of the issuing protocol with respect to the same number s_{0i}, or with respect to different s_{0i}'s. In the former case, Proposition 10 still holds, independent of whether the executions of the issuing protocol are performed sequentially or in parallel by $\mathcal{S}$; but this case is hardly of practical interest. So we will now study the latter case, in which issuing protocols are executed by $\mathcal{S}$ with respect to different blinding-invariant numbers.

When $\mathcal{S}$ sees to it that it performs executions of the issuing protocol with respect to *different* blinding-invariant numbers only *sequentially*, Proposition 10 provides fairly solid evidence that not even a *conspiracy* of receivers will be able to blind the presumed blinding-invariant predicate of one of their secret keys. The motivation for this is that essentially no cooperation is possible between different receivers while performing executions of the issuing protocol. This seems to justify the following conjecture.

Conjecture 11. *If S makes sure that it never performs two executions of the issuing protocol in parallel in case they involve different blinding-invariant numbers, then the defined certificate issuing scheme is a restrictive blind signature scheme.*

It is stressed that it cannot hurt if $\mathcal{S}$ performs executions of the issuing protocol in parallel that pertain to the *same* blinding-invariant number.

In many practical applications the simple measure of not running executions of the issuing protocol in parallel, in case they pertain to different blinding-invariant parts, will certainly suffice. The interested reader is referred to [5] for a detailed discussion.

3.6 Parallel Executions of the Issuing Protocol

Conjecture 11 is false if $\mathcal{S}$ performs executions of the issuing protocol *in parallel* when *different* blinding-invariant numbers are involved. Let s_{0i} be the blinding-invariant number for $\mathcal{R}_i$, and $s_{0j} \neq s_{0i} \bmod v$ that for $\mathcal{R}_j$; the corresponding "not-yet-blinded" public keys are h_i and h_j. In its simplest form (leaving out additional computations that need to be performed to completely blind the certified key pair, in order to prevent unduly obscuring of the description), the attack on the two parallel executions of the issuing protocol is the following:

(**Step 1 for $\mathcal{R}_i$**) $\mathcal{S}$ generates at random a number $w_i \in \mathbb{Z}_v$, and sends $a_i := w_i^v$ to $\mathcal{R}_i$.

(**Step 1 for $\mathcal{R}_j$**) $\mathcal{S}$ generates at random a number $w_j \in \mathbb{Z}_v$, and sends $a_j := w_j^v$ to $\mathcal{R}_j$.

(**Cooperation between $\mathcal{R}_i$ and $\mathcal{R}_j$**) $\mathcal{R}_i$ and $\mathcal{R}_j$ compute $h_k := g^{s_{0k}}$ for an arbitrary number s_{0k} of their choice (s_{0k} need not be in $\mathbb{Z}_v$; any number in $\mathbb{N}$ will do). They then compute $c_k := \mathcal{H}(h_k, a_i a_j)$.

(**Step 2 for $\mathcal{R}_i$**) $\mathcal{R}_i$ sends $c_i := c_k (s_{0j} - s_{0k})(s_{0j} - s_{0i})^{-1} \bmod v$ to $\mathcal{S}$.

(**Step 2 for $\mathcal{R}_j$**) $\mathcal{R}_j$ sends $c_j := c_k (s_{0k} - s_{0i})(s_{0j} - s_{0i})^{-1} \bmod v$ to $\mathcal{S}$.

(**Step 3 for $\mathcal{R}_i$**) $\mathcal{S}$ sends $r_i := (xy^{s_{0i}})^{c_i} w_i$ to $\mathcal{R}_i$.

(**Step 3 for $\mathcal{R}_j$**) $\mathcal{S}$ sends $r_j := (xy^{s_{0j}})^{c_j} w_j$ to $\mathcal{R}_j$.

$\mathcal{R}_i$ and $\mathcal{R}_j$ accept if and only if

$$r_i^v (h\,h_i)^{-c_i} = a_i \quad \text{and} \quad r_j^v (h\,h_j)^{-c_j} = a_j.$$

If this verification holds, then $\mathcal{R}_i$ and $\mathcal{R}_j$ compute

$$r_k := r_i r_j h^{-((c_i + c_j) \operatorname{div} v)} g^{-((c_i s_{0i} + c_j s_{0j}) \operatorname{div} v)} g^{c_k s_{0k} \operatorname{div} v}.$$

Proposition 12. *If $\mathcal{R}_i$ and $\mathcal{R}_j$ accept, then*

$$(s_{0k}, 1),\ h_k,\ (r_k, c_k)$$

is a certified key pair.

Proof. See [5]. □

In case the application really demands that $\mathcal{S}$ can securely run executions of the issuing protocol in parallel, without any restrictions, we must therefore alter the issuing protocol. The following minor adjustment is believed to suffice for this purpose. Rather than revealing s_{0i} to $\mathcal{R}_i$ *before* the execution of the issuing protocol, $\mathcal{S}$ does not make it known until it has received c of $\mathcal{R}_i$ in Step 2; instead, $\mathcal{S}$ only makes $h_i = g^{s_{0i}}$ known initially. Furthermore, $\mathcal{S}$ chooses s_{0i} at random from $\mathbb{Z}_v$: computing s_{0i} from h_i then requires breaking the *Discrete Log* assumption in $\mathbb{Z}_n^*$. This modification does not in any way prevent $\mathcal{R}_i$ from performing the necessary computations in Step 2, since only h_i needs to be known to $\mathcal{R}_i$.

The interested reader is referred to [4] for a detailed argument of why the modified issuing protocol is believed to be secure under parallel executions. Caution should be excersised, though, since the provided argument only considers one particular type of attack. Readers are therefore strongly encouraged to try to break the modified issuing protocol in other ways.

This concludes the exposition of the general technique for designing restrictive blind secret-key certificate issuing protocols. Although the description has been based on the Guillou-Quisquater signature scheme, enough handles have been provided throughout to easily apply the technique to any of the other Fiat-Shamir type signature schemes that can be subjected to the technique of Ohta and Okamoto for designing ordinary blind signature issuing protocols.

4   Relation to Blind Signature Issuing Protocols

Contrary to restrictive blind issuing protocols for public-key certificates, restrictive blind issuing protocols for secret-key certificates are *not* a particular case of "ordinary" blind signature issuing protocols. Consider a triple consisting of a secret key, a matching public key, and a certificate on the public key. The receiver in the certificate issuing protocol can completely blind the public key and the certificate, but not part of the secret key. If the certificate would be a public-key certificate, then the protocol would indeed be a particular case of an ordinary blind signature scheme; the public key is the message and the certificate is the signature on the message, and the pair is blinded.

However, if the certificate is a secret-key certificate, it is by definition *not* a digital signature on the public key (the extreme opposite is true: pairs consisting of a public key and a matching secret-key certificate can be generated by anyone with exactly the same probability distribution); the *secret key* is the message, and the certificate is the signature on the message. But the message *cannot* be blinded, by the very definition of a restrictive blind signature issuing protocol; only the signature can.

5   Conclusion

A variety of privacy-protecting signature transporting mechanisms can be obtained by combining the new restrictive blind signature issuing protocols with an appropriate showing protocol between the receiver of the issued triple and a third party. One particularly interesting such signature transporting mechanism is an untraceable off-line electronic cash system, first studied by Chaum, Fiat and Naor [11]. In [1] I introduced the most efficient and versatile untraceable off-line cash system known to that date. As will be appreciated, the new issuing protocols can be combined fairly straightforwardly with the techniques developed in [1] for achieving prior restraint of double-spending with fall-back to traceability after the fact. By basing the restrictive blind secret-key certificate issuing protocol on the Schnorr signature scheme, the resulting cash system is considerably more efficient than the system in [1]. The interested reader is referred to [5], and to [2] for practical optimizations. In [7] the application to Internet payments is discussed.

In [6] general techniques are described for designing showing protocols that can be combined with restrictive blind signature issuing protocols in order to design general privacy-protecting signature transporting mechanisms (also known as credential mechanisms, first studied by Chaum [10]). Here, again, the use of restrictive blind issuing protocols instead of a cut-and-choose issuing protocol enables significant improvements in terms of efficiency, functionality, and provability of security properties of the credential mechanisms. Instead of using Chaum's technique of encoding different types of credentials by using different signatures schemes, in the new credential mechanisms credentials are encoded into the blinding-invariant parts of secret keys; this enables the holder of a set of credentials to prove a variety of predicates of his credentials without providing additional information.

As the reader may have noticed, we have never used the fact that S may know the factorization of the modulus n. The sole reason for not having done so is that in Fiat-Shamir type signature schemes that are based on the Discrete Log assumption, such as the Schnorr signature scheme, the signer also does not have such trapdoor information at its disposal, and our goal was to explain a generally applicable technique. However, if we allow S to know (and make use of) the factorization of n, a powerful technique becomes available, which enables the issuer in the credential mechanisms of [6] to update credentials without needing to know their current values. This updating technique is not possible in credential mechanisms based on cut-and-choose issuing protocols.

In sum, the demonstrated technique for designing efficient secret-key certificate issuing protocols has a direct bearing on the design of efficient and versatile privacy-protecting credential mechanisms.

References

1. Brands, S., "Untraceable Off-Line Cash in Wallet with Observers," Advances in Cryptology – CRYPTO '93, Lecture Notes in Computer Science, no. 773, Springer-Verlag, pp. 302–318. An extended pre-print appeared as: "An efficient off-line electronic cash system based on the representation problem," Centrum voor Wiskunde en Informatica (CWI), Report CS-R9323, March 1993. Available by anonymous ftp from: ftp.cwi.nl:/pub/CWIreports/AA/CS-R9323.ps.Z.

2. Brands, S., "Off-line Cash Transfer by Smart Cards," Proceedings of the First Smart Card Research and Advanced Application Conference, Lille (France), Oct. 1994, pp. 101–117. See also: Centrum voor Wiskunde en Informatica (CWI), Report CS-R9455, September 1994. Available by anonymous ftp from: ftp.cwi.nl:/pub/CWIreports/AA/CS-R9455.ps.Z.

3. Brands, S., manuscript (1993) part (i): "Secret-Key Certificates," Centrum voor Wiskunde en Informatica (CWI), Technical Report, Februari 1995. Submitted for publication.

4. Brands, S., manuscript (1993) part (ii): "Restrictive Blinding of Secret-Key Certificates," Centrum voor Wiskunde en Informatica (CWI), Technical Report, Februari 1995.

5. Brands, S., manuscript (1993) part (iii): "Off-Line Electronic Cash Based on Secret-Key Certificates," Proceedings of the Second International Symposium of Latin American Theoretical Informatics (LATIN '95), Valparaíso, Chili, April 3–7, 1995. See also: Centrum voor Wiskunde en Informatica (CWI), Report CS-R9506, Januari 1995. Available by anonymous ftp from: ftp.cwi.nl:/pub/CWIreports/AA/CS-R9506.ps.Z.

6. Brands, S., manuscript (1993) part (v): "Privacy-protecting Digital Credentials Based on Restrictive Blinding," submitted for publication. Preprint available on request.

7. Brands, S., "Electronic Cash on the Internet," Proceedings of the Internet Society 1995 Symposium on Network and Distributed System Security, San Diego, California, Februari 16-17, 1995.

8. Brickell, E., McCurley, K., "An Interactive Identification Scheme Based on Discrete Logarithms and Factoring," Journal of Cryptology, Vol. 5, No. 1 (1992), pp. 29–39.

9. Chaum, D., "Blind Signatures for Untraceable Payments," Advances in Cryptology – CRYPTO '82, Lecture Notes in Computer Science, Springer-Verlag, pp. 199–203.

10. Chaum, D., "Security without identification: transaction systems to make big brother obsolete," CACM Vol. 28, No. 10, October 1985, pp. 1030–1044.

11. Chaum, D., Fiat, A., Naor, M., "Untraceable electronic cash," Advances in Cryptology – CRYPTO '88, Lecture Notes in Computer Science, no. 403, Springer-Verlag, pp. 319–327.

12. Feige, U., Shamir, A., "Witness Indistinguishable and Witness Hiding Protocols," Proceedings of the 22nd Annual ACM Symposium on the Theory of Computing, 1990, pp. 416–426.

13. Feige, U., Fiat, A., Shamir, A., "Zero-Knowledge Proofs of Identity," Journal of Cryptology, Vol. 1, No. 2 (1988), pp. 77–94.

14. Fiat, A., Shamir, A., "How to prove yourself: practical solutions to identification and signature problems," Advances in Cryptology – CRYPTO '86, Lecture Notes in Computer Science, Springer-Verlag, pp. 186-194.

15. Goldwasser, S., Micali, S., Rivest, R., "A digital signature scheme secure against adaptive chosen message attack," SIAM Journal on Computing, Vol. 17 No. 2 (1988), pp. 281–308.

16. Guillou, L., Quisquater, J.-J., "A Practical Zero-Knowledge Protocol Fitted to Security Microprocessor Minimizing Both Transmission and Memory," Advances in Cryptology – EUROCRYPT '88, Lecture Notes in Computer Science, no. 330, Springer-Verlag, pp. 123-128.

17. Okamoto, T., "Provably Secure and Practical Identification Schemes and Corresponding Signature Schemes," Advances in Cryptology – CRYPTO '92, Lecture Notes in Computer Science, no. 740, Springer-Verlag, pp. 31–53.

18. Okamoto, T., Ohta, K., "Divertible Zero-Knowledge Interactive Proofs and Commutative Random Self-Reducibility," Advances in Cryptology – EUROCRYPT '89, Lecture Notes in Computer Science, no. 434, Springer-Verlag, pp. 481 496.

19. Rivest, R., Shamir, S., Adleman, L., "A method for obtaining digital signatures and public-key cryptosystems," Communications of the ACM, Feb. 1978, pp. 120-126.

20. Schnorr, C, "Efficient Signature Generation by Smart Cards," Journal of Cryptology, Vol. 4, No. 3 (1991), pp. 161-174.

Towards Fast Correlation Attacks on Irregularly Clocked Shift Registers

Jovan Dj. Golić [*]

Information Security Research Centre, Queensland University of Technology
GPO Box 2434, Brisbane Q 4001, Australia
School of Electrical Engineering, University of Belgrade
Email: golic@fit.qut.edu.au

Abstract. A theoretical framework for fast correlation attacks on irregularly clocked linear feedback shift registers (LFSRs) based on a recently established linear statistical weakness of decimated LFSR sequences is developed. When the LFSR feedback polynomial is not known, methods for the statistical weakness detection and the feedback polynomial reconstruction are proposed. When the LFSR feedback polynomial is known, an iterative procedure for fast LFSR initial state reconstruction given an observed keystream sequence is introduced. The procedure is based on appropriately defined parity-check sums and consists in iterative recomputation of the posterior probabilities for unknown elements of the decimation sequence. A convergence condition in terms of the numbers of the parity-check sums needed for successful reconstruction and the required polynomial computational complexity indicate that the proposed fast correlation attack may be realistic, especially in the constrained clocking case. The number of the feedback polynomial multiples of relatively low weight and not too large degree thus proves to be critical for the security of irregularly clocked LFSRs.

1 Introduction

Clock-controlled linear feedback shift registers (LFSRs) have become important building blocks for keystream generators in stream cipher applications, because they are known to produce sequences of long period and high linear complexity, see [12], [5], and [17]. They are also immune to fast correlation attacks [13, 19, 20, 18] on additively noised LFSR sequences. They have even been proposed as the keystream generators themselves, see [4], [14]. A clock-controlled shift register is a LFSR that is irregularly clocked according to a decimation sequence which defines the number of symbols to be deleted before the next output symbol is produced. The decimation sequence is itself a pseudorandom sequence produced by a clock-control generator, for example, by another LFSR, as is proposed in [4]. Irregular clocking is called constrained if the number of consecutive deletions is limited and unconstrained otherwise, see [9]. The secret key is assumed to

---

[*] This research was supported in part by the Science Fund of Serbia, grant #0403, through the Institute of Mathematics, Serbian Academy of Arts and Sciences.

L.C. Guillou and J.-J. Quisquater (Eds.): Advances in Cryptology - EUROCRYPT '95, LNCS 921, pp. 248-262, 1995.
© Springer-Verlag Berlin Heidelberg 1995

control the LFSR initial state, the initial state and the structure of the clock-control generator, and, possibly the LFSR feedback polynomial as well. The objective is to reconstruct the feedback polynomial and the initial state of the clock-controlled LFSR given an observed segment of the keystream sequence.

When the LFSR feedback polynomial is known, a divide and conquer attack on the unknown decimation sequence based on the linear consistency test is proposed in [21], where the required length of the observed keystream sequence for the attack to be successful is also estimated. A similar attack for multiplexed sequences based on the collision test is independently suggested in [1], see also [2]. On the other hand, divide and conquer correlation attacks on the LFSR initial state are also possible if the observed decimated sequence is sufficiently long. Namely, the embedding correlation attack is in [23] and [10] analyzed for the constrained clocking case and in [10] for the unconstrained clocking case. The statistically optimal probabilistic correlation attack in the unconstrained clocking case is analyzed in [10] too. The attacks imply the exhaustive search over all possible decimation sequences and over all possible LFSR initial states, respectively. Although a divide and conquer effect is achieved, the computational complexity remains exponential.

The first problem to be considered in this paper is the LFSR feedback polynomial reconstruction given a known segment of the keystream sequence. We develop an approach based on a recently found linear statistical weakness of irregularly decimated LFSR sequences [11]. The weakness and the corresponding correlation coefficients are further analyzed in more detail in Section 2. In Section 3, it is then shown how to detect the statistical weakness when the feedback polynomial is not known, how to reconstruct the feedback polynomial by exhaustive search over appropriately defined shrunk polynomials, and how to reconstruct the feedback polynomial in a fast iterative way. The necessary length of the keystream sequence and the computational complexity are estimated in all the cases.

The second problem, dealt with in Section 4, is the fast reconstruction of the LFSR initial state based on an observed keystream sequence, assuming that the feedback polynomial is known. The linear statistical weakness [11] is again the starting point, and the method that we propose is based on the parity-check sums corresponding to the shrunk polynomials of specially defined polynomial multiples of the feedback polynomial. The algorithm consists in iterative recomputation of the posterior probabilities for unknown elements of the decimation sequence and is conceptually similar to the iterative probabilistic decoding algorithms used in fast correlation attacks on additively noised LFSR sequences [13, 22, 3, 15, 16]. Instead of the binary additive noise we deal with the integer decimation noise. By using the analogy with the problem considered in [16], we give a convergence condition in terms of the numbers of the parity-check sums needed for successful reconstruction. This condition along with the required polynomial computational complexity shows that the proposed fast correlation attack may be realistic, particularly in the constrained clocking case.

2 Linear Statistical Weakness

Consider a clock-controlled shift register as a keystream generator consisting of a binary linear feedback shift register (LFSR) that is irregularly clocked according to a nonnegative integer decimation sequence which defines the number of symbols to be deleted before the next output symbol is produced, see [12], [5]. It is thus assumed that the number of clocks per each output symbol is a positive integer. The decimation sequence is itself generated in a pseudorandom manner by a clock-control generator, for example, by another LFSR, as is proposed in [4]. More precisely, if $X = \{x_t\}_{t=0}^{\infty}$ denotes a regularly clocked LFSR sequence and $D = \{d_t\}_{t=0}^{\infty}$ a decimation sequence, then the output sequence $Y = \{y_t\}_{t=0}^{\infty}$ is defined as a decimated sequence

$$y_t = x\left(t + \sum_{i=0}^{t} d_i\right), \quad t \geq 0. \tag{1}$$

The secret key is assumed to control the LFSR initial state and the initial state and the structure of the clock-control generator. It may in addition be assumed that the LFSR feedback polynomial is also defined by the secret key, as in [4]. The objective is to reconstruct the initial state of the clock-controlled LFSR given an observed segment of the keystream sequence, provided that the feedback polynomial is known. If the feedback polynomial is not known, then the first objective is to determine this polynomial.

Since the decimation sequence is not known, it is reasonable to assume an appropriate probabilistic model based on partial or complete knowledge of the structure of the clock-control generator. Let thus D be a sequence of independent identically distributed nonnegative integer random variables with a probability distribution $\mathcal{P} = \{P(d)\}_{d \in \mathcal{D}}$ where $\mathcal{D}$ is the set of integers with positive probability. The deletion rate [10] is then defined as $\bar{p} = \frac{\bar{d}}{1+\bar{d}}, \bar{d} = \sum_{d \in \mathcal{D}} dP(d)$. The unknown LFSR initial state is assumed to be chosen uniformly at random. Let us now briefly review and then further analyze in more detail the linear statistical weakness of irregularly clocked LFSR sequences pointed out in [11]. Let the LFSR have length r and the feedback polynomial $f(z) = 1 + \sum_{i=1}^{r} f_i z^i = 1 + \sum_{k=1}^{w} z^{i_k}$, $1 \leq i_1 < \ldots < i_w = r$, where $W = w + 1$ is the weight of $f(z)$. Then a *shrunk* polynomial of $f(z)$ is defined as a polynomial of the form $\hat{f}(z) = 1 + \sum_{i=1}^{\hat{r}} \hat{f}_i z^i = 1 + \sum_{k=1}^{w} z^{\hat{i}_k}$, where $1 \leq \hat{i}_1 < \ldots < \hat{i}_w = \hat{r}$ and $\hat{i}_k - \hat{i}_{k-1} \leq i_k - i_{k-1}, 1 \leq k \leq w$, with $\hat{i}_0 = i_0 = 0$. The weight of $\hat{f}(z)$ is the same as the weight of $f(z)$ and its degree $\hat{r}$ is not bigger than the degree r. The degrees are equal if and only if $\hat{f}(z) = f(z)$. According to the LSCA method [8], it is shown in [11] that the linear equation

$$y_t + \sum_{k=1}^{w} y_{t-\hat{i}_k} = 0 \tag{2}$$

holds with probability $(1 + c)/2$ in the decimated sequence for any $t \geq \hat{r}$, where the corresponding correlation coefficient c depends on the probability distribution $\mathcal{P} = \{P(d)\}_{d \in \mathcal{D}}$. For simplicity, consider the geometric distribution $P(d) = p^d(1 - p), d \geq 0$, which corresponds to the case of independent deletions with probability $p > 0$. This is exactly the model to be used for the shrinking generator [4] where also $p = 1/2$. An arbitrary probability distribution $\mathcal{P}$ can be approximated by the geometric distribution by setting $p = \bar{p}$ where $\bar{p}$ is the deletion rate. The correlation coefficient [11] is in this case given by

$$c = p^{r - \hat{r}}(1 - p)^{\hat{r}+1} \prod_{k=1}^{w} \left(\frac{\Delta_k}{\hat{\Delta}_k} \right) \tag{3}$$

where $\Delta_k = i_k - i_{k-1} - 1$ and $\hat{\Delta}_k = \hat{i}_k - \hat{i}_{k-1} - 1, 1 \leq k \leq w$. Equation (3) has a clear combinatorial meaning in terms of the probability of decimation sequences. More precisely, the correlation coefficient is equal to the probability of the event that the bits satisfying the feedback polynomial in the shift register sequence remain undeleted in such a way that they satisfy the shrunk feedback polynomial in the decimated sequence. It is assumed that the conditional correlation coefficient is equal to one when the event occurs and to zero otherwise.

Consequently, the error sequence $\{e_t\}$ produced by applying the feedforward linear transform defined by $\hat{f}$ to the keystream sequence $\{y_t\}$, $e_t = y_t + \sum_{k=1}^{w} y_{t-\hat{i}_k}, t \geq \hat{r}$, is regarded as a sequence of nonbalanced identically distributed binary random variables with the correlation coefficient c to the constant zero variable. The variables are not independent. In order to detect the statistical weakness in the error sequence, one can apply a simple chi-square statistical test. For a sequence of length n with n_0 zeros and n_1 ones, the value of the chi-square statistic is given by

$$\chi^2 = \frac{(n_0 - n_1)^2}{n} = n \, \acute{c}^2 \tag{4}$$

where $\acute{c} = (n_0 - n_1)/n$ denotes an estimate of c. The error sequence can thus be distinguished from a purely random binary sequence with error probability less than about 10^{-3}, if n, which is approximately equal to the length of the observed keystream sequence, is equal to $10/c^2$ or larger. The same holds for the amount of computation needed.

The correlation coefficient depends on the chosen shrunk polynomial and is maximized if

$$\hat{\Delta}_k = \hat{\Delta}_k^{opt} \overset{\text{def}}{=} \lfloor (1 - p)(\Delta_k + 1) \rfloor, \quad 1 \leq k \leq w. \tag{5}$$

The maximum value denoted as c_f is given by

$$c_f = (1 - p)^{w+1} \prod_{k=1}^{w} p^{\Delta_k - \hat{\Delta}_k^{opt}} (1 - p)^{\hat{\Delta}_k^{opt}} \left(\frac{\Delta_k}{\hat{\Delta}_k^{opt}} \right). \tag{6}$$

The multiplicative terms in (6) are the central terms of the corresponding binomial distributions. If $2 \leq \hat{\Delta}_k^{opt} \leq \Delta_k - 2$, then the kth multiplicative term in (6) is by Stirling's formula well approximated as $(2\pi p(1-p)\Delta_k)^{-1/2}$. If $\hat{\Delta}_k^{opt}$ is equal either to Δ_k or to $\Delta_k - 1$, that is, if $p\Delta_k < 2 - p$, then the kth multiplicative term is lower-bounded by $(1-p)^{\Delta_k}$.

Let us analyze c_f in two extreme cases, when p is relatively large and when p is relatively small. The case when p is very close to 1 can be treated analogously, but is not of practical interest. Suppose first that none of r/w, Δ_k, $p\Delta_k$, and $(1-p)\Delta_k$ is very small. Then we have

$$c_f \simeq (1-p)\left(\frac{2\pi p}{1-p}\right)^{-\frac{w}{2}} \left(\prod_{k=1}^{w} \Delta_k\right)^{-\frac{1}{2}}. \tag{7}$$

The smallest magnitude of c_f, which is the worst case for cryptanalysis, is then obtained when the feedback taps are approximately equidistant and is equal to

$$c_f \simeq (1-p)\left(\frac{2\pi p}{1-p}\frac{r}{w}\right)^{-\frac{w}{2}}. \tag{8}$$

The necessary length of the keystream sequence and the amount of computation needed to detect the weakness are then both upper-bounded by $(10/(1-p)^2)$ $(2\pi p/(1-p))^w((r-w)/w)^w$. Given w, the larger the values of r and p the smaller the correlation coefficient. Given r and p, there exists an optimal value of w that minimizes the correlation coefficient. Also, $c \to 1$ when $p \to 0$ and $c \to 0$ when $p \to 1$. For example, if $p = 1/2$ as is suggested for the shrinking generator [4], then the necessary length becomes $40 (6.28 (r-w)/w)^w$. Second, suppose that the deletion rate p is relatively small so that $p\Delta_k < 2 - p$ for each k. Then we obtain a lower bound

$$c_f \geq (1-p)^{r+1} \geq \left(1 - \frac{2w}{r}\right)^{r+1}. \tag{9}$$

The minimum length to detect the weakness is then at most $10/(1-p)^{2(r+1)}$ $\leq 10/(1 - \frac{2w}{r})^{2(r+1)}$.

Consequently, one may conclude that the linear statistical weakness is realistic and may be relatively easy to detect unless both r and w are relatively large. If w is large, then instead of f one should use a polynomial multiple of f whose weight is relatively low and whose degree is not too large, based on the well-known fact that an LFSR sequence generated by f satisfies the linear recursions determined by all the polynomial multiples of f as well. In any case, the necessary length of the keystream sequence for the weakness detection is much smaller than 2^r. The underlying assumption is that the feedback polynomial f is known.

The required length of the observed keystream sequence can be significantly reduced by using more than just one different shrunk polynomials of a given polynomial f. The chi-square statistic is then computed on the resulting concatenation of the error sequences as a whole. There are exactly $M_f = \prod_{k=1}^{w}(\Delta_k + 1)$ different shrunk polynomials $\hat{f}$ of a given polynomial f. The sum of the correlation coefficients (3) for all the shrunk polynomials of f is readily verified to be

$$\sum_{\hat{f}} \left(p^{r-\hat{r}} (1-p)^{\hat{r}+1} \prod_{k=1}^{w} \binom{\Delta_k}{\hat{\Delta}_k} \right) = (1-p)^{w+1} \qquad (10)$$

and is equal to the probability that any $w+1$ bits satisfying the linear recursion defined by the polynomial f in the regularly clocked LFSR sequence all remain in the keystream sequence as undeleted after the decimation. The average correlation coefficient over all the shrunk polynomials of f is hence $(1-p)^{w+1}/M_f$. However, the total correlation virtually remains the same, up to a multiplicative constant close to one, if the summation is carried out only over the shrunk polynomials close to the optimal one for which the correlation coefficient is equal to the maximum value c_f. A shrunk polynomial is regarded to be close to the optimal one if for each $1 \leq k \leq w$ the absolute difference $\hat{\Delta}_k - \hat{\Delta}_k^{opt}$ is at most equal to the standard deviation $(2\pi p(1-p)\Delta_k)^{1/2}$ of the binomial distribution multiplied by a positive constant close to 1. Since the corresponding average correlation coefficient $\bar{c}_f$ is then very close to c_f, there are approximately $(1-p)^{w+1}/c_f$ such shrunk polynomials. So, the required length of the keystream sequence to detect the weakness is thus reduced to $10/(c_f(1-p)^{w+1})$, whereas the amount of computation remains the same, that is, $10/c_f^2$, comparing with the single optimal shrunk polynomial case.

3 Feedback Polynomial Reconstruction

In this section, it is assumed that the LFSR feedback polynomial f is key-dependent and hence unknown, as is proposed in [4]. In order to achieve long period and good long-term statistical properties on a period, f is usually chosen to be primitive or irreducible. However, since f is time-invariant given a key, the statistical weakness analyzed in the previous section remains, but the amount of computation required to detect the weakness is in general increased. Moreover, since the correlation coefficient (3) depends on the assumed shrunk polynomial $\hat{f}$ given f and is maximized if (5) is satisfied, the estimate $\hat{c}$, see (4), of the correlation coefficient may be used as a statistic to reconstruct the feedback polynomial f. We will now consider more closely the following three cases: the weakness detection, the feedback polynomial reconstruction, and the fast reconstruction of the feedback polynomial.

In practical applications, the degree r of f is typically known with relatively small uncertainty. The first objective is to find a shrunk polynomial $\hat{f}$ of f that is close to an optimal one yielding the maximum value of the correlation coefficient (6). Its degree $\hat{r}$ is then close to $(1-p)r + pw$. For example, if $p = 1/2$, then $\hat{r} \simeq (r+w)/2$, which may not be large. For each assumed w, one should then check $\binom{\hat{r}-1}{w-1}$ possible candidates for an optimal shrunk polynomial. The best candidate is the one with the maximum value of the statistic $\hat{c}$ over all possible values of w. The weakness is detected if the chi-square value (4) corresponding to this maximum value is significant. Typically, some prior information about w

is avalaible, so that the search is not complete. For example, in hardware realizations w is not large. The required length of the keystream sequence depends on f and is increased to $10\hat{r}/c_f^2$, because the optimal shrunk polynomial is described by $\hat{r}$ bits and the necessary length for each bit of uncertainty to be resolved is $10/c_f^2$. Since f is not known, instead of c_f one can use a lower bound (8) for relatively large p, $p \geq w/r$, or a lower bound (9) for small p, $p < w/r$. The obtained maximum value of the estimate $\hat{c}$ should be consistent with the chosen lower bound. The amount of computation for each of the assumed shrunk polynomials is proportional to the keystream sequence length. Another possibility is to simultaneously check all the shrunk polynomials close to the candidate one, which reduces the required keystream sequence length, as is described in the previous section. The unknown standard deviation can be estimated by using the equidistant taps assumption.

The second objective is to reconstruct the feedback polynomial f. Once the best shrunk polynomial is determined, the most likely feedback polynomial is the one that satisfies (5). However, the decision on f based on the estimate $\hat{c}$ of the correlation coefficient (3) is not reliable. Namely, the number of the shrunk polynomial candidates that are close to the best one with respect to the chi-square statistic is generally not small. In this case, there are many close candidates for the best feedback polynomial. To distinguish between them, one should then use the appropriate polynomial multiples of these polynomials. More precisely, for every chosen candidate for f, find a number of polynomial multiples and for each of them compute an estimate $\hat{c}$ of the correlation coefficient by using different shrunk polynomials close to the optimal one. The polynomial multiples should preferably have relatively low degrees and low weights so as to obtain reliable estimates of the correlation coefficients on a keystream sequence of a given length. In addition, the polynomial multiples of different candidates for f should be sufficiently different so as to obtain only one or just a few candidates whose correlation coefficient estimates are consistent with the maximum value corresponding to (5). So, in general, the longer the observed keystream sequence, the smaller the number of the remaining candidates. However, additional precomputation is required to obtain suitable polynomial multiples. For a candidate polynomial g, this can be done by computing the residues modulo g of the polynomials of the form z^i. Note that finding low weight polynomial multiples of a feedback polynomial is crucial for the fast correlation attacks on the initial state of regularly clocked LFSRs distorted by additive binary noise, see [13, 19, 3].

The third objective is to reconstruct the feedback polynomial f in a fast way, by reducing the computational effort needed to find an optimal shrunk polynomial. The feedback polynomial reconstruction then goes along the same lines as in the second approach. First observe that the average value of the correlation coefficient (3) over all the shrunk polynomials with fixed $\hat{\Delta}_i$ is maximized if (5) holds for $k = i$. Instead of the simultaneous search through all the shrunk polynomials of a given degree and weight, one can then proceed iteratively by reconstructing the values $\hat{\Delta}_i$ one at a time, starting from $i = 1$, for example.

More precisely, the sum of the correlation coefficients (3) for all the shrunk polynomials $\hat{f}$ with fixed $\hat{\Delta}_i$ is easily seen to be

$$\sum_{\hat{f}:\hat{\Delta}_i} \left(p^{r-\hat{r}}(1-p)^{\hat{r}+1} \prod_{k=1}^{w} \binom{\Delta_k}{\hat{\Delta}_k} \right) = (1-p)^{w+1} p^{\Delta_i - \hat{\Delta}_i}(1-p)^{\hat{\Delta}_i} \binom{\Delta_i}{\hat{\Delta}_i}. \quad (11)$$

The total number of such shrunk polynomials is $M_f^{(i)} = \prod_{k\neq i}(\Delta_k + 1)$, and the number of them with the correlation coefficient close to the maximum value, given the constraint on $\hat{\Delta}_i$, is approximately $(1-p)^w/c_f^{(i)}$ where $c_f^{(i)}$ is defined as c_f without the ith multiplicative term in (6) corresponding to $\hat{\Delta}_i$, that is,

$$c_f^{(i)} = (1-p)^w \prod_{k\neq i} p^{\Delta_k - \hat{\Delta}_k^{opt}}(1-p)^{\hat{\Delta}_k^{opt}} \binom{\Delta_k}{\hat{\Delta}_k^{opt}}. \quad (12)$$

However, since f is not known, one should consider the average correlation coefficient over all $M_f^{(i)}$ shrunk polynomials, that is,

$$\bar{c}_f(\hat{\Delta}_i) = \frac{(1-p)^{w+1}}{\prod_{k\neq i}(\Delta_k + 1)} p^{\Delta_i - \hat{\Delta}_i}(1-p)^{\hat{\Delta}_i} \binom{\Delta_i}{\hat{\Delta}_i}. \quad (13)$$

The maximum value is achieved when $\hat{\Delta}_i$ satisfies (5) and is equal to $\bar{c}_f^{(i)} = c_f(1-p)^w/(M_f^{(i)}c_f^{(i)})$. This value would exactly be equal to c_f if the average was taken over the $(1-p)^w/c_f^{(i)}$ shrunk polynomials. Since f is not known, let $\underline{c}$ denote a lower bound on $c_f^{(i)}/(1-p)^w$, for any $1 \le i \le w$, that is given by $(2\pi p(1-p)(r-w)/(w-1))^{-(w-1)/2}$ for relatively large p, $p \ge (w-1)/(r-1)$, see (8), and by $(1-p)^{r-w}$ for small p, $p < (w-1)/(r-1)$, see (9). A lower bound $\underline{\bar{c}}$ on $\bar{c}_f^{(i)}$, for any $1 \le i \le w$, can be obtained similarly. Roughly speaking, one may use a lower bound $\underline{\bar{c}} = (1-p)^{w+1}\pi^{-1/2}((r-1)/(w-1/2))^{-w+1/2}$.

The reconstruction procedure for finding an optimal shrunk polynomial is iterative and is based on estimating the correlation coefficient (13) on a given keystream sequence. For any assumed value of $\hat{\Delta}_i$, the number of different shrunk polynomials used to compute the average correlation coefficient is $1/\underline{c}$. Initially, for any possible value of $\hat{\Delta}_1$, one picks at random $1/\underline{c}$ shrunk polynomials with fixed $\hat{\Delta}_1$, estimates the average correlation coefficient on a given keystream sequence, and finds the best candidate for $\hat{\Delta}_1$ as the value for which the estimate is maximal. In the next step, one finds the best candidate for $\hat{\Delta}_2$ in a similar way except that the $1/\underline{c}$ shrunk polynomials are now chosen at random so that $\hat{\Delta}_1$ is around the best candidate value obtained in the first step. The procedure is then continued iteratively in an analogous way by using the best candidates for the values of $\hat{\Delta}_i$ from previous iterations to obtain the best estimate in the current iteration. After w steps, one then proceeds to recompute the best candidates obtained in the previous round of w steps and repeats the procedure for a number of rounds until no significant improvement is observed. It remains to determine the required length of the keystream sequence. For a reliable estimate of the correlation coefficient (13) one needs $20 \log \hat{r} / \bar{\underline{c}}^2$ computations, so that the necessary length would then be $20 \underline{c} \log \hat{r} / \bar{\underline{c}}^2$ (the total bit uncertainty of w and $\hat{\Delta}_i$ is here approximated as $2 \log \hat{r}$). It is easy to see that in such a way no reduction in computation is achieved. However, the maint point to observe is that the reliable estimate of the correlation coefficient is not needed. What is needed for the convergence of the iterative procedure to an optimal shrunk polynomial is just a slight improvement on the unknown value of $\hat{\Delta}_i$ in each iteration step, where the first round is the most critical one. This is to a certain extent similar to iterative probabilistic decoding procedures used in fast correlation attacks [13, 19, 3, 15, 16]. Accordingly, the convergence criteria established in [3] and [16] lead us to anticipate that the required number of computations per each step is then just about $1/\bar{\underline{c}}$, so that the necessary length is then $\underline{c}/\bar{\underline{c}}$. The convergence of iterative probabilistic procedures is discussed in more detail in the next section.

4 Initial State Reconstruction

In this section, it is assumed that the LFSR feedback polynomial f of degree r is known, and the objective is a fast reconstruction of the unknown LFSR initial state and the unknown decimation sequence, given an observed segment of the keystream/decimated sequence. The problem is known to be very difficult, see [9]. A divide and conquer attack on the unknown decimation sequence based on the linear consistency test [21] proves to be successful if the observed decimated sequence is sufficiently long, also see the collision test [1, 2]. On the other hand, divide and conquer correlation attacks on the LFSR initial state are also possible if the observed decimated sequence is sufficiently long. Namely, the embedding correlation attack is analyzed in [23] and [10] for the constrained clocking case and in [10] for the unconstrained clocking case, whereas the statistically optimal probabilistic correlation attack in the unconstrained clocking case is analyzed in [10]. The attacks imply the exhaustive search over all possible decimation sequences and over all possible LFSR initial states, respectively. Despite a divide and conquer effect, the computational complexity remains exponential. Our ultimate goal is to examine whether fast correlation attacks are possible, with linear or polynomial complexity.

The starting point is the linear statistical weakness of the decimated LFSR sequences based on shrunk feedback polynomials discussed in Section 2. Our aim is to reconstruct the decimation sequence $\{d_t\}$ one term at a time based on appropriate locally applied shrunk polynomials. Note that the decimation sequence is synchronous with the decimated sequence $\{y_t\}$ and that d_t is the number of bits to be deleted before the bit y_t is produced. Alternatively, the regularly clocked LFSR sequence can be obtained from the decimated sequence by inserting d_t bits before y_t, for all observed t. The decimation sequence need not be reconstructed completely: essentially only slightly more than r consecutive terms at any point in time are required to be known to determine the LFSR initial state uniquely or almost uniquely. Of course, the known decimation sequence can then be used for an attack on the secret key specifying the clock-control generator. Moreover, if the clock-control generator is an easy to reconstruct scheme, for example, another LFSR, then the decimation sequence reconstruction may incorporate the structure of the clock-control generator as well.

In the assumed probabilistic model, the decimation sequence is regarded as a sequence of independent identically distributed nonnegative integer random variables with a probability distribution $\mathcal{P} = \{P(d)\}_{d \in \mathcal{D}}$. The prior information about d_t is thus determined by the assumed probability distribution. For a statistical decision on the individual term d_t, some local statistic that would necessarily involve the two consecutive terms y_{t-1} and y_t is needed. The statistic should convey information about the value of d_t, so that the posterior probability of d_t becomes significantly different from the prior one. As in the fast correlation attacks on additively noised LFSR sequences, where the decision is being made on the individual terms of the binary noise sequence, the local information is extracted from the parity-check sums corresponding to the polynomial multiples of the feedback polynomial. They represent the codewords of the dual code of the linear code formed by the truncated LFSR sequences, see [3]. In this case, however, the parity-check sums are defined by the shrunk polynomials of the appropriate polynomial multiples of the feedback polynomial.

Let $h(z) = 1 + \sum_{k=1}^{\omega} z^{j_k}$, $1 \leq j_1 < \ldots < j_\omega = m$, denote a polynomial multiple of degree m and weight $\Omega = \omega + 1$, and let $\hat{h}(z) = 1 + \sum_{k=1}^{\omega} z^{\hat{j}_k}$, $1 \leq \hat{j}_1 < \ldots < \hat{j}_\omega = \hat{m}$, denote a shrunk polynomial of $h(z)$. Let $\tau_k = j_k - j_{k-1} - 1$ and $\hat{\tau}_k = \hat{j}_k - \hat{j}_{k-1} - 1$, $1 \leq k \leq \omega$, where $j_0 = \hat{j}_0 = 0$. The main idea is to find and use the polynomial multiples $h(z)$ such that $\tau_i = d$, for some $1 \leq i \leq \omega$, to check whether $d_t = d$, $d \in \mathcal{D}$. For each such $h(z)$ only the different shrunk polynomials $\hat{h}(z)$ such that $\hat{\tau}_i = 0$ are then used. The parity-check sum involving the consecutive terms y_{t-1} and y_t corresponds to $\hat{h}(z)$ in an obvious way. Each $h(z)$ may be multiply used to check the same value d, for all those i such that $\tau_i = d$. Let thus $\mathcal{H}_d$ denote a set of the polynomial multiples for checking the value d. For each polynomial $h(z)$ in $\mathcal{H}_d$ for which $\tau_i = d$, we choose a set of shrunk polynomials $\hat{h}(z)$ such that $\tau_i = 0$ and whose correlation coefficients are close to the maximum value $p^d(1-p)c_h^{(i)}$ where $c_h^{(i)}$ is obtained from the maximum value c_h by deleting the ith multiplicative term corresponding to τ_i, see (12). There are approximately $(1-p)^\omega / c_h^{(i)}$ such polynomials altogether, with the total conditional correlation coefficient $(1-p)^\omega$, see (11), and with the average conditional correlation coefficient close to $c_h^{(i)}$, where the assumed condition is that $d_t = d$. This is a very important point: although the conditional correlation coefficient associated with individual shrunk polynomials is small, the total conditional correlation coefficient may not be small because the number of the shrunk polynomials that are close to optimal is large for each feedback polynomial multiple $h(z)$.

The shrunk polynomials of different $h(z)$ in $\mathcal{H}_d$ may coincide. The conditional correlation coefficient $c(\hat{h})$ for a given polynomial $\hat{h}(z)$ is then defined as the sum of the individual conditional correlation coefficients associated with all the polynomials from $\mathcal{H}_d$ for which $\hat{h}$ is close to an optimal shrunk polynomial. In this case, each shrunk polynomial is used only once. The conditional correlation coefficient $c(\hat{h})$ given $h(z)$ is given by the same expression as $c_h^{(i)}$, see (12), except that $\hat{\tau}_k$ is substituted for $\hat{\tau}_k^{opt}$. Since the shrunk polynomials used are close to optimal, $c_h^{(i)}$ may be a good approximation for $c(\hat{h})$ given $h(z)$ which can be further simplified in a way described in Section 2, see (7) and (9). This is especially the case if $\hat{\tau}_k$ is very close to $\hat{\tau}_k^{opt}$. If not, one can also use the normal or Poisson approximations to the binomial distribution. The underlying assumption is that the probability distribution $\mathcal{P}$ is geometric. Similar expressions can also be obtained for the case of constrained irregular clocking. Let $\hat{\mathcal{H}}_d$ denote a set of the so-obtained different shrunk polynomials of the polynomials in $\mathcal{H}_d$. A basic requirement is that the pairwise intersections between the sets $\mathcal{H}_d$, for different d, should be insignificant, so that by discarding just a few shrunk polynomials one could obtain distinct sets $\hat{\mathcal{H}}_d$. This in fact means that the correlation coefficients $c_h^{(i)}$ should not be close for polynomials $h(z)$ in different $\mathcal{H}_d$. More importantly, one should also check that for each polynomial in $\mathcal{H}_d$, the correlation coefficient $c_h^{(i)}$ is not close to the correlation coefficient $c_{h'}^{(i)}$ for any possible feedback polynomial multiple $h'(z)$ that can be used to check any other possible value of d. Preliminary experiments have shown that this is possible to achieve if the set of possible values of d is $\mathcal{D} = \{0,1\}$ which corresponds to constrained irregular clocking. In any case, finding the appropriate shrunk polynomials can be done in precomputation time if the feedback polynomial is known. The obtained shrunk polynomials are called the parity-check polynomials.

We proceed now by describing and analyzing an iterative statistical decision procedure based on the parity-check sums computed by using the assumed parity-check polynomials. The statistically optimal decision rule for individual random variables d_t is based on the posterior probabilities. Let $\hat{P}_t(d)$ denote the posterior probability that $d_t = d$ given a set of the parity-check sums defined by the parity-check polynomials in $\hat{\mathcal{H}}_d$. Let $c(\hat{h})$ be a conditional correlation coefficient associated with a parity-check polynomial $\hat{h}$, where the assumed condition is that $d_t = d$. The basic assumption, justified by the choice of the parity-check polynomials, is that the conditional correlation coefficient is zero or very close to zero when $d_t \neq d$. Then by assuming that the observations determined by the individual parity-check sums are independent, it is not difficult to see that

$$\frac{\hat{P}(d)}{1 - \hat{P}(d)} = \frac{P(d)}{1 - P(d)} \prod_{\hat{h} \in \hat{\mathcal{H}}_d} (1 + c(\hat{h}))^{1 - s(\hat{h})} (1 - c(\hat{h}))^{s(\hat{h})} \tag{14}$$

where $s(\hat{h})$ is the binary value of the parity-check sum defined by the parity-check polynomial $\hat{h}$, and the index t is omitted for simplicity. The posterior probability ratio (the *odds*) of the event $d_t = d$ is thus increased or decreased depending on the individual observations. The expression is similar to the one that is obtained for fast correlation attacks on additively noised LFSR sequences [16].

The iterative statistical decision procedure that we now propose is in fact conceptually similar to the iterative probabilistic decoding procedures proposed for the fast correlation attacks in [13, 22, 3, 15, 16]. The main idea for the iterative procedure is to use the posterior probabilities obtained by (14) in the previous step as the prior probabilities for the current iteration step. Of course, since the independence assumption for different d is not quite accurate, the posterior probabilities calculated by (14) should be normalized after each step. For this iterative procedure to work, we need an expression for the conditional correlation coefficient $c(\hat{h})$ in the case when the probability distribution $\mathcal{P} = \{P(d)\}_{d \in \mathcal{D}}$ is time-dependent, because the expression (3) only holds for identically distributed variables with the geometric distribution. It is assumed that the decimation random variables d_t are independent. The general expression is not included here for simplicity. The computational effort may be large, but most of it can be done in precomputation time allowing a certain degree of numerical approximation. However, the amount of computation can be drastically reduced by using various simplifications, which may even improve the effectiveness of the attack. For example, it is reasonable to approximate the updated probability distribution of the decimation sequence by a different geometric distribution for each multiplicative term in (3) with the corresponding estimate of the deletion rate. Thus we get

$$c(\hat{h}) = (1 - p_0) \prod_{k \neq i} p_k^{\tau_k - \hat{\tau}_k} (1 - p_k)^{\hat{\tau}_k + 1} \binom{\tau_k}{\hat{\tau}_k} \tag{15}$$

where the probabilities p_k, $0 \leq k \leq \omega$, $k \neq i$, are computed as the appropriate average values based on the updated probability distribution. Another interesting approach is to identify the terms d_t for which the posterior (updated) probability distribution is most different from the prior one, and then to use these posterior probability distributions as they are, without averaging, or even to make 'hard' decisions on these terms by using the maximum posterior probability decision rule. Other tricks may also be possible. The constrained clocking case, in which the set $\mathcal{D}$ of the possible values of d is upper-bounded or limited, can be treated analogously.

The iterative procedure is successful if it converges to a sequence close to the decimation sequence that has actually produced the given keystream sequence. The main question is whether the required number of the parity-check polynomials is realistic or not in view of the additional constraints imposed on the choice of these polynomials. As is also the case with the fast correlation attacks [13, 22, 3, 15, 16], the exact mathematical derivation of the conditions for success of the new attack does not seem to be tractable. However, we now show that a simplified analysis [16] can be adapted to deal with the new attack. The main result of the analysis in [16] is to show that the fast correlation attack on additively noised LFSR sequences is successful if *the odds improvement in the first iteration step is 'significant' in the most favourable case when all the parity-check sums are different from zero.* Here, the most favourable case is when the parity-check sums are all equal to zero. The odds improvement is then given by

$$\prod_{\hat{h} \in \hat{\mathcal{H}}_d} (1 + c(\hat{h})) \simeq 1 + \sum_{\hat{h} \in \hat{\mathcal{H}}_d} c(\hat{h}) \simeq 1 + \sum_{h \in \mathcal{H}_d} (1 - p)^\omega \qquad (16)$$

because the correlation coefficients are very small. If we heuristically assume that the significant initial improvement is 2, then we obtain a nice convergence condition

$$\sum_\omega N_{d,\omega} (1 - p)^\omega > 1 \qquad (17)$$

where $N_{d,\omega}$ denotes the number of the feedback polynomial multiples of weight $\omega + 1$ in $\mathcal{H}_d$. The condition should be satisfied for all the values of d in $\mathcal{D}$ whose probability is not very close to zero. This condition applies in general, where p is the deletion rate of the decimation sequence. If it were not for the constrained choice of the feedback polynomial multiples, this condition would be less restrictive than the corresponding condition [16] for additively noised LFSR sequences. Note that for each polynomial h in $\mathcal{H}_d$, there are about $(1 - p)^\omega / c_h^{(i)}$ parity-check polynomials in $\hat{\mathcal{H}}_d$, where $c_h^{(i)}$ is determined by (12). Therefore, the computational complexity to obtain the required parity-check sums corresponding to a polynomial h of weight $\omega + 1$ and degree m is for relatively large p, $p \geq (\omega - 1)/(m - 1)$, upper-bounded by $(2\pi p(1 - p)(m - \omega)/(\omega - 1))^{(\omega-1)/2}$, see Section 3. This is feasible even if m is large provided that ω is relatively small. In the constrained clocking case where $\mathcal{D} = \{0, 1\}$, the condition (17) seems to be very realistic. In general, the chances for success of the proposed fast correlation attack appear to be greater in the constrained clocking case than in the unconstrained one. From the cryptographic standpoint, it turns out that the number of the feedback polynomial multiples of relatively low weight and not too large degree should be relatively small with respect to (17). This is in accordance with a vague anticipation given in [4].

After the convergence process is completed or at any iteration step when the odds are significant, one can make 'hard' decisions on the values of d_t for some t. If the uncertainty reduction regarding the decimation sequence is large, one may then check the remaining needed values by exhaustive search, either by applying the linear consistency test [21] or by applying the embedding algorithm [23, 10]. In view of [21], one has to determine slightly more than r consecutive terms of the decimation sequence at any point in time, where r is the degree of the LFSR feedback polynomial. The LFSR state at some time t is then determined uniquely by solving the appropriate linear equations. The time t is not known if all the initial terms of the decimation sequence are not known. They can be determined by applying the embedding or the Levenshtein-like distance algorithms [23, 10].

5 Conclusion

A theory of fast correlation attacks on irregularly clocked LFSRs based on a recently determined linear statistical weakness of decimated LFSR sequences is developed. The weakness and the corresponding correlation coefficients are analyzed in more detail. When the LFSR feedback polynomial is not known, the statistical weakness detection and the feedback polynomial reconstruction are discussed and an iterative algorithm for fast feedback polynomial reconstruction is proposed. When the LFSR feedback polynomial is known, an iterative procedure for fast LFSR initial state reconstruction given an observed keystream sequence is introduced. The procedure is based on specially defined parity-check sums corresponding to the shrunk polynomials of appropriate feedback polynomial multiples, and consists in iterative recomputation of the posterior probabilities for unknown elements of the decimation sequence. By using the analogy with the well-known fast correlation attacks on additively noised LFSR sequences, a convergence condition which determines the numbers of the parity-check sums required for successful reconstruction is derived. This condition and the corresponding polynomial computational complexity show that the proposed fast correlation attack may be realistic, especially in the constrained clocking case. Extensive computer simulations are needed to support the theory and to specify the technical details.

References

1. R. J. Anderson, "Solving a class of stream ciphers," *Cryptologia*, 14(3):285–288, 1990.
2. R. J. Anderson, "Faster attack on certain stream ciphers," *Electr. Lett.*, 29(15): 1322–1323, July 1993.
3. V. Chepyzhov and B. Smeets, "On a fast correlation attack on stream ciphers," Advances in Cryptology – EUROCRYPT '91, *Lecture Notes in Computer Science*, vol. 547, D. V. Davies ed., Springer-Verlag, pp. 176–185, 1991.
4. D. Coppersmith, H. Krawczyk, and Y. Mansour, "The shrinking generator," Advances in Cryptology – CRYPTO '93, *Lecture Notes in Computer Science*, vol. 773, D. R. Stinson ed., Springer-Verlag, pp. 22–39, 1994.
5. J. Dj. Golić and M. V. Živković, "On the linear complexity of nonuniformly decimated PN-sequences," *IEEE Trans. Inform. Theory*, 34:1077–1079, Sep. 1988.
6. J. Dj. Golić and M. J. Mihaljević, "A generalized correlation attack on a class of stream ciphers based on the Levenshtein distance," *Journal of Cryptology*, 3(3):201–212, 1991.
7. J. Dj. Golić and S. V. Petrović, "A generalized correlation attack with a probabilistic constrained edit distance," Advances in Cryptology - EUROCRYPT '92, *Lecture Notes in Computer Science*, vol. 658, R. A. Rueppel ed., Springer-Verlag, pp. 472–476, 1993.
8. J. Dj. Golić, "Correlation via linear sequential circuit approximation of combiners with memory," Advances in Cryptology – EUROCRYPT '92, *Lecture Notes in Computer Science*, vol. 658, R. A. Rueppel ed., Springer-Verlag, pp. 113–123, 1993.
9. J. Dj. Golić, "On the security of shift register based keystream generators," Fast Software Encryption – Cambridge '93, *Lecture Notes of Computer Science*, vol. 809, R. J. Anderson ed., Springer-Verlag, pp. 90–100, 1994.
10. J. Dj. Golić and L. O'Connor, "Embedding and probabilistic correlation attacks on clock-controlled shift registers," *Pre-proceedings of Eurocrypt '94*, pp. 231–243, Perugia, Italy, 1994.

11. J. Dj. Golić, "Intrinsic statistical weakness of keystream generators," *Pre-proceedings of Asiacrypt '94*, pp. 72–83, Wollongong, Australia, 1994.

12. D. Gollmann and W. G. Chambers, "Clock controlled shift registers: a review," *IEEE J. Sel. Ar. Commun.*, 7(4):525–533, 1989.

13. W. Meier and O. Staffelbach, "Fast correlation attacks on certain stream ciphers," *Journal of Cryptology*, 1(3):159–176, 1989.

14. W. Meier and O. Staffelbach, "The self-shrinking generator," *Pre-proceedings of Eurocrypt '94*, pp. 201–210, Perugia, Italy, 1994.

15. M. J. Mihaljević and J. Dj. Golić, "A comparison of cryptanalytic principles based on iterative error-correction," Advances in Cryptology – EUROCRYPT '91, *Lecture Notes in Computer Science*, vol. 547, D. V. Davies ed., Springer-Verlag, pp. 527–531, 1991.

16. M. J. Mihaljević and J. Dj. Golić, "Convergence of a Bayesian iterative error-correction procedure on a noisy shift register sequence," Advances in Cryptology – EUROCRYPT '92, *Lecture Notes in Computer Science*, vol. 658, R. A. Rueppel ed., Springer-Verlag, pp. 124–137, 1993.

17. R. A. Rueppel, "Stream ciphers," in *Contemporary Cryptology: The Science of Information Integrity*, G. Simmons ed., pp. 65–134. New York: IEEE Press, 1991.

18. T. Siegenthaler, "Decrypting a class of stream ciphers using ciphertext only," *IEEE Trans. Comput.*, 34:81–85, Jan. 1985.

19. K. C. Zeng and M. Huang, "On the linear syndrome method in cryptanalysis," Advances in Cryptology – CRYPTO '88, *Lecture Notes in Computer Science*, vol. 403, S. Goldwasser ed., Springer-Verlag, pp. 469–478, 1990.

20. K. C. Zeng, C. H. Yang, and T. R. N. Rao, "An improved linear syndrome algorithm in cryptanalysis with applications," Advances in Cryptology – CRYPTO '90, *Lecture Notes in Computer Science*, vol. 537, A. J. Menezes and S. A. Vanstone eds., Springer-Verlag, pp. 34–47, 1991.

21. K. C. Zeng, C. H. Yang, and T. R. N. Rao, "On the linear consistency test (LCT) in cryptanalysis and its applications," Advances in Cryptology – CRYPTO '89, *Lecture Notes in Computer Science*, vol. 435, G. Brassard ed., Springer-Verlag, pp. 164–174, 1990.

22. M. V. Živković, "On two probabilistic decoding algorithms for binary linear codes," *IEEE Trans. Inform. Theory*, 37:1707–1716, Nov. 1991.

23. M. V. Živković, "An algorithm for the initial state reconstruction of the clock-controlled shift register," *IEEE Trans. Inform. Theory*, 37:1488–1490, Sep. 1991.

Large Period Nearly deBruijn FCSR Sequences (Extended Abstract)

Andrew Klapper

Dept. of Computer Science
University of Kentucky
Lexington, KY 40506-0046 USA
klapper@cs.engr.uky.edu

Mark Goresky

Dept. of Mathematics
Northeastern University
Boston, MA 03115 USA
goresky@nuhub.neu.edu

Abstract. Recently, a new class of feedback shift registers (FCSRs) was introduced, based on algebra over the 2-adic numbers. The sequences generated by these registers have many algebraic properties similar to those generated by linear feedback shift registers. However, it appears to be significantly more difficult to find maximal period FCSR sequences. In this paper we exhibit a technique for easily finding FCSRs that generate nearly maximal period sequences. We further show that these sequence have excellent distributional properties. They are balanced, and nearly have the deBruijn property for distributions of subsequences.

Index Terms – **Binary sequences, feedback with carry shift registers, deBruijn property, 2-adic numbers.**

1 Introduction

Pseudorandom sequences with a variety of statistical properties, such as large period, high linear span, and good statistical distributions, are important in many areas of communications and computing, such as cryptography, spread spectrum communications, error correcting codes, and Monte Carlo integration. Thus devices for generating sequences with such good properties are basic tools for the design of stream ciphers (as well as for other applications). While such properties alone are insufficient to make sequences useful for encryption, they are initial minimal requirements. Once we can generate sequences with these properties, various techniques can be used to further scramble sequences making them suitable for encryption, while perhaps retaining good statistical properties.

One class of sequences that has many nice properties is the class of linear feedback shift register (LFSR) sequences. Maximal period LFSR sequences (or m-sequences) are known to have large period and a balance of zeros and ones, and to become deBruijn sequences when a single zero is inserted [3]. These properties, as well as the availability of algebraic tools for their analysis, have led to their use in a number of constructions of key stream generators. Examples

L.C. Guillou and J.-J. Quisquater (Eds.): Advances in Cryptology - EUROCRYPT '95, LNCS 921, pp. 263-273, 1995.

include nonlinear feedforward functions [6], nonlinear combining functions [14], and clock controlled shift registers [2].

Recently a new class of binary sequence generators, *feedback with carry shift registers* (or FCSRs) has been described by Klapper and Goresky [8, 9]. They have many of the nice algebraic properties of LFSR sequences and, it is hoped, will serve as building blocks for stream ciphers in much the same way that LFSR sequences have in the past. In this paper we study some of the basic statistical properties of FCSR sequences. We show how to construct, in an effective manner, FCSRs with very large period. Previously described methods for doing this required the choice of a prime number q for which 2 is a primitive root. Unfortunately, there is no known effective way of testing this condition, nor is it even known whether there are infinitely many such primes. Here we show that if p is a prime number such that 2 is a primitive root modulo p and modulo p^2, then for any positive integer e, using $q = p^e$ as the connection integer of a FCSR results in an output sequence with period $\phi(q) = p^e - p^{e-1}$. The condition that 2 be primitive modulo p^2 is known to hold whenever 2 is primitive modulo p for $p < 2 \cdot 10^{10}$. We also give an explicit procedure for finding the initial settings of FCSRs with this period.

We further show that the sequences so constructed have excellent statistical properties in the sense that they are nearly deBruijn sequences. Recall that a deBruijn sequence is a sequence bfa of period N such that every sequence of length $\log(N)$ (N must be a power of 2) occurs precisely once in each period of bfa. In other words, the numbers of occurrences of any two subsequences of length $\log(N)$ are equal. For sequences whose period is not a power of 2, the best we can hope for in this regard is that the numbers of occurrences of any two sequences differ by at most one. This, in fact, is the case when q is prime and 2 is a primitive root modulo q [8]. When q is a power of a prime and 2 is a primitive root modulo q, we show that the numbers of occurrences of any two subsequences differ by at most two. This holds for subsequences of *any* length.

Finally, we consider an arithmetic, or "with carry," analog of the crosscorrelation of two sequences. We show that for any two decimations of a FCSR sequence of the type described above, the arithmetic correlations are identically zero, except when the two sequences coincide.

2 Feedback with Carry Shift Registers

In this section we review the operation of FCSRs and recall their basic algebraic properties. See [8, 9] for details. Let q be an odd positive integer, and let $q + 1$ have the binary expansion $q+1 = \sum_{i=1}^{r} q_i 2^i$ with $q_i \in \{0, 1\}$. For convenience we also let $q_0 = -1$, so $q = \sum_{i=0}^{r} q_i 2^i$. The coefficients $q_1, \cdots, q_r$ are to be thought of as the taps on a feedback register. We can think of q as giving a recurrence with carry on the output sequence of this register.

Definition 1. The FCSR with connection integer q is a feedback register with r bits of storage plus additional memory for carry. If the contents of the register at any given time are $(a_{r-1}, a_{r-2}, \ldots, a_1, a_0)$ and the memory is m, then the operation of the shift register is defined as follows:

A1. Form the integer sum $\sigma = \sum_{k=1}^{r} q_k a_{r-k} + m$.

A2. Shift the contents one step to the right, outputting the rightmost bit a_0.

A3. Place $a_r = \sigma \pmod 2$ into the leftmost cell of the shift register

A4. Replace the memory m with $(\sigma - a_r)/2$.

Such a register outputs an infinite binary sequence $\mathbf{a} = (a_0, a_1, a_2, \cdots)$. The analysis of FCSR sequences employs the 2-adic number associated with $\mathbf{a}$, i.e., the power series with indeterminate replaced by 2, $\alpha = \sum_{i=0}^{\infty} a_i 2^i$. This 2-adic number plays a role similar to that of the generating function in linear feedback shift register theory. See [11] for background on 2-adic numbers. For convenience from time to time we speak of α as being the output of a FCSR, or as being periodic, or as having any other property that should more properly be attributed to the sequence $\mathbf{a}$.

The following facts are known about FCSRs and their output sequences [8, 9].

1. A binary sequence $\mathbf{a}$ is eventually periodic if and only if its associated 2-adic number α is a rational number c/q. It is strictly periodic if and only if moreover $-q < c \le 0$.

2. If $\mathbf{a}$ is the output sequence of a FCSR, and α is the associated 2-adic number, then $\mathbf{a}$ is eventually periodic and $\alpha = c/q$ where q is the connection number of a FCSR that outputs $\mathbf{a}$. In this case we can write

$$c = \sum_{i=0}^{r-1} \sum_{j=0}^{r-i-1} q_i a_j 2^{i+j} - m 2^r, \tag{1}$$

where m is the initial state of the extra memory.

3. Conversely, every eventually periodic binary sequence $\mathbf{a}$ whose associated 2-adic number can be written $\alpha = c/q$ for integers c, q, with q odd, is the output of a FCSR with connection number q.

4. Suppose $\mathbf{a}$ is the output sequence of a FCSR with connection integer q. Let γ be the inverse of 2 modulo q. Then there exists $A \in \mathbf{Z}/(q)$ such that for every i, $a_i = (A\gamma^i \bmod q) \bmod 2$. This composition of mod operations means first reduce modulo q to a number between 0 and $q - 1$, then reduce the result modulo 2.

5. Adding b to the initial memory changes α by $-b2^r/q$.

As a consequence of the exponential representation of FCSR sequences, it is apparent that the period of the output of a FCSR with connection number q can be no more than the cardinality of the multiplicative group of the integers modulo q, $(\mathbf{Z}/(q))^*$.

Definition 2. An ℓ-sequence is a FCSR sequence with maximum possible period $T = |(\mathbf{Z}/(q))^*|$.

An ℓ-sequence is analogous to an m-sequence in LFSR theory. Such a sequence is generated by connection numbers q for which 2 is a *primitive root*. The best we can hope is that the period is $q - 1$. This occurs when q is prime and 2 is a primitive root modulo q. The search for primes q such that 2 is a primitive root is related to a large body of contemporary number theory. It is believed that there are infinitely many primes q with this property [5]. However, finding such primes (and even finding large primes at all) is problematic.

In this paper we consider two fundamental questions about FCSR sequences:

1. How can we guarantee the output sequence has large period?
2. What are the statistical properties of large period FCSR sequences?

The first question can be divided into two parts:

1. How can we guarantee that the 2-adic number c/q has large period? Equivalently, how can we guarantee that 2 is a primitive root modulo q and c is relatively prime to q?
2. Given a rational number c/q, how can we efficiently construct the initial loading of a FCSR that outputs c/q?

3 Finding ℓ-Sequences for Prime Powers

In this section we give a method for generating ℓ-sequences based on FCSRs whose connection numbers are prime powers. Note that if q is not a prime power, then $\mathbf{Z}/(q)^*$ is not a cyclic group, so 2 cannot be primitive and we can have no ℓ-sequences. The following fact is well known, but we include a brief proof for completeness.

Theorem 3. *Let q be a power of a prime p, say $q = p^e$. If 2 is a primitive root modulo p^2, then 2 is a primitive root modulo q as well.*

Proof: The order of the multiplicative group of integers modulo q is $\phi(q) = p^{e-1}(p - 1)$. Thus $2^{p^{e-1}(p-1)} \equiv 1 \bmod q$, and we must show that there is no prime number $t > 1$ dividing $p^{e-1}(p - 1)$ such that $2^{p^{e-1}(p-1)/t} \equiv 1 \bmod q$. We do so by induction on e. That is, we may assume that the order of 2 modulo p^{e-1} is $p^{e-2}(p - 1)$. Also note that if 2 is a primitive root modulo p^2, it must also be a primitive root modulo p.

Suppose t divides $p - 1$. From the fact that $2^p \equiv 2 \bmod p$, it follows that $2^{p-1}/t \equiv 1 \bmod p$, contradicting the primitivity of 2 modulo p.

Thus we may assume $t = p$. Thus p^e divides $2^{p^{e-2}(p-1)} - 1$, but by induction p^{e-1} does not divide $2^{p^{e-3}(p-1)} - 1$. Also, p^{e-2} divides $2^{p^{e-3}(p-1)} - 1$. Thus $2^{p^{e-3}(p-1)} = 1 + p^{e-2}y$ for some y relatively prime to p. But then

$$2^{p^{e-2}(p-1)} - 1 = (1 + p^{e-2}y)^p - 1 \equiv p^{e-1}y \bmod p^e$$

when $e \geq 3$. Thus p^e does not divide $2^{p^{e-2}(p-1)} - 1$, a contradiction. $\square$

Suppose we have a prime p for which 2 is primitive. Checking whether 2 is primitive modulo p^2 is quite easy. Suppose this is not the case. The order of the multiplicative group $\mathbf{Z}/(p^2)^*$ is $p(p-1)$, so for some divisor $t \neq 1$ of $p(p-1)$, the order of 2 is $p(p-1)/t$. That is, p^2 divides $2^{p(p-1)/t} - 1$. We may assume t is prime. Thus either $t = p$ or t is a divisor of $p - 1$.

If t is a divisor of $p-1$, then p is a divisor of $2^{p(p-1)/t} - 1$. But $2^p \equiv 2 \bmod p$, so p divides $2^{(p-1)/t} - 1$. This contradicts the assumption that 2 is primitive modulo p. Thus $t = p$, so p^2 divides $2^{p-1} - 1$. It follows that to check whether 2 is a primitive root modulo p^2, it suffices to check whether p^2 divides $2^{p-1} - 1$.

Thus we can find ℓ-sequences as follows: choose a small prime p for which 2 is primitive (for small p this can be checked easily); check that p^2 does not divide $2^{p-1} - 1$; choose $e > 0$ and let $q = p^e$; choose an integer c relatively prime to p, with $0 < c < q$; and construct the FCSR with connection integer q and output $-c/q$. This gives us a FCSR whose output is strictly periodic with period $|\mathbf{Z}/(q)^*| = p^e - p^{e-1}$. In the next section we discuss the construction of the initial loading of the FCSR for a given $-c/q$.

One can ask about the abundance of primes p for which 2 is a primitive root modulo p^2. Hardy and Wright point out that the condition that p^2 divides $2^{p-1} - 1$ holds for only two primes p less than $3 \cdot 10^7$ [4, p. 73], and by computer search Bombieri has extended this limit to $2 \cdot 10^{10}$ [1]. (The two primes are 1093 and 3511.) In both cases 2 is not primitive modulo p. Thus for a large number of primes, we need only check the primitivity of 2 modulo p. In fact, it is not known whether there are *any* primes p such that 2 is primitive modulo p but not modulo p^2, though there is no compelling reason to believe there are no such primes.

4 Initial Loading of a FCSR

In this section we describe how an initial loading can be chosen for a FCSR that guarantees the output will be purely periodic and will have the maximum period for the given connection number.

It has been shown that, for a given rational number c/q, the initial loading for an FCSR that gives output c/q can be found by the following procedure [8, 9].

B1. Set $m_{-1} = c$.
B2. For each $i = 0, 1, \ldots, r - 1$ compute the following numbers:

$$\sigma_i = \sum_{k=0}^{i-1} q_{i-k} a_k + m_{i-1} \in \mathbf{Z} \tag{2}$$

$$a_i = \sigma_i \ (\bmod \ 2) \in \mathbf{Z}/(2) \tag{3}$$

$$m_i = \frac{\sigma_i - a_i}{2}. \tag{4}$$

If we use the initial loading $(a_{r-1}, a_{r-2}, \ldots, a_1, a_0)$ and initial memory $m_{r-1} \in R$, then the resulting FCSR outputs the 2-adic expansion of c/q. If c is relatively prime to q, then the period of the sequence is $T = \mathrm{ord}_q(2)$. However if c and q have a common factor then the period may be smaller but at least it will divide $\mathrm{ord}_q(2)$. Thus for $q = p^e$ with 2 primitive modulo p^2 (and hence also modulo q), if we randomly choose c, check that $\gcd(c, p) = 1$, and then find an initial loading using the above procedure, we will find an initial loading that gives a period $p^e - p^{e-1}$ strictly period output sequence. The expected number of random choices of c needed to achieve this is $p/(p-1)$, since the probability that c is relatively prime to p is $(p-1)/p$.

Alternatively, we may want more control over the initial setting of the register. We can choose the initial contents of the register, then attempt to find an initial value of the memory that gives the desired output sequence. We proceed as follows.

1. Randomly choose bits $a_0, \cdots, a_{r-1} \in \{0, 1\}$.
2. Compute

$$z = \sum_{i=0}^{r-1} \sum_{j=0}^{r-i-1} q_i a_j 2^{i+j} \tag{5}$$

$$= \sum_{k=0}^{r-1} \sum_{i=0}^{k} q_i a_{k-i} 2^k. \tag{6}$$

3. Let

$$m = \left\lceil \frac{z}{2^r} \right\rceil.$$

4. Check $\gcd(2^r m - z, q) = 1$. If so, use $a_0, \cdots, a_{r-1}$ as the initial loading, and m as the initial memory. If not, repeat (1) – (4). In some cases $\lfloor (z+q)/2^r \rfloor = \lceil z/2^r \rceil + 1$ and can also be tried as the the initial memory.

To see that this gives a maximal period purely periodic sequence for q, it suffices to check that $0 \leq 2^r m - z < q$, since the output from the q-FCSR with these initial values is $z - 2^r m/q$. But this follows immediately from the choice of m.

The big question is the time complexity. First observe that in any given repetition of (1) – (4), the probability of success is at least $(p^e - p^{e-1})/p^e = (p-1)/p$. Thus the expected number of trials is only $p/(p-1)$.

The most costly part of this algorithm is step (2). This can be done quickly using a divide and conquer algorithm similar to divide and conquer multiplication. For $r - 1$ bit integers q and a, we define the operation

$$\mathrm{semimult}(q, a, r) = \sum_{k=0}^{r-1} \sum_{i=0}^{k} q_i a_{k-i} 2^k.$$

If we write $q = q' + 2^{[r/2]}q''$ and $a = a' + 2^{[r/2]}a''$, then we have

semimult$(q, a, r) =$

$$q'a' + 2^{[r/2]}(\text{semimult}(q', a'', r - [r/2]) + \text{semimult}(q'', a', r - [r/2])).$$

Thus the time it takes to compute semimult(q, a, r) satisfies a recurrence

$$T(r) = 2T([r/2]) + S([r/2]) + \mathcal{O}(r),$$

where $S(r)$ is the time it takes to multiply two r bit numbers. Thus $T(r) \leq S(r) + cr$ for some constant c. Furthermore, if we use the Schönhage-Strassen algorithm [15], then $S(r) = \mathcal{O}(r \log r \log \log r)$. This can be improved to $S(r) \sim r \log r$ using Pollard's nonasymptotic algorithm for $r < 2^{37}$ on a 32 bit machine or $r < 2^{70}$ on a 64 bit machine [13].

Finally, observe that $\gcd(2^r m - z, q) = 1$ if and only if $\gcd(2^r m - z, p) = 1$. This can be checked using the Euclidean algorithm in $\mathcal{O}(r \log(p)^2)$ bit operations.

In summary, a desired initial loading can be found in less than expected $2r \log r + \mathcal{O}(r \log(p)^2)$ time for $r < 2^{37}$ on a 32 bit machine, or $r < 2^{70}$ on a 64 bit machine.

5 Distributional Properties

In this section we show that the sequences constructed above have excellent distributional properties. First we note that they are balanced.

Proposition 4. *Let q be a power of a prime p, say $q = p^e$, and suppose that 2 is primitive modulo q. Let **a** be any maximal period FCSR sequence, generated by a FCSR with connection integer q. The number of zeros and the number of ones in one period of **a** are equal.*

Furthermore we can consider higher order distributions. We show next that these sequences are close to having the deBruijn property that each subsequence of length log of the period occurs exactly once in each period. We show that for any two such subsequences, their numbers of occurrences can differ by at most two.

Theorem 5. *Let q be a power of a prime p, say $q = p^e$, and suppose that 2 is primitive modulo q. Let s be any nonnegative integer, and let A and B be s bit subsequences. Let **a** be any maximal period, purely periodic FCSR sequence, generated by a FCSR with connection integer q. Then the numbers of occurrences of A and B in **a** with their starting positions in a fixed period of **a** differ by at most 2.*

Proof: The purely periodic FCSR sequences with connection integer q are precisely the 2-adic expansions of rational numbers $-x/q$, with $0 \leq x < q$ [8, 9]. Such a sequence has maximum period if and only if p does not divide x. Since 2 is primitive modulo q, the cyclic shifts of bfa correspond to the set of all rational numbers $-x/q$, with $0 \leq x < q$. Thus an s bit subsequence A occurs in bfa if and only if it occurs as the first s bits in the 2-adic expansion of some rational numbers $-x/q$ with $0 \leq x < q$ and p not dividing x. Two rational number $-x_1/q$ and $-x_2/q$ have the same first s bits if and only if $-x_1/q \equiv -x_2/q \bmod 2^s$, if and only if $x_1 \equiv x_2 \bmod 2^s$. Thus we want to count the number of x with a given first s bits, $0 \leq x < q$, and x not divisible by p.

Let $2^r < q < 2^{r+1}$. If $s > r$, there are either zero or one such x, so the result follows. Thus we may assume $s \leq r$.

We first count the number of x with the first s bits fixed and $0 \leq x < q$, ignoring the divisibility condition. If $A = a_0, \cdots, a_{s-1}$, we let $\alpha = \sum_{i=0}^{s-1} a_i 2^i$. Let $q = \sum_{i=0}^{r} q_i 2^i$, and $q' = \sum_{i=0}^{s-1} q_i 2^i$. If $\alpha < q'$, then every choice of $a_s, \cdots, a_r$ with $\sum_{i=s}^{r} a_i 2^i \leq \sum_{i=s}^{r} q_i 2^i$ gives a unique x in the right range. If $\alpha \geq q'$, then every choice of $a_s, \cdots, a_r$ with $\sum_{i=s}^{r} a_i 2^i < \sum_{i=s}^{r} q_i 2^i$ gives a unique x in the right range. Thus for different choices of A, the numbers of such x differ by at most one.

Next we consider those x for which $0 \leq x < q$ and p divides x. That is, $x = py$ for some y, and $0 \leq y < q/p = p^{e-1}$. As above, $x_1 = py_1$ and $x_2 = py_2$ have the same first s bits if and only if the same is true of y_1 and y_2. The preceding paragraph shows that the numbers of such y for different choices of the first s bits differ by at most one. But if $x = py$, then $y \equiv A \bmod 2^s$ if and only if $x \equiv pA \bmod 2^s$, so for any B and C, the number of xs divisible by p with first s bits equal to B differs from the number of xs divisible by p with first s bits equal to C by at most 1. We have

$$|\{x : 0 \leq x < q, p \nmid x, \wedge x \equiv \alpha \bmod 2^s\}|$$
$$= |\{x : 0 \leq x < q \wedge x \equiv \alpha \bmod 2^s\}| - |\{x : 0 \leq x < q, p | x, \wedge x \equiv \alpha \bmod 2^s\}|.$$

As α varies the two terms on the right hand side vary by at most one from their values for any fixed choice of α. Thus the difference varies by at most 2. □

It is easy to check that the difference can be as large as 2.

6 Arithmetic Correlations

Traditionally, the shifted cross-correlations of two sequences have been used as a measure of the extent to which the sequences are independent. These values are small if corresponding bits in one sequence are as likely to be equal as they are to be different. In the case of FCSR sequences, it appears quite difficult to compute cross-correlations in the usual sense. There is, however, an arithmetic (or "with carry") analog of the cross-correlation. This has been studied previously in the case of autocorrelation functions by Mandelbaum [12].

Definition 6. Let $\mathbf{a}$ and $\mathbf{b}$ be two eventually periodic sequences with period N, and let $0 \leq \tau < N$. Let $\mathbf{b}^\tau$ be the sequence formed by shifting $\mathbf{b}$ by τ positions, $\mathbf{b}_i^\tau = \mathbf{b}_{i+\tau}$. Then the *shifted arithmetic cross-correlation* $\Theta_{\mathbf{a},\mathbf{b}}(\tau)$ of $\mathbf{a}$ and $\mathbf{b}$ is the difference between the number of zeros and the number of ones in a complete period of the periodic part of the sequence formed by subtracting $\mathbf{b}^\tau$ from $\mathbf{a}$ *with carry*. When $\mathbf{a} = \mathbf{b}$, the cross-correlation is called the *autocorrelation* of $\mathbf{a}$.

This corresponds to forming the 2-adic numbers α and β associated with $\mathbf{a}$ and $\mathbf{b}$, computing $\gamma = \alpha - 2^{-\tau}\beta$, and taking the difference between the number of coefficients that are zero and the number of coefficients that are one in a single period of γ. Even when $\mathbf{a}$ and $\mathbf{b}$ are purely periodic, there may be a transient prefix before γ becomes periodic. However, in this case the purely periodic part of γ is guaranteed to begin after at most N bits.

In the case of m-sequences and standard correlations the cross-correlations must be at least one in absolute value, simply because the periods of the sequences are odd. In general, the larger a family of sequences, the larger the maximum cross-correlations in the family. Remarkably, we can exhibit families of sequences in which all arithmetic correlations are identically zero. This generalizes a result of Mandelbaum showing that shifted autocorrelations of ℓ-sequences based on prime connection integers are identically zero [12]. Recall that a sequence $\mathbf{b}$ is a *k-fold decimation* of a sequence $\mathbf{a}$ if $\mathbf{b}$ is formed by taking every kth term of $\mathbf{a}$. That is, $b_i = a_{ki}$.

Theorem 7. *Let $\mathbf{a}$ be an ℓ-sequence based on connection integer $q = p^e$, p prime. Let k and m be integers that are relatively prime to the period $p^e - p^{e-1}$ of $\mathbf{a}$. Let $\mathbf{b}$ and $\mathbf{c}$ be k-fold and m-fold decimations of $\mathbf{a}$, respectively. Let τ be any shift. If $\mathbf{c}$ is a shift of $\mathbf{b}$, then there is one value of τ for which $\Theta_{\mathbf{b},\mathbf{c}}(\tau) = p^e - p^{e-1}$. In all other cases (whether or not $\mathbf{c}$ is a shift of $\mathbf{b}$), $\Theta_{\mathbf{b},\mathbf{c}}(\tau) = 0$.*

One can see, for example, that if we choose $q = 25$, then the decimations of the sequence of bits in the 2-adic expansion of $-1/q$ give eight cyclically distinct sequences of period 20 with ideal pairwise arithmetic correlations. In the classical theory of cross-correlations, any family of five or more sequence with this period must have maximum cross-correlation at least 5.

7 Conclusions

We have demonstrated that a large class of FCSR sequences are ℓ-sequences. This means that their periods are exponentially larger than the amount of initial information (taps on the register, initial register contents, and initial memory) required to generate the sequences. We have further shown that these sequences have excellent statistical properties, being nearly deBruijn sequences.

The picture for FCSRs whose connection number is not a prime power is more complicated. If $q = \prod p_i^{e_i}$, then the cardinality of $\mathbf{Z}/(q)$ is $\prod_{i=1}^{k} p_i^{e_i-1}(p_i - 1)$. However, this group is the product of cyclic groups of order $p_i^{e_i-1}(p_i - 1)$, so the order of its maximal order element is the least common multiple of the $p_i^{e_i-1}(p_i - 1)$. Since each of the $p_i - 1$ are even, the order of 2 cannot be the order of the full group. By the Chinese remainder theorem, the order of 2 modulo q is the least common multiple of the orders of 2 modulo $p_i^{e_i}$, $i = 1, \cdots, k$. Thus if the p_i are chosen so that 2 is primitive modulo each p_i, then its order modulo q is

$$\mathrm{lcm}\{p_i^{e_i-1}(p_i - 1)\} = \prod_{i=1}^{k} p_i^{e_i-1}\mathrm{lcm}\{p_i - 1\}.$$

In fact, if we choose the p_i so that the greatest common divisor of any two of the $p_i - 1$ is 2, then the order of 2 modulo q is $\prod_{i=1}^{k} p_i^{e_i-1}\mathrm{lcm}\{p_i - 1\}/2^{k-1}$. This is the largest we can hope for for the period of a FCSR whose connection number has k distinct prime factors.

The question of distributional properties of general FCSR sequences also remains. It would be nice to have bounds of the form in Section 5 in more general cases, or even just in the case described in the preceding paragraph.

Various extensions of the notion of FCSR have been suggested, based on complete valued fields other than the 2-adic numbers [10, 7]. The questions discussed in this paper can be asked in these settings as well.

Finally, now that we have established that certain FCSR sequences have good statistical properties, it remains to show that they can be modified (say with nonlinear feedforward functions, or by using nonlinear combiners) so they have large linear span and large 2-adic span. This would give us sequences with good statistics and resistance to the Berlekamp-Massey and 2-adic rational approximation algorithms, and thus good candidates for use in stream ciphers.

References

1. E. Bombieri, personal communication.
2. D. Gollman, Pseudo Random Properties of Cascade Connections of Clock Controlled Shift Registers, *Advances in Cryptology, Proceedings of Eurocrypt 84*, ed. T. Beth, N. Cot, and I. Ingemarsson, *Springer-Verlag LNCS* vol. 209, 1985, pp. 93-98.
3. S. Golomb, *Shift Register Sequences*. Aegean Park Press, Laguna Hills CA, 1982.
4. G. Hardy and E. Wright, *An Introduction to the Theory of Numbers*. Oxford University Press, Oxford UK, 1979.
5. C. Hooley, On Artin's conjecture. *J. Reine Angew. Math.* vol. 22, 1967 pp. 209-220.
6. E. L. Key, "An Analysis of the structure and complexity of nonlinear binary sequence generators," *IEEE Trans. Info. Theory*, vol. IT-22 no. 6, pp. 732-736, Nov. 1976.
7. A. Klapper, Feedback with Carry Shift Registers over Finite Fields, *Proceedings of Leuven Algorithms Workshop*, Leuven, Belgium, December, 1994.
8. A. Klapper and M. Goresky, Feedback Shift Registers, Combiners with Memory, and Arithmetic Codes, *Univ. of Kentucky, Dept. of Comp. Sci. Tech. Rep. No. 239-93*.
9. A. Klapper and M. Goresky, 2-Adic Shift Registers, *Fast Software Encryption: Proceedings of 1993 Cambridge Security Workshop*, ed. R. Anderson, *Springer-Verlag LNCS*, vol. 809, 1994, pp. 174-178.
10. A. Klapper, and M. Goresky, Feedback Registers Based on Ramified Extensions of the 2-Adic Numbers, to appear. *Proceedings, Eurocrypt 1994*, Perugia, Italy,

11. N. Koblitz, *p-Adic Numbers, p-Adic Analysis, and Zeta Functions.* Graduate Texts in Mathematics Vol. 58, Springer Verlag, N.Y. 1984.

12. D. Mandelbaum, Arithmetic codes with large distance. *IEEE Trans. Info. Theory,* vol. IT-13, 1967 pp. 237-242.

13. J. Pollard The Fast Fourier Transform in a Finite Field, *Math. Comp.,* vol. 25, 1971, pp. 365-374.

14. R. Rueppel, *Analysis and Design of Stream Ciphers.* Springer Verlag, New York, 1986.

15. A. Schönhage and V. Strassen, Schnelle Multiplikation Grosser Zahlen, *Computing,* vol. 7, 1971, pp. 281-292.

On Nonlinear Resilient Functions
(Extended Abstract)

Xian-Mo Zhang[1] and Yuliang Zheng[2]

[1] The University of Wollongong, Wollongong, NSW 2522, Australia
xianmo@cs.uow.edu.au
[2] Monash University, Melbourne, VIC 3199, Australia
yzheng@fcit.monash.edu.au

Abstract. This paper studies resilient functions which have applications in fault-tolerant distributed computing, quantum cryptographic key distribution and random sequence generation for stream ciphers. We present a number of methods for synthesizing resilient functions. An interesting aspect of these methods is that they are applicable both to linear and to nonlinear resilient functions. Our second major contribution is to show that every linear resilient function can be transformed into a large number of nonlinear resilient functions with the same parameters. As a result, we obtain resilient functions that are highly nonlinear and have a high algebraic degree.

1   Introduction

A (n, m, t)-resilient function is an n-input m-output function F with the property that it runs through every possible output m-tuple an equal number of times when t arbitrary inputs are fixed and the remaining $n - t$ inputs runs through all the 2^{n-t} input tuples once. The concept was introduced by Chor *et al* in [5] and independently, by Bennett *et al* in [1]. It turned out that (balanced) correlation immune functions introduced by Siegenthaler [20] is a special case of resilient functions. Areas where resilient functions find their applications include fault-tolerant distributed computing, quantum cryptographic key distribution and random sequence generation for stream ciphers.

Researchers have concentrated themselves on linear resilient functions, with only one exception being the work by Stinson and Massey [22]. The two researchers' aim was solely to disprove a conjecture that if there exists a nonlinear resilient function then there exists a linear resilient function with the same parameters which was posed in [5], rather than to explore cryptographic merits of nonlinear resilient functions. Recent advances in cryptanalysis, in particular the discovery of the linear cryptanalytic attack [12], have shown the vital importance of nonlinear functions in data encryption and one-way hashing algorithms. With the further revelation of the potential power of the linear attack, we might see its serious implications on the security of many other cryptographic routines, including those employing resilient functions. A relevant but earlier development is the best affine approximation (BAA) attack proposed by Ding, Xiao and Shan

L.C. Guillou and J.-J. Quisquater (Eds.): Advances in Cryptology - EUROCRYPT '95, LNCS 921, pp. 274-288, 1995
© Springer-Verlag Berlin Heidelberg 1995

in [6]. It has been shown in their book that the BAA attack can successfully break a number of types of key stream generators that employ a combining or filtering function which, though correlation immune, has a *low* nonlinearity. Success of these attacks clearly shows a need to investigate highly nonlinear resilient functions.

The rest of the paper is organized as follows: Section 2 introduces basic definitions. It also reviews important properties of resilient functions, as well as previous work in the area. Section 3 presents a number of methods for constructing new resilient functions from old. Some of them significantly generalize methods known previously. An exceptional feature of these methods is that they can be applied *both to linear and to nonlinear* resilient functions. Section 4 shows how to turn a known resilient function into a new one. As a result we can obtain a large number of highly nonlinear resilient functions from a linear one. Some miscellaneous results on resilient functions, including a discussion on algebraic degree, are included in Section 5, and the paper is closed by some concluding remarks in Section 6.

2 Preliminaries

The vector space of n tuples of elements from $GF(2)$ is denoted by V_n. These vectors, in ascending alphabetical order, are denoted by $\alpha_0, \alpha_1, \ldots, \alpha_{2^n-1}$. As vectors in V_n and integers in $[0, 2^n - 1]$ have a natural one-to-one correspondence, it allows us to switch from a vector in V_n to its corresponding integer in $[0, 2^n - 1]$, and vice versa.

Let f be a (Boolean) function from V_n to $GF(2)$ (or simply, a function on V_n). The *sequence* of f is defined as $((-1)^{f(\alpha_0)}, (-1)^{f(\alpha_1)}, \ldots, (-1)^{f(\alpha_{2^n-1})})$, while the *truth table* of f is defined as $(f(\alpha_0), f(\alpha_1), \ldots, f(\alpha_{2^n-1}))$. f is said to be *balanced* if its truth table assumes an equal number of zeros and ones. We call $h(x) = a_1 x_1 \oplus \cdots \oplus a_n x_n \oplus c$ an *affine function*, where $x = (x_1, \ldots, x_n)$ and $a_j, c \in GF(2)$. In particular, h will be called a *linear function* if $c = 0$. The sequence of an affine (linear) function will be called an *affine (linear) sequence*.

The algebraic degree $deg(f)$ of a function f is the size of the longest term in the algebraic normal form representation of the function. The *Hamming weight* of a vector v, denoted by $W(v)$, is the number of ones in v. Let f and g be functions on V_n. Then $d(f, g) = \sum_{f(x) \neq g(x)} 1$, where the addition is over the reals, is called the *Hamming distance* between f and g. Let $\varphi_0, \ldots, \varphi_{2^{n+1}-1}$ be the affine functions on V_n. Then $N_f = \min_{i=0,\ldots,2^{n+1}-1} d(f, \varphi_i)$ is called the *nonlinearity* of f. It is well-known that the nonlinearity of f on V_n satisfies $N_f \leq 2^{n-1} - 2^{\frac{1}{2}n-1}$. An extensive investigation of highly nonlinear balanced functions has been carried out in [17].

Algebraic degree and nonlinearity can also be defined for mappings or tuples of Boolean functions. Let $F = (f_1, \ldots, f_m)$ be a function from V_n to V_m (where each f_i is a function on V_n). The algebraic degree of F, denoted by $deg(F)$, is defined as the minimum among the algebraic degrees of all nonzero linear combinations of the component functions of F, namely,

$$deg(F) = \min_g \{deg(g)|g = \bigoplus_{j=1}^{m} c_j f_j\}.$$

Similarly the nonlinearity of F, denoted by N_F, is defined as

$$N_F = \min_g \{N_g|g = \bigoplus_{j=1}^{m} c_j f_j\}.$$

This definition regarding N_F was first introduced by Nyberg in [13].

$F = (f_1, \ldots, f_m)$ is said to be linear if all its component functions are linear, and to be nonlinear otherwise. If F is linear, then $deg(F) = 1$ and $N_F = 0$. The converse, however, is not always true.

2.1 Properties of Resilient Functions

In this sub-section we summarize a number of facts regarding resilient functions. Though most of these results are either previously known in, for instance, [1, 5, 3], or can be proven easily, they are collected here with the intention to help the reader in understanding our results to be presented in the coming sections. We start with a formal definition of a resilient function.

Definition 1. Let $F = (f_1, \ldots, f_m)$ be a function from V_n to V_n, where $n \geq m \geq 1$, and let $x = (x_1, \ldots, x_n) \in V_n$.

1. F is said to be *unbiased* with respect to a fixed subset $T = \{j_1, \ldots, j_t\}$ of $\{1, \ldots, n\}$, if for every $(a_1, \ldots, a_t) \in V_t$

$$(f_1(x), \ldots, f_m(x))|_{x_{j_1} = a_1, \ldots, x_{j_t} = a_t}$$

   runs through all the vectors in V_m each 2^{n-m-t} times while $(x_{i_1}, \ldots, x_{i_{n-t}})$ runs through V_n once, where $t \geq 0$, $\{i_1, \ldots, i_{n-t}\} = \{1, \ldots, n\} - \{j_1, \ldots, j_t\}$ and $i_1 < \cdots < i_{n-t}$.
2. F is said to be a (n, m, t)-resilient function if F is unbiased with respect to every $T \subseteq V_n$ with $|T| = t$. The parameter t is called the *resiliency* of the function.

Obviously, $n - m \geq t$ holds for each (n, m, t)-resilient function.

Resilient functions are closely related to correlation immune functions introduced by Siegenthaler [20]. As was noticed by Stinson and co-workers, a $(n, 1, t)$-resilient function is the same as a *balanced* tth-order correlation immune Boolean function. We will come back to this issue shortly.

The following lemma is helpful in understanding the relationship between a resilient function and its component functions. It has been called *XOR Lemma* and expressed in terms of independence of random variables in [5, 1]. Here we follow the version described in [19].

Lemma 2. *A function* $(f_1, \ldots, f_m)$, *where each* f_i *is a function on* V_n *and* $n \geq m$, *is unbiased, namely, it runs through all the vectors in* V_m *each* 2^{n-m} *times while* x *runs through* V_n *once, if and only if each nonzero linear combinations of* $f_1, \ldots, f_m$ *are balanced.*

Hence we have

Lemma 3. *Let* $F = (f_1, \ldots, f_m)$ *be a function from* V_n *to* V_n, *where* n *and* m *are integers with* $n \geq m \geq 1$ *and each* f_j *is a function on* V_n. *Then* F *is unbiased with respect to* $T = \{j_1, \ldots, j_t\}$, *a fixed subset of* $\{1, \ldots, n\}$, *if and only if every nonzero linear combination of* $f_1, \ldots, f_m$, $f(x) = \bigoplus_{j=1}^{m} c_j f_j(x)$, *is unbiased (i.e., balanced) with respect to* $T = \{j_1, \ldots, j_t\}$, *where* $x = (x_1, \ldots, x_n) \in V_n$.

As an immediate consequence, we have

Theorem 4. *Let* $F = (f_1, \ldots, f_m)$ *be a function from* V_n *to* V_m, *where* n *and* m *are integers with* $n \geq m \geq 1$ *and each* f_j *is a function on* V_n. *Then* F *is·a* (n, m, t)-*resilient function if and only if every nonzero linear combination of* $f_1, \ldots, f_m$, $f(x) = \bigoplus_{j=1}^{m} c_j f_j(x)$, *is a* $(n, 1, t)$-*resilient function, where* $x = (x_1, \ldots, x_n) \in V_n$.

It follows from Theorem 4 that if $F = (f_1, \ldots, f_m)$ is a (n, m, t)-resilient function, then $G = (f_1, \ldots, f_s)$ is a (n, s, t)-resilient function for each integer $1 \leq s \leq m$.

Theorem 4 shows that each (n, m, t)-resilient function gives $2^m - 1$ distinct balanced tth-order correlation immune functions on V_n. It also indicates that we can study (n, m, t)-resilient functions, including their properties and constructions, through investigating the correlation immune characteristics of their component functions.

To facilitate our investigations, we introduce the following lemma.

Lemma 5. *A function* f *on* V_n *is unbiased with respect to* $T = \{j_1, \ldots, j_t\}$, *a fixed subset of* $\{1, \ldots, n\}$, *if and only if for each linear function* $\varphi(x) = c_{j_1} x_{j_1} \oplus \cdots \oplus c_{j_t} x_{j_t}$ *on* V_n, *where* $x = (x_1, \ldots, x_n)$, $f(x) \oplus \varphi(x)$ *is balanced.*

Proof. First we consider the simplest case where $T = \{1, \ldots, t\}$. Let $(a_1, \ldots, a_t)$ be an arbitrary but fixed vector in V_t. Then

$$(f(x) \oplus \varphi(x))|_{x_1 = a_1, \ldots, x_t = a_t} = f(a_1, \ldots, a_t, x_{t+1}, \ldots, x_n) \oplus \varphi(a_1, \ldots, a_t, x_{t+1}, \ldots, x_n).$$

Now suppose that f is unbiased with respect to $T = \{1, \ldots, t\}$. Then

$$f(a_1, \ldots, a_t, x_{t+1}, \ldots, x_n)$$

is balanced. Note that $\varphi(a_1, \ldots, a_t, x_{t+1}, \ldots, x_n)$ is a constant. Thus

$$(f(x) \oplus \varphi(x))|_{x_1 = a_1, \ldots, x_t = a_t}$$

is balanced. As $(a_1, \ldots, a_t)$ is arbitrary, $f(x) \oplus \varphi(x)$ is a balanced function on V_n.

Conversely, suppose that $f(x) \oplus \varphi(x)$ is balanced for an arbitrary $\varphi(x) = c_1 x_1 \oplus \cdots \oplus c_t x_t$.

Let $\xi_{a_1 \cdots a_t}$ be the sequence of $f(a_1, \ldots, a_t, x_{t+1}, \ldots, x_n)$. By Lemma 1 of [16],

$$\xi = \xi_{0 \cdots 0}, \xi_{0 \cdots 1}, \cdots, \xi_{1 \cdots 1}$$

is the sequence of $f(x_1, \ldots, x_n)$.

Recall that a $(1, -1)$-matrix H of order m is called a *Hadamard* matrix if $HH^t = mI_m$, where H^t is the transpose of H and I_m is the identity matrix of order m [15]. A Sylvester-Hadamard matrix of order 2^n, denoted by H_n, is generated by the following recursive relation

$$H_0 = 1, \quad H_n = \begin{bmatrix} H_{n-1} & H_{n-1} \\ H_{n-1} & -H_{n-1} \end{bmatrix}, \quad n = 1, 2, \ldots . \tag{1}$$

Now let L be the sequence of φ. Then L is a row of H_n. Since $H_n = H_t \times H_{n-t}$, where $\times$ denotes the Kronecker product, we have $L = \ell' \times \ell''$, where ℓ' is a row of H_t and ℓ'' is a row of H_{n-t}. Write $\ell' = (d_0, \ldots, d_{2^t-1})$. Then $L = (d_0 \ell'', \ldots, d_{2^t-1} \ell'')$ and hence

$$\langle \xi, L \rangle = d_0 \langle \xi_{0 \cdots 0}, \ell'' \rangle + d_1 \langle \xi_{0 \cdots 01}, \ell'' \rangle + \cdots, + d_{2^t-1} \langle \xi_{1 \cdots 01}, \ell'' \rangle$$

Since $f(x) \oplus \varphi(x)$ is balanced, $\langle \xi, L \rangle = 0$. Note that $\ell' = (d_0, \ldots, d_{2^t-1})$, a row or column of H_t, is also the sequence of $\varphi'(x') = c_1 x_1 \oplus \cdots \oplus c_t x_t$. A fact with H_t is that the rows (columns) of H_t comprises all the linear sequences (see Lemma 2 of [16]). Then from (2),

$$\langle \xi_{0 \cdots 0}, \ell'' \rangle, \langle \xi_{0 \cdots 01}, \ell'' \rangle, \cdots, \langle \xi_{1 \cdots 11}, \ell'' \rangle) H_t = (0, 0, \cdots, 0).$$

As H_t has inverse, we have

$$\langle \xi_{0 \cdots 0}, \ell'' \rangle = \langle \xi_{0 \cdots 01}, \ell'' \rangle = \cdots, \langle \xi_{1 \cdots 11}, \ell'' \rangle) = 0.$$

Rewrite $\varphi(x) = \varphi(x') \oplus \varphi(x'')$, where $x' \in V_t$ and $x'' \in V_{n-t}$. Now ℓ' is the sequence of φ' while ℓ'' is the sequence of ℓ''. Note that $\varphi'(x') = c_1 x_1 \oplus \cdots \oplus c_t x_t$. Thus $\varphi'' = 0$ and $\ell'' = (1, \ldots, 1)$. As a result, $\langle \xi_{a_1 \cdots a_t}, \ell'' \rangle = 0$, which implies that $\xi_{a_1 \cdots a_t}$ is balanced and hence $f(a_1, \ldots, a_t, x_{t+1}, \ldots, x_n)$ is balanced, where $(a_1, \ldots, a_t)$ is an arbitrary vector in V_t. This shows that f is unbiased with respect to $T = \{1, \ldots, t\}$.

For the more general case where $T = \{j_1, \ldots, j_t\}$, set

$$f(x_1, \ldots, x_n) = g(x_{j_1}, \ldots, x_{j_t}, x_{j_{t+1}}, \ldots, x_{j_n}),$$

where $\{j_1, \ldots, j_t\} = T$ and $\{x_{j_t}, x_{j_{t+1}}, \ldots, x_{j_n}\} = \{1, \ldots, n\} - T$. Also set

$$x_{j_1} = y_1, \ldots, x_{j_n} = y_n. \tag{2}$$

Thus

$$g(x_{j_1}, \ldots, x_{j_t}, x_{j_{t+1}}, \ldots, x_{j_n}) = g(y_1, \ldots, y_t, y_{t+1}, \ldots, y_n).$$

Now write $\psi(y) = \psi(y_1, \ldots, y_n) = c_1 y_1 \oplus \cdots \oplus c_t y_t$, where $y = (y_1, \ldots, y_n)$. Obviously $\psi(y_1, \ldots, y_n) = \varphi(x_1, \ldots, x_n)$. Hence $f(x) \oplus \varphi(x) = g(y) \oplus \psi(y)$. Clearly, f is unbiased with respect to $\{j_1, \ldots, j_t\}$ if and only if g is unbiased with respect to $\{1, \ldots, t\}$, and by the above discussions, if and only if $g(y) \oplus \psi(y) = f(x) \oplus \varphi(x)$ is balanced. $\square$

A corollary of Lemma 5 is

Corollary 6. f is a $(n, 1, t)$-resilient function if and only if for each linear function $\varphi(x) = c_1 x_1 \oplus \cdots \oplus c_n x_n$ with $W(c_1, \ldots, c_n) \leq t$, $f(x) \oplus \varphi(x)$ is balanced.

From this corollary and Theorem 4 it follows

Corollary 7. F is a (n, m, t)-resilient function if and only if it is a (n, m, s)-resilient function for each $0 \leq s \leq t$.

Now we go back to correlation immune functions. Work by Xiao and Massey provides us with an equivalent definition of the concept [10]:

Definition 8. A function f on V_n is said to be tth-order correlation immune if for each linear function $\varphi(x) = c_1 x_1 \oplus \cdots \oplus c_n x_n$ with $1 \leq W(c_1, \ldots, c_n) \leq t$, $f(x) \oplus \varphi(x)$ is balanced.

As $W(c_1, \ldots, c_n) = 0$ is excluded, the definition covers both balanced and non-balanced correlation immune function, although stream ciphers prefer balanced to non-balanced functions.

Comparing the definition with Corollary 6, it becomes clear that a balanced tth-order correlation immune function is indeed identical to a $(n, 1, t)$-resilient function.

Having presented essential facts on resilient functions, next we consider transformations on the coordinates of a resilient function. Unlike nonlinearity and algebraic degree, the resiliency of functions is not invariant under a nonsingular linear transformation on the coordinates. This can be seen from the following example.

Let $f(x) = x_1 \oplus x_2 \oplus \cdots \oplus x_n$, where $x = (x_1, \ldots, x_n)$. Then f is a $(n, 1, n-1)$-resilient function. Now let B be a matrix of order n over $GF(2)$ satisfying $(x_1, x_2, \ldots, x_{n-1}, x_n)B = (x_2, x_3, \ldots, x_{n-1}, \bigoplus_{j=1}^{n} x_j)$. Set $g(x) = f(xB^{-1})$. Then $g(x) = x_n$ whose resiliency is zero.

Another issue is in relation to the transformation of the component functions, namely output, of a resilient function. This will be discussed in detail in Section 4, where we show an important result regarding invariant properties of resilient functions under transformations of (output) component functions.

2.2 Related Work

The concept of a resilient function was introduced in [5, 1]. The equivalence between linear resilient functions and linear error correcting codes was established also in [5, 1], while the equivalence between resilient functions and large sets of orthogonal arrays was proved in [21]. Two upper bounds on resiliency which are the best known so far were derived in [7, 3]. In [22] Stinson and Massey disproved the conjecture that if there exists a nonlinear resilient function then there exists a linear resilient function with the same parameters. The nonlinear resilient functions they constructed were based on the (nonlinear) Kerdock and Preparata codes [11]. Some linear resilient functions achieving an upper bound on resiliency can be found in [7, 3]. Resilient functions which are symmetric were studied in [5, 8], while non-binary resilient functions were examined in [9].

Soon after the concept of a correlation immune function was introduced by Siegenthaler [20], Xiao and Massey gave an equivalent definition in [10]. These were followed by [4, 18] where various methods for constructing correlation immune functions were presented.

3   Constructing New Resilient Functions from Old

Constructing new resilient functions from old ones is an interesting problem that has many practical implications. There are two opposite directions in relation to this problem, these being constructing "large" ones from "small" ones and "small" ones from "large" ones. Due to the close relationship between resilient functions and error correcting codes, in particular the equivalence between linear codes and linear resilient functions as was revealed in [5, 1], numerous techniques can be borrowed from the theory of error correcting codes to construct new resilient functions from old. These techniques have been further enriched by Stinson's work on the equivalence between resilient functions and large sets of orthogonal arrays [21]. Some concrete examples on constructing new from old can be found in [3].

The main purpose of this section is to present a number of methods for directly synthesizing large resilient functions from small ones. A distinctive feature of these methods is that they are applicable both to linear and to nonlinear resilient functions.

We start with correlation immune functions. Let f_i be a $(n_i, 1, t_i)$-resilient function, $i = 1, 2$. Then $f_1(x) \oplus f_2(y)$ is a $(n_1 + n_2, 1, t_1 + t_2 + 1)$-resilient function, where $x \in V_{n_1}$ and $y \in V_{n_2}$. To show that this is correct, let φ be a linear function on $V_{n_1 + n_2}$ defined by

$$\varphi(x, y) = c_1 x_1 \oplus \cdots \oplus c_{n_1} x_{n_1} \oplus d_1 y_1 \oplus \cdots \oplus d_{n_2} y_{n_2},$$

where $x = (x_1, \ldots, x_{n_1})$, $y = (y_1, \ldots, y_{n_2})$, $c_j, d_i \in GF(2)$. Suppose that

$$W(c_1, \ldots, c_{n_1}, d_1, \ldots, d_{n_2}) \leq t_1 + t_2 + 1.$$

Then either $W(c_1, \ldots, c_{n_1}) \leq t_1$ or $W(d_1, \ldots, d_{n_2}) \leq t_2$. By Corollary 6, either $f_1(x) \oplus \varphi_1(x)$ or $f_2(y) \oplus \varphi_2(y)$ is balanced, where $\varphi_1(x) = c_1 x_1 \oplus \cdots \oplus c_{n_1} x_{n_1}$ and $\varphi_2(y) = d_1 y_1 \oplus \cdots \oplus d_{n_2} y_{n_2}$. Note that the sum of two functions with disjoint variables is balanced if one of the two functions is balanced (for a simple proof see

Lemma 9 of [16]). Hence $f_1(x) \oplus f_2(y) \oplus \varphi(x, y) = [f_1(x) \oplus \varphi_1(x)] \oplus [f_2(y) \oplus \varphi_2(y)]$ is balanced. Again by Corollary 6, $f_1(x) \oplus f_2(y)$ is a $(n_1+n_2, 1, t_1+t_2+1)$-resilient function.

By induction, we have the following result.

Lemma 9. *Let f_i be a $(n_i, 1, t_i)$-resilient function, $i = 1, \ldots, s$. Then $f_1(x) \oplus \cdots \oplus f_s(y)$ is a $(\sum_{j=1}^{s} n_j, 1, s - 1 + \sum_{j=1}^{s} t_j)$-resilient function, where $x \in V_{n_1}, \ldots, y \in V_{n_s}$.*

As an application of Lemma 9, we can combine known resilient functions to obtain a new one. First we show that *if $F = (f_1, \ldots, f_m)$ is a (n, m, t)-resilient function, then $G(x, y, z) = (F(x) \oplus F(y), F(y) \oplus F(z))$ is a $(3n, 2m, 2t + 1)$-resilient function, where $x, y, z \in V_n$.*

To prove that G is a $(3n, 2m, 2t + 1)$-resilient function, we first note that $f_1(x) \oplus f_1(y), \ldots, f_m(x) \oplus f_m(y), f_1(y) \oplus f_1(z), \ldots, f_m(y) \oplus f_m(z)$ comprise all the $2m$ component functions of G. Consider a nonzero linear combination of these $2m$ component functions

$$f(x, y, z) = \bigoplus_{j=1}^{m} c_j(f_j(x) \oplus f_j(y)) \oplus \bigoplus_{j=1}^{m} d_j(f_j(y) \oplus f_j(z)),$$

where either $(c_1, \ldots, c_m) \neq (0, \ldots, 0)$ or $(d_1, \ldots, d_m) \neq (0, \ldots, 0)$.

Note that

$$f(x, y, z) = \bigoplus_{j=1}^{m} c_j f_j(x) \oplus \bigoplus_{j=1}^{m} (c_j \oplus d_j) f_j(y) \oplus \bigoplus_{j=1}^{m} d_j f_j(z).$$

By Theorem 4, $\bigoplus_{j=1}^{m} c_j f_j(x)$ is a $(n, 1, t)$-resilient function when $(c_1, \ldots, c_m) \neq (0, \ldots, 0)$. Similarly, $\bigoplus_{j=1}^{m} d_j f_j(z)$ is a $(n, 1, t)$-resilient function when $(d_1, \ldots, d_m) \neq (0, \ldots, 0)$, and $\bigoplus_{j=1}^{m} (c_j \oplus d_j) f_j(y)$ is a $(n, 1, t)$-resilient function when $(c_1 \oplus d_1, \ldots, c_m \oplus d_m) \neq (0, \ldots, 0)$.

Since either $(c_1, \ldots, c_m) \neq (0, \ldots, 0)$ or $(d_1, \ldots, d_m) \neq (0, \ldots, 0)$, at least two hold among $(c_1, \ldots, c_m) \neq (0, \ldots, 0)$, $(d_1, \ldots, d_m) \neq (0, \ldots, 0)$ and $(c_1 \oplus d_1, \ldots, c_m \oplus d_m) \neq (0, \ldots, 0)$. By Lemma 9, when two hold $f(x, y, z)$ is a $(3n, 1, 2t+1)$-resilient function, while when three hold it is a $(3n, 1, 3t+2)$-resilient function. By Theorem 4, $G(x, y, z)$ is indeed a $(3n, 2m, 2t + 1)$-resilient function.

It was first observed in [5] that $g(x_1, \ldots, x_{3h}) = (x_1 \oplus \cdots \oplus x_{2h}, x_{h+1} \oplus \cdots \oplus x_{3h})$ is a linear $(3h, 2, 2h - 1)$-resilient function. We can view this function as being obtained from $f(x_1, \ldots, x_h) = x_1 \oplus \cdots \oplus x_h$, which is a $(h, 1, h - 1)$-resilient function, by using the technique described above. Conversely we can also regard our technique as a significant generalization of the idea underling the construction of $g(x_1, \ldots, x_{3h}) = (x_1 \oplus \cdots \oplus x_{2h}, x_{h+1} \oplus \cdots \oplus x_{3h})$.

Now applying the same technique to the resulting function G itself, we obtain a $(3^2 n, 2^2 m, 2^2(1 + t) - 1)$-resilient function. In general repeating the technique for k times, $k = 1, 2, \ldots$, we obtain a $(3^k n, 2^k m, 2^k(1 + t) - 1)$-resilient function from a (n, m, t)-resilient function.

The technique can also be generalized in other directions. In particular, it is easy to prove that if $F = (f_1, \ldots, f_m)$ is a (n, m, t)-resilient function, then $G(x, y, z, u) = (F(x) \oplus F(y), F(y) \oplus F(z), F(z) \oplus F(u))$ is a $(4n, 3m, 2t + 1)$-resilient function, where $x, y, z, u \in V_n$. Again by iterating the technique, we can construct from a (n, m, t)-resilient function a $(4^k n, 3^k m, 2^k(1 + t) - 1)$-resilient function for all $k = 1, 2, \ldots$.

To summarize the discussions, we have

Lemma 10. *Given a (n, m, t)-resilient function, there is an iterative method to construct a $((h + 1)^k n, h^k m, 2^k(1 + t) - 1)$-resilient function for all $h = 2, 3, \ldots$ and $k = 1, 2, \ldots$.*

As another application of Lemma 9, we give the following result.

Corollary 11. *Let $F = (f_1, \ldots, f_m)$ be a (n_1, m, t_1)-resilient function and $G = (g_1, \ldots, g_m)$ a (n_2, m, t_2)-resilient function. Then $P(z) = F(x) \oplus G(y) = (f_1(x) \oplus g_1(y), \ldots, f_m(x) \oplus g_m(y))$ is a $(n_1 + n_2, m, t_1 + t_2 + 1)$-resilient function, where $z = (x, y)$, $x \in V_{n_1}$ and $y \in V_{n_2}$.*

Proof. Consider an arbitrary nonzero linear combination of the component functions of $P(z)$, say

$$p(z) = \bigoplus_{j=1}^m c_j [f_j(x) \oplus g_j(y)] = \bigoplus_{j=1}^m c_j f_j(x) \oplus \bigoplus_{j=1}^m c_j g_j(y).$$

By Theorem 4, $\bigoplus_{j=1}^m c_j f_j(x)$ is a t_1-resilient function, while $\bigoplus_{j=1}^m c_j g_j(y)$ is a t_2-resilient function. Hence by Lemma 9, $p(z)$ is a $t_1 + t_2 + 1$-resilient function. As $p(z)$ is arbitrary, again by Theorem 4, $P(z)$ is a $(n_1 + n_2, m, t_1 + t_2 + 1)$-resilient function. $\square$

A special case of the technique indicated in Corollary 11, namely when both F and G are linear, has been employed by Bierbrauer, Gopalakrishnan and Stinson in proving their Theorem 7 in [3].

The following result is concerned with placing resilient functions in parallel.

Corollary 12. *Let $F = (f_1, \ldots, f_{m_1})$ be a (n_1, m_1, t_1)-resilient function and $G = (g_1, \ldots, g_{m_2})$ be a (n_2, m_2, t_2)-resilient function. Then*

$$P(z) = (f_1(x), \ldots, f_{m_1}(x), g_1(y), \ldots, g_{m_2}(y))$$

is a $(n_1 + n_2, m_1 + m_2, \rho)$-resilient function, where $z = (x, y)$, $x \in V_{n_1}$, $y \in V_{n_2}$, and $\rho = \min\{t_1, t_2\}$.

Proof. Consider an arbitrary nonzero linear combination of the component functions of $P(z)$

$$p(z) = \bigoplus_{j=1}^{m_1} c_j f_j(x) \oplus \bigoplus_{j=1}^{m_2} d_j g_j(y).$$

As $(c_1, \ldots, c_{m_1}, d_1, \ldots, d_{m_2})$ is a nonzero vector, without loss of generality, we can assume that $(c_1, \ldots, c_{m_1}) \neq (0, \ldots, 0)$. For any λ_1-subset $\{j_1, \ldots, j_{\lambda_1}\} \subseteq \{1, \ldots, n_1\}$ and any λ_2-subset $\{i_1, \ldots, i_{\lambda_2}\} \subseteq \{1, \ldots, n_2\}$, where $\lambda_1 + \lambda_2 = \rho$, and any $a_1, \ldots, a_{\lambda_1}, b_1, \ldots, b_{\lambda_2} \in GF(2)$, by Theorem 4, and the fact that the sum of two functions with disjoint variables is balanced if one of the two functions is balanced (Lemma 9 of [16]), $\bigoplus_{j=1}^{m_1} c_j f_j(x)|_{x_{j_1}=a_1, \ldots, x_{j_{\lambda_1}}=a_{\lambda_1}}$ is balanced. Thus

$$\bigoplus_{j=1}^{m_1} c_j f_j(x)|_{x_{j_1}=a_1, \ldots, x_{j_{\lambda_1}}=a_{\lambda_1}} \oplus \bigoplus_{j=1}^{m_2} d_j g_j(y)|_{y_{i_1}=b_1, \ldots, b_{i_{\lambda_2}}=b_{\lambda_2}}$$

is balanced. It follows from Theorem 4 that

$$P(z) = (f_1(x), \ldots, f_{m_1}(x), g_1(y), \ldots, g_{m_2}(y))$$

is a $(n_1 + n_2, m_1 + m_2, \rho)$-resilient function. □

4 Transforming Linear Resilient Functions to Nonlinear Ones

Recall that a resilient function is said to be *linear* if its component functions are all linear, and said to be *nonlinear* otherwise. When the concept of resilient functions was introduced, it was conjectured that if there exists a *nonlinear* resilient function with certain parameters, then there exists a *linear* resilient function with the same parameters [5, 1]. This conjecture was disproved by Stinson and Massey [22]. In particular, they showed that there exists an infinite class of nonlinear resilient functions for which there do not exist linear resilient functions with the same parameters. They used nonlinear error correcting codes in their proof. In this section we investigate this topic in a slightly different direction. In particular we show that by permuting the output m-tuples (i.e., all 2^m vectors in V_m), instead of only re-ordering the m component functions of a (n, m, t)-resilient function, we can obtain $2^m!$ distinct (n, m, t)-resilient functions. A consequence of this result is that the *converse* of the conjecture in [5, 1] is true, namely if there exists a *linear* resilient function with certain parameters, then there exists a *nonlinear* resilient function with the same parameters.

Here is the main result in this section.

Theorem 13. *Let F be a (n, m, t)-resilient function and G be a permutation on V_m. Then $P = G \circ F$, namely $P(x) = G(F(x))$, is also a (n, m, t)-resilient function.*

Proof. Since F is a (n, m, t)-resilient function, for each $\{j_1, \ldots, j_t\} \subseteq \{1, \ldots, n\}$ and $a_1, \ldots, a_t \in GF(2)$,

$$F(x)|_{x_{j_1}=a_1, \ldots, x_{j_t}=a_t}$$

runs through all the vectors in V_m each 2^{n-m-t} times while $(x_{i_1}, \ldots, x_{i_{n-t}})$ runs through V_n once, where $\{i_1, \ldots, i_{n-t}\} = \{1, \ldots, n\} - \{j_1, \ldots, j_t\}$ and $i_1 < \cdots < i_{n-t}$. As G is a permutation on V_m,

$$P(x)|_{x_{j_1} = a_1, \ldots, x_{j_t} = a_t} = G(F(x))|_{x_{j_1} = a_1, \ldots, x_{j_t} = a_t}$$

runs through all the vectors in V_m each 2^{n-m-t} times while $(x_{i_1}, \ldots, x_{i_{n-t}})$ runs through V_n once. It immediately follows that P is a (n, m, t)-resilient function. $\boxminus$

Note that the total number of different permutations on V_m is $2^m!$ which is far larger than $m!$. The latter is the number of ways to re-order the m component functions. New resilient functions generated using these permutations are all different. To prove this, let G_1 and G_2 be two different permutations on V_m. We want to prove that $G_1 \circ F \neq G_2 \circ F$. Suppose for contradiction that $G_1 \circ F = G_2 \circ F$. Then $F = G_1^{-1} \circ G_2 \circ F$. As F is unbiased, for each $\beta \in V_m$, there exist 2^{n-m} different vectors $\alpha \in V_n$ such that $F(\alpha) = \beta$. This causes $\beta = G_1^{-1} \circ G_2(\beta)$. As β is arbitrary, $G_1^{-1} \circ G_2$ must be the identity permutation on V_m, which contradicts the fact that $G_1 \neq G_2$. Thus we have proved the following:

Corollary 14. *Given a (n, m, t)-resilient function, Theorem 13 produces $2^m!$ distinct (n, m, t)-resilient function.*

Now we describe an example to show applications of Theorem 13. It is easy to verify that

$$F(x_1, x_2, x_3, x_4, x_5, x_6) = (x_1 \oplus x_2 \oplus x_3, \quad x_3 \oplus x_4 \oplus x_5, \quad x_5 \oplus x_6 \oplus x_1)$$

is a linear $(6, 3, 2)$-resilient function. Consider a permutation G on V_3 defined by

$$G(u_1, u_2, u_3) = (u_1 \oplus u_3 \oplus u_2 u_3, \quad u_1 \oplus u_2 \oplus u_1 u_3, \quad u_2 \oplus u_3 \oplus u_1 u_2).$$

By Theorem 13, $P = G \circ F$ is also a $(6, 3, 2)$-resilient function.

Note that all component functions of the resulting resilient function P are quadratic. The rest of this section is devoted to this direction, namely converting linear resilient functions to nonlinear ones. We also show how to calculate the nonlinearity of a resulting nonlinear resilient function. The following lemma will be used in the discussions.

Lemma 15. *Let g be a function on V_m whose nonlinearity is N_g. Let $n \geq m$ and B be an $n \times m$ matrix over $GF(2)$ whose rank is m. Set $h(x_1, \ldots, x_n) = g((x_1, \ldots, x_n)B)$. Then the nonlinearity N_h of h, a function on V_n, satisfies $N_h = 2^{n-m} N_g$, and the algebraic degree of h is the same as that of g.*

Proof. First we note that this lemma is a generalization of the following result: Let $h(x_1, \ldots, x_n) = g(x_1, \ldots, x_k)$. Then h, a function on V_n, satisfies $N_h = 2^{n-m} N_g$. A proof for this special case can be found in, for instance, [18].

To prove this lemma, we append to B an $n \times (n - m)$ matrix C so that $A = [B, C]$ is a nonsingular matrix of order n over $GF(2)$. Set $(u_1, \ldots, u_n) = (x_1, \ldots, x_n)A$. Now define a function on V_n, say g^*, as follows

$$g^*(u_1, \ldots, u_n) = g(u_1, \ldots, u_m).$$

Then $N_{g^*} = 2^{n-m}N_g$, and g^* and g share the same algebraic degree. On the other hand, from the construction of h,

$$h(x_1, \ldots, x_n) = g((x_1, \ldots, x_n)B) = g^*((x_1, \ldots, x_n)A).$$

By noting the fact that the nonlinearity and algebraic degree of a function are invariant under a nonsingular linear transformation on coordinates, we have $N_h = N_{g^*} = 2^{n-m}N_g$, and that h has the same algebraic degree as that of g^*, which is the same as that of g. $\qquad \square$

Now we prove a significant result on constructing new resilient functions from old, linear ones.

Theorem 16. *Let F be a linear (n, m, t)-resilient function and G be a permutation on V_m whose nonlinearity is N_G. Then $P = G \circ F$ is a (n, m, t)-resilient function and*

(i) the nonlinearity N_P of P satisfies $N_P = 2^{n-m}N_G$.
(ii) the algebraic degree of P is the same as that of G.

Proof. As F is a linear resilient function, it can be written as $F(x_1, \ldots, x_n) = (x_1, \ldots, x_n)B$ where B is an $n \times m$ matrix of rank m over $GF(2)$ and $(x_1, \ldots, x_n) \in V_n$. The theorem follows immediately from Lemma 15. $\qquad \square$

We turn our attention back to the nonlinear $(6, 3, 2)$-resilient function constructed above. It is easy to verify that the nonlinearity of each nonzero linear combination of the component functions of G is 2. By Theorem 16, the nonlinearity of P is 16, and as we have seen, the algebraic degree of P is indeed 2.

Theorem 16 implies that highly nonlinear resilient functions can be constructed from linear resilient functions by applying highly nonlinear permutations in the transforming process. A number of highly nonlinear permutations which are based on polynomials on a finite field have been shown in [14, 2]. In particular, it is shown in [14] that the nonlinearity of a permutation G based on the inverse function on $GF(2^m)$ satisfies $N_G \geq 2^{m-1} - 2^{\frac{1}{2}m}$ and the algebraic degree of G is $m - 1$. Hence the following is proved:

Corollary 17. *If there exists a linear (n, m, t)-resilient function, then there exists a nonlinear (n, m, t)-resilient function P whose nonlinearity satisfies $N_P \geq 2^{n-1} - 2^{n-\frac{1}{2}m}$ and whose algebraic degree is $m - 1$.*

Another important implication of Theorem 16 is that from each linear resilient function, we can derive a large number nonlinear resilient functions with the same parameters. This, together with the result by Stinson and Massey [22], shows that it is more affluent in nonlinear resilient functions than in linear resilient functions, in terms of either the numbers or the parameters.

5 Remarks on Algebraic Degree

In his pioneering work [20], Siegenthaler showed, by a lengthy argument, that the algebraic degree of a balanced correlation immune function, i.e., a $(n, 1, t)$-resilient function, is at most $n - t - 1$, except for the case when $t = n - 1$. Here we show that the proof can substantially shortened by employing Theorem 1 on Page 372 of [11].

Let f be a $(n, 1, t)$-resilient function. As f is a function on V_n, by Theorem 1 on Page 372 of [11], it can be expressed in the algebraic normal form, namely

$$f(x_1, \ldots, x_n) = \bigoplus_{a_1, \ldots, a_n \in GF(2)} g(a_1, \ldots, a_n) x_1^{a_1} \cdots x_n^{a_n},$$

where

$$g(a_1, \ldots, a_n) = \bigoplus_{(b_1, \ldots, b_n) \subset (a_1, \ldots, a_n)} f(b_1, \ldots, b_n),$$

and by $(b_1, \ldots, b_n) \subset (a_1, \ldots, a_n)$ we mean that if $b_j = 1$ then $a_j = 1$.

Consider the coefficient of the term $x_1 \cdots x_{n-t}$, that is

$$\bigoplus_{b_1, \ldots, b_{n-t} \in GF(2)} f(b_1, \ldots, b_{n-t}, 0, \ldots, 0). \tag{3}$$

Since f is a $(n, 1, t)$-resilient function, (3) becomes zero, except for $n - t = 1$ in which case (3) becomes one. By the same reasoning, we can see that the coefficient of every term of algebraic degree $n - t$ is zero. This proves that the algebraic degree of f is at most $n - t - 1$.

By noting our Theorem 4, we have

Corollary 18. *The algebraic degree of a (n, m, t)-resilient function is at most $n - t - 1$, except for the case when $t = n - 1$.*

Recall that it is easy to construct linear $(n, n - 1, 1)$-resilient functions from linear error correcting codes. Using Corollaries 14 and 17, we obtain $2^{n-1}!$ distinct $(n, n - 1, 1)$-resilient functions, a large number of which have a nonlinearity of at least $2^{n-1} - 2^{\frac{n+1}{2}}$ and whose algebraic degree is $n - 2$.

It should be noted, however, that due to Corollary 18, applying Theorem 16 to a *nonlinear* $(n, n - 1, 1)$-resilient function does not always yield a function that has a higher algebraic degree.

In [7] Friedman proved that the resiliency t of a (n, m, t)-resilient function is bounded from above by

$$B_1 = \lfloor \frac{2^{m-1} n}{2^m - 1} \rfloor - 1.$$

Theorem 3 of [3] gives another upper bound

$$B_2 = 2 \lfloor \frac{2^{m-2}(n + 1)}{2^m - 1} \rfloor - 1. \tag{4}$$

As shown in [3] a linear $(2^m - 1, m, 2^{m-1} - 1)$-resilient function can be obtained from a simplex code. This function achieves the upper bound on resiliency (5). Applying Corollaries 14 and 17 to this resilient function, we obtain $2^m!$ distinct $(2^m - 1, m, 2^{m-1} - 1)$ resilient functions, some of which have a non-linearity of at least $2^{2^m-2} - 2^{2^m-1-\frac{1}{2}m}$ and whose algebraic degree is $m - 1$. All the resulting functions achieve the upper bound on resiliency indicated in (5).

6 Conclusion

Main results of this paper are related to the construction of nonlinear resilient functions. Of particular importance to practical applications is the method for transforming linear resilient functions into nonlinear ones. Currently we are in the process of extending in various directions the results reported in this paper.

Acknowledgment

We would like thank G. Stinson for kindly providing us with his papers. This project was supported in part by the Australian Research Council (ARC) under the reference number A49232172 and by the Australian Telecommunications and Electronics Research Board (ATERB) under the reference numbers C010/058 and N069/412.

This work was done while the second author was with the University of Wollongong.

References

1. BENNETT, C. H., BRASSARD, G., AND ROBERT, J. M. Privacy amplification by public discassion. *SIAM J. Computing 17* (1988), 210–229.
2. BETH, T., AND DING, C. On permutations against differential cryptanalysis. In *Advances in Cryptology - EUROCRYPT'93* (1994), vol. 765, Lecture Notes in Computer Science, Springer-Verlag, Berlin, Heidelberg, New York, pp. 65–76.
3. BIERBRAUER, J., GOPALAKRISHNAN, K., AND STINSON, D. R. Bounds on resilient functions and orthogonal arrays. In *Advances in Cryptology - CRYPTO'94* (1994), vol. 839, Lecture Notes in Computer Science, Springer-Verlag, Berlin, Heidelberg, New York, pp. 247–256.
4. CAMION, P., CARLET, C., CHARPIN, P., AND SENDRIER, N. On correlation-immune functions. In *Advances in Cryptology - CRYPTO'91* (1991), vol. 576, Lecture Notes in Computer Science, Springer-Verlag, Berlin, Heidelberg, New York, pp. 87–100.
5. CHOR, B., GOLDREICH, O., HÅSTAD, J., FRIEDMAN, J., RUDICH, S., AND SMOLENSKY, R. The bit extraction problem or *t*-resilient functions. *IEEE Symposium on Foundations of Computer Science 26* (1985), 396–407.
6. DING, C., XIAO, G., AND SHAN, W. *The Stability Theory of Stream Ciphers*, vol. 561 of *Lecture Notes in Computer Science*. Springer-Verlag, Berlin, Heidelberg, New York, 1991.

7. FRIEDMAN, J. On the bit extraction problem. *Proc. 33rd IEEE Symp. on Foundations of Computer Science* (1992), 314–319.

8. GOPALAKRISHNAN, K., HOFFMAN, D. G., AND STINSON, D. R. A note on a conjecture concerning symmetric resilient functions. *Information Processing Letters 47* (1993), 139–143.

9. GOPALAKRISHNAN, K., AND STINSON, D. R. Three characterizations of non-binary correlation-immune and resilient functions. Submitted to *Designs, Codes and Cryptograpy*, 1994.

10. GUO-ZHEN, X., AND MASSEY, J. L. A spectral characterization of correlation-immune combining functions. *IEEE Transactions on Information Theory 34*, 3 (1988), 569–571.

11. MACWILLIAMS, F. J., AND SLOANE, N. J. A. *The Theory of Error-Correcting Codes*. North-Holland, Amsterdam, New York, Oxford, 1977.

12. MATSUI, M. Linear cryptanalysis method for DES cipher. In *Advances in Cryptology - EUROCRYPT'93* (1994), vol. 765, Lecture Notes in Computer Science, Springer-Verlag, Berlin, Heidelberg, New York, pp. 386–397.

13. NYBERG, K. On the construction of highly nonlinear permutations. In *Advances in Cryptology - EUROCRYPT'92* (1993), vol. 658, Lecture Notes in Computer Science, Springer-Verlag, Berlin, Heidelberg, New York, pp. 92–98.

14. NYBERG, K. Differentially uniform mappings for cryptography. In *Advances in Cryptology - EUROCRYPT'93* (1994), vol. 765, Lecture Notes in Computer Science, Springer-Verlag, Berlin, Heidelberg, New York, pp. 55–65.

15. SEBERRY, J., AND YAMADA, M. Hadamard matrices, sequences, and block designs. In *Contemporary Design Theory: A Collection of Surveys*, J. H. Dinitz and D. R. Stinson, Eds. John Wiley & Sons, Inc, 1992, ch. 11, pp. 431–559.

16. SEBERRY, J., AND ZHANG, X.-M. Highly nonlinear 0-1 balanced functions satisfying strict avalanche criterion. In *Advances in Cryptology - AUSCRYPT'92* (1993), vol. 718, Lecture Notes in Computer Science, Springer-Verlag, Berlin, Heidelberg, New York, pp. 145–155.

17. SEBERRY, J., ZHANG, X. M., AND ZHENG, Y. Nonlinearly balanced boolean functions and their propagation characteristics. In *Advances in Cryptology - CRYPTO'93* (1994), vol. 773, Lecture Notes in Computer Science, Springer-Verlag, Berlin, Heidelberg, New York, pp. 49–60.

18. SEBERRY, J., ZHANG, X. M., AND ZHENG, Y. On constructions and nonlinearity of correlation immune functions. In *Advances in Cryptology - EUROCRYPT'93* (1994), vol. 765, Lecture Notes in Computer Science, Springer-Verlag, Berlin, Heidelberg, New York, pp. 181–199.

19. SEBERRY, J., ZHANG, X. M., AND ZHENG, Y. Relationships among nonlinearity criteria. Presented at *EUROCRYPT'94*, 1994.

20. SIEGENTHALER, T. Correlation-immunity of nonlinear combining functions for cryptographic applications. *IEEE Transactions on Information Theory IT-30 No. 5* (1984), 776–779.

21. STINSON, D. R. Resilient functions and large sets of orthogonal arrays. *Congressus Numerantium 92* (1993), 105–110.

22. STINSON, D. R., AND MASSEY, J. L. An infinite class of counterexamples to a conjecture concerining non-linear resilient functions. to appear in *Journal of Cryptology*, 1994.

Combinatorial Bounds for Authentication Codes with Arbitration

Kaoru KUROSAWA   *and*   Satoshi OBANA

Department of Electrical and Electronic Engineering,
Faculty of Engineering, Tokyo Institute of Technology
2-12-1 O-okayama, Meguro-ku, Tokyo 152, Japan
E-mail kkurosaw@ss.titech.ac.jp

Abstract. Unconditionally secure authentication codes with arbitration (A^2-codes) protect against deceptions from the transmitter and the receiver as well as that from the opponent.

In this paper, we present combinatorial lower bounds on the cheating probabilities for A^2-codes in terms of the number of source states, that of the whole messages and that of messages which the receiver accepts as authentic for each source state. Previously, only entropy based lower bounds were known. Our bounds for the model without secrecy are tight because the A^2-codes given by Johansson meet our bounds with equality.

1 Introduction

In the model of unconditionally secure authentication codes (A-codes) [1], there are three participants, a transmitter, a receiver and an opponent. The opponent tries to cheat the receiver by impersonation attack and substitution attack. This model has been studied extensively so far. Lower bounds on the cheating probabilities based on entropy were given by [2, 3]. Combinatorial lower bounds were given by [4, 5, 6, 7, 8, 9]. In this model, the transmitter and the receiver are both honest and trust each other. However, it is not always the case that the two parties want to trust each other.

Inspired by this problem, Simmons introduced an extended model, A^2-code model, in which there is a fourth person, an arbiter [10, 11]. In this model, caution is taken against deception of the transmitter and the receiver as well as that of the opponent. The arbiter has access to all key information of the transmitter and the receiver, and solves disputes between them. We denote by E_R the set of keys of the receiver and by E_T denotes the set of keys of the transmitter, respectively.

In this model, there are essentially five different kinds of cheatings, impersonation by the opponent, substitution by the opponent, impersonation by the transmitter, impersonation by the receiver and substitution by the receiver. Denote these cheating probabilities by P_I, P_S, P_T, P_{R_0} and P_{R_1}. Johansson showed an entropy based lower bound on these five cheating probabilities [12]. By assuming $\max(P_I, P_S, P_T, P_{R_0}, P_{R_1}) = 1/q$, he also showed a lower bound on the size of keys in terms of q [12]. Recently, Kurosawa showed a more tight lower bound

L.C. Guillou and J.-J. Quisquater (Eds.): Advances in Cryptology - EUROCRYPT '95, LNCS 921, pp. 289-300, 1995.
© Springer-Verlag Berlin Heidelberg 1995

for a larger set of source states by assuming $P_I = P_S = P_T = P_{R_0} = P_{R_1} = 1/q$ [15]. However, combinatorial lower bounds on the cheating probabilities are not known. (The structure of A^2-codes is not well known.)

In this paper, we present combinatorial lower bounds on $P_I, P_S, P_T, P_{R_0}, P_{R_1}$, $|E_R|$ and $|E_T|$. First, we show the following bounds for a A^2-code model without secrecy. Let $|S|$ denote the number of source states and $|M|$ denote that of messages, respectively. Assume that each $f \in E_R$ accepts c messages for each source state s. Let $l \stackrel{\triangle}{=} |M|/c|S|$. Then $P_I \geq 1/l$ and $P_T \geq (c-1)|S|/(|M|-|S|)$. If $P_I = 1/l$, then $P_S \geq 1/l$. For a separable case, if $P_I = 1/l$, then $P_T \geq 1/l$. If $P_I = P_S = P_T = 1/l$, then $|E_R| \geq c|S|(l-1)+1$. Similar bounds are obtained for P_{R_0}, P_{R_1} and $|E_T|$. Our bounds on P_I, P_S, P_T, P_{R_0} and P_{R_1} are tight because the A^2-codes given by Johansson [13, 14] (in which $c = l = q$) meet our bounds with equality.

Further, we show such combinatorial lower bounds for general A^2-codes.

2 Preliminaries

2.1 Authentication code (A-code)

In the model of A-codes, there are three participants, a transmitter T, a receiver R and an opponent O. The transmitter T and the receiver R share a common encoding rule e. On input a source state s, T computes a message $m = e(s)$ and sends m to R. R accepts or rejects m based on e. An A-code is called an A-code without secrecy if a source state is uniquely determined from a message m. It is possible that more than one message can be used to communicate a particular source state; this is called splitting. Defining

$$M(e, s) \stackrel{\triangle}{=} \{m \mid e(s) = m\}$$

splitting means $|M(e, s)| > 1$. If $|M(e, s)| = 1$ for $\forall e$ and $\forall s$, the A-code is called an A-code without splitting.

We assume independent probability distributions on source states and on encoding rules, respectively. In the impersonation attack, the opponent O sends a message m to the receiver. O succeeds if m is accepted by the receiver as authentic. The impersonation attack probability P_I is defined by

$$P_I \stackrel{\triangle}{=} \max_{m \in M} \Pr[\text{R accepts } m] \tag{2.1}$$

In the substitution attack, O observes a message m that is transmitted by T and substitutes m with another message $\hat{m}$. O succeeds if $\hat{m}$ is accepted by the receiver as authentic. For no splitting, the substitution attack probability P_S is defined by

$$P_S \stackrel{\triangle}{=} \sum_{m \in M} \Pr(M = m) \max_{\hat{m} \neq m} \Pr[\text{R accepts } \hat{m} | \text{R accepts } m] \tag{2.2}$$

For splitting, the maximum is taken over $\hat{m}$ such that the source state conveyed by $\hat{m}$ is different from that of m. Let $S \triangleq \{s\}, E \triangleq \{e\}$ and $M \triangleq \{m\}$. We denote an A-code by (S, E, M).

Proposition 1. *[4] In an A-code without splitting, $P_I \geq |S|/|M|$. The equality holds if and only if* $\Pr[\text{R accepts } m] = |S|/|M|$ *for $\forall m$.*

Proposition 2. *[7] In an A-code without splitting and without secrecy, if $P_I = |S|/|M|$, then $P_S \geq |S|/|M|$.*

Definition 3. An orthogonal array $OA(l, k, \lambda)$ is a $\lambda l^2 \times k$ array of l symbols such that, in any two columns of the array, every one of the possible l^2 pairs of symbols occurs in exactly λ rows.

Proposition 4. *Suppose we have an A-code without splitting and without secrecy such that $P_I = P_S = |S|/|M| = 1/l$.*

1. *$|E| \geq |S|(l-1) + 1$. The equality occurs if and only if the incidence matrix of E is an orthogonal array $OA(l, |S|, \lambda)$, where $\lambda = (|S|(l-1)+1)/l^2$ and each $e \in E$ is used with equal probability [8].*
2. *Also, $|E| \geq l^2$ [7].*

Proposition 5. *[9] In a splitting A-code, let $M(e) \triangleq \{m \mid e \text{ accepts } m\}$. Then,*

$$P_I \geq \min_{e \in E} \frac{|M(e)|}{|M|} \qquad P_S \geq \min_{e \in E} \frac{|M(e)| - \max_{s \in S} |M(e, s)|}{|M| - \min_{s \in S} |M(e, s)|}$$

2.2 Authentication code with arbitration (A^2-code)

We denote an A^2-code by (S, M, E_R, E_T), where $S = \{s\}$ is a set of source states, $M = \{m\}$ is a set of messages, $E_R = \{f\}$ is a set of the receiver's decoding rules and $E_T = \{e\}$ is a set of the transmitter's encoding rules.

The selection of e and f may be done in several ways. One choice is to let the receiver R choose his f and then secretly pass this on to the arbiter. In this case, the arbiter constructs e and passes this on to the transmitter T. Another choice is to do the other way around and the third approach is to let the arbiter construct both rules. In any case, on input s, T sends m such that $m = e(s)$ to R. R accepts m iff $f(m)$ is valid. The arbiter accepts m as authentic iff e can generate m.

In this model, there are five different kinds of attacks.

I, Impersonation by the opponent. The cheating probability P_I is defined in the same way as eq.(2.1). S, Substitution by the opponent. The cheating probability P_S is defined in the same way as eq.(2.2). T, Impersonation by the transmitter. The transmitter sends a message to the receiver and denies having sent it. The transmitter succeeds if the message is accepted by the receiver as authentic and if the message is not one of the messages that the transmitter could have generated due to his encoding rule. This cheating probability P_T is defined as follows

$$P_T \stackrel{\triangle}{=} \max_{e \in E_T} \max_{m \in M} \Pr[\text{R accepts } m \text{ and } m \text{ is not generated by } e | \text{T has } e] \quad (2.3)$$

R_0, Impersonation by the receiver. The receiver claims to have received a message from the transmitter. The receiver succeeds if the message could have been generated by the transmitter due to his encoding rule. This cheating probability P_{R_0} is defined by

$$P_{R_0} \stackrel{\triangle}{=} \max_{f \in E_R} \max_{m \in M} \Pr[\text{Arbiter (or T) accepts } m | \text{R has } f \in E_R] \quad (2.4)$$

R_1, Substitution by the receiver. The receiver receives a message from the transmitter but claims to have received another message. The receiver succeeds if this other message could have been generated by the transmitter due to his encoding rule. This cheating probability P_{R_1} is defined by

$$P_{R_1} \stackrel{\triangle}{=} \max_{f \in E_R} \sum_{m \in M} \Pr(m)$$

$$\max_{\hat{m} \neq m} \Pr[\text{Arbiter (or T) accepts } \hat{m} | \text{R has } f \text{ and T sends } m] \quad (2.5)$$

Let

$$E_R \circ E_T \stackrel{\triangle}{=} \{(e, f) \mid \Pr[\text{T has } e \in E_T \text{ and R has } f \in E_R] > 0\}$$

For A^2-codes, Johansson showed an information theoretic bound such as follows.

Proposition 6. *[12] In an A^2-code without splitting*

$$\begin{aligned} P_I &\geq 2^{-\inf I(E_R;M)} & P_{R_0} &\geq 2^{-\inf I(E_T;M|E_R)} \\ P_S &\geq 2^{-\inf I(E_R;M'|M)} & P_{R_1} &\geq 2^{-\inf I(E_T;M'|E_R,M)} \\ P_T &\geq 2^{-\inf I(E_R;M|E_T)} \end{aligned}$$

where $I(X;Y)$ denotes the mutual entropy of X and Y.

Proposition 7. *[12]*

$$\begin{aligned} |E_R| &\geq (P_I P_S P_T)^{-1} & |E_R \circ E_T| &\geq (P_I P_S P_T P_{R_0} P_{R_1})^{-1} \\ |E_T| &\geq (P_I P_S P_{R_0} P_{R_1})^{-1} \end{aligned}$$

3 Combinatorial bounds for A^2-codes without secrecy

In this section, we present combinatorial lower bounds for A^2-codes without splitting and without secrecy. To derive our bounds, we develop three techniques. The first technique is a reduction of an A^2-code to an A-code. The second one is a restriction of messages which is described by the following Theorem.

Theorem 8. *For an A-code (S, E, M), consider a subcode $(S, E, \hat{M})$ such that $\hat{M} \subseteq M$. Then, for $\forall m \in \hat{M}$,*

$$\Pr[R \text{ accepts } m \text{ in the original } A\text{-code}]$$
$$= \Pr[R \text{ accepts } m \text{ in the subcode}]$$

The third technique is given by the following Theorem. This technique will be a basic Theorem for A-codes.

Theorem 9. *In an A-code without splitting (S, E, M), if $\Pr[R \text{ accepts } m] = 1/l$ for $\forall m$, then $|M| = |S|l$.*

Proof. Let $X = \{x_{ij}\}$ be the incidence matrix of E. That is,

$$x_{ij} = \begin{cases} 1 & \text{if } e_i \in E \text{ accepts } s_j \in S \\ 0 & \text{otherwise} \end{cases}$$

Then, $\Pr[R \text{ accepts } m_j] = \sum_i \Pr[e = e_i] x_{ij}$. If $\Pr[R \text{ accepts } m_j] = 1/l$ for $\forall m_j$, then

$$|M|/l = \sum_j \Pr[R \text{ accepts } m_j] = \sum_j \sum_i \Pr[e = e_i] x_{ij}$$
$$= \sum_i \Pr[e = e_i] \sum_j x_{ij} = \sum_i \Pr[e = e_i] |S|$$
$$= |S|$$

Hence, $|M| = l|S|$. $\qquad\qquad\qquad\qquad\qquad\qquad\qquad\qquad\qquad$ □

For an A^2-code (S, M, E_R, E_T), we define a decoding rule of the receiver $f \in E_R$ as follows.

$$f(m) = \begin{cases} s \in S & \text{if } m \text{ is accepted as } s. \\ \text{reject} & \text{if } m \text{ is rejected.} \end{cases}$$

Let

$$M(f, s) \overset{\triangle}{=} \{m \mid f(m) = s\}$$
$$M(f) \overset{\triangle}{=} \{m \mid f \text{ accepts } m\} \; (= \bigcup_{s \in S} M(f, s))$$
$$M_s \overset{\triangle}{=} \{m \mid f(m) = s, f \in E_R\} \; (= \bigcup_{f \in E_R} M(f, s))$$
$$E_R(e) \overset{\triangle}{=} \{f \mid \Pr[R \text{ has } f | T \text{ has } e] > 0\}$$

3.1 Lower bound on P_I and P_S

To prevent the impersonation attack of the receiver, it must be that $|M(f, s)| > 1$. That is, from a view point of the receiver, A^2-codes must be splitting. To derive our bounds, we assume the following assumption.

Assumption 10. $|M(f, s)| = c$ for $\forall f \in E_R$ and $\forall s \in S$.

That is, each $f \in E_R$ accepts c messages for each $s \in S$. For each source state $s_i \in S$, define

$$S_i \overset{\triangle}{=} \{s_{i1}, s_{i2}, \ldots, s_{ic}\}, \qquad \hat{S} \overset{\triangle}{=} S_1 \cup S_2 \cup \cdots$$

Then, we can consider an A-code without splitting $(\hat{S}, E_R, M)$ which corresponds to the original A^2-code (S, M, E_R, E_T) in a natural way. Clearly, $|\hat{S}| = c|S|$.

Theorem 11. Under assumption 10, $P_I \geq c|S|/|M|$. The equality holds if and only if

$$\Pr[R \text{ accepts } m] = c|S|/|M| \text{ for } \forall m \in M$$

Proof. It is clear that P_I of the A^2-code is equal to the impersonation attack probability for the splitting A-code $(\hat{S}, E_R, M)$. Apply proposition 1 to $(\hat{S}, E_R, M)$. $\qquad \square$

Assumption 12. $P_I = c|S|/|M| = 1/l$.

Theorem 13. Under assumption 10 and 12, $|M_s| = cl$ for $\forall s \in S$.

Proof. From assumption 12 and theorem 11,

$$\Pr[R \text{ accepts } m] = 1/l \text{ for } \forall m \in M \qquad (3.1)$$

We consider a subcode of $(\hat{S}, E_R, M)$ such that the set of messages is restricted to M_s. This is an A-code without splitting in which the number of messages is $|M_s|$ and that of source states is c (from assumption 10). Even for this restricted A-code, we have

$$\Pr[R \text{ accepts } m] = 1/l \text{ for } \forall m \in M_s$$

from Theorem 11. Then, from Theorem 9, $|M_s| = cl$ $\qquad \square$

Theorem 14. Under assumption 10 and 12,

$$P_S \geq 1/l \qquad (3.2)$$

Proof. P_S is defined by

$$P_S \overset{\triangle}{=} \sum_{m' \in M} \Pr[T \text{ sends } m'] \max \Pr[R \text{ accepts } m | R \text{ accepts } m'] \qquad (3.3)$$

Let $m' \in M_{s_i}$. Then, the maximum is taken over $m \in M \backslash M_{s_i}$. For $m' \in M_{s_i}$, define

$$P'_I(m') \triangleq \max_{m \in M \backslash M_{s_i}} \Pr[\text{R accepts } m | \text{R accepts } m']. \tag{3.4}$$

Then, this is the impersonation attack probability against an A-code without splitting $(\hat{S} \backslash S_i, E_R, M \backslash M_{s_i})$, where the probability distribution on E_R is conditioned by the fact that R accepts m'. Then, from proposition 1,

$$P'_I(m') \geq \frac{|\hat{S} \backslash S_i|}{|M \backslash M_{s_i}|}.$$

Now

$$|\hat{S} \backslash S_i| = |\hat{S}| - |S_i| = c|S| - c = c(|S| - 1), \qquad |M \backslash M_{s_i}| = |M| - |M_{s_i}| = cl|S| - cl$$

from assumption 12 and Theorem 13. Therefore, we have $P'_I(m') \geq 1/l$ for $\forall m'$. Then, we have eq.(3.2) from eq.(3.3) and eq.(3.4). $\qquad \square$

3.2 Lower bound on P_T

Let $h \triangleq \min_{e \in E_T} |E_R(e)|$.

Theorem 15. *Under assumption 10, $P_T \geq \max\{(c - 1)|S|/(|M| - |S|), 1/h\}$.*

Proof. It is easy to see that $P_T \geq 1/h$.
Suppose that the transmitter T has an encoding rule $e \in E_T$. Let $m_i = e(s_i)$ for $i = 1, 2, \ldots, |S|$. T must send m to R such that $m \neq m_i$ for $\forall i$ for cheating. Consider the following subcode of $(\hat{S}, E_R, M)$ such as follows. Let X be the incidence matrix of $E_R(e)$ which is a $|E_R(e)| \times |M|$ binary matrix. Remove the columns corresponding to $\{m_i | i = 1, 2, \cdots, |S|\}$ from this matrix. Then, we obtain an incidence matrix of an A-code without splitting $(S', E_R(e), M \backslash \{m_i\})$, where $|S'| = |\hat{S}| - |S| = c|S| - |S| = (c-1)|S|$. Theorem 11 holds for this subcode. The best strategy of the transmitter is at least as good as the impersonation attack against this modified A-code. Then from proposition 1, we have

$$P_T \geq \max_{e \in E_T} \frac{|S'|}{|M \backslash \{m_i\}|} = \frac{(c - 1)|S|}{|M| - |S|}$$

$\qquad \square$

Corollary 16. *Under assumption 10 and 12,*

$$P_T \geq \max\{(c - 1)/(lc - 1), 1/h\}. \quad \text{Further, if } c = l = q, \text{ then} \quad P_T \geq \frac{1}{q + 1}$$

Next, we consider a separable case.

Assumption 17. *For $\forall s$, M_s can be grouped into $A^s_1, A^s_2, \ldots$ in such a way that $|A^s_i \cap M(f, s)| = 1$ for $\forall s$, $\forall A^s_i$ and $\forall f \in E_R$. An A^2-code given by [14] satisfies this assumption.*

Lemma 18. *Under assumption 10, 12 and 17, $|A_i^s| = l$ and $|\{A_i^s\}| = c$ for $\forall s$.*

Proof. The proof is almost the same as that of Theorem 13. We consider a subcode of $(\hat{S}, E_R, M)$ such that the set of messages is restricted to A_i. In this restricted A-code, the number of messages is $|A_i|$ and that of source states is 1 from assumption 17. From Theorem 11, assumption 12 and Theorem 11, even for this restricted A-code, $\Pr[\text{R accepts } m] = 1/l$ for $\forall m \in A_i$. Then, from Theorem 9, $|A_i^s| = 1 \times l = l$. From Theorem 13, $|\{A_i^s\}| = |M_s|/|A_i| = c$ $\qquad\square$

Theorem 19. *Under assumption 10, 12 and 17, $P_T \geq \max\{1/l, 1/h\}$.*

Proof. It is clear that $P_T \geq 1/h$.
Suppose that the transmitter has $e \in E_T$. Let $m_i = c(s_i)$ for $i = 1, 2, \ldots, |S|$. For simplicity, suppose that $m_i \in A_1^{s_i}$ for $i = 1, 2, \ldots, |S|$. As in the proof of Theorem 15, we consider an A-code without splitting $(S', E_R(e), M \setminus \{A_i^{s_i}\})$ such that $|S'| = (c-1)|S|$. Then, as in that proof, we have

$$P_T \geq \max_{e \in E_T} \frac{|S'|}{|M \setminus \{A_i^{s_i}\}|}$$

From assumption 12, and lemma 18,

$$|M \setminus \{A_i^{s_i}\}| \;=\; |M| - \sum |A_1^{s_i}| \;=\; |M| - l|S| \;=\; lc|S| - l|S|$$

Then, $P_T \geq (c-1)|S|/(lc|S| - l|S|) = 1/l$ $\qquad\square$

3.3 Lower bound on P_{R_0} and P_{R_1}

Let $E_T(f) \overset{\triangle}{=} \{e \mid \Pr[\text{T has } e | \text{R has } f] > 0\}$. Suppose that R has f. Then, R knows that T has some $e \in E_T(f)$. Consider an A-code $(S, E_T(f), M(f))$. It is an A-code without splitting and without secrecy because the original A^2-code is so. Let $P_I(f)$ and $P_S(f)$ denote the impersonation attack probability and the substitution attack probability, respectively. Then it is easy to see that

$$P_{R_0} = \max_{f \in E_R} P_I(f) \qquad\qquad P_{R_1} = \max_{f \in E_R} P_S(f)$$

(Remember that the arbiter accepts m as authentic iff e can generate m.) From assumption 10, $|M(f)| = c|S|$. Now, from proposition 1, we have Theorem 20.

Theorem 20. *Under assumption 10,*

1. $P_{R_0} \geq 1/c$
2. *If $P_{R_0} = 1/c$, then $P_{R_1} \geq 1/c$.*

3.4 Tightness

Corollary 21. *Suppose that $c = l = q$. Then*

1. $P_I \geq 1/q$, $P_T \geq 1/(q+1)$, $P_{R_0} \geq 1/q$.
2. If $P_I = 1/q$, then $P_S \geq 1/q$, If $P_{R_0} = 1/q$, then $P_{R_1} \geq 1/q$.
3. Under assumption 17, if $P_I = 1/q$, then $P_T \geq 1/q$.

Corollary 21 is tight because all the bounds are satisfied with equality by the A^2-codes given by Johansson [13, 14].

3.5 Lower bound on $|E_R|, |E_R \circ E_T|$ and $|E_T|$

In this subsection, we show more tight lower bounds on $|E_R|, |E_R \circ E_T|$ and $|E_T|$ than proposition 7.

Assumption 22. $P_I = P_S = P_T = c|S|/|M| = 1/l$

Theorem 23. *Under assumption 10, 17 and 22, $|E_R| \geq c|S|(l-1)+1$. The equality holds if and only if the incidence matrix of E_R is an orthogonal array $OA(l, c|S|, \lambda)$ where $\lambda = (c|S|(l-1)+1)/l^2$ and each $f \in E_R$ is used with equal probability.*

Proof. From assumption 17, we can consider that our A-code without splitting $(\hat{S}, E_R, M)$ is without secrecy. Remember that $|\hat{S}| = c|S|$. For this A-code without secrecy, let $\hat{P}_I$ and $\hat{P}_S$ be the impersonation attack probability and the substitution attack probability, respectively. Then, clearly $\hat{P}_I = P_I = c|S|/|M|$. From assumption 22, $\hat{P}_S = \max\{P_S, P_T\} = c|S|/|M|$. Now, from proposition 4, we have this Theorem. $\qquad\square$

Remark. From proposition 7, we have another bound such that $|E_R| \geq l^3$. If $c|S| \geq l^2 + l + 1$, Theorem 23 is more tight than this bound.

Assumption 24. $P_{R_0} = P_{R_1} = 1/c$

Theorem 25. *Under assumption 10 and 24, $E_T(f) \geq \max\{c^2, |S|(c-1)+1\}$.*

Proof. From proposition 4. $\qquad\square$

Theorem 26. *Under assumption 10 and 24,*

$$|E_R \circ E_T| \geq |E_R| \times \max\{c^2, |S|(c-1)+1\}.$$

Proof. From Theorem 25. $\qquad\square$

From Theorem 19, if $P_T = 1/l$, then $l \leq h = \min_e |E_R(e)|$

Assumption 27. *For $\forall e \in E_T, |E_R(e)| = l$*

Theorem 28. *Under assumption 10, 24 and 27,*

$$|E_T| \geq \frac{|E_R|}{l} \times \max\{c^2, |S|(c-1)+1\}.$$

Proof. From assumption 27, $l \times |E_T| = |E_R \circ E_T|$. Then, from Theorem 26, Theorem 28 holds. $\qquad\square$

4 Combinatorial bounds for general A^2-codes

In this section, we show combinatorial bounds for general A^2-codes without splitting.

Theorem 29.

$$P_I \geq \min_{f \in E_R} \frac{|M(f)|}{|M|} \qquad\qquad P_S \geq \min_{f \in E_R} \frac{|M(f)| - \max_{s \in S} |M(f,s)|}{|M| - \min_{s \in S} |M(f,s)|}$$

Proof. We consider a splitting A-code (S, E_R, M) which corresponds to the original A^2-code (S, M, E_R, E_T) in a natural way (where $|M(f,s)| > 1$). From proposition 5, we have this theorem. $\qquad\square$

Theorem 30. $\qquad\qquad P_T \geq \min_{f \in E_R} \frac{|M(f)| - |S|}{|M| - |S|}$

The proof is almost the same as that of theorem 15.

Theorem 31. $\qquad\qquad P_{R_0} \geq \max_{f \in E_R} \frac{|S|}{|M(f)|}$

Proof. We consider an A-code without splitting $(S, E_T(f), M(f))$ as shown in subsection 3.3. From proposition 1, we have this theorem. $\qquad\square$

Theorem 32. $\qquad\qquad P_{R_1} \geq \max_{f \in E_R} \frac{|S| - 1}{|M(f)| - \min_{s \in S} |M(f,s)|}$

Proof. Fix $f \in E_R$ and $m \in M(f)$ arbitrarily. For $\forall m' \in M(f) - M(f, f(m))$, let

$$P_{m'} \stackrel{\triangle}{=} \Pr[\text{Arbiter accepts } m' \mid \text{R has } f \text{ and T sends } m]$$

Then

$$P_{m'} = \frac{\sum_{\{e \mid m, m' \in M(e)\}} \Pr[E_T = e | E_R = f] \Pr[S = f(m)]}{\sum_{\{e \mid m \in M(e)\}} \Pr[E_T = e | E_R = f] \Pr[S = f(m)]}$$

Now, we have

$$\sum_{m' \in M(f)} P_{m'}$$

$$= \frac{\sum_{m' \in M(f)} \sum_{\{e \mid \exists s: m' = e(s), m \in M(e)\}} \Pr[E_T = e | E_R = f] \Pr[S = f(m)]}{\sum_{\{e \mid m \in M(e)\}} \Pr[E_T = e | E_R = f] \Pr[S = f(m)]}$$

$$= \frac{\sum_s \sum_{m' \in M(f)} \sum_{\{e \mid m' = e(s), m \in M(e)\}} \Pr[E_T = e | E_R = f] \Pr[S = f(m)]}{\sum_{\{e \mid m \in M(e)\}} \Pr[E_T = e | E_R = f] \Pr[S = f(m)]}$$

$$= \frac{\sum_s \sum_{\{e \mid m \in M(e)\}} |\{m' \mid m' = e(s)\}| \Pr[E_T = e | E_R = f] \Pr[S = f(m)]}{\sum_{\{e \mid m \in M(e)\}} \Pr[E_T = e | E_R = f] \Pr[S = f(m)]}$$

$$= \frac{\sum_s \sum_{\{e \mid m \in M(e)\}} \Pr[E_T = e | E_R = f] \Pr[S = f(m)]}{\sum_{\{e \mid m \in M(e)\}} \Pr[E_T = e | E_R = f] \Pr[S = f(m)]}$$

$$= |S|$$

Further,

$$\sum_{m'\in M(f,f(m))} P_{m'}$$

$$= \frac{\sum_{m'\in M(f,f(m))} \sum_{\{e \mid \exists s: m'=e(s), m\in M(e)\}} \Pr[E_T = e|E_R = f]\Pr[S = f(m)]}{\sum_{\{e \mid m\in M(e)\}} \Pr[E_T = e|E_R = f]\Pr[S = f(m)]}$$

$$= \frac{\sum_{m'\in M(f,f(m))} \sum_{\{e \mid m'=e(f(m)), m\in M(e)\}} \Pr[E_T = e|E_R = f]\Pr[S = f(m)]}{\sum_{\{e \mid m\in M(e)\}} \Pr[E_T = e|E_R = f]\Pr[S = f(m)]}$$

$$= \frac{\sum_{\{e \mid m\in M(e)\}} |\{m' \mid m' = e(f(m))\}| \Pr[E_T = e|E_R = f]\Pr[S = f(m)]}{\sum_{\{e \mid m\in M(e)\}} \Pr[E_T = e|E_R = f]\Pr[S = f(m)]}$$

$$= \frac{\sum_{\{e \mid m\in M(e)\}} \Pr[E_T = e|E_R = f]\Pr[S = f(m)]}{\sum_{\{e \mid m\in M(e)\}} \Pr[E_T = e|E_R = f]\Pr[S = f(m)]}$$

$$= 1$$

Hence,
$$\sum_{m'\in M(f)\backslash M(f,f(m))} P_{m'} = \sum_{m'\in M(f)} P_{m'} - \sum_{m'\in M(f,f(m))} P_{m'} = |S| - 1$$

Therefore, there exists $\hat{m}' \in M(f)\backslash M(f, f(m))$ such that

$$P_{\hat{m}'} \geq \frac{|S| - 1}{|M(f)| - |M(f, f(m))|}$$

Thus

$$P_{R_1} \triangleq \max_{f\in E_R} \sum_{m\in M} \Pr(m) \max_{\hat{m}\neq m} \Pr[\text{Arbiter accepts } \hat{m}|\text{R has } f \text{ and T sends } m]$$

$$\geq \max_{f\in E_R} P_{\hat{m}'}$$

$$\geq \max_{f\in E_R} \frac{|S| - 1}{|M(f)| - \min_{s\in S} |M(f, s)|}$$

$\square$

5 Another definition of P_{R_0} and P_{R_1}

We can define P_{R_0} and P_{R_1} in a different way from eq.(2.4) and eq.(2.5). The alternative definitions are

$$P_{R_0} \triangleq \sum_{f\in E_R} \Pr(f) \max_{m\in M} \Pr[\text{Arbiter (or T) accepts } m|\text{R has } f \in E_R]$$

$$P_{R_1} \triangleq \sum_{f\in E_R} \Pr(f) \sum_{m\in M} \Pr(m)$$
$$\max_{\hat{m}\neq m} \Pr[\text{Arbiter (or T) accepts } \hat{m}|\text{R has } f \text{ and T sends } m]$$

For the model without secrecy, the lower bounds on the above P_{R_0} and P_{R_1} are the same as those on the original P_{R_0} and P_{R_1}. For the general model, we obtain the following bound.

Theorem 33.

$$P_{R_0} \geq \min_{f \in E_R} \frac{|S|}{|M(f)|} \qquad\qquad P_{R_1} \geq \min_{f \in E_R} \frac{|S| - 1}{|M(f)| - \min_{s \in S} |M(f, s)|}$$

References

1. G.J.Simmons, "A survey of Information Authentication", in *Contemporary Cryptology, The science of information integrity*, ed. G.J.Simmons, IEEE Press, New York, 1992.

2. G.J.Simmons, "Authentication theory/coding theory", Proceedings of Crypto'84, Lecture Notes in Computer Science, LNCS 196, Springer Verlag, pp.411–431 (1985).

3. E.F.Brickell, "A few results in message authentication", Congresus Numerantium, vol.43, pp.141–154 (1984).

4. G.J.Simmons, "Message authentication: a game on hypergraphs", Congresus Numerantium, vol.45, pp.161–192 (1984).

5. J.L.Massey, *Cryptography – a selective survey*, in *Digital Communications*, North Holland (pub.), pp.3–21, (1986).

6. D.R.Stinson, "Some constructions and bounds for authentication codes", Journal of Cryptology, Vol.1, 1988, pp.37–51, (1988).

7. D.R.Stinson, "The combinatorics of authentication and secrecy codes", Journal of Cryptology, Vol.2, no 1, 1990, pp.23–49, (1990).

8. D.R.Stinson, "Combinatorial Characterization of Authentication Codes", Proceedings of Crypto'91, Lecture Notes in Computer Science, LNCS 576, Springer Verlag, pp62–72 (1992).

9. Marijke De Soete, "New Bounds and Constructions for Authentication/Secrecy Codes with Splitting", Journal of Cryptology, Vol.3, no 3, 1991, pp173–186 (1991).

10. G.J.Simmons, "Message Authentication with Arbitration of Transmitter/Receiver Disputes", Proceedings of Eurocrypt'87, Lecture Notes in Computer Science, LNCS 304, Springer Verlag, pp.150–16 (1987)

11. G.J.Simmons, "A Cartesian Product Construction for Unconditionally Secure Authentication Codes that Permit Arbitration", Journal of Cryptology, Vol.2, no.2, 1990, pp.77–104 (1990).

12. Thomas Johansson, "Lower Bounds on the Probability of Deception in Authentication with Arbitration", In *Proceedings of 1993 IEEE International Symposium on Information Theory*, San Antonio, USA, January 17–22, 1993, pp.231.

13. Thomas Johansson, "Lower Bounds on the Probability of Deception in Authentication with Arbitration", submitted to IEEE Trans. on IT (private communication).

14. Thomas Johansson, "On the construction of perfect authentication codes that permit arbitration", Proceedings of Crypto'93, Lecture Notes in Computer Science, LNCS 773, Springer Verlag, pp.341-354 (1993).

15. K.Kurosawa, "New bound on authentication code with arbitration", Proceedings of Crypto'94, Lecture Notes in Computer Science, LNCS 899, Springer Verlag, pp.140–149 (1994).

New Hash Functions for Message Authentication

Hugo Krawczyk

IBM T.J. Watson Research Center
Yorktown Heights, NY 10598
(hugo@watson.ibm.com)

Abstract. We show that Toeplitz matrices generated by sequences drawn from small biased distributions provide hashing schemes applicable to secure message authentication. This work extends our previous results from Crypto'94 [4] where an authentication scheme based on Toeplitz matrices generated by linear feedback shift registers was presented.

Our new results have as special case the LFSR-based construction but extend to a much wider and general family of sequences, including several simple and efficient constructions with close to optimal security. Examples of the new constructions include Toeplitz matrices generated by the Legendre symbols of consecutive integers modulo a prime (of size significantly shorter than required by public-key modular arithmetic) as well as other algebraic constructions. The interest of these schemes extends beyond the proposed cryptographic applications to other uses of universal hashing (including other cryptographic applications).

1 Introduction

In *Crypto'94*, we introduced [4] a new scheme for hash functions suitable for message authentication in the symmetric key model. The scheme uses a linear feedback shift register sequence for the generation of a Boolean Toeplitz matrix that in turn is used to hash the message using matrix-vector multiplication. When combined with a one-time pad encryption of the hash value this schemes gives (provable) unconditional security.

The efficiency and simplicity of that construction makes it attractive conceptually as well as for practical use. In this paper we generalize the above scheme so that it can be used with a variety of different "weakly random" sequences as an alternative to LFSRs. We prove that any sequence taken from an ε-*biased distribution* of sequences can be used to generate a Toeplitz matrix that has all the security properties for authentication that the original LFSR-based construction had (and essentially the same strength of a completely random matrix!).

ε-*biased* distributions (introduced by Naor and Naor [7]) on bit sequences of length ℓ are characterized by the property that for any Boolean vector $\alpha \neq 0$ of length ℓ and for a sequence s chosen from that distribution, the probability of (α, s) (i.e., the scalar product modulo 2 of α and s) being 1 deviates from $1/2$ by at most ε. (See Section 2.3 for a formal definition). These sequences prove to be useful for replacement of true randomness in different applications (see [7]). The advantage of these sequences over purely random ones is that they allow for easy generation,

L.C. Guillou and J.-J. Quisquater (Eds.): Advances in Cryptology - EUROCRYPT '95, LNCS 921, pp. 301-310, 1995.
© Springer-Verlag Berlin Heidelberg 1995

simplicity and short description. All these properties translate in our case to a variety of attractive schemes for secure message authentication: simple, efficient, requiring short keys and short authentication tags.

In addition to LFSRs, which are a particular case of ε-biased sequences, this class of sequences includes several constructions [7, 1] from error correcting codes as well as other algebraic sequences; e.g., the sequence of Legendre symbols (or quadratic character) of consecutive integers modulo a prime (of size significantly shorter than required by public-key modular arithmetic).

Before stating our main result, we recall that in the setting of message authentication, we use Toeplitz matrices to first hash the message and then encrypt the hash value using a one-time pad (or stream cipher); the resultant encrypted hash value is the authentication tag for the message. Therefore, the communicating parties need to share a description of a particular hashing matrix as their shared secret key, and need to append to each transmitted message the corresponding authentication tag. This implies the need for a family of matrices that can be generated efficiently out of a short seed (the secret key), result in short hash values (the authentication tag) and still have the required security properties. The schemes we construct in this paper out of ε-biased sequences satisfy all these requirements with essentially the same security as provided by a completely random matrix (but while the later may require millions of random bits to describe the function, the ε-biased approach can do with keys in the order of 100 bits).

Next, we state our main theorem in terms of *otp-secure* hash functions, a notion defined in [4] (see Section 2.2 below). Informally, a family of hash functions is δ-otp-secure if the probability of an adversary to defeat the authentication is no more than δ where a message is authenticated by encrypting its hash value under a one-time pad (otp).

Main Theorem: *Let T_S be a family of Toeplitz matrices corresponding to sequences selected from an ε-biased distribution S. Then the hashing scheme that uses multiplication of the message by the Toeplitz matrix as the message hash is $\left(\frac{1}{2^n} + \varepsilon\right)$-otp-secure, where n is the length of the hash output.*

By using constructions of ε-biased sequences introduced by Alon et al. [1], we get explicit realizations of the above theorem. In particular, we show δ-otp-secure constructions where δ can be as close as desired to the optimal value 2^{-n} (this is optimal since the adversary can always guess the n-bit authentication tag with probability 2^{-n}). As an example, for any values of n and m, we present schemes that authenticate messages of length m with authentication tags of length n, and have security $\delta = 2^{-n+1}$. The shared key for describing such an authentication function is of length $2 \cdot (n + \log(n + m))$ (i.e., increases only logarithmically with the message which is typically much longer than n). These results are presented in Section 4.

Our proofs use as basic tools a characterization theorem from [4] that determines sufficient (and necessary) conditions for a family of hash functions to be secure for message authentication; and Discrete Fourier analysis for proving that ε-biased sequences induce Toeplitz hashing with the stated properties.

The interest of our work extends beyond the proposed cryptographic applications to other uses of universal hashing (including other cryptographic applications).

Our work extends the work of Krawczyk [4] that in turn follows the approach introduced by Carter and Wegman [10] of basing message authentication on hash functions. The same approach was followed by several authors, e.g., [8, 2]. We refer to [4] for a more complete survey of the relevant works.

ORGANIZATION: Section 2 presents the technical background and basic notions used throughout the paper. Section 3 presents the proof of the Main Theorem. Section 4 describes explicit constructions and their properties for message authentication.

2 Technical Background

In this section we introduce the main notions used throughout the paper as well as the technical background and tools.

2.1 Toeplitz matrices

Toeplitz matrices are characterized by having *fixed diagonals*. More precisely, each left-to-right diagonal is fixed, i.e., if $k - i = l - j$ for any indices $1 \leq i, k \leq n$, $1 \leq j, l \leq m$, then $A_{i,j} = A_{k,l}$. See Figure 1 for an example. Notice that an $n \times m$ Toeplitz matrix is fully described by its first column and first row (i.e., by $n + m - 1$ elements).

Notation: *To any given sequence s of $n + m - 1$ bits we associate an $n \times m$ Toeplitz matrix T_s, where the elements of s determine the first column and first row of T_s, and therefore the whole matrix. We map the first n elements of s into the first column of T_s starting from the bottom (i.e., $T_s(n,1) = s_1, \ldots, T_s(1,1) = s_n$) and then the last m bits of s into the first row of T_s (i.e., $T_s(1,1) = s_n, T_s(1,2) = s_{n+1}, \ldots, T_s(1,m) = s_{n+m-1}$). See Figure 1. We say that s generates T_s.*

$$\begin{pmatrix} 0\,1\,1\,0\,1\,0\,0\,1 \\ 0\,0\,1\,1\,0\,1\,0\,0 \\ 1\,0\,0\,1\,1\,0\,1\,0 \\ 1\,1\,0\,0\,1\,1\,0\,1 \end{pmatrix}$$

Fig. 1. The 4×8 Toeplitz matrix $T_{11001101001}$

Toeplitz matrices of dimension $n \times m$ can be used to hash messages of length m by multiplying the message (seen as a column vector) by the matrix. The resultant hash value has length n. It is well-known that the family of Toeplitz matrices T_s with s chosen at random constitutes a strongly universal$_2$ family of hash functions (see [6]). One of the main results in [4] is to show that when the sequence s is generated out of only n random bits using a random irreducible LFSR, the resultant family is still almost universal (or ε-balanced in the terminology of [4] − see Section 2.2 below). This is especially important when, as it is the case in practice, $n << m$. Here we show that this result extends to *any* family of sequences s which are ε-biased distributed. Explicit constructions are presented in Section 4.

2.2 ε-balanced hashing and authentication

Following the approach initiated by Carter and Wegman [10] (and developed by several authors – e.g., [8, 2, 4]) we study the message authentication model in which the communicating parties share a specific hash function h_k, chosen out of a family of hash functions H, and use this particular function h_k to authenticate multiple messages. The function h_k is described by a (secret) key k shared by the parties. Each message to be authenticated is first hashed using h_k and the resultant hash value encrypted by xor-ing it with a secret one-time pad known only to the legitimate parties. We use m to denote the length of the message M, and n for the length of the resultant hash value $h_k(M)$. Notice that all the functions described here can work, in principle, with arbitrary and variable length messages (typically, the security of the authentication degrades logarithmically with the length of messages). The output length n is the same for all messages and can be thought of as a security parameter.

The task of an adversary A that tries to break the authentication is to intercept a message M, sent between the legitimate parties, together with its legitimate authentication tag $h_k(M) \oplus r$ and replace it with another message M' (which may depend on M and other messages exchanged between the parties in the past [1]) for which A can produce the legal authentication tag $h_k(M') \oplus r$. Notice that we assume that A does not know k or r. (The above is called a *substitution* attack; an *impersonation* attack where the adversary initiates the sending of a fake message without the legitimate parties having a communication between them trivially has the minimal probability 2^{-n} to succeed because of the use of one-time pads).

The following definitions and theorem are from [4].

Definition 1. A family H of hash functions is called ε-*otp-secure* if for any message M, and for h_k chosen according to the distribution on H, no adversary succeeds in the above scenario with probability larger than ε.

Definition 2. A family of hash functions H with a probability distribution attached to it is called ε-*balanced* if

$$\forall M \neq 0, \ \forall b, \ Prob_h(h(M) = b) \leq \varepsilon$$

where the probability is taken for h chosen according to the distribution on H.

The relation between the above two definitions is given by the following theorem.

Theorem 3 [4]. *If H is a family of linear functions relative to the bitwise XOR operation then H is ε-otp-secure if and only if H is ε-balanced.*

Our goal is to prove that the families of functions that we build in this paper (i.e., Toeplitz hashing generated by small biased sequences) are secure for use in message authentication, namely, they are ε-otp-secure for a very small ε. Since these functions use (Boolean) matrix multiplication for hashing then they are linear relative to XOR, and by virtue of the above theorem our task reduces to prove that they are ε-balanced for small ε. This proof is presented in Section 3.

---

[1] we assume a *chosen message attack* in which messages exchanged between the legitimate parties can be chosen by the adversary

2.3 ε-biased distributions

ε-biased distributions were introduced by Naor and Naor [7] (following the work of Vazirani [9]) as a tool for constructing small sample spaces, or more generally as a tool for replacement of truly random sequences with more "compact" and easier to generate sequences. This approach works in a variety of applications as shown in [7] and subsequent works. In this paper we show how ε-biased distributions are useful for generating efficient and short-key authentication functions. More precisely, we prove that Toeplitz matrices generated out of ε-biased sequences preserve essentially the same properties for hashing and authentication as completely random matrices.

Definition 4. Let S be a distribution on sequences of length ℓ. Let (α, s) denote the scalar product modulo 2 of $\alpha \in \{0, 1\}^\ell$ and $s \in \{0, 1\}^\ell$. Then,

1. S is said *to pass the linear test α with bias ε* if $|Prob((\alpha, s) = 1) - \frac{1}{2}| \leq \varepsilon$ (the probability taken over the choice of s from the distribution S).
2. S is said to be an *ε-biased distribution* if it passes *all* linear tests $\alpha \neq 0$ with bias ε.

The case $\varepsilon = 0$ corresponds to the uniform distribution; therefore, ε-biased distributions can be viewed as approximations to the uniform distribution. However, even for very small ε, there may be significant distinctions between ε-biased and uniform distributions. In the negative side, ε-biased distributions can be weakly random (such is the case, e.g., of LFSR sequences that are exponentially small biased but can be predicted from a very short prefix, or the set of all binary strings with Hamming weight divisible by three, proven to be exponentially small biased in [3]). Fortunately, there is a positive side. First there exist some very simple and efficient constructions for ε-biased sequences. Examples of such constructions, due to Alon et al. [1], are presented in Section 4. Second, in some applications these weakly random sequences can replace truly random bits with the advantage of simplicity and efficiency of generation. The results in this paper are an example of such an application.

Multiple tests If S is an ε-biased distribution then by definition S passes all (non-zero) linear tests with bias at most ε. Now, let $\alpha_1, \alpha_2, \ldots, \alpha_k$ be k elements from $\{0, 1\}^\ell$ and $b_1, b_2, \ldots, b_k$ be k bits. What can we say about the *simultaneous* probability that $(\alpha_1, s) = b_1, (\alpha_2, s) = b_2, \ldots, (\alpha_k, s) = b_k$? Notice that if the set of α_i's is linearly independent and s is chosen uniformly from $\{0, 1\}^\ell$ then the above probability would be 2^{-k}. The following theorem states that for s drawn from the ε-biased distribution S this probability is at most $2^{-k} + \varepsilon$.

Theorem 5. *Let S be an ε-biased distribution on sequences of length ℓ. Let A be a $k \times \ell$ matrix of (full) rank k, and let b be a column vector of length k then*

$$Prob_s(A \cdot s = b) \leq \frac{1}{2^k} + \varepsilon$$

for s taken from the distribution S.

Notice that this is an extension of Definition 4, which corresponds to the particular case of $k = 1$. In other words, the power of passing single linear tests extends to the power of passing multiple simultaneous tests.

As we will show in Section 3, the proof of our Main Theorem reduces to the above theorem.

Naor and Naor [7] prove this property using a result by Vazirani [9, 1] that connects between ε-biased distributions and k-wise independence; here we present a direct proof using Discrete Fourier analysis (and get Vazirani's Theorem as a corollary). We present the proof of Theorem 5 in Section 3.2.

2.4 Discrete Fourier Transform

In this section we bring some minimal technical background and known facts on discrete Fourier transform, and its relationship to ε-biased distributions, that we use in our proof of Theorem 5

The set of real valued functions defined on the Boolean domain $\{0,1\}^\ell$, i.e., functions $f : \{0,1\}^\ell \to \mathbf{R}$, has an orthonormal basis composed of the following functions χ_σ, $\sigma \subseteq \{1, \cdots, \ell\}$:

$$\chi_\sigma(x_1, \cdots, x_\ell) = \begin{cases} +1 & \text{if } \sum_{i \in \sigma} x_i \text{ is even} \\ -1 & \text{if } \sum_{i \in \sigma} x_i \text{ is odd} \end{cases}$$

In other words, each real valued function over $\{0,1\}^\ell$ can be written as a (unique) linear combination of the functions χ_σ. The respective coefficients are denoted by $\hat{f}(\sigma)$, i.e., $f = \sum_\sigma \hat{f}(\sigma)\chi_\sigma$. This representation is called the *(discrete) Fourier transform* of the function f.

Property 1 *For Boolean f (i.e., $f : \{0,1\}^\ell \to \{0,1\}$), the coefficients $\hat{f}(\sigma)$ have the following special form:*

$$\hat{f}(\sigma) = Pr[f(x) = \oplus_{i \in \sigma} x_i] - Pr[f(x) \neq \oplus_{i \in \sigma} x_i] = 2 \cdot Pr[f(x) = \oplus_{i \in \sigma} x_i] - 1$$

where $x = (x_1, x_2, \ldots, x_\ell)$ is chosen uniformly at random.

We need the following definition and property of the L_1-norm of the function f.

Definition 6. Let f be a function from $\{0,1\}^\ell$ to the real numbers. Define $L_1(f) = \sum_\sigma |\hat{f}(\sigma)|$.

Property 2 *Let f and g be functions from $\{0,1\}^\ell$ to the real numbers. Then, $L_1(fg) \leq L_1(f)L_1(g)$.*

The connection between ε-biased distributions and Fourier transform is given by the following Lemma. Notice that probability distributions over the set of strings $\{0,1\}^\ell$ are real valued functions and therefore their Fourier transform is well-defined.

Lemma 7. *Let μ be an ε-biased probability distribution over $\{0,1\}^\ell$. Then, for every subset $\sigma \subset \{1 \ldots \ell\}$, $|\hat{\mu}(\sigma)| \leq \varepsilon 2^{-\ell}$.*

The following lemma due to Kushilevitz and Mansour [5] relates ε-biased distributions and the norm $L_1(f)$, and plays a central role in the proof of our Main Theorem.

Lemma 8 [5].

$$|E_U[f] - E_\mu[f]| \le \varepsilon L_1(f)$$

where E denotes expectation, U is the uniform distribution and μ is an ε-biased distribution.

3 Main Result

This section is devoted to prove the Main Theorem of this paper (see the Introduction), namely, that Toeplitz matrices generated by ε-biased sequences constitute a family of $(2^{-n} + \varepsilon)$-otp-secure hash functions. By Theorem 3, this task reduces to proving the following result.

Theorem 9. Let S be distribution on sequences of length $n+m-1$ bits. Let $\{T_s\}_s$ be the set of $n \times m$ Toeplitz matrices generated by the elements $s \in S$. If the distribution S is ε-biased then the family $\{T_s\}_s$ with the distribution induced by S is a $(2^{-n} + \varepsilon)$-balanced family of hash functions.

3.1 Proof of Theorem 9

Following the definition of ε-balanced hash functions, we need to show that, for any given vector b of length n and message $M \ne 0$ of length m, the probability that $T_s \cdot M = b$ is at most $(2^{-n} + \varepsilon)$. This probability is taken over the distribution on T_s induced by choosing s from the ε-biased distribution S.

We start by transforming the representation of the problem. Notice that in a Toeplitz matrix each row is shifted (to the right) relative to the previous row, with a new element set to the first position of the row. This allows as to "swap" the roles of the sequence s and the message M in the following way.

We generate a new $n \times (n+m-1)$ matrix A_M which is cyclic and is defined by its first row containing $n-1$ zeros and then the entries of the vector M (viewed now as a row vector). Each new row in A_M is defined as a cyclic shift from right to left relative to the previous row (e.g., the second row of A_M contains the vector M but this time prepended by $n-2$ zeros and appended with one zero). It is easy to see that the following relationship, where the sequence s is represented by a $n+m-1$ column vector, holds. (See Figure 2 for an illustration of $A_M \cdot s$).

Lemma 10.

$$T_s \cdot M = b \quad \text{if and only if} \quad A_M \cdot s = b .$$

Therefore, the proof of Theorem 9 (and then of the Main Theorem) reduces to prove that for any $M \ne 0$, $Prob(A_M \cdot s = b) \le 2^{-n} + \varepsilon$, for s chosen according to the distribution S. But due to the special form of A_M this is a special case of Theorem 5 with $k = n$ and $\ell = n+m-1$. We conclude the proof of our main result by proving Theorem 5.

3.2 Proof of Theorem 5

We start by defining the following function.

$$f_{A,b}(s) = \begin{cases} 1 \text{ if } A \cdot s = b \\ 0 \text{ otherwise} \end{cases}$$

This definition can be specialized to $f_{\alpha,\beta}(s)$ where α is a vector of length ℓ and β a bit.

Lemma 11. $L_1(f_{\alpha,\beta}) = 1$

Proof. The proof uses Property 1 and the definition of the norm L_1. Details are omitted.

Lemma 12. $L_1(f_{A,b}) \leq 1$

Proof. Using Property 2 and the fact that $f_{A,b} = \prod_{i=1}^{k} f_{\alpha_i,b_i}$, where $\alpha_1, \ldots, \alpha_k$ are the rows of the matrix A and $b = (b_1, \ldots, b_k)^T$, we get

$$L_1(f_{A,b}) \leq \prod_{i=1}^{k} L_1(f_{\alpha_i,b_i}) = 1$$

where the last equality is derived from Lemma 11.

Now, we can conclude the proof of the theorem. Notice that, by the definition of $f_{A,b}$, for any distribution on the sequences s we have the expectation $E(f_{A,b}) = Prob(A \cdot s = b)$. For the uniform distribution this gives $E_U(f_{A,b}) = Prob_U(A \cdot s = b) = 2^{-k}$, where the last equality uses the fact that A has full rank k. Then, using Lemma 8 and Lemma 12, we get:

$$\left| Prob_S(A \cdot s = b) - 2^{-k} \right| = |E_S(f_{A,b}) - E_U(f_{A,b})| \leq \epsilon \cdot L_1(f_{A,b}) \leq \epsilon.$$

That is, $Prob_S(A \cdot s = b) \leq 2^{-k} + \epsilon$, which proves the theorem.   ·♡

4   Explicit Constructions

Our Main Theorem tells us that Toeplitz hashing generated out of ϵ-biased sequences, and combined with a one-time pad, provides a secure message authentication scheme in which no adversary can break the system with chance better than $2^{-n} + \epsilon$. (Naturally, if the one-time pad is generated using a pseudorandom generator or stream cipher the security of the whole authentication scheme reduces to that of the pseudorandom generator.)

The advantage of using ϵ-biased sequences (to define the Toeplitz matrices) as opposed to purely random bits is that the former can be generated efficiently out of a short random seed (typically, in the order of 100 bits while a pure random Toeplitz matrix would require millions of random bits in order to hash a few megabits of information). In particular, this implies a short shared authentication key for specifying a particular matrix. Moreover, the process of generation of the whole matrix is efficient and the resultant authentication tags short. Also, let us stress that in practical applications there is no need to simultaneously generate or work with the whole matrix, but only small portions of it (e.g., one column at a time).

Here we present several examples of explicit constructions of ε-biased sequences due to Alon et al. [1], with each of these constructions translating, using our results, into efficient authentication schemes. The first example is intended to show how the result of [4] about the quality of LFSR-based Toeplitz hashing for authentication can be derived as a special case of our results. We use ℓ to denote the length of the ε-biased sequences in these constructions and r to denote the number of random bits required to generate a sequence. In our setting of $n \times m$ Toeplitz matrices we have $\ell = n + m - 1$, and r is the length of the key that determines the specific shared matrix.

LFSR CONSTRUCTION. Sequences are determined by a random seed of length $r/2$ and an irreducible polynomial over $GF(2)$ of degree $r/2$. The sequence of ℓ bits is generated using an LFSR loaded initially with the above seed and the connections corresponding to the irreducible polynomial.

LEGENDRE SYMBOL CONSTRUCTION. Let p be a fixed prime number of length r. The sequences s are generated out of a random number $x \in \{0, \ldots, p - 1\}$. For $1 \le i \le \ell$, s_i is defined as 1 if $x + i$ is a quadratic residue modulo p or 0 otherwise. (We note that the prime p is not part of the secret key but a public value).

SCALAR PRODUCT CONSTRUCTION. Sequences s are determined by two random elements $x, y \in GF(2^{r/2})$. For $1 \le i \le \ell$, s_i is defined as the scalar product (x^i, y) where x^i is the i-th power of x as an element of $GF(2^{r/2})$.

Theorem 13 [1]. *The above three constructions produce ε-biased distributions on sequences of length ℓ with $\varepsilon = \frac{\ell}{2^{r/2}}$. Each sequence is generated out of r initial bits.*

Combining this with our Main Theorem we get the following result.

Theorem 14. *The families of $n \times m$ Toeplitz matrices resulting from each of the above three constructions by putting $l = n + m - 1$ constitute δ-otp-secure hashing schemes with $\delta = \frac{1}{2^n} + \frac{n+m-1}{2^{r/2}}$ and with each matrix described by a key of length r.*

Notice that for any fixed value of n, one can choose r such that δ gets as close as desired to the optimal value of 2^{-n}. In particular, for $r = 2n + 2\log(n + m - 1)$ one gets $\delta = 2^{-n+1}$ (notice that r increases only logarithmically with the size of messages). By choosing $r = 2n$ one gets $\delta = \frac{n+m}{2^n}$. In particular, for the LFSR construction the later parameters coincide exactly with the construction in [4]. Interestingly enough, the general bound proved here results in a just slightly larger bound than the $m/2^n$ bound derived in [4] specifically for the LFSR construction.

Finally, we remark that additional constructions of ε-biased distributions exist, some of them based on techniques from error correcting codes. In particular, using dual BCH codes. See [7, 1].

310

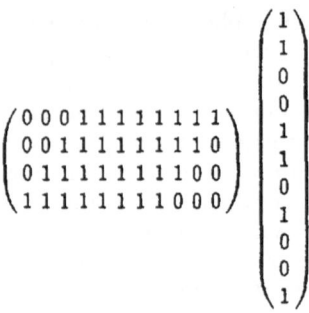

Fig. 2. The product $A_M \cdot s$ corresponding to $M = 1^8$ and $s = 11001101001$

Acknowledgments: I thank Oded Goldreich for helpful conversations regarding small biased distributions.

References

1. Noga Alon, Oded Goldreich, Johan Hastad, and Rene Peralta. Simple constructions of almost k-wise independent random variables. In 31th *Annual Symposium on Foundations of Computer Science, St. Louis, Missouri*, pages 544–553, October 1990.
2. Bierbrauer J., Johansson T., Kabatianskii G., and Smeets, B., "On Families of Hash Functions via Geometric Codes and Concatenation", *Crypto'93*
3. G. Even. *Construction of small probability spaces for simulation.* M.Sc. thesis, Dept. of Computer Science, Technion, August 1991.
4. Krawczyk, H., "LFSR-based Hashing and Authentication", *Advances in Cryptology – CRYPTO 94 Proceedings, Lecture Notes in Computer Science Vol. 839, Springer-Verlag, Y. G. Desmedt, ed* 1994, pp. 129-139.
5. E. Kushilevitz and Y. Mansour. "Learning decision trees using the Fourier spectrum", *SIAM Journal on Computing* 22(6) 1331-1348, December 1993.
6. Mansour, Y., Nisan, N., and Tiwari, P., "The Computational Complexity of Universal Hash Functions", *Theoretical Computer Science*, 107(1):121–133, 1993.
7. Joseph Naor and Moni Naor. Small bias probability spaces: efficient construction and applications. *SIAM Jour. on Computing*, Vol. 22, No. 4, 1993, pp. 838-856.
8. Stinson, D.R., "Universal hashing and authentication codes", *Proc. of Crypto'91*, pp. 74-85.
9. Vazirani, U.V., "Randomness, Adversaries and Computation", Ph.D. Thesis, EECS, UC Berkeley, 1986.
10. Wegman, M.N., and Carter, J.L., "New Hash Functions and Their Use in Authentication and Set Equality", *JCSS*, 22, 1981, pp. 265-279.

A^2−codes from universal hash classes

Jürgen Bierbrauer
Department of Mathematical Sciences
Michigan Technological University
HOUGHTON, MI 49931 (USA)

Abstract

We describe a general method to construct codes for unconditional authentication with arbitration (A^2-codes), which protect not only against outside opponents but also against certain types of frauds from the receiver and transmitter. The constructions are based on orthogonal arrays and universal hash classes. The idea is to construct A^2-codes out of pairs of A-codes. The hitherto known examples are special cases of the construction. Along the way we also construct new universal hash classes.

1 Introduction

It is the purpose of the traditional theory of unconditional authentication (A-codes) to protect the *transmitter* and *receiver* from deception by an outside opponent. It is assumed that transmitter and receiver trust each other. Simmons [9, 10] extended this model to include protection against certain frauds by transmitter and receiver. This model uses an *arbiter*, who distributes partial keys to transmitter and receiver and decides in cases of controversy between transmitter and receiver. It is assumed that the arbiter is trustworthy. The corresponding systems have been termed A^2-codes by Simmons [10].

Let us first consider A-codes: The basic ingredients in their construction are universal hash classes, in particular $\epsilon - ASU_2$ classes.

Definition 1 *Let $\epsilon > 0$. A multiset $\mathcal{A}$ of b functions from a k-set S to a v-set E is $\epsilon-$ almost strongly universal$_2$ (short: ASU_2) if*

1. *for every $u \in S$ and $x \in E$ the number of elements of $\mathcal{A}$ mapping $u \mapsto x$ is b/v,*

2. *for every pair $u_1, u_2 \in S, u_1 \neq u_2$, and every pair $x_1, x_2 \in E$ the number of elements of $\mathcal{A}$ affording the operation $u_1 \mapsto x_1, u_2 \mapsto x_2$ is $\leq \epsilon \cdot b/v$.*

L.C. Guillou and J.-J. Quisquater (Eds.): Advances in Cryptology - EUROCRYPT '95, LNCS 921, pp. 311-318, 1995.
© Springer-Verlag Berlin Heidelberg 1995

It is easy to see that $\epsilon \geq 1/v$. The special case $\epsilon = 1/v$ is known as an orthogonal array of strength 2, denoted as $OA_\lambda(2, k, v)$. Here $b = \lambda \cdot v^2$.

In *authentication without secrecy* it is the case that secrecy either cannot be tolerated or is provided by different means. Technically this means that every message sent through the public channel has to correspond to a unique source state. An A-code may be described as an array $\mathcal{A}$ whose columns correspond to the source states, whose entries are the messages sent through the channel, and whose rows correspond to the keys. We may describe $\mathcal{A}$ equivalently as a set of mappings (the rows) : $S \longrightarrow M$ from source states to messages. The mappings f have to be injective. If secrecy has to be excluded, then this translates into the property that no entry occurs in more than one column. In particular, if $\mid S \mid = k$ and v different entries occur in each column, then $\mid M \mid = k \cdot v$. Simmons [10] uses the term *Cartesian* in this case, Stinson [11] calls these structures *perfectly disclosed*. In the opposite case, when the set of entries is the same for each column, let us speak of ·a *(perfectly) enclosed* array. Universal hash classes are perfectly enclosed arrays. We shall come back to the relation between enclosed and disclosed arrays in the next section.

We describe the setup of A^2-codes: The transmitter has a set E_T of mappings : $S \longrightarrow M$. These are injective and form a perfectly disclosed array. We can therefore write $M = S \times E$ and write each $F \in E_T$ in the form $F(s) = (s, f(s))$. The mappings f form then the perfectly enclosed array $en(E_T)$ of mappings : $S \longrightarrow E$. Put $\mid S \mid = k, \mid E \mid = v$. The receiver has a set E_R of mappings $G : M \longrightarrow \{0, 1\}$. The set of admissible pairs $E_R \circ E_T$ is chosen such that for every $(G, F) \in E_R \circ E_T$ and every source state $s \in S$ we have $G(F(s)) = 0$. The receiver, upon seeing the message m, will accept it as authentic if $G(m) = 0$. At each unit of time the arbiter distributes a pair of encoding rules F to the transmitter and G to the receiver. Here the pair $(G, F) \in E_R \circ E_T$ is chosen according to the uniform distribution. The threats against which the system yields protection (in probability) are the following: impersonation by the opponent, substitution by the opponent, impersonation by the receiver, substitution by the receiver, impersonation by the transmitter. Let $p_I, p_S, p_{R_0}, p_{R_1}, p_T$ be the corresponding probabilities of success with the corresponding attacks when using an optimal strategy. Bounds on these probabilities are obtained from (and are in fact equivalent to) combinatorial properties of $en(E_T), E_R$ and $E_R \circ E_T$. For a detailed discussion see for example [6]. Our construction is based on the idea of choosing $en(E_T)$ as a pair $(\mathcal{A}_1, \mathcal{A}_2)$ of ASU_2 universal hash classes (A-codes, that is) and of letting E_R be determined by another such universal hash class, which is closely related to $\mathcal{A}_2$ (in fact by $E_R = \overline{\mathcal{A}_2}$ in the notation of Lemma 1). The choice of $E_R \circ E_T$ will be more or less forced. Our A^2-codes are thus obtained from pairs $(\mathcal{A}_1, \mathcal{A}_2)$ of A-codes. This construction contains earlier examples given by T.Johansson and K.Kurosawa as special cases ([6, 8]).

2 Enclosed and disclosed arrays

We wish to point out in this section that the distinction between enclosed and disclosed arrays is mathematically meaningless. This is illustrated by showing that a famous geometrical authentication code as constructed by Gilbert, McWilliams and Sloane [4] is simply the extended 2-dimensional Reed-Solomon code in disguise.

Definition 2 *Let $\mathcal{A}$ be a multiset of mappings from the k-set S to the set M of entries. Alternatively we use the array notation introduced above. For each column $s \in S$ let $A(s)$ be the set of different entries in column s. Assume that $\mid A(s) \mid = v$ for every s.*

- *Fix a v-set E and bijections $\phi_s : A(s) \longrightarrow E$ for every column s and define $\mathcal{A}' = \{f' \mid f \in \mathcal{A}\}$, where*

$$f'(s) = \phi_s(f(s)).$$

Call $\mathcal{A}'$ an enclosure of $\mathcal{A}$ and write $\mathcal{A}' = en(\mathcal{A})$.

- *Fix a family E_s of pairwise disjoint v-sets, fix bijections $\phi_s : A(s) \longrightarrow E_s$ for every column s and define an array $\mathcal{A}'' = \{f'' \mid f \in \mathcal{A}\}$, where*

$$f''(s) = \phi_s(f(s)).$$

Call $\mathcal{A}''$ a disclosure of $\mathcal{A}$ and write $\mathcal{A}'' = dis(\mathcal{A})$.

It is an elementary but important fact that the set of probabilities of deception is not changed if the entries in some column of $\mathcal{A}$ are permuted or even replaced by different entries. In this sense every array $\mathcal{A}$ is equivalent to $en(\mathcal{A})$ and to $dis(\mathcal{A})$. We find it more natural to handle and to construct enclosed $\mathcal{A}$-codes. In case of authentication without secrecy one should simply use the corresponding disclosure. $\epsilon - ASU_2$ classes are $\mathcal{A}$-codes in enclosed form.

There is a vast literature on so-called *geometric authentication systems*. They come by construction in disclosed form. If one considers the corresponding (and, as we know, formally equivalent) enclosed arrays, it turns out that these are either well-known or uninteresting. We give just one example:

The first geometric authentication scheme was proposed in [4]: Let q be a prime-power, $\mathcal{P}$ a projective plane of order q. Fix a line g_0 of $\mathcal{P}$. Construct a perfectly disclosed array $\mathcal{A}$ with q^2 rows in the following way:

- the columns are indexed by the points on g_0.

- the entries are indexed by the lines $\neq g_0$.

- the rows are indexed by the affine points ($\in \mathcal{P} - g_0$).

If P is the source state, Q the key, then the corresponding message is the line PQ. We see that $\mathcal{A}$ is perfectly disclosed. In fact $\mathcal{A}(P)$ is the set of all lines $\neq g_0$ through P. As lines of $\mathcal{P}$ intersect in precisely one point, it is clear that the enclosure $en(\mathcal{A})$ is an array with only q different entries and the property that every ordered pair of entries occurs in every ordered pair of different columns precisely once. Thus $en(\mathcal{A})$ is an orthogonal array $OA_1(2, q+1, q)$. We have rediscovered the 2-dimensional extended Reed-Solomon codes.

3 JK-systems

We present a method of construction for A^2–codes based on $\epsilon - ASU_2$-classes of hash functions (A–codes). We call these structures JK–systems in honour of Thomas Johansson and Kaoru Kurosawa who first studied special cases of this construction. We need the following recursive construction for $\epsilon - ASU_2$ classes of hash functions:

Lemma 1 *Let q be a prime-power, $\mathcal{A}$ an $\epsilon - ASU_2$ class of b hash functions : $S \longrightarrow \mathbb{F}_q$. We define a class $\overline{\mathcal{A}}$ of $q \cdot b$ hash functions : $S \times \mathbb{F}_q \longrightarrow \mathbb{F}_q$ as follows: The functions of $\overline{\mathcal{A}}$ are pairs $(f, z), f \in \mathcal{A}, z \in \mathbb{F}_q$, the value of (f, z) at (s, u) is $f(s) + z \cdot u$.*
Then $\overline{\mathcal{A}}$ is an $\epsilon - ASU_2$ class of hash functions.

This is easily proved. Consider the special case when $\mathcal{A}$ is an orthogonal array of strength 2 (equivalently $\epsilon = 1/q$). Then $\mathcal{A}$ has parameters $OA_\lambda(2, k, q)$, where $\mid S \mid = k, b = q^2 \cdot \lambda$. Repeated application of the lemma yields orthogonal arrays $OA_{\lambda \cdot q^i}(2, k \cdot q^i, q)$ for every natural number i. Let in particular $\mathcal{A}$ be the 2-dimensional extended Reed-Solomon code. This is an $OA_1(2, q+1, q)$. Our construction yields $OA_{q^{n-1}}(2, q^n + q^{n-1}, q)$ for every n. This is not an achievement however as the simplex code (the dual of the Hamming code) is an $OA_{q^{n-1}}(2, q^n + q^{n-1} + \ldots + q + 1, q)$.
More interesting is an application to the $\frac{k}{q} - ASU_2$ classes of hash functions as constructed in [7]. In [2] we generalized this to $\frac{k}{q^m} - ASU_2$ classes of q^{n+m} hash functions from a q^{kn}-set to a q^m-set, whenever $n \geq m$. Here we obtain a generalization in a different direction:

Proposition 1 *For every prime-power q and natural numbers k, i there is a $\frac{k}{q} - ASU_2$ class of q^{2+i} hash functions from a q^{k+i}-set to a q-set.*

We come to our basic construction:

Definition 3 *Let q be a prime-power, let $\mathcal{A}_i$ be an $\epsilon_i - ASU_2-class$ of hash functions : $S \longrightarrow \mathbb{F}_q, i = 1, 2$. Assume that $\mathcal{A}_2$ is a vector space over $\mathbb{F}_q$ and $\mathcal{A}_1 \subseteq \mathcal{A}_2$. The $JK-system$ $JK(\mathcal{A}_1, \mathcal{A}_2, q)$ consists of $\mathcal{A}_1, \mathcal{A}_2$ and the set $\mathcal{T}$ of admissible triples. Here a triple (f_1, f_2, g), where $f_i \in \mathcal{A}_i, g = (f_2', z) \in \overline{\mathcal{A}_2}$ is admissible if*

$$f_2 = f_2' + z \cdot f_1.$$

$JK-$systems are thus constructed using pairs $\mathcal{A}_1, \mathcal{A}_2$ of $\epsilon_i - ASU_2-$class of hash functions. As discussed in the Introduction each such universal hash class is equivalent to a traditional authentication code, yielding security against intruders but no protection against frauds by insiders. The $JK-$system uses the properties of $\mathcal{A}_1$ to protect against certain frauds by the receiver. The properties of $\mathcal{A}_2$ yield protection against frauds by the outside opponent. Finally, the choice of admissible triples guarantees protection against the transmitter using a signature different from his own. All this will be discussed in greater detail below.

We remark that the construction is not as flexible as it might appear. In fact, as $\mathcal{A}_2$ is linear, we have $\epsilon_2 = 1/q$. Its generator matrix consists of k projectively inequivalent columns. This means that $\mathcal{A}_2$ is a shortened simplex code. Given $k = \mid S \mid$, an $\mathcal{A}_2$ of minimum dimension may be obtained as follows: Let n be minimal such that $k \leq q^n + q^{n-1} + \ldots + q + 1 = (q^{n+1} - 1)/(q-1)$. Then $\mathcal{A}_2$ may be chosen as a shortening of the $(n+1)-$dimensional Simplex code.

The system $JK(\mathcal{A}_1, \mathcal{A}_2, q)$ is used as an A^2-code in the following way: at each unit of time the arbiter chooses an admissible triple $(f_1, f_2, g = (f_2', z))$ at random (according to the uniform distribution on the set of admissible triples). He distributes the partial keys: (f_1, f_2) to the transmitter, $g = (f_2', z)$ to the receiver, via secure channels. The following is the legitimate use of the system by transmitter and receiver: the transmitter, wishing to send source state $s \in S$, encodes it into the message $m = (s, f_1(s), f_2(s)) = F(s)$ and sends this via the public channel. The receiver, upon seeing a message (s, x, y), accepts it as authentic if $g(s, x) = y$. Thus $G(s, x, y) = g(s, x) - y$ in the notation of the introduction. If $m = (s, f_1(s), f_2(s))$ is unaltered, then $g(s, f_1(s)) = f_2'(s) + z \cdot f_1(s) = f_2(s)$ as the triple was admissible. We determine the basic parameters of this A^2-code:

Theorem 1 *If $JK(\mathcal{A}_1, \mathcal{A}_2, q)$ is used as an A^2-code as above, where $\mathcal{A}_1$ is an $\epsilon_1 - ASU_2$ class of b_1 hash functions and $\mathcal{A}_2$ is obtained by shortening the $(n+1)-dimensional$ Simplex code, where $k \leq (q^{n+1}-1)/(q-1)$, then the sizes of the transmitter's and receiver's key space are $\mid E_T \mid = b_1 \cdot q^{n+1}, \mid E_R \mid = q^{n+2}$, the number of admissible triples is $\mid E_R \circ E_T \mid = q^{n+2}b_1$. The probabilities of deception are*

$$p_I = p_{R_0} = p_T = 1/q, p_S \leq 1/q, p_{R_1} \leq \epsilon_1.$$

proof: The statements concerning the sizes of the key spaces are obvious. We concentrate on the more interesting of the probabilities of deception: p_S (substitution by the opponent), p_{R_1} (substitution by the receiver) and p_T (impersonation by the transmitter):

Substitution by the opponent: The opponent observes a legitimate message $(s, f_1(s), f_2(s))$, changes this into (s', x, y) for some $s' \neq s$. He succeeds if this is accepted as authentic by the receiver. Security is based on the properties of $\mathcal{A}_2$ only. So let us assume that the opponent guessed correctly the values of $x = f_1(s')$ and of z. He has to guess the value of $f_2(s', f_1(s')) = f_2'(s') + z \cdot f_1(s')$, equivalently of $f_2'(s')$, based on the knowledge of $f_2(s) = f_2'(s) + z \cdot f_1(s)$, equivalently of $f_2'(s)$. As $s \neq s'$ and as f_2' is chosen from $\mathcal{A}_2$ according to the uniform distribution, we have $p_S \leq 1/q$.

Substitution by the receiver: The receiver R, upon seeing a legitimate message $(s, f_1(s), f_2(s))$, claims having received (s', x, y) instead for some $s' \neq s$. He succeeds if $x = f_1(s')$. Clearly he will use $y = g(s', x)$. Security is based on the properties of $\mathcal{A}_1$. It follows from the basic rule $f_2 = f_2' + z \cdot f_1$ that knowledge of the partial key (f_2', z) will give R no information on f_1. He thus has to guess $f_1(s')$ on the basis of his knowledge of $f_1(s)$ only. It follows $p_{R_1} \leq \epsilon_1$.

Impersonation by the transmitter: The transmitter T sends a message (s, x, y), where $x \neq f_1(s)$ (i.e. using a signature different from his own). In order to succeed he has to guess the value of $g(s, x) = f_2'(s) + z \cdot x$. As T does know $f_2'(s) + z \cdot f_1(s) = f_2(s)$ we see that T has to guess equivalently the value of $z \cdot (x - f_1(s))$. However, as $x \neq f_1(s)$, this is equivalent to guessing z. As T has no a priori information on z, this shows $p_T = 1/q$. ∎

The A^2–codes which have hitherto been proposed are special cases of this construction: Johansson's construction I is $JK(\mathcal{A}, \mathcal{A}, q^m)$, where $\mathcal{A} = OA_1(2, q^n, q^m), n \leq m$ is a shortened Reed-Solomon code. His construction II is $JK(\mathcal{A}, \mathcal{A}, q)$, where $\mathcal{A} = OA_{q^{n-1}}(2, q^n, q)$ is a member of a well-known family of OA of strength 2. We have seen above that the number of columns can be further increased. Kurosawa's first construction coincides with Johansson's construction II. Kurosawa's second construction is much more interesting. Here $\mathcal{A}_2$ is $OA_{q^{n-1}}(2, q^n, q)$ as before, but $\mathcal{A}_1$ is the (non linear) $t/q - ASU_2$ class of q^2 functions considered above.

4   The Wegman&Carter-effect

It is a fundamental observation, due to Wegman and Carter (see [3, 12]) that the number of rows (functions) of an $\epsilon - ASU_2$ class of functions from a k-set into a v-set may be dramatically decreased if ϵ, instead of equaling the theoretical minimum $1/v$ (this is the case of orthogonal arrays of strength 2), is allowed to be slightly larger. The latest developments in this area may be found in [2]. We use JK-systems $JK(\mathcal{A}_1, \mathcal{A}_2, q)$ to show that the same effect occurs for A^2−codes. Unfortunately this is only partially true. We remarked already that $\mathcal{A}_2$ has to be an orthogonal array of strength 2. We can however reduce b_1 by choosing $\mathcal{A}_1$ to be an $\epsilon_1 - ASU_2$ class of hash functions with ϵ slightly bigger than $1/q$. Kurosawa's second system is a first example. In his case $\epsilon_1 = t/q$ for an integer $t > 1$. We present a large class of JK-systems, where $\mathcal{A}_1$ is obtained by *Stinson-composition* from a Reed-Solomon code and an orthogonal array (see [1]). The technical problem is to choose the $\mathcal{A}_i$ such that $\mathcal{A}_1 \subseteq \mathcal{A}_2$.

Let $\mathbb{F}_Q$ be the ground field, where $Q = q^m$ for some prime-power q. Let $\mid S \mid = q^{nt}$ for some t and $n \geq m$. Assume further that $m \mid nt$. An explicit description of the $\epsilon_1 - ASU_2$ class of hash functions : $S \longrightarrow \mathbb{F}_Q$ obtained via Stinson composition as promised above is the following (see [1]):

We have $b_1 = q^{2n+m}$. Write the elements (functions) of $\mathcal{A}_1$ as triples (α, β, γ), where $\alpha, \beta \in \mathbb{F}_{q^n}, \gamma \in \mathbb{F}_{q^m}$. Write the elements of $\mathbb{F}_{q^{nt}}$ as polynomials $p(X)$ of degree $\leq t - 1$, with coefficients in $\mathbb{F}_{q^n}$. Function (α, β, γ) maps $p(X)$ into $\Phi(p(\alpha) \cdot \beta) + \gamma$, where $\Phi : \mathbb{F}_{q^n} \longrightarrow \mathbb{F}_{q^m}$ is a surjective $\mathbb{F}_q$−linear mapping. We have seen in [1] that $\mathcal{A}_1$ is an $\epsilon_1 - ASU_2$ class, where $\epsilon_1 < \frac{1}{Q} + \frac{t-1}{q^n}$.

Let $q^{nt} = Q^l$. We claim that $\mathcal{A}_1 \subseteq \mathcal{A}_2$, where $\mathcal{A}_2 = OA_{Q^{l-1}}(2, Q^l, Q)$ is one of the aforementioned orthogonal arrays: its functions are given by pairs $(x, z), x \in \mathbb{F}_{Q^l}, x \in \mathbb{F}_Q$, mapping $u \in \mathbb{F}_{Q^l}$ into $\phi(u \cdot x) + z$. Here $\phi : \mathbb{F}_{Q^l} \longrightarrow \mathbb{F}_Q$ is a surjective $\mathbb{F}_Q$−linear mapping. The reader can easily convince himself that this does define an orthogonal array with the given parameters. We want to show that each function (α, β, γ) of $\mathcal{A}_1$ also is a function $(x, z) \in \mathcal{A}_2$. This means that x, z have to be chosen such that

$$\Phi(p(\alpha) \cdot \beta) + \gamma = \phi(p(X) \cdot x) + z$$

for every element $p(X) = u \in \mathbb{F}_{Q^l} = \mathbb{F}_{q^{nt}}$. Naturally we will choose $z = \cdot \gamma$. Consider an $\mathbb{F}_{q^n}$−linear surjective mapping $\rho : \mathbb{F}_{q^{nt}} \longrightarrow \mathbb{F}_{q^n}$ (for example the trace) such that ϕ is the concatenation of ρ and Φ. It suffices then to find x such that the following is satisfied:

$$p(\alpha) \cdot \beta = \rho(p(X) \cdot x).$$

The left side defines an $\mathbb{F}_{q^n}$−linear mapping from $\mathbb{F}_{q^{nt}}$ to $\mathbb{F}_{q^n}$. It can therefore be represented in the required fashion. We arrive at the following:

Theorem 2 *Let q be a prime-power, t, n, m natural numbers such that $m \leq n$ and $m \mid nt, \mid S \mid = q^{nt}$. Then there is $JK(\mathcal{A}_1, \mathcal{A}_2, q^m)$ such that $\epsilon_1 < q^{-m} + \frac{k-1}{q^n}, b_1 = q^{2n+m}, b_2 = q^{nt+m}$.*

In particular we have $\epsilon_1 < 2/q^m$ provided $t \leq q^{n-m} + 1$.

We see that it would be very desirable to find a more general construction such that $\mathcal{A}_2$ could be chosen as an $\epsilon_2 - ASU_2$ class of hash functions for arbitrary ϵ_2. This would allow us to also reduce the value of b_2.

References

[1] J.Bierbrauer: *Universal hashing and geometric codes,* to appear in *Designs, Codes and Cryptography.*

[2] J.Bierbrauer, T.Johansson, G.Kabatianskii,B.Smeets: *On families of hash functions via geometric codes and concatenation,* Proceedings of CRYPTO'93, Lecture Notes in Computer Science 773(1994),331-342.

[3] J.L.Carter,M.N.Wegman: *Universal Classes of Hash Functions,* J.Computer and System Sci. 18(1979), 143-154.

[4] E.N.Gilbert,F.J.MacWilliams,N.J.A.Sloane: *Codes which detect deception,* Bell System Technical Journal 53(1974),405-424.

[5] T.Johansson:*Lower bounds on the probability of deception in authentication with arbitration,* in *Proceedings of the 1993 IEEE International Symposium on Information Theory,* San Antonio, USA, page 231.

[6] T.Johansson: *On the construction of perfect authentication codes that permit arbitration,* Proceedings of CRYPTO'93, Lecture Notes in Computer Science 773(1994), 343-354.

[7] T.Johansson,G.Kabatianskii,B.Smeets: *On the relation between A-codes and codes correcting independent errors,* Proceedings Eurocrypt 93, 1-11.

[8] K.Kurosawa: *New bound on authentication code with arbitration,* Proceedings of CRYPTO'94, Lecture Notes in Computer Science 839(1994), 140-149.

[9] G.J.Simmons: *Message authentication with arbitration of transmitter/receiver disputes,* Proceedings Eurocrypt 87, 151-165.

[10] G.J. Simmons: *A cartesian product construction for unconditionally secure authentication codes that permit arbitration,* Journal of Cryptology 2 (1990),77-104.

[11] D.R.Stinson: The combinatorics of authentication and secrecy codes, Journal of Cryptology 2 (1990),23-49.

[12] M.N.Wegman,J.L.Carter: *New hash functions and their use in authentication and set equality,* J.Computer and System Sci. 22(1981),265-279.

A New Identification Scheme
Based on
the Perceptrons Problem

David POINTCHEVAL
David.Pointcheval@ens.fr

Laboratoire d'Informatique
École Normale Supérieure
45, rue d'Ulm
F-75230 PARIS Cedex 05

Abstract. Identification is a useful cryptographic tool. Since zero-knowledge theory appeared [3], several interactive identification schemes have been proposed (in particular Fiat-Shamir [2] and its variants [8, 5, 4], Schnorr [9]). These identifications are based on number theoretical problems. More recently, new schemes appeared with the peculiarity that they are more efficient from the computational point of view and that their security is based on $\mathcal{NP}$-complete problems: PKP (Permuted Kernels Problem) [10], SD (Syndrome Decoding) [12] and CLE (Constrained Linear Equations) [13].

We present a new $\mathcal{NP}$-complete linear problem which comes from learning machines: the Perceptrons Problem. We have some constraints, m vectors X^i of $\{-1, +1\}^n$, and we want to find a vector V of $\{-1, +1\}^n$ such that $X^i \cdot V \geq 0$ for all i.

Next, we provide some zero-knowledge interactive identification protocols based on this problem, with an evaluation of their security. Eventually, those protocols are well suited for smart card applications.

1 Introduction

An interactive identification protocol involves two persons Alice and Bob. Alice wants to prove interactively that she is really Alice. She has a public key which everybody knows, and a secret key associated to her public key. She is the only one who knows the secret key, and nobody can compute it. To prove her identity, Alice proves that she knows a secret key associated to her public key. In general, the public key is a problem, a difficult problem, and the secret key is a solution of this problem.

Recently, the zero-knowledge theory showed that we can prove the knowledge of a solution of a problem without revealing anything about this solution. The verifier learns nothing but the conviction that the prover knows a solution. The

The first efficient zero-knowledge protocols were based on number theoretical problems (Fiat-Shamir [2] and its variants [8, 5, 4], Schnorr [9]). They have two major disadvantages:

L.C. Guillou and J.-J. Quisquater (Eds.): Advances in Cryptology - EUROCRYPT '95, LNCS 921, pp. 319-328, 1995.
© Springer-Verlag Berlin Heidelberg 1995

- The hardness of the problems used (factorization and discrete logarithm) is not proved. Moreover, efficient algorithms and computers threaten them.
- Arithmetic operations are very expensive (modular multiplications, modular exponentiations).

Since 1989, new schemes have appeared, which rely on $\mathcal{NP}$-complete problems, and require only operations over small numbers or even on bits: PKP (Permuted Kernels Problem) [10], SD (Syndrome Decoding) [12] or CLE (Constrained Linear Equations) [13].

This paper introduces another linear scheme based on the Perceptrons Problem, an $\mathcal{NP}$-complete problem, which seems to be well suited for smart card applications.

2 Problems

2.1 The Perceptrons Problem

The following problem appears in physics and the study of the Ising's perceptrons, and in artificial intelligence with the learning machines. We call it *The Perceptrons Problem*.

Notation 1. We call an ε-vector (or matrix) a vector (or matrix) which components are either -1 or $+1$.

Definition 2. The Perceptrons Problem **PP**:

Input   :   an ε-matrix A of size $m \times n$.
Problem :   find an ε-vector Y of size n such that $AY \geq 0$.

It is easy to show that this problem is difficult to solve, even to approximate in the sense of Papadimitriou and Yannakakis [6]. Proofs can be found in [7].

2.2 The Permuted Perceptrons Problem

It is possible to design a zero-knowledge identification protocol with every $\mathcal{NP}$-complete problem provided one-way hash functions are granted. But in order to get one quicker and easier, we will take a variant of this problem:

Definition 3. The Permuted Perceptrons Problem **PPP**

Input   :   an ε-matrix A of size $m \times n$,
               a multiset S of nonnegative integers, of size m.
Problem :   find an ε-vector Y of size n such that
               $\{\{(AV)_j | j = \{1, \ldots, m\} \}\} = S$.

It is clear that a solution for **PPP** is a solution for **PP**. So the Permuted Perceptrons Problem is more difficult to solve than the original Perceptrons Problem.

3 Size of the problem

Secure values of m and n will be precised later with the efficiency of the attacks. But we will see that efficiency of the attacks will depend on the number of solutions of the instance.

So we will first evaluate the number of solutions of an average instance (A, S) of the Permuted Perceptrons Problem of size $m \times n$ when we know there exists at least one solution V.

- On one hand, we can count the solutions for the Perceptrons Problem instance associated. Let $PP(m, n)$ be the average number of solutions for the Perceptrons Problem instance A of size $m \times n$ (it is a big formula).
- On the other hand, we have to evaluate the probability to obtain a given multiset S with the components of the product of A and a given solution of **PP**: $P_{m,n,S}$
  With every multiset S, this probability is less than the probability to obtain the multiset Σ which elements have a Gaussian distribution
  (*i.e.* $|\Sigma|_j = mp_{n,j}$, where $p_{n,j}$ is the probability for two ε-vectors of size n,
  X and Y, to be such that $|X \cdot Y| = j$: $p_{n,j} = 2^{-n+1} \binom{n}{\frac{n+j}{2}} \simeq \sqrt{\frac{8}{\pi n}} e^{-\frac{j^2}{2n}}$
  when $n = j \bmod 2$)

$$P_{m,n,S} \le m! \prod_{j=1}^{n} \frac{p_{n,j}^{mp_{n,j}}}{(mp_{n,j})!}$$

Then the number of solutions for an average instance of **PPP** is less than $PP(m, n) \times P_{m,n,\Sigma}$.

Against the most efficient attack, we will need the smallest number of solutions. Then we will choose m and n such that $PP(m, n) \times P_{m,n,\Sigma} \le 1$

As a consequence, we see that for interesting sizes ($100 < m < 200$), we can approximatively take $n \approx m + 16$.

4 Making the problem practical

4.1 How to get the keys ?

For cryptographic purposes, we only use instances with a known solution. We also want all instances with at least one solution may appear. To get such an instance, we firstly choose an arbitrary ε-vector V, of size n, which will be the solution of the future instance, and an ε-matrix A of size $m \times n$. We next modify it in the following way:

- If $(AV)_i < 0$, we replace the ith row of A by its opposite value.
- If $(AV)_i \ge 0$, then we don't change this row.

Then, we compute $S = \{\{(AV)_i | i = 1, \ldots, m\}\}$. Consequently, (A, S) is an instance of the Permuted Perceptrons Problem, with V as a solution. In addition, with this method, all the "good instances" may appear, with a probability which is proportional to the number of solutions of the **PP** instance.

4.2 Finite field

Also for cryptographic use, we need to bound the size of the numbers used in order to store them on a constant number of bits. So we will work in the finite field with p elements. If we bound the components of the product by a nonnegative odd integer t (i.e. if $T \in \{1, 3, \ldots, t\}^m$), we can prove that if n, p and t are such that $2p > n + t$ we have:

$$AV = T \iff AV = T \bmod p$$

5 Possible attacks

We tried several attacks against **PP** and **PPP** in order to evaluate the security of a possible protocol. But since there is no algebraic structure in those problems, no manipulation of the matrix will leave the problem unchanged (manipulation like Gaussian elimination, used in the past against PKP, CLE or any problem based on error correcting codes will not help here). So, it seems that only (more or less intelligent) exhaustive search or probabilistic attacks would succeed.

5.1 The majority vector

The first attack against **PPP** which comes to mind is the *majority vector* M:
for all j, $\begin{cases} M_j = +1 & \text{if } \#\{i | A_{i,j} = +1\} > \frac{n}{2} \\ M_j = -1 & \text{otherwise} \end{cases}$

Theorem 4. *For an $m \times n$-instance constructed as shown below, with solution V and $m \leq n$, the average Hamming distance between V and M, is roughly*

$$n \cdot \left(\frac{1}{2} - \frac{1}{\pi}\sqrt{\frac{m}{n}}\right) \approx 0.2n.$$

Firstly, we can change 20% of the components of M, and trying the products, but there are $\binom{n}{0.20n}$ such possibilities.

We can already fix a bound for n (and m) to overtake the usual work factor of 2^{64}: $n \geq 95$.

To improve this attack, we could arrange these changes beginning with components of M which values are litigious (i.e. $\#\{i | A_{i,j} = +1\}$ near $\frac{n}{2}$). But surprisingly enough, some components which seem to be very good are false: on average, 80% of the components will have to be handled. So, this improvement doesn't modify the bound on n.

5.2 Simulated annealing

Because of the inefficiency of the previous attack, we tried the well-known probabilistic algorithm that comes from artificial intelligence, known as *simulated annealing* [11]. This attack tries to minimize a function, *Energy*, defined on a finite metric space, in a probabilistic way. Simulated annealing algorithms are an improvement of gradient descent algorithms. Whereas gradient descent algorithms can converge to a local minimum and stay there, simulated annealing algorithms try to go away with some small random perturbations. These perturbations which may be important at the beginning have to decrease to zero.

Such an algorithm can be efficient only if the *Energy* function is roughly "continuous" (the difference between the images of two neighbors is bounded by a small number). For this reason, simulated annealing doesn't seem to be well suited for **PPP** because of the multiset, but it should be perfect for **PP** with

$$E(V) = \frac{1}{2} \sum_{i=1}^{m} (|(AV)_i| - (AV)_i)$$

This algorithm turned out to be the most efficient. We have carried out many tests on square matrices $(m = n)$, and on some other sizes, and during a day, we can find a solution for any instance of **PP** which size is less than about 200.

Those attacks have been running for a few months and we never find a solution for **PPP** for sizes greater than 71.

If we suppose that each of those solutions can appear with the same probability, we have to repeat this attack about $0.7 \times \#\{$solutions of **PP**$\}$ rounds, in order to find the good solution with probability 0.5.

Then, we can evaluate the work factor of such an attack with probability of success equal to 0.5:

size	number of solutions **PP**	time for a solution (seconds)	time Solution $Pr = 1/2$ (seconds)	Work factor* 2^n elementary operations
101×117	$4.7 \ 10^9$	85	399.10^9	64
121×137	$8.7 \ 10^{10}$	130	11.10^{12}	68
151×167	$3.7 \ 10^{12}$	180	666.10^{12}	74

* work factor estimated using a 60-70 MIPS processor speed

Then, we can say that even small sizes are secure enough. In addition, whatever the probabilistic attack, it will not be able to differentiate the good solution of **PPP** from any solution of **PP**. So, even if we supposed a quick attack for **PP** (an $\mathcal{NP}$-complete problem) which would need only 1 second to find a solution for a 141×157-sized instance, the work factor would remain above 2^{64}.

6   Practical values

With those results, we can suggest $m = 101$ and $n = 117$ as a secure size of the problem. So, in the average instance, $|S|_1 = 15$, $|S|_3 = 14$, etc.

$$\Pr_{Instances} [Max\ S > 33] < .3$$

Then we can suppose that there is no greater number than 33, (*i.e.* $t = 33$). We must take $p > 75$, and then we take $p = 127$ (it will optimize the probability from a cheater to be rejected).

7   Protocols

Common data: some integers p, n, t such that $2p > t + n$ and a collision-free, random hash function H.

Let A be a matrix of size $m \times n$, a Perceptrons Problem instance, with V as a solution. Let S be the multiset of the components of AV.
Public key : (A, S)
Secret key : V

•The prover selects
- a random permutation P over $\{0, \ldots, m - 1\}$ (to mix the rows of A.)
- a random signed permutation Q over $\{0, \ldots, n - 1\}$
  (to mix the columns of A, and to multiply them randomly by $+1$ or -1.)
- a random vector W of $\mathbb{F}_p^n$

7.1   Three pass identification protocol (3p zk)

1. The prover computes $A' = PAQ$, $V' = Q^{-1}V$, $R = W + V'$
   and $h_0 = H(P|Q)$, $h_1 = H(W)$, $h_2 = H(R)$, $h_3 = H(A'W)$, $h_4 = H(A'R)$
   and sends $(h_0, h_1, h_2, h_3, h_4)$ to the verifier.
2. The verifier randomly selects c in $\{0, \ldots, 3\}$ and sends c to the prover.
3. The prover sends: 4. The verifier checks:

if $c = 0 : (P, Q, W)$ $h_0 = H(P|Q)$, $h_1 = H(W)$ and $h_3 = H(PAQW)$.

if $c = 1 : (P, Q, R)$ $h_0 = H(P|Q)$, $h_2 = H(R)$ and $h_4 = H(PAQR)$.

if $c = 2 : (A'W, A'V')$ $h_3 = H(A'W)$, $h_4 = H(A'W + A'V')$
                            and $\{\{(A'V')_i\}\} = S$.

if $c = 3 : (W, V')$ $h_1 = H(W)$, $h_2 = H(W + V')$
                            and $V' \in \{-1, +1\}^n$.

7.2   Properties

***Theorem 5.** The 3p zk protocol is an Interactive Proof System for **PPP**.*

Lemma 6. *Assume that some probabilistic polynomial-time adversary is accepted with probability greater than* $\left(\dfrac{3}{4}\right)^r + \epsilon$ *after* r *rounds, then there exists a polynomial-time probabilistic machine which extracts the secret key* S *from the public data or outputs collisions for the commitment function, with overwhelming probability.*

Proof. Consider the tree $T(\omega)$ of all 4^k executions corresponding to all possible questions of the verifier over k rounds when the adversary has a fixed random tape ω.

$$\alpha = \Pr_{\omega}[T(\omega) \text{ has a vertex with 4 sons}]$$

It is clear that $\alpha \geq \epsilon$, and by resetting the adversary $\frac{1}{\epsilon}$ times, one finds, with constant probability, an execution tree with a vertex having 4 sons. Repeating again, this probability can be made very close to one.

A vertex with 4 sons corresponds to a situation where 5 commitments h_0, h_1, h_2, h_3 and h_4 have been made and where the adversary can provide answers to the 4 possible queries of the verifier.

Consider answers :

$$H(P_0|Q_0) = h_0 = H(P_1|Q_1)$$
$$H(W_0) \quad = h_1 = H(W_3)$$
$$H(R_1) \quad = h_2 = H(W_3 + V_3')$$

$$H(P_0 A Q_0 W_0) = h_3 = H(Y_2)$$
$$H(P_1 A Q_1 R_1) = h_4 = H(Y_2 + Z_2)$$

Unless we have found a collision for the hash function H, we can consider

$$P = P_0 = P_1 \qquad R \;=\; R_1 \;=\; W_3 + V_3' \;=\; W + V'$$
$$Q = Q_0 = Q_1 \qquad Y \;=\; Y_2 \;=\; P_0 A Q_0 W_0 = PAQW$$
$$W = W_0 = W_3 \qquad Y + Z = Y_2 + Z_2 = P_1 A Q_1 R_1 = PAQR$$

such that $V' \in \{-1,+1\}^n$ and $\{\{Z_i\}\} = S$.

so $Y + Z = PAQR = PAQW + Z = PAQW + PAQV'$, then $Z = PAQV'$.

Let $V = QV'$ ($V \in \{-1,+1\}^n$), then $Z = PAV$, Consequently $\{\{(AV)_i\}\} = S$.

7.3 Five pass identification protocol (5p zk)

1. The prover computes $A' = PAQ$, $V' = Q^{-1}V$
   and $h_0 = H(P|Q)$, $h_1 = H(W|V')$, $h_2 = H(A'W|A'V')$
   and sends (h_0, h_1, h_2) to the verifier.
2. The verifier randomly selects k in $\mathbb{F}_p^*$ and sends k to the prover.
3. The prover computes $R = kW + V'$ and $h_3 = H(R)$, $h_4 = H(A'R)$
   and sends (h_3, h_4) to the verifier.
4. The verifier randomly selects c in $\{0,1,2\}$ and sends c to the prover.

5. The prover sends: 6. The verifier checks:

if $c = 0 : (P, Q, R)$	$h_0 = H(P|Q)$, $h_3 = H(R)$ and $h_4 = H(PAQR)$
if $c = 1 : (A'W, A'V')$	$h_2 = H(A'W|A'V')$, $h_4 = H(kA'W + A'V')$
	and $\{\{(A'V')_i\}\} = S$.
if $c = 2 : (W, V')$	$h_1 = H(W|V')$, $h_3 = H(kW + V')$
	and $V' \in \{-1,+1\}^n$

7.4 Properties

Theorem 7. *The **5p zk** protocol is an Interactive Proof System for **PPP**.*

Lemma 8. *Assume that some probabilistic polynomial-time adversary is accepted with probability greater than* $\left(\dfrac{2p-1}{3(p-1)}\right)^{r} + \epsilon$ *after r rounds, then there exists a polynomial-time probabilistic machine which extracts the secret key S from the public data or outputs collisions for the commitment function, with overwhelming probability.*

Using the idea of resettable simulation [3], in the random oracle model [1], it can be shown that both protocols are zero-knowledge. Alternatively, one has to assume specific statistical independance properties for the hash function.

7.5 Light versions

A light version (**3p light** and **5p light**) of those protocols, which reduces the number of required rounds, can be designed.

Five pass identification protocol (5p light)
The initialization is the same as in the previous schemes.
1. The prover computes $A' = PAQ$, $V' = Q^{-1}V$
   and $h_0 = H(P|Q)$, $h_1 = H(W|A'W|V'|A'V')$
   and sends (h_0, h_1) to the verifier.
2. The verifier randomly selects k in $\mathbb{F}_p^*$ and sends k to the prover.
3. The prover computes $R = kW + V'$ and $h_2 = H(R|A'R)$
   and sends h_2 to the verifier.
4. The verifier randomly selects c in $\{0, 1\}$ and sends c to the prover.
5. The prover sends: 6. The verifier checks:
   if $c = 0 : (P, Q, R)$ $h_0 = H(P|Q)$, $h_2 = H(R|PAQR)$
   if $c = 1 : (W, V', A'W, A'V')$ $\{\{(A'V')_i\}\} = S,$
                                 $V' \in \{-1, +1\}^n$
                                 $h_2 = H(X|Y)$ with
                                     $X = kW + V'$
                                     $Y = kA'W + 2T + U$

These light protocols are no longer zero-knowledge. However, the information released appears quite small. In fact, in the case $c = 1$, the verifier learns two vectors and their images by A', then he can deduce something about A', and then about P and Q. As he knows V', he theoretically learns a fraction of bit of V. Since the permutations P and Q are different at each round, the given information seems unusable.

8 Performances

The performances of this scheme are similar to those of the already existing linear ones:

	SD Stern	CLE Stern	PKP Shamir	**PPP** 3p ZK	**PPP** 5p ZK
matrix size	256×512	24×24	37×64	101×117	
over the field	$\mathbb{F}_2$	$\mathbb{F}_{16}$	$\mathbb{F}_{251}$	$\mathbb{F}_2$	
best known attack complexity	2^{68}	2^{52}	$> 2^{100}$	2^{64}	
Number of rounds	35	20	20	48	35
public key (bits)	256	80	296	144	
secret key (bits)	512	80	384	117	
bits sent by round	954	824	832	896	1040
global transmission rate (kbytes)	4.08	2.01	2.03	5.25	4.44

- As we can see in the figure, with a secret key more secure than in some other schemes (work factor of the attack greater than 2^{64}), and with a probability of 10^{-6} for a cheater to be accepted, an identification requires less than 4.5 kbytes of communication between the prover and the verifier (to be compared with the 2 kbytes for PKP and CLE, and the 4 kbytes for SD). And we can improve them by the use of hash trees.
- Moreover, all the operations are no more than additions and subtractions between small integers (less than one byte) even modulo 2. They are well suited to a very minimal environment of 8-bit processors.
- If we use a common matrix M, stored in the seed of a pseudo-random generator, and if the keys are:

  secret key : a random ε-vector V of size n (less than 15 bytes)

  public key : the ε-vector L such that $L_i(MV)_i \geq 0$

        and the multiset $S = \{\{L_i(MV)_i\}\}$ (18 bytes)

  It is very few bytes if we compare them with PKP or SD. But we should not forget that as for PKP, SD and CLE, this scheme is not *identity based*. It means that public keys have to be certified by an authority.
- They require only simple operations so the program is very small. Moreover, the size of the data (common data and keys) are tiny. As a consequence, little EEPROM is needed.
- Very few temporary computations have to be stored so they require very little RAM.

9 Conclusion

We have defined a new identification scheme which is very easy to implement on every kind of smart card because of its very simple operations and the small size of the data. We welcome attacks from readers.

Acknowledgements

I would like to thank Louis Granboulan for the results about the majority vector, and Jacques Stern for fruitful discussions.

References

1. Bellare, M., Rogaway, P.: Random oracles are practical: a paradigm for designing efficient protocols. In Proceedings of the 1st ACM Conference on Computer and Communications Security (1993) pp. 62–73.
2. Fiat, A., Shamir, A.: How to prove yourself: practical solutions of identification and signature problems. In Advances in Cryptology – Proceedings of CRYPTO '86 (1987) vol. Lecture Notes in Computer Science 263 Springer-Verlag pp. 186–194.
3. Goldwasser, S., Micali, S., Rackoff, C.: Knowledge complexity of interactive proof systems. In Proceedings of the 17th ACM Symposium on the Theory of Computing STOC (1985) pp. 291–304.
4. Ohta, K., Okamoto, T.: A modification of the fiat-shamir scheme. In Advances in Cryptology – Proceedings of CRYPTO '88 (1989) vol. Lecture Notes in Computer Science 403 Springer-Verlag pp. 232–243.
5. Ong, H., Schnorr, C.: Fast signature generation with a fiat shamir-like scheme. In Advances in Cryptology – Proceedings of EUROCRYPT '90 (1991) vol. Lecture Notes in Computer Science Springer-Verlag pp. 432–440.
6. Papadimitriou, C., Yannakakis, M.: Optimization, approximation, and complexity classes. Journal of Computer and Systems Sciences **43** (1991) pp. 425–440.
7. Pointcheval, D.: Les réseaux de neurones et leurs applications cryptographiques. Tech. rep. Laboratoire d'Informatique de l'École Normale Supérieure Février 1995. LIENS-95-2.
8. Quisquater, J., Guillou, L.: A practical zero-knowledge protocol fitted to security microprocessor minimizing both transmission and memory. In Advances in Cryptology – Proceedings of EUROCRYPT '88 (1989) vol. Lecture Notes in Computer Science 330 Springer-Verlag pp. 123–128.
9. Schnorr, C.: Efficient identification and signatures for smart cards. In Advances in Cryptology – Proceedings of CRYPTO '89 (1990) vol. Lecture Notes in Computer Science 435 Springer-Verlag pp. 235–251.
10. Shamir, A.: An efficient identification scheme based on permuted kernels. In Advances in Cryptology – Proceedings of CRYPTO '89 (1990) vol. Lecture Notes in Computer Science 435 Springer-Verlag pp. 606–609.
11. Skubiszewski, M.: Optimisation par recuit simulé : mise en œuvre matérielle de la machine de Boltzmann, application à l'étude des suites synchronisantes. PhD thesis Université d'Orsay juin 1993.
12. Stern, J.: A new identification scheme based on syndrome decoding. In Advances in Cryptology – proceedings of CRYPTO '93 (1994) vol. Lecture Notes in Computer Science 773 Springer-Verlag pp. 13–21.
13. Stern, J.: Designing identification schemes with keys of short size. In Advances in Cryptology – proceedings of CRYPTO '94 (1994) vol. Lecture Notes in Computer Science 839 Springer-Verlag pp. 164–173.

Fast RSA-type Schemes
Based on Singular Cubic Curves
$y^2 + axy \equiv x^3 \pmod{n}$

Kenji Koyama

NTT Communication Science Laboratories
2-2, Hikaridai, Seika-cho, Soraku-gun, Kyoto, 619-02 Japan
E-mail: koyama@cslab.kecl.ntt.jp

Abstract

This paper proposes fast RSA-type public-key schemes based on singular cubic curves $y^2 + axy = x^3$ over the ring Z_n. The x and y coordinates of a 2 log n-bit long plaintext/ciphertext are transformed to a log n-bit long shadow plaintext/ciphertext by isomorphic mapping. Decryption is carried out by exponentiating this shorter shadow ciphertext over Z_n. The decryption speed of the proposed schemes is about 2.0 times faster than that of the RSA scheme for a K-bit long message if $\lceil K/\log n \rceil$ is even. We prove that breaking each of the proposed schemes is computationally equivalent to breaking the RSA scheme in one-to-one communication circumstances. We also prove that the proposed schemes have the same security as the RSA scheme against the Hastad attack when linearly related plaintexts are encrypted in broadcast applications.

1   Introduction

In 1991, an RSA-type scheme over elliptic curves, i.e., non-singular cubic curves, was presented by Koyama, Maurer, Okamoto and Vanstone [4]. This scheme, the KMOV scheme for short, is more secure than the RSA scheme [9] against the Hastad attack [2] [6]. The decryption speed of the KMOV scheme, however, is 5.8 times slower than that of the RSA scheme even if rapid computational techniques are used [5].

By changing the base from elliptic curves to singular cubic curves, this paper proposes faster RSA-type schemes based on curves $E_n : y^2 + axy \equiv x^3 \pmod{n}$. The x and y coordinates of a 2 log n-bit long plaintext/ciphertext are transformed to a log n-bit long shadow plaintext/ciphertext by isomorphic mapping. Decryption is carried out by exponentiating this shorter shadow ciphertext over Z_n instead of a sequential addition of the points over singular cubic curves E_n. The decryption speed of the proposed schemes is about 2.0 times faster than that of the RSA scheme for a K-bit long message if $\lceil K/\log n \rceil$ is even. We prove that breaking each of the proposed schemes is computationally equivalent to breaking the RSA scheme. This equivalence in security is guaranteed under usual one-to-one communication circumstances. We also prove that the proposed schemes have the same security as the RSA scheme against the Hastad attack when linearly related plaintexts are encrypted in broadcast applications.

L.C. Guillou and J.-J. Quisquater (Eds.): Advances in Cryptology - EUROCRYPT '95, LNCS 921, pp. 329-340, 1995.
© Springer-Verlag Berlin Heidelberg 1995

The organization of this paper is as follows. Section 2 mentions singular cubic curves over a finite field and a finite ring. In Section 3, we describe new schemes. The efficiency of the proposed and other schemes is discussed in Section 4. The security of the proposed schemes is discussed in Section 5. Section 6 concludes this paper.

2 Singular Cubic Curves

Let F_p be a finite field with p elements and F_p^* be a multiplicative group of F_p, where $p\ (> 3)$ is a prime.

Definition 1 ([3][7]) *A non-singular part of a singular cubic curve, denoted by $E_p(a, b)$, is defined as the set of solutions $(x, y) \in F_p \times F_p$ to Eq.(1), excluding a singular point $(0, 0)$ and including the point at infinity $\mathcal{O}$.*

$$y^2 + axy = x^3 + bx^2 \quad over\ F_p, \quad a, b \in F_p \tag{1}$$

An addition "$\oplus$" on $E_p(a, b)$ is given by the chord-and-tangent law similar to that for elliptic curves.

The sum (x_3, y_3) of (x_1, y_1) and (x_2, y_2) in F_p is computed as

$$\begin{cases} x_3 = & \lambda^2 + a\lambda - b - x_1 - x_2, \\ y_3 = & \lambda(x_1 - x_3) - y_1, \end{cases} \tag{2}$$

where

$$\lambda = \begin{cases} \dfrac{y_2 - y_1}{x_2 - x_1} & \text{if } (x_1, y_1) \neq (x_2, y_2), \\ \dfrac{3x_1^2 + 2bx_1 - ay_1}{2y_1 + ax_1} & \text{if } (x_1, y_1) = (x_2, y_2). \end{cases}$$

Note that $E_p(a, b)$ is a group. Operation $\otimes$ is defined as follows.

$$k \otimes (x, y) = \overbrace{(x, y) \oplus \cdots \oplus (x, y)}^{k\ times} \quad over\ E_p(a, b).$$

A group $E_p(a, b)$ is isomorphic to F_p^*. The isomorphic relationship is generally described in [3] and [7] for curves $(y - \alpha x)(y - \beta x) = x^3$ over F_p^*, where $\alpha, \beta \in F_p^*$, which is equivalent to equation (1) with $a = -\alpha - \beta \bmod p$, $b = -\alpha\beta \bmod p$. When $b = 0$, we can put $\alpha = 0$ and $\beta = -a(\neq 0)$, and the simplified relationship is carried out explicitly in the following theorem.

Theorem 1 *The mapping $\omega : E_p(a, 0) \to F_p^*$ defined by*

$$\omega : \mathcal{O} \mapsto 1, \quad (x, y) \mapsto 1 + \frac{ax}{y} = \frac{x^3}{y^2}$$

is a group isomorphism. The group isomorphism mapping $\omega^{-1} : F_p^* \to E_p(a, 0)$ is defined by

$$\omega^{-1} : 1 \mapsto \mathcal{O}, \quad v \mapsto \left(\frac{a^2 v}{(v - 1)^2}, \frac{a^3 v}{(v - 1)^3} \right).$$

Hence, an order of $E_p(a,0)$, denoted by $\#E_p(a,0)$, is $p-1$.

Let $Z_n = \{0,1,\cdots,n-1\}$ and Z_n^* be a multiplicative group of Z_n. A non-singular part of a singular cubic curve over Z_n is defined as follows.

Definition 2 *Let n be a product of primes p,q (> 3). A non-singular part of a singular cubic curve, denoted by $E_n(a,b)$, is defined as the set of solutions $(x,y) \in Z_n \times Z_n$ to Eq.(3), excluding a singular point $(0,0)$ and including the point at infinity $\mathcal{O}$.*

$$y^2 + axy = x^3 + bx^2 \quad \text{over } Z_n, \quad a,b \in Z_n. \tag{3}$$

An addition on $E_n(a,b)$ is defined by the chord-and-tangent law. Although the addition is not always defined, the probability for such a case is negligibly small for large p and q. By Theorem 1 and the Chinese Remainder Theorem, the following theorem holds.

Theorem 2 *For (x_i, y_i) and (x_1, y_1) satisfying $(x_i, y_i) = i \otimes (x_1, y_1)$ over $E_n(a,0)$, we have*

$$1 + \frac{ax_i}{y_i} \equiv \left(1 + \frac{ax_1}{y_1}\right)^i \pmod{n},$$

i.e.,

$$\frac{x_i^3}{y_i^2} \equiv \left(\frac{x_1^3}{y_1^2}\right)^i \pmod{n}.$$

The following theorem is a base of a pair of an encryption and a decryption of public-key cryptosystems over $E_n(a,0)$.

Theorem 3 *Let n be a product of primes p,q (> 3) and $N = \text{lcm}(p-1, q-1)$. For any integer k satisfying $k \equiv 1 \pmod{N}$, we have*

$$(x,y) = k \otimes (x,y) \quad \text{over } E_n(a,0)$$

with the overwhelming probability for large p and q.

3 New RSA-type Schemes Based on $E_n(a,0)$

We can construct RSA-type public-key schemes over singular cubic curves $E_n(a,b)$ with a message-dependent variable a and a fixed constant b. Considering the computational efficiency among variants of instances of these schemes, we put $b = 0$. We propose two new RSA-type schemes over $E_n(a,0)$: scheme 1 and scheme 2. These proposed schemes can be used in both secret communications and digital signatures. For simplicity, we describe protocols of secret communications.

The security of the proposed schemes is based on the difficulty of factoring n, which is a product of large primes p and q. Let a plaintext (m_x, m_y) be an integer pair, where $m_x, m_y \in Z_n^*$ and $m_x^3 \not\equiv m_y^2 \pmod{n}$. A concept of RSA-type schemes based on isomorphism over singular cubic curves is shown in Figure 1. This figure also includes a flow diagram of scheme 1. In scheme 1, the encryption is carried out over $E_n(a,0)$ along the path from plaintext (m_x, m_y) to ciphertext (c_x, c_y). In scheme 2, the encryption is carried out over Z_n^* along the path from plaintext (m_x, m_y) to shadow ciphertext c via shadow plaintext m. Although the decryption of naive cryptosystems based on cubic curves is computed directly from (c_x, c_y) to (m_x, m_y) over E_n $(= E_p \times E_q)$ in the left half of Figure 1, the decryptions for schemes 1 and 2 are carried out over F_p^* and F_q^* because decryption over F_p^* and F_q^* is faster than that over $E_p(a, 0)$ and $E_q(a, 0)$.

Note that for the original RSA scheme, the encryption and decryption are carried out between (shadow) plaintext m and (shadow) ciphertext c in Z_n^*, more exactly in Z_n, in the right half of Figure 1.

$$E_n \; (= E_p \times E_q) \quad \text{Isomorphism} \quad Z_n^* \; (= F_p^* \times F_q^*)$$

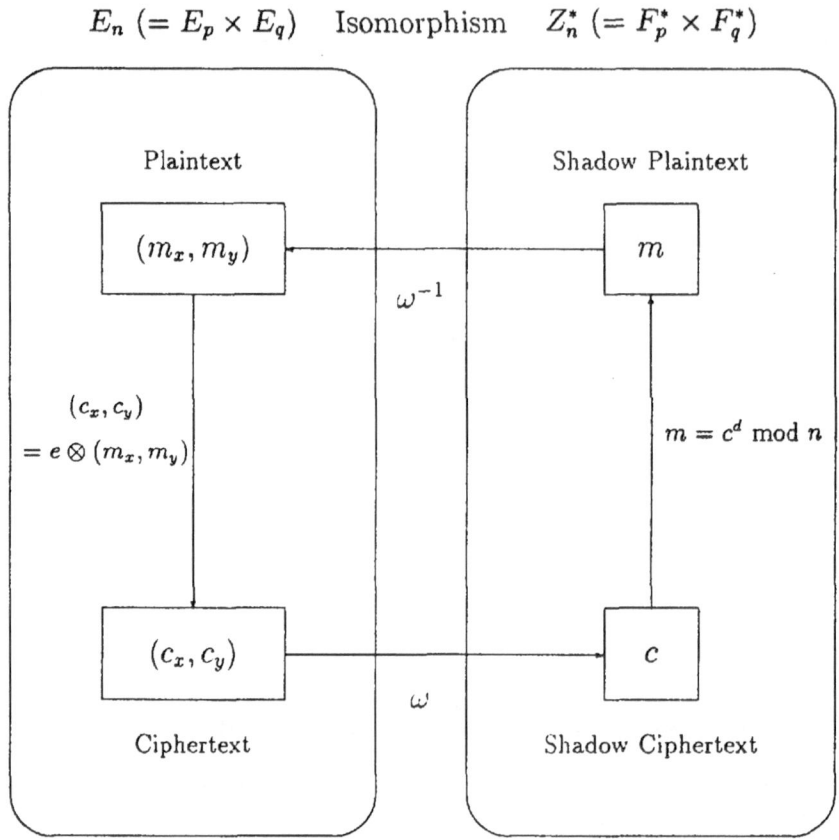

Fig.1 Concept of RSA-type schemes and a flow of scheme 1

3.1 Key Generation of Scheme 1 and Scheme 2

A key generation procedure is common for scheme 1 and scheme 2.

Receiver R chooses two large primes p and q. Let $n = pq$ and $N = \text{lcm}(p-1, q-1)$. R determines an integer e satisfying $\gcd(e, N) = 1$. Decryption keys d_p and d_q are computed from encryption key e as $d_p = \frac{1}{e} \bmod (p-1)$ and $d_q = \frac{1}{e} \bmod (q-1)$, respectively. R's public keys are e and n. R's secret keys are p, q, d_p and d_q.

3.2 Scheme 1

Encryption

Sender S encrypts plaintext (m_x, m_y) with the receiver's public keys e and n as

$$(c_x, c_y) = e \otimes (m_x, m_y) \text{ over } E_n(a, 0),$$

where $a = \dfrac{m_x^3 - m_y^2}{m_x m_y} \bmod n$, and sends a ciphertext (c_x, c_y) to receiver R.

Remark

· Plaintext condition such that $m_x, m_y \in Z_n^*$ and $m_x^3 \not\equiv m_y^2 \pmod{n}$ holds true with overwhelming probability for large primes p and q and uniformly distributed integers m_x and m_y.

Decryption

Receiver R decrypts ciphertext (c_x, c_y) with secret keys p, q, d_p and d_q. First, R computes $c_{xp} = c_x \bmod p$, $c_{yp} = c_y \bmod p$ and shadow ciphertext $c_p = \dfrac{c_{xp}^3}{c_{yp}} \bmod p$ by using the isomorphic mapping ω in Theorem 1. R computes shadow plaintext m_p as

$$m_p = c_p^{d_p} \bmod p = \left(\frac{c_{xp}^3}{c_{yp}^2} \right)^{d_p} \bmod p. \tag{4}$$

R computes $(m_{xp}, m_{yp}) \in E_p(a_p, 0)$ with $a_p = \dfrac{c_{xp}^3 - c_{yp}^2}{c_{xp} c_{yp}} \bmod p$ by using the isomorphic mapping ω^{-1} in Theorem 1 as

$$m_{xp} = \frac{a_p^2 m_p}{(m_p - 1)^2} \bmod p, \quad m_{yp} = \frac{m_{xp} a_p}{(m_p - 1)} \bmod p.$$

R computes $(m_{xq}, m_{yq}) \in E_q(a_q, 0)$ in the same way. Finally, R obtains (m_x, m_y) by combining (m_{xp}, m_{yp}) and (m_{xq}, m_{yq}) via the Chinese Remainder Theorem.

Remarks

· By the isomorphic mappings in Theorem 1, computing $d_p \otimes (c_{xp}, c_{yp})$ over $E_p(a_p, 0)$ corresponds to computing $(c_{xp}^3 / c_{yp}^2)^{d_p}$ over F_p^*. The decryption of scheme 1 corresponds to the path from (c_x, c_y) to (m_x, m_y) via c and m.

· Since $m_{xp}, m_{yp} \in F_p^*$ and $m_{xp}^3 \not\equiv m_{yp}^2 \pmod{p}$, we have $m_p \neq 1$.

3.3 Scheme 2

Encryption

Sender S encrypts plaintext (m_x, m_y) with the receiver's public keys e and n as

$$c = \left(\frac{m_x^3}{m_y^2} \right)^e \bmod n,$$

$$a = \frac{m_x^3 - m_y^2}{m_x m_y} \bmod n,$$

and sends a pair (c, a) of shadow ciphertext c and the corresponding variable a to receiver R.

Remark

· The length of the transmitted message in scheme 2 is the same as that in scheme 1, which is 2 log n bits.

Decryption

Receiver R decrypts shadow ciphertext (c, a) with secret keys p, q, d_p and d_q. First, R computes $c_p = c \bmod p$ and shadow plaintext m_p from c_p, d_p and p as

$$m_p = c_p^{d_p} \bmod p. \tag{5}$$

R computes $(m_{xp}, m_{yp}) \in E_p(a_p, 0)$ with $a_p = a \bmod p$ by using the isomorphic mapping ω^{-1} in Theorem 1 as

$$m_{xp} = \frac{a_p^2 m_p}{(m_p - 1)^2} \bmod p, \quad m_{yp} = \frac{m_{xp} a_p}{(m_p - 1)} \bmod p.$$

R computes $(m_{xq}, m_{yq}) \in E_q(a_q, 0)$ in the same way. Finally, R obtains (m_x, m_y) by combining (m_{xp}, m_{yp}) and (m_{xq}, m_{yq}) via the Chinese Remainder Theorem.

Remarks

· The decryption of scheme 2 corresponds to the path from c to (m_x, m_y) via m.

· Computations of c_p and a_p in the decryption of scheme 2 need less time than that of scheme 1 because divisions of $c_p = c_{xp}^3/c_{yp}^2$ and $a_p = (c_{xp}^3 - c_{yp}^2)/c_{xp}c_{yp}$ can be avoided.

4 Efficiency

4.1 Comparison of Proposed Schemes and Other Schemes

Since encryption key e can be set as a small value and decryption keys d_p, d_q are large enough such that $\log d_p \approx \log p$, $\log d_q \approx \log q$, we focus on the decryption procedure. We evaluate the average number of modular multiplications for decryption. Here, we assume $\log p \approx \log q$.

In the proposed schemes, i.e., scheme 1 and scheme 2, the dominant computations involve equations (4) and (5). They require $1.5 \log p$ multiplications modulo p on average. Including the $1.5 \log q$ multiplications modulo q, the decryption of each of the proposed schemes requires about $3 \log p$ modular multiplications.

The block size for the RSA scheme is $\log n$ bits, and that for the proposed schemes is $2 \log n$ bits. The number of modular multiplications in the new schemes and previously proposed schemes are shown in Table 1. We define "speed ratio"; the bigger the speed ratio is, the faster the decryption speed is. Let the decryption speed ratio of the RSA scheme be normalized to 1.0. When a K-bit long message is given, the speed ratios for the KMOV scheme and the new schemes are determined as $0.085r$ and r, respectively, where $r = s/\lceil \frac{s}{2} \rceil$ and $s = \lceil K/\log n \rceil$. Note that $1.0 \leq r \leq 2.0$. When integer s is even, the speed ratios for the KMOV scheme and the new schemes are fixed as 0.17 and 2.0, respectively. If message length K is uniformly distributed, the probability that s is even is $1/2$. If message length K is predetermined such that $K = 2 \log n$, then integer s is always even. For the Demytko scheme based on elliptic curves [1], its speed ratio is always fixed as 0.14 because the block size is $\log n$. These results are summarized in Table 1. We can observe that the decryption speed of the proposed schemes is about 2.0 times faster than that of the RSA scheme for a K-bit long message if $\lceil K/\log n \rceil$ is even.

Table 1: Efficiency of decryptions

Cryptosystems	Block size	No. of mod. multi.	Speed ratio ($\lceil K/\log n \rceil$ is even)
RSA	$\log n$	$3 \log p$	1.0
KMOV	$2 \log n$	$35 \log p$	0.17
Demytko	$\log n$	$22 \log p$	0.14
New schemes	$2 \log n$	$3 \log p$	2.0

Nowadays, the RSA scheme with 512 bits modulus n (block size) is practically used for key distributions and digital signatures. In this standard RSA scheme, eight DES keys can be distributed in one block. In the new schemes with 1024 bits block size, 16 DES keys can be distributed at the same decryption speed and the same security level.

4.2 Encryption Efficiency of Scheme 1 and Scheme 2

Although the dominant computations involve the decryptions in scheme 1 and scheme 2, we evaluate their encryption efficiency to compare thses schemes. We focus on pure encryption procedures excluding isomorphic mapping procedures. Let $|e|$ be the bit-length of encryption key e. A possible minimum value of e is 3, and $|e| = 2$. It is clear that the encryption of scheme 2 requires $1.5|e|$ multiplications modulo n on average. In scheme 1, computing the multiples of a point on curve E_n can be performed in affine coordinates (2) or homogeneous coordinates. A point (x, y) on the affine plane is equivalent to a point (X, Y, Z) on the projective plane, where $x = X/Z$, $y = Y/Z$. When we put $b = 0$, the addition formula in affine coordinates can be rewritten in homogeneous coordinates as equations (7) and (8) in the Appendix. The revised formulae with minimum number of multiplications are equations (9) and (10) in the Appendix. In the addition formula in homogeneous coordinates, contrary to that in affine coordinates, the divisions in Z_n in each addition over E_n can be avoided. Each elementary addition over E_n is calculated using addition, subtraction, multiplication and division in Z_n. For simplicity, addition, subtraction and special multiplication by a small constant were neglected for the comparison. In affine coordinates, each non-doubling addition requires three multiplications and one division in Z_n, and each doubling requires six multiplications and one division in Z_n. In homogeneous coordinates, each non-doubling addition requires 26 multiplications in Z_n, and each doubling requires 26 multiplications in Z_n. Let ℓ be the ratio of the computation amount of division in Z_n to that of multiplication in Z_n. Consequently, the encryption of scheme 1 based on affine coordinates requires $(7.5 + 1.5\ell)|e|$ multiplications in Z_n on average. That based on homogeneous coordinates requires $39|e| + \ell$ multiplications in Z_n on average. Since $1.5|e| < (7.5 + 1.5\ell)|e|$ and $1.5|e| < 39|e| + \ell$, the encryption of scheme 2 is faster than that of scheme 1. In particular, encryption efficiency of scheme 1 differs by the implemented coordinates. For example, when $e = 3$ and $e = 21$, the encryption in homogeneous coordinates is faster than that in affine coordinates if and only if $\ell > 31.5$ and $\ell > 24.2$, respectively.

5 Security

5.1 Security in One-to-one Communication

We show a theorem about the security relationship between the proposed schemes and the RSA scheme.

Theorem 4 *Breaking each of the proposed schemes is computationally equivalent to breaking the RSA scheme. That is, the following sentences are equivalent.*

  (i) There is an efficient algorithm A1 such that for all $c_x, c_y \in Z_n^*$, $(c_x, c_y) \in E_n(a, 0)$, if $(c_x, c_y) = e \otimes (m_x, m_y)$ over $E_n(a, 0)$, then $A1(c_x, c_y, e, n) = (m_x, m_y)$.

  (ii) There is an efficient algorithm A2 such that for all $c, a \in Z_n^*$, $(m_x, m_y) \in E_n(a, 0)$, if $c = \left(\dfrac{m_x^3}{m_y^2} \right)^e \bmod n$, then $A2(c, a, e, n) = (m_x, m_y)$.

  (iii) There is an efficient algorithm B such that for all $c \in Z_n^*$, if $c = m^e \bmod n$, then $B(c, e, n) = m$.

Proof: First, the equivalence between (i) and (iii) is shown as follows.
(i) $\Rightarrow$ (iii)
Assuming algorithm A1 is given, algorithm B is defined as follows.
Input: c, e, n
Step 1: Choose $a \in Z_n^*$ randomly.
Step 2: Compute $(c_x, c_y) \in E_n(a, 0)$ from c, a and n by using isomorphic mapping, without knowing factors of n as

$$c_x = \frac{a^2 c}{(c-1)^2} \bmod n, \quad c_y = \frac{a^3 c}{(c-1)^3} \bmod n.$$

Step 3: Compute $(m_x, m_y) = A1(c_x, c_y, e, n)$.
Step 4: Compute $m = 1 + \frac{a m_x}{m_y} \bmod n$
Output: m
If algorithm A1 requires $O(T)$ bit-operations, then algorithm B requires $O(T + (\log n)^3)$ bit-operations, and is polynomially reducible from algorithm A1.

(iii) $\Rightarrow$ (i)
Assuming algorithm B is given, algorithm A1 is defined as follows.
Input: $(c_x, c_y), e, n$
Step 1: Compute $a = \dfrac{c_x^3 - c_y^2}{c_x c_y} \bmod n$.
Step 2: Compute $c = 1 + \dfrac{a c_x}{c_y}$.
Step 3: Compute $m = B(c, e, n)$.
Step 4: Compute $(m_x, m_y) \in E_n(a, 0)$ from m, a and n by using isomorphic mapping, without knowing factors of n as

$$m_x = \frac{a^2 m}{(m-1)^2} \bmod n, \quad m_y = \frac{a^3 m}{(m-1)^3} \bmod n.$$

Output: (m_x, m_y)
If algorithm B requires $O(T)$ bit-operations, then algorithm A1 requires $O(T + (\log n)^3)$ bit-operations, and is polynomially reducible from algorithm B.

Next, the equivalence between (ii) and (iii) is shown as follows.

(ii) $\Rightarrow$ (iii)

Assuming algorithm $A2$ is given, algorithm B is defined as follows.

Input: c, e, n

Step 1: Choose $a \in Z_n^*$ randomly.

Step 2: Compute $(m_x, m_y) = A2(c, a, e, n)$.

Step 3: Compute $m = 1 + \frac{am_x}{m_y} \bmod n$

Output: m

If algorithm $A2$ requires $O(T)$ bit-operations, then algorithm B requires $O(T + (\log n)^3)$ bit-operations, and is polynomially reducible from algorithm $A2$.

(iii) $\Rightarrow$ (ii)

Assuming algorithm B is given, algorithm $A2$ is defined as follows.

Input: c, a, e, n

Step 1: Compute $m = B(c, e, n)$.

Step 2: Compute $(m_x, m_y) \in E_n(a, 0)$ from m, a and n,

without knowing factors of n as

$$m_x = \frac{a^2 m}{(m-1)^2} \bmod n, \quad m_y = \frac{a^3 m}{(m-1)^3} \bmod n.$$

Output: (m_x, m_y)

If algorithm B requires $O(T)$ bit-operations, then algorithm $A2$ requires $O(T + (\log n)^3)$ bit-operations, and is polynomially reducible from algorithm B.

The above theorem is concerning on usual passive attacks. Consider possibility of active known-plaintext attacks. Assume that an attacker knows a value of m_y in addition to the values of c_x, c_y, e and n. The attacker aims at obtaining m_x by solving cubic congruence $m_x^3 - m_y^2 \equiv am_xm_y \bmod n$ with known m_y and $a = \frac{c_x^3 - c_y^2}{c_xc_y} \bmod n$. However, it seems difficult to obtain m_x if breaking the RSA scheme is difficult. On the other hand, assume that an attacker knows a value of m_x in addition to the values of c_x, c_y, e and n. The attacker aims at obtaining m_y by solving quadratic congruence $m_x^3 - m_y^2 \equiv am_xm_y \bmod n$ with known m_x and $a = \frac{c_x^3 - c_y^2}{c_xc_y} \bmod n$. However, it seems difficult to obtain m_y if breaking the Rabin scheme [8] is difficult. Note that breaking the Rabin scheme (i.e., factoring n) is more difficult than the breaking the RSA scheme in a usual sense. Thus, additive information on m_x or m_y seems useless for cryptanalysis.

5.2 Security in Broadcast Applications

In broadcast applications, the original RSA scheme is not secure if encryption key e is small. Let e and n_i be public keys of the original RSA scheme for a receiver R_i $(1 \leq i \leq k)$. The common plaintext m is encrypted as $c_i = m^e \bmod n_i$ $(1 \leq i \leq k)$ for k receivers. If $k \geq e$, then the system of congruences $c_i \equiv m^e (\bmod n_i)$ $(1 \leq i \leq e)$ can be transformed into the equation $c = m^e$, where c is the combined ciphertext from c_i via the Chinese Remainder Theorem. Hence, the plaintext m can be computed as $m = c^{1/e}$ over the real field. Even if known terms like "user ID" are included in the plaintexts such that $m_i = \alpha_i m + \beta_i$, where α_i and β_i are publicly known, Hastad [2] showed that similar attacks aimed at obtaining m can be successful by solving a set of k congruences of polynomials $\sum_{j=0}^{u} t_{ij} m^j \equiv 0 \bmod n_i$. The inequality condition for a successful attack is given by

$$\prod_{i=1}^{k} n_i > n_s^{u(u+1)/2}(k+u+1)^{(k+u+1)/2} 2^{(k+u+1)^2/2}(u+1)^{u+1},$$

where $n_s = \min(n_i)$. This condition is the most sensitive to the degree u of the obtained set of congruences of polynomials. In the RSA scheme with the linearly related plaintexts $m_i = \alpha_i m + \beta_i$, the system of congruences in m with degree e can be obtained in broadcast applications. In the KMOV scheme over elliptic curves, the system of congruences in m_x with degree e^2 can be obtained in broadcast applications. Thus, it was shown in [6] that the KMOV scheme is more secure than the original RSA scheme against the Hastad attack.

We evaluate the security of the new schemes (i.e., scheme 1 and scheme 2) in broadcast applications, in which the plaintext is purely common or linearly related. First, consider scheme 1. There is a recursive formula for computing x_i such that $(x_i, y_i) = i \otimes (x_1, y_1)$ over $E_n(a, 0)$, where $(x_1, y_1) \in E_n(a, 0)$ is the initial point:

$$x_{2i} = \frac{x_i^2}{4x_i + a^2} \bmod n, \qquad x_{2i+1} = \frac{x_{i+1}^2 x_i^2}{x_1(x_{i+1} - x_i)^2} \bmod n. \tag{6}$$

Using Eq. (6), ciphertext c_x in scheme 1 is expressed by

$$c_x = \frac{m_x^e}{h_e(m_x)} \bmod n,$$

where m_x is a plaintext and $h_i(m_x)$ is recursively defined as

$$h_1(m_x) = 1,$$

$$h_{2i}(m_x) = 4m_x^i h_i(m_x) + a^2(h_i(m_x))^2 \bmod n \ (i \geq 1),$$

$$h_{2i+1}(m_x) = (h_{i+1}(m_x) - m_x h_i(m_x))^2 \bmod n \ (i \geq 1).$$

Since the degree of $h_i(m_x)$ is $i-1$, and m_x^i and $h_i(m_x)$ are relatively prime polynomials, the system of congruences in m_x with degree e can be obtained as $m_x^e - c_x h_e(m_x) \equiv 0 \ (\bmod \ n)$. Thus, it is shown that scheme 1 has the same security as the RSA scheme when linearly related plaintexts are encrypted in broadcast applications. It is also shown that scheme 2 has the same security as the RSA scheme when linearly related plaintexts are encrypted in broadcast applications. Note that the RSA scheme with a purely common plaintext generates a simpler monomial m^e than a set of polynomials with degree e. Thus, the new schemes are more secure than the RSA scheme when purely common plaintexts are encrypted in broadcast applications.

We show numerical examples. When modulus n_i is 512 bits long and $e = 5$, the Hastad attack is applicable if more than 16 ciphertexts are obtained for the new schemes and the RSA scheme with linearly related plaintexts. When modulus n_i is 512 bits long and $e = 19$, the Hastad attack is applicable if more than 282 ciphertexts are obtained for the new schemes and the RSA scheme with linearly related plaintexts. When modulus n_i is 512 bits long and $e \geq 21$, the Hastad attack is *not* applicable for the new schemes and the RSA scheme with linearly related plaintexts. Note that when modulus n_i is 512 bits long and $e \geq 5$, the Hastad attack is *not* applicable for the KMOV scheme.

6 Conclusion

We have proposed fast RSA-type schemes over $E_n(a, 0)$. For a $2 \log n$-bit long message, the decryption speed of the proposed schemes is about 2.0 times faster than that of the RSA scheme. We have proved that breaking the proposed scheme is equivalent to breaking the RSA scheme.

Acknowledgement

We thank Hidenori Kuwakado and Yukio Tsuruoka for their valuable discussions.

References

[1] N. Demytko: "A new elliptic curves based analogue of RSA", Advances in Cryptology - Eurocrypt'93, LNCS **765** pp. 40-49 (1992).

[2] J. Hastad: "On using RSA with low exponent in a public key network", Advances in Cryptology - Crypto'85, LNCS **218** pp. 403–408 (1985).

[3] D. Husemöller: "Elliptic Curves", Springer-Verlag (1987).

[4] K. Koyama, U. M. Maurer, T. Okamoto and S. A. Vanstone: "New public-key schemes based on elliptic curves over the ring Z_n", Advances in Cryptology - Crypto'91, LNCS **576** pp. 252-266 (1991).

[5] K. Koyama and Y. Tsuruoka: "A signed binary window method for fast computing over elliptic curves", Advances in Cryptology - Crypto'92, LNCS **740** pp. 345-357 (1992).

[6] H. Kuwakado and K. Koyama: "On the security of RSA-type cryptosystems over elliptic curves against the Hastad attack", IEE Electronics Letters, Vol.30, No.22, pp. 1843-1844 (1994).

[7] A. J. Menezes: "Elliptic Curve Public Key Cryptosystems", Kluwer Academic Publishers (1993).

[8] M. Rabin: "Digital signatures and public-key cryptosystems", MIT/LCS/TR-21, (1979).

[9] R. L. Rivest, A. Shamir and L. Adleman: "A method for obtaining digital signatures and public-key cryptosystems", Comm. of the ACM, 21, 2, pp. 120–126 (1978).

Appendix: Addition Formula for Singular Cubic Curves

For singular cubic curves $y^2 + axy = x^3$, the addition: $(x_3, y_3) = (x_1, y_1) \oplus (x_2, y_2)$ is given by chord-and-tangent law. The addition formula in affine coordinates is shown in equation (2). The addition formula in homogeneous coordinates is as follows.

Non-doubling Addition Formula for $(X_1, Y_1, Z_1) \neq (X_2, Y_2, Z_2)$

$$
\begin{cases}
\begin{aligned}
X_3 = {} & X_2{}^4 Z_1{}^4 + 2 X_1 X_2{}^3 Z_1{}^3 Z_2 + a X_2{}^2 Y_2 Z_1{}^4 Z_2 + X_2 Y_2{}^2 Z_1{}^4 Z_2 \\
& - a X_2{}^2 Y_1 Z_1{}^3 Z_2{}^2 - 2 a X_1 X_2 Y_2 Z_1{}^3 Z_2{}^2 - 2 X_2 Y_1 Y_2 Z_1{}^3 Z_2{}^2 \\
& - X_1 Y_2{}^2 Z_1{}^3 Z_2{}^2 - 2 X_1{}^3 X_2 Z_1 Z_2{}^3 + 2 a X_1 X_2 Y_1 Z_1{}^2 Z_2{}^3 \\
& + X_2 Y_1{}^2 Z_1{}^2 Z_2{}^3 + a X_1{}^2 Y_2 Z_1{}^2 Z_2{}^3 + 2 X_1 Y_1 Y_2 Z_1{}^2 Z_2{}^3 \\
& + X_1{}^4 Z_2{}^4 - a X_1{}^2 Y_1 Z_1 Z_2{}^4 - X_1 Y_1{}^2 Z_1 Z_2{}^4 \\
Y_3 = {} & X_2{}^3 Y_2 Z_1{}^4 - 2 X_2{}^3 Y1 Z_1{}^3 Z_2 - a X_2 Y_2{}^2 Z_1{}^4 Z_2 - Y_2{}^3 Z_1{}^4 Z_2 \\
& + 3 X_1 X_2{}^2 Y_1 Z_1{}^2 Z_2{}^2 - 3 X_1{}^2 X_2 Y_2 Z_1{}^2 Z_2{}^2 + 2 a X_2 Y_1 Y_2 Z_1{}^3 Z_2{}^2 \\
& + a X_1 Y_2{}^2 Z_1{}^3 Z_2{}^2 + 3 Y_1 Y_2{}^2 Z_1{}^3 Z_2{}^2 + 2 X_1{}^3 Y_2 Z_1 Z_2{}^3 \\
& - a X_2 Y_1{}^2 Z_1{}^2 Z_2{}^3 - 2 a X_1 Y_1 Y_2 Z_1{}^2 Z_2{}^3 - 3 Y_1{}^2 Y_2 Z_1{}^2 Z_2{}^3 \\
& - X_1{}^3 Y_1 Z_2{}^4 + a X_1 Y_1{}^2 Z_1 Z_2{}^4 + Y_1{}^3 Z_1 Z_2{}^4 \\
Z_3 = {} & X_2{}^3 Z_1{}^4 Z_2 - 3 X_1 X_2{}^2 Z_1{}^3 Z_2{}^2 + 3 X_1{}^2 X_2 Z_1{}^2 Z_2{}^3 - X_1{}^3 Z_1 Z_2{}^4
\end{aligned}
\end{cases}
\tag{7}
$$

Doubling Formula for $(X_1, Y_1, Z_1) = (X_2, Y_2, Z_2)$

$$
\begin{cases}
\begin{aligned}
X_3 = {} & 9 a X_1{}^5 Z_1 + 18 X_1{}^4 Y_1 Z_1 + a^3 X_1{}^4 Z_1{}^2 - 6 a^2 X_1{}^3 Y_1 Z_1{}^2 \\
& - 24 a X_1{}^2 Y_1{}^2 Z_1{}^2 - 16 X_1 Y_1{}^3 Z_1{}^2 - a^4 X_1{}^2 Y_1 Z_1{}^3 \\
& - 3 a^3 X_1 Y_1{}^2 Z_1{}^3 - 2 a^2 Y_1{}^3 Z_1{}^3 \\
Y_3 = {} & -27 X_1{}^6 + 45 a X_1{}^4 Y_1 Z_1 + 36 X_1{}^3 Y_1{}^2 Z_1 + 2 a^3 X_1{}^3 Y_1 Z_1{}^2 \\
& - 15 a^2 X_1{}^2 Y_1{}^2 Z_1{}^2 - 24 a X_1 Y_1{}^3 Z_1{}^2 - 8 Y_1{}^4 Z_1{}^2 \\
& - a^4 X_1 Y_1{}^2 Z_1{}^3 - a^3 Y_1{}^3 Z_1{}^3 \\
Z_3 = {} & a^3 X_1{}^3 Z_1{}^3 + 6 a^2 X_1{}^2 Y_1 Z_1{}^3 + 12 a X_1 Y_1{}^2 Z_1{}^3 + 8 Y_1{}^3 Z_1{}^3
\end{aligned}
\end{cases}
\tag{8}
$$

By introducing moderate intermediate variables, addition formulae (7) and (8) can be revised to minimize the number of multiplications:

Revised Non-doubling Addition Formula

$$
\begin{cases}
X_3 = {} & H\{Z_1 Z_2 (T + X_1 X_2 K) - M - Q\}, \\
Y_3 = {} & L(M - Q) - Z_1 Z_2 \{GT + 3H(X_2^2 Y_1 Z_1 + X_1^2 Y_2 Z_2)\}, \\
Z_3 = {} & Z_1 Z_2 H^3,
\end{cases}
\tag{9}
$$

where $H = X_2 Z_1 - X_1 Z_2$, $G = Y_2 Z_1 - Y_1 Z_2$, $K = X_2 Z_1 + X_1 Z_2$, $L = Y_2 Z_1 + Y_1 Z_2$, $M = X_2^3 Z_1^3$, $Q = X_1^3 Z_2^3$, $T = G(aH + G)$.

Revised Doubling Formula

$$
\begin{cases}
X_3 = {} & Z_1 A [X_1 \{9V + Z_1 (a^2 I - 8Y_1 J)\} - a^2 C^2], \\
Y_3 = {} & -27 V^2 + C [9VD + Z_1 \{B^3 + a(X_1 EF - a^2 CJ)\}], \\
Z_3 = {} & Z_1^3 A^3,
\end{cases}
\tag{10}
$$

where $A = aX_1 + 2Y_1$, $B = aX_1 - 2Y_1$, $C = Y_1 Z_1$, $D = 5aX_1 + 4Y_1$, $E = aX_1 - 12Y_1$, $F = aX_1 + 3Y_1$, $I = X_1^2 - aC$, $J = aX_1 + Y_1$, $V = X_1^3$.

Relationships among the Computational Powers of Breaking Discrete Log Cryptosystems

Kouichi Sakurai[1][*] and Hiroki Shizuya[2]

[1] Dept. of Computer Science and Communication Engineering, Kyushu University,
812-81 Japan. e-mail:sakurai@csce.kyushu-u.ac.jp
[2] ECIP & GSIS, Tohoku University, 980-77 Japan.
e-mail:shizuya@ecip.tohoku.ac.jp

Abstract. We investigate the complexity of breaking cryptosystems of which security is based on the discrete logarithm problem. We denote the algorithms of breaking the Diffie-Hellman's key exchange scheme by DH, the Bellare-Micali's non-interactive oblivious transfer scheme by BM, the ElGamal's public-key cryptosystem by EG, the Okamoto's conference-key sharing scheme by CONF, and the Shamir's 3-pass key-transmission scheme by 3PASS, respectively. We show a relation among these cryptosystems that

$$3PASS \leq_m^{FP} CONF \leq_m^{FP} EG \equiv_m^{FP} BM \equiv_m^{FP} DH,$$

where $\leq_m^{FP}$ denotes the polynomial-time functionally many-to-one reducibility, i.e. a function version of the $\leq_m^P$ -reducibility. We further give some condition in which these algorithms have equivalent difficulty. Namely,

1. If the complete factorization of $p-1$ is given, i.e. if the the discrete logarithm problem is a certified one, then these cryptosystems are equivalent w.r.t. expected polynomial-time functionally Turing reducibility.
2. If the underlying group is the Jacobian of an elliptic curve over $\mathbf{Z}_p$ with a prime order, then these cryptosystems are equivalent w.r.t. polynomial-time functionally many-to-one reducibility.

We also discuss the complexity of several languages related to those computing problems.

1 Introduction

1.1 Motivation

The discrete logarithm problem, DLP for short, is the problem that on input $y, g \in G$, outputs an integer x such that $y = g^x$, where G is some finite group with efficiently computable group law. A cryptosystem based on DLP is secure if the DLP is hard to solve. A typical DLP is the case where $G = \mathbf{Z}_p^*$ with p prime.

---

[*] A part of this work was done while the first author was working for Mitsubishi Electric Corp.

L.C. Guillou and J.-J. Quisquater (Eds.): Advances in Cryptology - EUROCRYPT '95, LNCS 921, pp. 341-355, 1995.
© Springer-Verlag Berlin Heidelberg 1995

In 1976, Diffie and Hellman [DH76] first proposed a key exchange scheme that is secure if the DLP over $\mathbf{Z}_p^*$ is hard to solve. A lot of cryptosystems based on DLP have been proposed to construct a public-key cryptosystem, an oblivious transfer protocol, a key-transmission scheme, a zero-knowledge proof of possession of information, and so on. It is clear that all these cryptosystems would no longer be secure if there were an efficient algorithm to solve the DLP, but no such algorithm is known to exist (see, e.g. [COS86, Odl84]). However, it is worth noting that the converse does not generally hold, i.e. it is not known that a polynomial-time algorithm to crack one of these cryptosystems implies feasibility of the DLP. Recently, a great progress has been made by Maurer toward the equivalence of the DLP and breaking the Diffie-Hellman scheme [Ma94], but the equivalence is not known to hold without assumption. Therefore, in general, all these cryptosystems could be breakable without solving the DLP.

In this paper, instead of studying whether there exists a cracking algorithm for cryptosystems without breaking DLP, we investigate the relation among such cryptosystems. Let S_1 and S_2 be two cryptosystems both based on some DLP. Our interest is whether S_1 remains secure even if a polynomial-time algorithm to break S_2 has been found, and vice versa. Although such discussion appears to be essential to clarify the security level of the cryptosystem, we know little about that, surprisingly.

1.2 Summary of Results

Let us denote the problems of breaking the Diffie-Hellman's key exchange scheme by DH, the Bellare-Micali's non-interactive oblivious transfer scheme [BeMi89] by BM, the ElGamal's public-key cryptosystem [ElG85] by EG, the Okamoto's conference-key sharing scheme [Oka88] by CONF and the Shamir's 3-pass key-transmission scheme [SRA79, Riv90] by 3PASS, respectively.

We first show a relation among these cryptosystems that

$$\text{3PASS} \leq_m^{\text{FP}} \text{CONF} \leq_m^{\text{FP}} \text{EG} \equiv_m^{\text{FP}} \text{BM} \equiv_m^{\text{FP}} \text{DH},$$

where $\leq_m^{\text{FP}}$ denotes the polynomial-time functionally many to one reducibility. We further gives some condition in which these algorithms have equivalent w.r.t. certain reductions. Namely,

1. If the complete factorization of $p-1$ is given, i.e. if the the discrete logarithm problem is a certified one, then these cryptosystems are equivalent w.r.t. expected polynomial-time functionally Turing reduction, i.e., $\text{3PASS} \equiv_T^{\text{FEP}}$ $\text{CONF} \equiv_T^{\text{FEP}} \text{EG} \equiv_m^{\text{FP}} \text{BM} \equiv_m^{\text{FP}} \text{DH}.$
2. If the underlying group is the Jacobian of an ordinary elliptic curve over $\mathbf{Z}_p$ with a prime order, then these cryptosystems are equivalent w.r.t. polynomial-time functionally many-to-one reduction, i.e. $\text{3PASS} \equiv_m^{\text{FP}} \text{CONF} \equiv_m^{\text{FP}} \text{EG} \equiv_m^{\text{FP}} \text{BM} \equiv_m^{\text{FP}} \text{DH}.$

We will also investigate the complexity of languages associated with these problems. Let L_{3PASS} be the language associated with 3PASS defined as

$$L_{3\text{PASS}} = \{((A,B,C,p),s) \mid 3\text{PASS}(A,B,C,p) = s\},$$

i.e. its membership problem is to recognize that the s is a correct answer to the instance (A,B,C,p) of 3PASS. Although $L_{3\text{PASS}}$ is not known to be in $\mathcal{P}$ or $\mathcal{BPP}$, we show that if $L_{3\text{PASS}}$ is in $\mathcal{P}$, there is a probabilistic polynomial-time algorithm that reduces DH to 3PASS. Thus, if $L_{3\text{PASS}}$ is in $\mathcal{P}$, all the problems to crack these cryptosystems become equivalent.

In the same way, let L_{DH} be the language associated with DH defined as

$$L_{\text{DH}} = \{((A,B,g,p),C) \mid \text{DH}(A,B,g,p) = C\}.$$

Although L_{DH} is not known to be in $\mathcal{P}$ or $\mathcal{BPP}$ as observed in [B93], we show that L_{DH} is random self-reducible in the sense of [TW87], and therefore L_{DH} is in $\mathcal{PZK}$, the class of languages that have perfect zero-knowledge proof systems.

1.3 Computational Complexity, Communication Complexity, and Cryptographic Mechanisms

Complexity assumption is an important measure of security of cryptographic protocols. In general, a protocol with sophisticated mechanism requires stronger complexity assumption.

Impagliazzo and Rudich [IR89], in fact, presented evidence that secure secret key agreement protocols require stronger complexity assumption than the existence of one-way permutations.

The protocol with sophisticated mechanism also requires a number of interactions. Rudich [Rud91] constructed an oracle relative to which secret agreement can be done in k passes, but not in $k - 1$, and showed that there exists a 3-pass system based on an assumption which seems to be weaker than the existence of trapdoor functions.

We should note that the schemes discussed in this paper achieve different mechanisms and require different number of interactions. Thus, the results of this paper reveal relationships among computational complexity assumption, round complexity, and mechanisms in cryptographic protocols based on the discrete logarithms.

2 Preliminaries

2.1 Cryptosystems based on DLP

We give a brief review of the cryptosystems considered in this paper. All those are based on the discrete logarithm problem (DLP). To avoid complicated generalization of DLP defined over an generic finite group, we restrict ourselves to the case where the underlying group is $\mathbf{Z}_p^*$ with p prime. Thus, the DLP is now the problem that on input y, g, p, outputs x such that $y \equiv g^x \pmod{p}$. Here g does not necessarily generate $\mathbf{Z}_p^*$. For notational convenience, we will write simply g^x rather than $g^x \bmod p$, etc.

We will refer to Alice and Bob as two parties, respectively, that follow the scheme and communicate with each other.

Diffie-Hellman's Key Exchange Scheme [DH76]

Alice and Bob agree on p and the base $g \in \mathbf{Z}_p^*$ before starting their communication. Alice picks a randomly from $\mathbf{Z}_{p-1}$, computes $A = g^a$, and sends A to Bob. Bob picks b randomly from $\mathbf{Z}_{p-1}$, computes $B = g^b$, and sends B to Alice. Alice computes $C = B^a$ and Bob computes $C = A^b$.

Bellare-Micali's Non-Interactive Oblivious Transfer Scheme [BeMi89]

Alice and Bob agree on p and the base $g \in \mathbf{Z}_p^*$ and some $C \in \mathbf{Z}_p^*$. Bob picks $i \in \{0,1\}$ at random, $x_i \in \mathbf{Z}_{p-1}$, and sets $\beta_i = g^{x_i}$ and $\beta_{1-i} = C \cdot (g^{x_i})^{-1}$. Bob publishes (β_0, β_1) as his public key whereas he keeps (i, x_i) as his secret key. Suppose Alice wants to send Bob one of the strings (s_0, s_1) in an oblivious transfer manner. Alice picks at random $y_0, y_1 \in \mathbf{Z}_{p-1}$ and sends $\alpha_0 = g^{y_0}, \alpha_1 = g^{y_1}$ to Bob. Alice then computes $\gamma_0 = \beta_0^{y_0}$ and $\gamma_1 = \beta_1^{y_1}$, and sends $r_0 = s_0 \oplus \gamma_0$ and $r_1 = s_1 \oplus \gamma_1$ to Bob, where $\oplus$ designates the bitwise addition mod 2.

On receiving α_0 and α_1, Bob uses his secret key to compute $\alpha_i^{x_i} = \gamma_i$. He then computes $\gamma_i \oplus r_i = s_i$.

ElGamal's Public-Key Cryptosystem [ElG85]

Bob sets $g \in \mathbf{Z}_p^*$ as the base, picks $x \in \mathbf{Z}_{p-1}$ at random, and computes $y = g^x$. Bob publishes y, g, p as his public key whereas he keeps x as his secret key. Suppose Alice wants to send a string m to Bob. Alice picks $r \in \mathbf{Z}_{p-1}$ at random, computes $C_1 = g^r, C_2 = my^r$ and sends (C_1, C_2) to Bob. On receiving (C_1, C_2), Bob uses his secret key to compute $m = C_2/(C_1)^x$.

Okamoto's Conference-Key Sharing Scheme [Oka88]

Alice and Bob agree on p and the base $g \in \mathbf{Z}_p^*$ before starting their communication. Alice picks a randomly from $\mathbf{Z}_{p-1}^*$, computes $A = g^a$, and sends A to Bob. Bob picks b randomly from $\mathbf{Z}_{p-1}$, computes $B = A^b$, and sends B to Alice. Alice computes $C = B^{a^{-1}}$ and Bob computes $C = g^b$.

We shall note that the established key depends only on Bob's randomness b. Thus Bob can decide the value of the key g^b by himself although Bob can not send directly a message. This property has an advantage over Diffie-Hellman's key exchange scheme in the case of a conference-key sharing scheme for multiple users [Oka88].

Shamir's 3-Pass Message Transmission Scheme [SRA79]

This is also called the Massey-Omura's cryptosystem (see, e.g. [Kob87b]), and originally proposed as a tool for mental poker by Shamir et al. [SRA79, Riv90]. Alice and Bob agree on p before their communication. Suppose Alice wants to send a string (message) s to Bob. Alice picks $a \in \mathbf{Z}_{p-1}^*$ at random, computes $A = s^a$, and sends A to Bob. On receiving A, Bob picks $b \in \mathbf{Z}_{p-1}^*$ at random, computes $C = A^b$, and sends C to Alice. On receiving C, Alice uses her secret a to compute $B = C^{a^{-1}}$ and sends B to Bob. On receiving B, Bob uses his secret b to compute $s = B^{b^{-1}}$.

Remark.

Shamir's 3-pass key transmission scheme is useful not only for secret message transferring but also for an oblivious transfer [Ra81]. An oblivious transfer is a protocol satisfying the following three conditions.

1. Alice can send any message m_0 or m_1.
2. Bob gets only one of message m_0 or m_1.
3. Alice cannot know which message, m_0 or m_1 Bob obtains.

However, certain attacks (on the third condition above) were pointed out (e.g. [Co85]). Shamir et al. [SRA79, Ra81] applied the protocol above into shuffling cards together among two parties in an electronic poker game. Thus, we consider that the Shamir's 3-pass has a more sophisticated mechanism than an oblivious transfer. The protocol is as follows:

Before starting the protocol, A (Alice) and B (Bob) agree on a prime p.

1. For two message m_0 and m_1, A randomly picks $a \in \mathbf{Z}^*_{p-1}$, computes $\alpha_0 = m_0{}^a$ and $\alpha_1 = m_1{}^a$, and sends (α_0, α_1) to B.
2. B picks $e \in \{0, 1\}$ and randomly selects $b \in \mathbf{Z}^*_{p-1}$, then computes $\beta_0 = \alpha_0{}^b$, and sends β to A.
3. A computes $\gamma = \beta^{a^{-1}}$, and sends it to B.
4. B obtains m_e by computing $\gamma^{b^{-1}}$.

2.2 Definitions of Problems

We give the formal definitions of the problems to crack the cryptosystems considered in this paper. These problems will be formalized as something like functions from some tuple of Σ^*'s to Σ^*, where Σ^* is the set of all possible strings over the finite alphabet $\Sigma = \{0, 1\}$.

$\mathrm{DLP}(y, g, p)$ is the problem that on input p prime and $y, g \in \mathbf{Z}^*_p$, outputs $x \in \mathbf{Z}_{p-1}$ such that $y = g^x$ if such an x exists.

$\mathrm{DH}(A, B, g, p)$ is the problem that on input p prime and $A, B, g \in \mathbf{Z}^*_p$, outputs $C \in \mathbf{Z}^*_p$ such that $C = g^{ab}$, $A = g^a$ and $B = g^b$ if such a C exists.

$\mathrm{BM}((\alpha_0, \alpha_1), (r_0, r_1), C, (\beta_0, \beta_1), g, p)$ is the problem that on input p prime and $\alpha_0, \alpha_1, r_0, r_1, C, \beta_0, \beta_1, g \in \mathbf{Z}^*_p$ with $\beta_0 \beta_1 = C$, outputs one of (s_0, s_1) such that $s_i = \gamma_i \oplus r_i$, $\gamma_i = g^{x_i y_i}$, $\alpha_i = g^{y_i}$, $\beta_i = g^{x_i}$ if such an s_i exists ($i = 0$ or 1).

$\mathrm{EG}(C_1, C_2, y, g, p)$ is the problem that on input p prime and $C_1, C_2, y, g \in \mathbf{Z}^*_p$, outputs $m \in \mathbf{Z}^*_p$ such that $C_2 = m g^{xr}$, $y = g^x$, $C_1 = g^r$ if such an m exists.

$\mathrm{CONF}(A, B, g, p)$ is the problem that on input p prime and $A, B, g \in \mathbf{Z}^*_p$, outputs $C \in \mathbf{Z}^*_p$ such that $A = g^a$ where $a \in \mathbf{Z}^*_{p-1}$, $B = A^b$ where $b \in \mathbf{Z}_{p-1}$ and $C = g^b$ if such an C exists.

$\mathrm{3PASS}\ (A, B, C, p)$ is the problem that on input p prime and $A, B, C \in \mathbf{Z}^*_p$, outputs s such that $A = s^a$, $B = s^b$, $C = s^{ab}$ and $a, b \in \mathbf{Z}^*_{p-1}$ if such an s exists.

The functions above always returns a correct answer if there is a solution to the query. However, there are no mentions of the behavior in the case when there is no solution to the query. However, we consider stronger functions which output $\perp$ if there are no solutions, where $\perp$ is the special string to designate the status that the function has no returnable value (Theorem 5).

2.3 Reducibility

In order to compare the relative complexity of different functions, we use the concept of *reducibility*. Intuitively a function f is reducible to another function g if the value of the first function f is computed by an algorithm which uses an algorithm for the second function g as a subroutine. We will consider three types of such reducibilities based on the types of subroutines.

Definition 1. A function f is polynomial-time functionally Turing reducible to a function g (in symbols $f \leq_T^{\mathrm{FP}} g$) if a polynomial-time oracle Turing machine with access to values of g can compute f. Regarding the complexity of such a algorithm we suppose that the cost of one calling the oracle B is just one step.

Definition 2. A function f is *expected* polynomial-time functionally Turing reducible to a function g (in symbols $f \leq_T^{\mathrm{FEP}} g$) if an expected polynomial-time oracle Turing machine with access to values of g can compute f. (NOTE: This paper says that a machine M is *expected polynomial-time* if there exists an $e > 0$ such that, for all $x \in \{0,1\}^*$, the expectation, taken over the infinite bit sequences r, of $(t_M(x,r))^e$ is bounded above by $|x|$ (i.e., $E((t_M(x,r))^e) \leq |x|$).)

Definition 3. A function f is polynomial-time functionally many-one reducible to a function g (in symbols $f \leq_m^{\mathrm{FP}} g$) if there exists a pair of polynomial-time computable functions h_1, h_2 such that for every input string x, $f(x) = h_2(g(h_1(x)))$.

3 Main Results

3.1 Relationships among the Cryptosystems

We first show the following relation among these cryptosystems.

Theorem 4. 3PASS $\leq_m^{\mathrm{FP}}$ CONF $\leq_m^{\mathrm{FP}}$ EG $\equiv_m^{\mathrm{FP}}$ BM $\equiv_m^{\mathrm{FP}}$ DH $\leq_m^{\mathrm{FP}}$ DLP.

Proof. Since it is clear that DH $\leq_m^{\mathrm{FP}}$ DLP, we show that 3PASS $\leq_m^{\mathrm{FP}}$ CONF, CONF $\leq_m^{\mathrm{FP}}$ EG, EG $\equiv_m^{\mathrm{FP}}$ DH, and BM $\equiv_m^{\mathrm{FP}}$ DH.

3PASS $\leq_m^{\mathrm{FP}}$ CONF:
Let $(A, B, C, p) = (s^a, s^b, s^{ab}, p)$ be an instance of 3PASS.

$$3\mathrm{PASS}(A, B, C, p) = \mathrm{CONF}(C, A, B, p) = \mathrm{CONF}((s^b)^a, (s^b)^{b^{-1}a}, s^b, p) = (s^b)^{b^{-1}} = s.$$

CONF $\leq_m^{FP}$ EG:
Let $(A, B, C, p) = (g^a, g^{ab}, g, p)$ be an instance of CONF.

$$\text{CONF}(A, B, C, p) = 1/\text{EG}(g, 1, g^{ab}, g^b, p).$$

3PASS $\leq_m^{FP}$ EG:
Let $(A, B, C, p) = (s^a, s^b, s^{ab}, p)$ be an instance of 3PASS. First we compute (C_1, C_2, y, g, p), an instance of EG, by (CA, ABC, CB, C, p). Here,

$$
\begin{aligned}
(C_1, C_2, y, g, p) &= (CA, ABC, CB, C, p) \\
&= (s^{a(b+1)}, s^{ab+a+b}, s^{b(a+1)}, s^{ab}, p) \\
&= ((s^{ab})^{b^{-1}(b+1)}, s^{ab+a+b}, (s^{ab})^{a^{-1}(a+1)}, s^{ab}, p).
\end{aligned}
$$

Thus,

$$
\begin{aligned}
m &= \text{EG}(CA, ABC, CB, C, p) \\
&= s^{ab+a+b} / (s^{ab})^{a^{-1}b^{-1}(a+1)(b+1)} \\
&= s^{ab+a+b} / s^{ab+a+b+1} = s^{-1}.
\end{aligned}
$$

This implies that the oracle returns $m = s^{-1}$. Therefore, we get $s = m^{-1}$, which is computed in time polynomial in $|p|$.

EG $\leq_m^{FP}$ DH:
This is a trivial reduction. Let $(C_1, C_2, y, g, p) = (g^r, mg^{xr}, g^x, g, p)$ be an instance of EG. Since the oracle DH returns g^{xr} to the query (C_1, y, g, p), m is immediately computed by $m = C_2/g^{xr}$.

DH $\leq_m^{FP}$ EG [Oka94]:
Let $(A, B, g, p) = (g^a, g^b, g, p)$ be an instance of DH. g^{ab} is the inverse of the answer of the oracle EG to the query $(A, 1, B, g, p)$.

BM $\equiv_m^{FP}$ DH:
It is not hard to see that BM $\leq_m^{FP}$ DH because DH returns $\gamma_i = g^{x_i y_i}$ to the query $(\alpha_i, \beta_i, g, p)$, and s_i is computed by $s_i = \gamma_i \oplus r_i$. Conversely, for $(A, B, g, p) = (g^a, g^b, g, p)$, an instance of DH, we let

$$((\alpha_0, \alpha_1), (r_0, r_1), C, (\beta_0, \beta_1), g, p) = ((A, g^u), (0, 0), t, (B, tB^{-1}), g, p),$$

where u, t are picked randomly from $\mathbf{Z}_{p-1}$ and $\mathbf{Z}_p^*$, respectively. Since we set $r_0 = 0$, the oracle $\text{BM}((A, g^u), (0, 0), t, (B, tB^{-1}), g, p)$ returns $s_0 = r_0 \oplus \gamma_0 = 0 \oplus g^{ab} = g^{ab}$.
This completes the proof. ∎

We give a simple alternative proof for 3PASS $\leq_m^{FP}$ DH. It holds that for an instance $(A, B, C, p) = (s^a, s^b, s^{ab}, p)$ of 3PASS,

$$\text{DH}(A, B, C, p) = \text{DH}((s^{ab})^{b^{-1}}, (s^{ab})^{a^{-1}}, s^{ab}, p) = (s^{ab})^{a^{-1}b^{-1}} = s.$$

We do not know if EG $\leq_m^{FP}$ 3PASS. However, if we consider more stronger cracking algorithms which answer the special symbol "⊥" when there is no solution to the instance, we obtain a further result. Consider the following function:

3PASS$^\star(A, B, C, p)$ is the problem that on input p prime and $A, B, C \in \mathbf{Z}_p^*$, outputs s such that $A = s^a$, $B = s^b$, $C = s^{ab}$ and $a, b \in \mathbf{Z}_{p-1}^*$ if such an s exists. Otherwise it outputs $\perp$.

Theorem 5. DH $\leq_T^{\text{FEP}}$ 3PASS*.

Proof. Let $(A, B, g, p) = (g^a, g^b, g, p)$ be an instnace of DH. For randomly picked $u, v \in \mathbf{Z}_{p-1}$, an instance of 3PASS* is computed by
$$(Ag^u, Bg^v, g, p) = (g^{a+u}, g^{b+v}, g, p).$$
If both $a + u$ and $b + v$ happen to be in $\mathbf{Z}_{p-1}^*$, the oracle 3PASS$^\star(g^{a+u}, g^{b+v}, g, p)$ returns $g^{(a+u)(b+v)}$ because

$$g^{a+u} = \left(g^{(a+u)(b+v)}\right)^{(b+v)^{-1}},$$
$$g^{b+v} = \left(g^{(a+u)(b+v)}\right)^{(a+u)^{-1}},$$
$$g = \left(g^{(a+u)(b+v)}\right)^{(a+u)^{-1}(b+v)^{-1}}.$$

Thus $g^{ab} = g^{(a+u)(b+v)}/(A^v B^u g^{uv})$.
However, if either $a + u$ or $b + v$ is not in $\mathbf{Z}_{p-1}^*$, the oracle 3PASS$^\star(g^{a+u}, g^{b+v}, g, p)$ returns some s or $\perp$. We show that if 3PASS$^\star(g^{a+u}, g^{b+v}, g, p)$ returns s, then $s = g^{(a+u)(b+v)}$. If the oracle returns s, it satisfies that for some $\alpha, \beta \in \mathbf{Z}_{p-1}^*$,
$$s^\alpha = g^{a+u}, \quad s^\beta = g^{b+v}, \quad s^{\alpha\beta} = g.$$
Thus, over $\mathbf{Z}_{\text{ord}(g)}$,
$$r\alpha = a + u, \quad r\beta = b + v, \quad r\alpha\beta = 1,$$
where $\text{ord}(g)$ designates the order of g, and r is an element in $\mathbf{Z}_{\text{ord}(g)}$ such that $s = g^r \bmod p$. Then, we have that $(a + u)(b + v) = r^2\alpha\beta = r(r\alpha\beta) = r$. Thus, $s = g^r = g^{(a+u)(b+v)}$.

Conversely, if no such r, α, β exist, the oracle returns $\perp$. Therefore, another $u, v \in \mathbf{Z}_{p-1}$ should be picked, and this is repeated until the oracle returns a string other than $\perp$.

To summarize, the Algorithm 1 named DHto3PASS solves DH using the oracle 3PASS*.

Now we estimate how many times the while-statement is repeated. The probability ρ that the oracle returns a string other than $\perp$ to a query is greater than the probability that both $a + u$ and $b + v$ are in $\mathbf{Z}_{p-1}^*$. Thus, $\rho \geq (\varphi(p-1)/(p-1))^2$, where φ is the Euler's totient function. Since $\varphi(n) \leq \ln(2) \cdot n/\ln(2n)$ for a positive integer n [Rib88], the expected number of repetition of the while-statement is less than $(\ln(2(p-1))/\ln(2))^2$, which is bounded by a polynomial in $|p|$. Thus, DH reduces to 3PASS* in probabilistic polynomial-time. This completes the proof. ∎

```
% Algorithm 1
% DHto3PASS
input A, B, g, p
s := ⊥
while (s = ⊥) do
    pick u, v ∈ Z_{p-1} at random
    A' := Ag^u; B' := Bg^v
    s := 3PASS*(A', B', g, p)
end while
C := s/(A^v B^u g^{uv})
output C
end
```

Remark. The Algorithm 1 above does not give the answer "⊥" even when the input of DH has no solution. So, we do not know if DH* $\leq_T^{\text{FEP}}$ 3PASS*. However, we can obtain a polynomial-time reduction from DH* to 3PASS* with one-sided error by terminating the algorithm DHto3PASS within a suitable step, as shown in Algorithm 2.

```
% Algorithm 2
% DHto3PASS with one-sided error
input A, B, g, p
s := ⊥; C := ⊥; i := 1
T := q(|p|) % some polynomial in |p|
while ([s = ⊥] ∧ [i ≤ T]) do
        pick u, v ∈ Z_{p-1} at random
        A' := Ag^u; B' := Bg^v
        s := 3PASS*(A', B', g, p)
        i := i + 1
end while
if s ≠ ⊥ then C := s/(A^v B^u g^{uv})
output C
end
```

We do not know if DH ≡ DH* nor 3PASS ≡ 3PASS* because there are no known efficient algorithms to check the answers of these cracking algorithms DH and 3PASS. Nevertheless, we show that DH is reducible to 3PASS over some special discrete logarithms.

3.2 The Case of Certified Discrete Logarithms

First we show that if the complete factorization of $p - 1$ is given and the base is a generator of $\mathbf{Z}_p^*$, i.e. if the discrete logarithm problem is a certified one, there is a probabilistic polynomial-time algorithm that solves DH using 3PASS as an oracle. This reduces DH to 3PASS, and the above reductions become equivalent.

Theorem 6. *If the complete factorization of $p - 1$ with p prime is given and the base g is a generator of $\mathbf{Z}_p^*$,*

$$\text{DH} \leq_T^{\text{FEP}} \text{3PASS}.$$

Proof. In the proof of Theorem 5, we have shown that DH* reduces to 3PASS*, where 3PASS* is an algorithm which returns a special symbol "⊥" if and only if there is no solution. Now we consider a weaker algorithm which returns any polynomially bounded string instead of ⊥. However, this happens if either $a + u$ or $b + v$ is not in $\mathbf{Z}_{p-1}^*$. Thus, if we restrict ourselves to the query such that both $a + u$ and $b + v$ are in $\mathbf{Z}_{p-1}^*$, and if the instance of DH is appropriate, then the answer from the oracle is always correct. Therefore, we modify the algorithm DHto3PASS as shown in Algorithm 3.

Here, $d = \textbf{true}$ if and only if both X and Y are generators of $\mathbf{Z}_p^*$, which implies that both $a + u$ and $b + v$ are in $\mathbf{Z}_{p-1}^*$. The expected number of repetition of the while-statement is bounded by $(\ln(2(p-1))/\ln(2))^2$, which is also bounded by a polynomial in $|p|$. ∎

```
% Algorithm 3
% DHto3PASS for Certified DLP
input A, B, g, p = p_0^{e_0} p_1^{e_1} ··· p_k^{e_k} + 1
d := false
while (d = false) do
    pick u, v ∈ Z_{p-1} at random
    X := Ag^u; Y := Bg^v
            k
    d := [ ⋀  X^{(p-1)/p_i} ≠ 1]∧
           i=0
            k
         [ ⋀  Y^{(p-1)/p_i} ≠ 1]
           i=0
end while
s := 3PASS(X, Y, g, p)
C := s/(A^v B^u g^{uv})
output C
end
```

den Boer [dB88] showed that the Diffie-Hellman problem is as strong as the discrete logarithms for certain primes. It is remarkable that Maurer [Ma94] made this result stronger to cover generic cyclic groups. Let $\varphi(N)$ be the order of the group $\mathbf{Z}_N^*$.

Theorem 7 [dB88], see also [Ma94]. *If $\varphi(p-1)$ is smooth, i.e., it consists of small prime factors with respect to a fixed polynomial in $q(|p|)$, then $\mathrm{DLP} \leq_T^{\mathrm{FEP}} \mathrm{DH}$.*

We should note that our reductions keep the modulus, then the following is induced.

Corollary 8. *Suppose that $\varphi(p-1)$ is smooth, i.e., it consists of small prime factors with respect to a fixed polynomial in $q(|p|)$. If the complete factorization of $p-1$ with p prime is given and the base g is a generator of $\mathbf{Z}_p^*$, then*
$$\mathrm{3PASS} \equiv_T^{\mathrm{FEP}} \mathrm{CONF} \equiv_T^{\mathrm{FEP}} \mathrm{EG} \equiv_T^{\mathrm{FEP}} \mathrm{BM} \equiv_T^{\mathrm{FEP}} \mathrm{DH} \equiv_T^{\mathrm{FEP}} \mathrm{DLP}.$$

3.3 The Case of Elliptic Discrete Logarithms

Next we consider these cryptosystems based on the elliptic-curve discrete logarithm problem [Kob87a, Mil85], denoted by EDLP.

Here we briefly review the EDLP. Let $C(a,b)_p$ be an elliptic curve defined over $\mathbf{Z}_p$, where p prime $\neq 2, 3$, with parameters $a, b \in \mathbf{Z}_p$, that is

$$C(a,b)_p = \{(x,y) \in \mathbf{Z}_p \times \mathbf{Z}_p \mid [y = x^3 + ax + b] \wedge [a, b \in \mathbf{Z}_p] \wedge$$
$$[4a^3 + 27b^2 \not\equiv 0 \pmod{p}]\} \cup \{O\},$$

where O is the point at infinity. The Jacobian of $C(a,b)_p$, which happens to be the same as $C(a,b)_p$, forms an abelian group. The EDLP is the problem that on input a point $Q \in C(a,b)_p$ and the base point $P \in C(a,b)_p$, outputs m such that

$Q = mP$ if such an m exists. Here, we denote by mP the m-time addition of the point P. The order of $C(a,b)_p$, denoted by $\#C$, is computed in time polynomial in $|p|$ [Sch85]. The order is bounded as $-2\sqrt{p} \leq \#C(a,b)_p - (p+1) \leq 2\sqrt{p}$.

The elliptic curve $C(a,b)_p$ defined over $\mathbf{Z}_p$ is said to be supersingular if and only if $\#C(a,b) = p + 1$. Nonsupersingular elliptic curves are called ordinary. Thus an elliptic curve group with prime order is ordinary and simple, where by a simple group we mean that there is no non-trivial normal subgroup in $C(a,b)_p$. If $C(a,b)_p$ is supersingular, the EDLP reduces in probabilistic polynomial-time to a discrete logarithm problem over the multiplicative group of a certain extension field of $\mathbf{Z}_p$ [MOV91]. However, no such reduction algorithm is known to exist for elliptic-curve groups with prime order [Miy91].

It is not hard to see all the cryptosystems considered in this paper can actually be constructed over $C(a,b)_p$ as analogues of those over $\mathbf{Z}_p^*$, and the reductions shown in Theorem 4 also hold for the EDLP-based systems. Let DH_E (resp. BM_E, EG_E, CONF_E, $\mathrm{3PASS}_E$) designate the EDLP-based DH (resp. BM, EG, CONF, 3PASS) problem. We have the following theorem.

Theorem 9. *If the cryptosystems are based on the discrete logarithm problem whose underlying group is the Jacobian of an elliptic curve defined over $\mathbf{Z}_p$ with prime order, then*

$$\mathrm{3PASS}_E \equiv_m^{\mathrm{FP}} \mathrm{CONF}_E \equiv_m^{\mathrm{FP}} \mathrm{EG}_E \equiv_m^{\mathrm{FP}} \mathrm{BM}_E \equiv_m^{\mathrm{FP}} \mathrm{DH}_E.$$

Proof. As Theorem 4, it is easily seen that $\mathrm{3PASS}_E \leq_m^{\mathrm{FP}} \mathrm{CONF}_E \leq_m^{\mathrm{FP}} \mathrm{EG}_E \equiv_m^{\mathrm{FP}} \mathrm{BM}_E \equiv_m^{\mathrm{FP}} \mathrm{DH}_E$. Thus, it suffices to show that $\mathrm{DH}_E \leq_m^P \mathrm{3PASS}_E$. Let E be an elliptic curve defined over $\mathbf{Z}_p$ with p prime $\neq 2,3$, and let $\#E = q$ with q prime. For an instance $(A,B,P,E,p) = (aP,bP,P,E,p)$ of DH_E, if $A \neq O$ and $B \neq O$, then both a and b are units in $\mathbf{Z}_q$. This is because E is simple. Thus, the oracle $\mathrm{3PASS}_E$ always returns the correct answer to a query (A,B,P,E,p). Hence, $\mathrm{DH}_E \leq_m^p \mathrm{3PASS}_E$. ∎

There are few known research on the distribution of the prime-order elliptic curves over all elliptic curves. A construction of the prime-order elliptic curves is studied also in [Miy91], and finding more efficient algorithms to construct such ordinary elliptic curves is an interesting future topic. Thus, the previously known merit of ordinary elliptic curves over $\mathbf{Z}_p$ is just that it is immune from the attack by [MOV91]. Our theorem above is based on another interesting property of ordinary prime-order elliptic curves over $\mathbf{Z}_p$ that any non-zero element has the inverse.

3.4 Languages Associated with the Cryptosystems

We return to the cryptosystems based on DLP defined over $\mathbf{Z}_p^*$.

Associated with the problems $\mathbf{Q}$, we define the language $L_{\mathbf{Q}}$ by

$$L_{\mathbf{Q}} = \{(x,y)| \ \mathbf{Q}(x) = y\},$$

where $\mathbf{Q}$ is one of DLP, DH, BM, EG, CONF, or 3PASS. The problem to decide membership in $L_{\mathbf{Q}}$ is to recognize that y is an answer to the instance x of $\mathbf{Q}$. Clearly, these languages are in $\mathcal{NP} \cap$ co-$\mathcal{NP}$. Indeed, L_{DLP} is in $\mathcal{P}$. However, it is not known that one of L_{DH}, L_{BM}, L_{EG}, L_{CONF}, or L_{3PASS} is in $\mathcal{P}$ or $\mathcal{BPP}$. The same observation on L_{DH} can also be found in [B93]. Thus, there may be a reduction sequence among these languages which is different from the reductions given in Theorem 4, though, at the moment, no reductions among L_{DH}, L_{BM}, L_{EG}, L_{CONF}, and L_{3PASS} are known.

One connection to the reductions among the cracking problems is shown in the following.

Theorem 10. L_{3PASS} *is not in* $\mathcal{P}$ *unless* DH $\leq_T^{\text{FEP}}$ 3PASS.

Proof. We show the contraposition. That is, if L_{3PASS} is in $\mathcal{P}$, DH $\leq_T^{\text{FEP}}$ 3PASS. The algorithm DHto3PASS in the proof of Theorem 5 can be modified, if L_{3PASS} is in $\mathcal{P}$, as shown in Algorithm 4. This completes the proof. ∎

```
% Algorithm 4
input A, B, g, p
d := false
while (d = false) do
    pick u, v ∈ Z_{p-1} at random
    A' := Ag^u; B' := Bg^v
    s := 3PASS(A', B', g, p)
    d := [((A', B', g, p), s) ∈ L_{3PASS}]
end while
C := s/(A^v B^u g^{uv})
output C
end
```

The theorem above gives a characterization of the complexity of L_{3PASS}. Also we obtain

Corollary 11. *If* L_{3PASS} *is in* $\mathcal{P}$, *then* 3PASS $\equiv_T^{\text{FEP}}$ CONF $\equiv_T^{\text{FEP}}$ EG $\equiv_m^{\text{FP}}$ BM $\equiv_m^{\text{FP}}$ DH.

Theorem 12. *The language* L_{DH} *is random self-reducible in the sense of [TW87].*

Proof. For an instance $((A, B, g, p), C)$, let $A' = Ag^r$, $B' = Bg^s$, and $C' = CA^s B^r g^{rs}$ to make another instance $((A', B', g, p), C')$, where r and s are randomly picked from $\mathbf{Z}_{p-1}$. Note that if $A = g^a$, $B = g^b$ and $C = g^{ab}$, then $A' = g^{a+r}$, $B' = g^{b+s}$, and $C' = g^{(a+r)(b+s)}$. Hence, the distribution of A' (resp. B', C') is exactly the same as that of A (resp. B, C). It is clear that if $((A', B'g, p), C')$ is in L_{DH}, so is $((A, B, g, p), C)$. This implies L_{DH} is random self-reducible. ∎

The theorem above implies that L_{DH} has a perfect zero-knowledge interactive proof.

3.5 Single-Use versus Multiple-Use in Cryptosystems

Consider the situation that we use the Shamir's 3-pass scheme for transferring the same message s polynomially many times. In such a case, an adversary can get much information than single use. We discuss the relative security between single-use and multiple-use in the cryptosystem. So, we formulate the following k-3PASS problem.

k-3PASS is the problem that on input p prime and $A_1, B_1, C_1, \ldots, A_k, B_k, C_k, \in$ $\mathbf{Z}_p^*$, s such that $A_j = s^{a_j}$, $B_j = s^{b_j}$, $C_j = s^{a_j b_j}$ and $a_j, b_j \in \mathbf{Z}_{p-1}^*$ $(j = 1, \ldots, k)$ if such an s exists.

We show that multiple-use is as secure as single use.

Theorem 13. _1-3PASS (= 3PASS)_ $\leq_m^{FP}$ _k-3PASS_

Proof. Let (A, B, C, p) be an instance of 1-3PASS. Pick $(u_1, v_1), \ldots, (u_k, v_k) \in$ $\mathbf{Z}_{p-1}^* \times \mathbf{Z}_{p-1}^*$ at random. Put

$$A_i = A^{u_i}, B_i = B^{v_i}, C_i = C^{u_i v_i} (1 \leq i \leq k).$$

Then, $((A_1, B_1, C_1, p), \ldots, (A_k, B_k, C_k, p))$ is an instance of k-3PASS. ∎

The theorem above suggests a role of the randomness of each party in the scheme. The same property holds in some other cryptosystems, namely k-EG and k-CONF defined as follows.

k-EG   is the problem that on input p prime and $C_{11}, C_{21}, \ldots, C_{1k}, C_{2k}, y, g \in$ $\mathbf{Z}_p^*$, outputs $m \in \mathbf{Z}_p^*$ such that $C_{2j} = mg^{xr}$, $y = g^x$, $C_{1j} = g^{r_j}$ $(j = 1, \ldots, k)$ if such an m exists.

k-CONF is the problem that on input p prime and $A_1, \ldots, A_k, B, g \in \mathbf{Z}_p^*$, outputs $C \in \mathbf{Z}_p^*$ such that $A = g^{a_j}$ where $a_j \in \mathbf{Z}_{p-1}^*$, $B = A_j^b$ where $b \in \mathbf{Z}_{p-1}$ $(j = 1, \ldots, k)$ and $C = g^b$ if such an C exists.

Theorem 14. _1-EG(= EG)_ $\leq_m^{FP}$ _k-EG_

Proof. Let (C_1, C_2, y, g, p) be an instance of 1-EG. We show that for any $k \leq q(|p|)$ with q polynomial, this can be transformed into an instance of k-EG in polynomial-time. First, pick $u_1, \ldots, u_k \in \mathbf{Z}_{p-1}$ at random. Then, put

$$C_{1i} = C_1 g^{u_i}, C_{2i} = C_2 y^{u_i} (1 \leq i \leq k).$$

Since $C_{1i} = g^{r+u_i}$ and $C_{2i} = mg^{x(r+u_i)}$, we now have an instance of k-EG as
$$((C_{11}, C_{21}, y, g, p), \ldots, (C_{1k}, C_{2k}, y, g, p)). ∎$$

Okamoto [Oka88] observed such a property in his scheme.

Theorem 15 [Oka88]. _1-CONF(= CONF)_ $\leq_m^{FP}$ _k-CONF_

4   Concluding Remarks

We have given the reductions among the problems to break some cryptosystems based on the discrete logarithms over $\mathbf{Z}_p^*$ (Theorem 4). Specifically, we have shown that these problems are equivalent under the stronger function model (Theorem 5), although none of them is known to be equivalent to the discrete logarithm problem itself.

We have also shown that the equivalence occurs if the discrete logarithm problem is a certified one over $\mathbf{Z}_p^*$ (Theorem 6), or if it is the elliptic-curve discrete logarithm problem associated with an ordinary elliptic curve defined over $\mathbf{Z}_p$ (Theorem 9). Therefore, if one cryptosystem is breakable, so are the others. This means that if one wants to crack one of the cryptosystems, there are several possible approaches to the algorithm for breaking the target cryptosystem. However, this also implies that one cryptosystem is as secure as the others, namely, the provable security of the cryptosystems. Although those theorems can be interpreted in two ways as above, it is true that they give an interesting aspect of the cryptosystems based on the certified discrete logarithm or the ordinary elliptic-curve discrete logarithm.

Further, we have defined some languages associated with those problems. We have pointed out that each language to recognize the correct answer of the problem is not known to be in $\mathcal{P}$, whereas the language corresponding to the discrete logarithm problem is in $\mathcal{P}$. Some questions remain open:

- Does L_{3PASS} reduce to L_{DH} with respect to $\leq_T^P$ -reducibility?
- Does L_{DH} reduce to L_{3PASS} with respect to $\leq_T^P$ -reducibility?
- Does L_{3PASS} have a perfect zero-knowledge interactive proof?

Acknowledgments

We would like to thank the following people. Toshiya Itoh pointed out a flaw of a mathematical formula in an earlier version of this paper. Kojiro Kobayashi gave us invaluable comments on the (non-)transitivity of randomized reducibilities. Tatsuaki Okamoto informed us of his conference-key sharing scheme discussed in his Ph.D thesis.

References

[B93]   Brands, S., "An efficient off-line electronic cash system based on the representation problem," CWI Technical Report CS-R9323 (Apr. 1993).

[BeMi89] Bellare, M. and S. Micali, "Non-interactive oblivious transfer and applications," in Advances in Cryptology – Crypto'89, Lecture Notes in Computer Science 435, pp.547-557, *Springer-Verlag*, Berlin (1990).

[Co85]   Coppersmith, D. "Cheating at mental poker," Advances in Cryptology – Crypto'85, Lecture Notes in Computer Science 218, *Springer-Verlag*, Berlin, pp.104-107 (1986).

[COS86]   Coppersmith, D., A. M. Odlyzko, and R. Schroeppel, "Discrete logarithms in $GF(p)$," Algorithmica 1, pp.1-15 (1986).

[dB88] den Boer, B., "Diffie-Hellman is as strong as discrete log for certain primes," Advances in Cryptology – Eurocrypt'88, Lecture Notes in Computer Science 403, *Springer-Verlag*, Berlin, pp.530-539 (1990).

[DH76] Diffie, W. and M. E. Hellman, "New directions in cryptography," IEEE Trans. Inform. Theory, IT-22, No.6, pp.644-654, (Nov. 1976).

[ElG85] ElGamal, T., "A public key cryptosystem and a signature scheme based on discrete logarithms," IEEE Trans. Inform. Theory, IT-31, No.4, pp.469-472, (July 1985).

[IR89] Impagliazzo, R. and Rudich, S., "Limits on the provable consequences of one-way permutations," *Proc. of 21st STOC*, pp.44-61 (1989).

[Kob87a] Koblitz, N., "Elliptic curve cryptosystems," Math. Comp., 48, pp.203-209 (1987).

[Kob87b] Koblitz, N., "A Course in Number Theory and Cryptography," GTM 114, *Springer-Verlag* (1987).

[Ma94] Maurer, U. M., "Towards the equivalence of breaking the Diffie-Hellman protocol and compuing discrete logarithms," Advances in Cryptology – Crypto'94, Lecture Notes in Computer Science 839, *Springer-Verlag*, Berlin, pp.271-281 (1994).

[Mil85] Miller, V., "Uses of elliptic curves in cryptography," Advances in Cryptology – Crypto'85, Lecture Notes in Computer Science 218, *Springer-Verlag*, Berlin, pp.417-426 (1986).

[Miy91] Miyaji, A., "On ordinary elliptic curve cryptosystems," in Advances in Cryptology – Asiacrypt'91, Lecture Notes in Computer Science 739, *Springer-Verlag*.

[MOV91] Menezes, A., T. Okamoto, and S. A. Vanstone, "Reducing elliptic logarithms to logarithms in a finite field," *Proc. of 23rd STOC*, pp.80-89 (1991).

[Odl84] Odlyzko, A. M., "Discrete logarithms in finite fields and their cryptographic significance," Advances in Cryptology – Eurocrypt'84, Lecture Notes in Computer Science 209, *Springer-Verlag*, Berlin, pp.224-314 (1985).

[Oka88] Okamoto, T., "Encryption and authentication schemes based on public-key systems" Ph.D. Thesis, *The University of Tokyo* (1988).

[Oka94] Okamoto, T., Personal communication via email (1994).

[Ra81] Rabin, M., "How to exchange secrets by oblivious transfer," Tech. Memo TR-81, Aiken Computation Laboratory, Harvard University, (1981).

[Rib88] Ribenboim, P., "The Book of Prime Number Records," Springer-Verlag (1988).

[Riv90] Rivest, R. L., "Cryptography," Chapter 13 of Handbook of Theoretical Computer Science, Vol.A, Algorithms and Complexity, edited by Jan van Leeuwen, *The MIT*, pp.717-755 (1990).

[Rud91] Rudich, S., "The use of interaction in public cryptosystems," Advances in Cryptology – Crypto'91, Lecture Notes in Computer Science 576, *Springer-Verlag*, Berlin, pp.242-251 (1992).

[SRA79] Shamir, A., R. L. Rivest, and L. Adleman, "Mental Poker," MIT/LCS, TM-125, (Feb. 1979).

[Sch85] Schoof, R., "Elliptic curves over finite field and the computation of square roots mod p," Math. Comp., 44, pp.483-494 (1985).

[TW87] Tompa, M. and H. Woll, "Random Self-Reducibility and Zero-Knowledge Interactive Proofs of Possession of Information," *Proc. of 28th FOCS*, pp.472-482 (1987).

Universal Hash Functions & Hard Core Bits

Mats Näslund

Royal Institute of Technology,
Dept. of Numerical Analysis and Computing Science,
S-100 44 Stockholm, Sweden
email: matsn@nada.kth.se

Abstract. In this paper we consider the bit-security of two types of universal hash functions: linear functions on $GF[2^n]$ and linear functions on the integers modulo a prime. We show individual security for all bits in the first case and for the $O(\log n)$ least significant bits in the second case. Both types of functions are shown to have $O(\log n)$ simultaneous secure bits. For the second type of functions, primes of length $\Omega(n)$ are needed.

Together with the Goldreich-Levin theorem, this shows that all the common types of universal hash functions provide so called hard-core bits.

1 Introduction

Most cryptographic protocols are based on the access to some source of random bits. Examples of such protocols are private key crypto systems, authentication schemes, commitment schemes etc. For practical purposes it is desirable to reduce the number of true random bits needed. Instead we would like to deterministically expand a short truly random sequence into a longer one that is "just as good" as a truly random sequence of the same length. In other words we would like to deterministically produce some "extra" bits that "look" totally random. So called *hard-core* bits can serve as these extra bits. Intuitively, a hard-core bit is a 0-1 function that cannot be approximated essentially better than simply guessing it.

Another common technique is to try to make "slightly random" sources more random looking. This can be achieved by means of *universal hash functions*, first introduced by Carter & Wegman in [2]. Such hash functions will map elements pairwise independently and the image of each element will be uniformly distributed. Universal hash functions have been used extensively in the construction of pseudo random number generators (PRG's), see for instance [5].

This paper is concerned with the relation between universal hash functions and hard-core bits. The first function that was shown to provide hard-core bits is the usual inner product taken modulo 2. This was done by Goldreich and Levin in [4]. This inner product was from [2] known to be a universal hash function. The natural question is therefore: Can we obtain hard-core bits from the other known universal hash functions as well? More generally, does *every* universal hash function have such hard bit(s)? Not very surprisingly we will answer the first question positively. Even if the second question also could be answered

L.C. Guillou and J.-J. Quisquater (Eds.): Advances in Cryptology - EUROCRYPT '95, LNCS 921, pp. 356-366, 1995 positively, some new proof technique seems to be needed and we must leave this an open problem.

After giving some notation we first go about hash functions given by linear functions in a finite field of characteristic 2. Here we show that for randomly chosen $a, x, b \in \mathrm{GF}[2^n]$, any single bit of the function $x \mapsto ax + b$ is a hard-core bit. Also, the $O(\log n)$ least significant bits are shown to be simultaneously hard-core. Next we study hash functions obtained as linear functions on the integers modulo a prime. Using an adaptation of the techniques used by Chor et al. in [1], [3], we are able to prove both individual and simultaneous hardness for $O(\log n)$ bits. Primes of length $\Omega(n)$ are needed though.

2 Preliminaries

The model of computation used is that of probabilistic Turing machines. We will only be interested in such machines that run in time polynomial in the length of the input, pptm's for short. The length of the input to such a machine is referred to by n. In general, we denote by $|y|$ the length of the binary string y. By $x \in_U S$ we mean an x chosen from the set S according to the uniform distribution. If S is a set, $\|S\|$ is the cardinality of S.

We call a function $g(n)$ *negligible* if for every constant $c > 0$ and for every sufficiently large n, $g(n) < n^{-c}$.

By a *one-way function* we mean a function f such that for every pptm, M, the probability that M on input $f(x)$ finds an $x' \in f^{-1}(x)$ is negligible. The probability is taken over $x \in_U \{0,1\}^n$ and M's random coin flips. For simplicity all one-way function in this paper are assumed to be *length-preserving*, i.e. $|f(x)| = |x|$.

Let H be an efficiently sampleable family of functions were each $h \in H$ is computable in deterministic polynomial time and maps $\{0,1\}^n \mapsto \{0,1\}^{l(n)}$, $l(n) \leq n$. Let f be a one-way function. An *approximation algorithm* for H is a pptm that on input $f(x)$ and the description of $h \in H$ tries to compute $h(x)$. We call H a family of *hard-core functions for f with security $s(n)$* if for all approximation algorithms A: $Pr[A(f(x), h) = h(x)] < 2^{-l(n)} + s(n)$. The probability is taken over $x \in_U \{0,1\}^n$, $h \in_U H$ and A's random choices. (When $l(n) = 1$ we have a family of hard-core predicates.) We will sometimes just say that H is $s(n)$-secure and if H is $s(n)$-secure for all non-negligible $s(n)$ we simply call H a family of hard-core functions. If indeed for some A and $s(n)$, $Pr[A(f(x), h) = h(x)] \geq 2^{-l(n)} + s(n)$ holds, we call A an $s(n)$-oracle for H.

The following fact (a version of the Goldreich-Levin theorem from [4]) will be useful:

Fact 1. *Let $\langle r, x \rangle$ denote the inner product, $\sum_{i=1}^{n} r_i x_i$. The family of functions: $b_r(x) = \langle x, r \rangle \pmod 2$ for $r \in_U \{0,1\}^n$ is a family of hard-core predicates for any one-way function.*

Generalizing, any one-way function has a family of hard-core functions, B, defined as follows: Let $k \in O(\log n)$ and let $r_1, r_2, \ldots, r_k \in_U \{0,1\}^n$. Then $B_{r_1, r_2, \ldots, r_k}(x) = b_{r_1}(x) \circ b_{r_2}(x) \circ \cdots \circ b_{r_k}(x)$, where $\circ$ means concatenation.

Finally, a family of *(strong) universal hash functions* (UHF's) is a set of functions, $H_{n,m}$, with each $h \in H_{m,n}$ such that $h : \{0,1\}^n \mapsto \{0,1\}^m$, $m \le n$, and for any $x_1 \ne x_2 \in \{0,1\}^n$ and any $y_1, y_2 \in \{0,1\}^m$:

$$Pr_{h \in H_{n,m}}[h(x_1) = y_1 \wedge h(x_2) = y_2] = 2^{-2m}.$$

Throughout this paper $y_{<k>}$ denotes the k least significant bits in y. For the special case $y_{<1>}$, we write $\text{lsb}(y)$.

3 Hash functions given by linear functions on $\text{GF}[2^n]$

3.1 Notation

We will here study the family $\{h_{A,B}(X) = A(t)X(t) + B(t) \mid A, B \in_U \text{GF}[2^n]\}$. As usual, $\text{GF}[2^n]$ is the field of 2^n elements. We assume that we have a representation of the field as $\mathbf{Z}_2[t]/Q(t)$ where Q is an irreducible polynomial of degree n, $\sum_{i=0}^n Q_i t^i$. We map elements $x \in \{0,1\}^n$ to $\text{GF}[2^n]$ in the natural way by

$$\phi(x_{n-1}x_{n-2}\cdots x_0) = \sum_{i=0}^n x_i t^i = X(t).$$

We will use small letters (such as x) when we refer to values as binary strings and capital letters (such as X) when we have polynomials.

The notion of lsb is not well defined in $\text{GF}[2^n]$. However it turns out that what bit we interpret as least significant really does not matter so let us for the moment define it as the constant term in the polynomials. Finally, note that the addition below is modulo 2.

3.2 Security of $\text{lsb}(A(t)X(t) + B(t))$

We aim to show

Theorem 2. *The family $\{\text{lsb}(h_{A,B}(X)) \mid A, B \in_U \text{GF}[2^n]\}$ is a family of hard-core predicates, i.e. it is n^{-c} secure for any $c > 0$.*

The idea behind the proof is simple and quite intuitive. We will set up a 1–1 correspondence between the usual inner-product bit mod 2 and the lsb in the above representation.

Observe that for any two polynomials $P(t), S(t)$ we have $\text{lsb}(P(t) + S(t)) = \text{lsb}(P(t)) + \text{lsb}(S(t))$ so that we, for the moment, can forget about $B(t)$ above. As before, let $\langle x, r \rangle$ be the inner product of the n-vectors r and x.

Lemma 3. *Given any $r \in \{0,1\}^n$ there is a unique polynomial $R(t) \in \text{GF}[2^n]$ such that for all $x \in \{0,1\}^n$ we have*

$$\langle x, r \rangle \pmod 2 = \text{lsb}(\phi(x)R(t)). \tag{1}$$

Furthermore, $R(t)$ can be found in polynomial time.

Proof. Let L_1 be the set of all linear functions $\{0,1\}^n \mapsto \{0,1\}$ and L_2 the set of all linear functions $GF[2^n] \mapsto \{0,1\}$. First note that distinct r's define distinct linear functions: $\langle x, r \rangle \pmod 2$ and distinct polynomials $R(t)$ define distinct linear functions $\mathsf{lsb}(R(t)X(t))$. Since $||L_1|| = ||L_2|| = 2^n$, all function in L_1 can be expressed as $\langle x, r \rangle \pmod 2$ for some $r \in \{0,1\}^n$ and all functions in L_2 can be written as $\mathsf{lsb}(R(t)X(t))$ for some $R(t) \in GF[2^n]$. We conclude that there is a bijection between the respective set of linear functions. Finally, given the values of $\langle x, r \rangle \pmod 2$ for n lineary independent x's we can in polynomial time, using standard linear algebra methods, find a corresponding polynomial $R(t) \in GF[2^n]$ such that (1) is satisfied for all $\phi(x) = X(t) \in GF[2^n]$. □

We can now reduce the problem of approximating the inner product to the problem of approximating $\mathsf{lsb}(h_{A,B}(X))$. Theorem 2 then follows from Fact 1:

Proof. (Of Theorem 2.) Let T be an $\delta(n)$-oracle for $\mathsf{lsb}(A(t)X(t) + B(t))$. On input $r \in_U \{0,1\}^n, y = f(x)$, do the following: Using Lemma 3 find the unique $R(t) \in GF[2^n]$ such that $\langle r, x \rangle \pmod 2 = \mathsf{lsb}(R(t)\phi(x))$ for all $x \in \{0,1\}^n$. Choose $B(t) \in_U GF[2^n]$ and run T on input (R, B, y). Suppose that T outputs γ. Output $\mathsf{lsb}(B(t)) + \gamma$.

If r is uniformly distributed in $\{0,1\}^n$, R will be uniformly distributed in $GF[2^n]$. Thus the success probability is exactly the same as that of T. The reduction is clearly polynomial time.

Summing up, if $\delta(n)$ is non-negligible, we now have an approximation algorithm for the inner product bit, also with non-negligible success probability, a contradiction to the Goldreich-Levin theorem. □

In fact there is nothing special about the least significant bit.

Corollary 4. *Any single fixed bit position (coefficient) in the family $\{h_{A,B}(X) \mid A, B \in_U GF[2^n]\}$ is a family of hard-core predicates.*

Proof. The proof is the same as before. Just note that each single bit of $R(t)X(t)$ is a linear function $GF[2^n] \mapsto \{0,1\}$. □

3.3 Simultaneous security

The same technique as above can be used to prove

Theorem 5. *Let c be a constant. The family $\{(A(t)X(t)+B(t))_{<c \log n>} \mid A, B \in_U GF[2^n]\}$ is a family of hard-core functions. In general any set of $O(\log n)$ bits constitute a family hard-core functions.*

The theorem will follow from the following two lemmas, the first being the well known XOR-lemma, see [6]. A function, $h : \{0,1\}^n \mapsto \{0,1\}^{l(n)}$, is called *length-regular* if $l(n)$ increases with n.

Lemma 6. *Let $\{h\}$ be a set family of length-regular function with $|h(x)| \in O(\log n)$. Then $\{h\}$ is a family of hard-core functions if, and only if, the exclusive-or of any non-empty subset of its bits is a family hard-core predicates.*

Next we show that the problem of approximating $\text{lsb}(A(t)X(t) + B(t))$ can be reduced (in polynomial time) to that of approximating the exclusive-or of any non-empty subset of the bits of $A(t)X(t) + B(t)$.

Lemma 7. *Let S be a non-empty subset of $\{0, 1, 2, \ldots, n-1\}$ and let $l_i(z)$ be the i:th bit of z. Given $R(t) \in GF[2^n]$ there is a unique polynomial $P(t) \in GF[2^n]$ such that for all $X(t) \in GF[2^n]$:*

$$\text{lsb}(R(t)X(t)) = \sum_{i \in S} l_i(P(t)X(t))$$

and where P can be found in time polynomial in n.

Proof. For each $i = 0, 1, \ldots, n-1$, $l_i(P(t)X(t))$ is a linear function $GF[2^n] \mapsto \{0, 1\}$. Being a sum (mod 2) of such linear functions, $\sum_{i \in S} l_i(P(t)X(t))$ is also a linear function. Arguing as before it will suffice to show that distinct $R(t)$ define distinct functions.

Assume that for some $R(t) \neq 0$ we have $\sum_{i \in S} l_i(R(t)X(t)) = 0$ for all $X(t) \in GF[2^n]$. Let $i_0 \in S$. Since we are working in a field there is some $W(t) \in GF[2^n]$ such that $R(t)W(t) = t^{i_0}$. By assumption $\sum_{i \in S} l_i(R(t)W(t)) = 0$ but $\sum_{i \in S} l_i(R(t)W(t)) = \sum_{i \in S} l_i(t^{i_0}) = 1$, a contradiction. We must conclude that if $\sum_{i \in S} l_i(R(t)X(t)) = 0$ is to hold for all $X(t) \in GF[2^n]$ then $R(t) = 0$ and hence, distinct $R(t)$ give distinct linear functions. $\qquad\square$

4 Hash functions given by linear functions in $\mathbb{Z}_p$

4.1 Notation

As usual, $\mathbb{Z}_p$ denotes the field of integers modulo a prime, p. By $\mathcal{P}_k$, $k > 0$, we mean the set of primes p of length n/k. Here, $n = |x|$, the security parameter of some one-way function $f(x)$. As in [1], [3] we devide $\mathbb{Z}_p$ into "positive" and "negative" elements. The positive being $\{1, 2, \ldots, \frac{p-1}{2}\}$ and the negative $\{\frac{p-1}{2} + 1, \frac{p-1}{2} + 2, \ldots, p-1\}$. Thus it is natural to define an absolute value for each $x \in \mathbb{Z}_p$ by $|x|_p = x$ if $x \leq \frac{p-1}{2}$ and $|x|_p = p - x$ otherwise. If $|y|_p \leq \frac{p}{\gamma}$, we say that y is (γ, p)-small.

We shall need a notion of "oddness" and "eveness" and so we define the parity of $x \in \mathbb{Z}_p$ by $\text{parity}(x, p) = \text{lsb}(|x|_p)$. In this way the odd/even concept agrees with the intuition both for positive and negative x.

The hash functions we study here is the set $\{h_{a,b,p}(x) = ax + b \pmod{p} \mid p \in_U \mathcal{P}_k, a, b \in_U \mathbb{Z}_p\}$ Note that this set is not totally universal. The least significant bit has a $\frac{1}{p}$ "preference" for attaining the value 0. However this deviation tends to zero exponentially in $|p|$. For large p we can for all practical purposes consider the set to be universal. No polynomial time algorithm can distinguish this distribution from that of a "totally" universal hash function.

4.2 Security of $ax + b$ (mod p)

In this section of the paper we show:

Theorem 8. *Let k be any positive constant, let $p \in_U \mathcal{P}_k$. The family*

$$\{ lsb(h_{a,b,p}(x)) \mid a, b \in_U \mathbf{Z}_p \}$$

is a family of hard-core predicates for any one-way function.

The idea behind the proof is to show that an n^{-c}-oracle for $lsb(h_{a,b,p}(x))$ can be used to retrieve x (mod p). We then repeat this process for several distinct p and combine these results using the Chinese remainder theorem to find x (mod 2^n).

For the first part we will use a modification of the ideas used by Chor et al. in [1], [3]. To get some motivation we review some of that work.

Previous work. Chor et al. in [1], [3] use a parity oracle to invert the RSA function by computing the gcd of $E_N(ax), E_N(bx)$ for random a, b. E_N is the RSA encryption function with composite modulus N. This can be done by observing that $E_N(ax) = E_N(a)E_N(x)$ and by using the well known bit-gcd procedure that only makes parity tests. We get a gcd of the form $E_N(lx)$ (with l known) and with probability $\frac{6}{\pi^2}$ for large N this gcd equals 1. Since $E_N(1) = 1$, x is easily found.

Using an lsb-oracle for parity. The first step is to convert an lsb-oracle to a parity-oracle. Suppose we have a known integer j and an unknown integer i and that we somehow are able to obtain information on $lsb(j)$ and $lsb(j + i)$. We can deduce the parity of i from this since the parity of i is 0 if and only if the lsb of j equals the lsb of $j + i$.

Is this true also in $\mathbf{Z}_p$? Not in general. If $j + i$ causes "overflow" mod p, the parity will be missrepresented. However it is easy to see that the probability of such overflow is $\frac{|i|_p}{p}$. Thus, if i is "small" mod p, we can deduce that

$$parity(i, p) = 0 \Leftrightarrow lsb(j) = lsb(j + i) \text{ for most } j.$$

(We shall shortly specify what is required to be considered small.) This gives a simple way to approximate $parity(ax + b \pmod{p})$ using a lsb-oracle, O_l: On input a, b, p choose $c, d \in_U \mathbf{Z}_p$ and ask O_l about $lsb(cx + d \pmod{p})$ and $lsb((a + c)x + (b + d) \pmod{p})$. Output 0 if O_l answers the same to both questions, 1 otherwise.

To improve performance we can choose many (c, d)-pairs and take a majority decision. However, calling O_l twice for each (c, d)-pair has its drawbacks which we have reason to return to later.

We next show how we can use a "very good" parity oracle and then show how to obtain such from a "fairly good" lsb-oracle.

$\left(\frac{1}{2} - \frac{\alpha}{n}\right)$**-oracles for parity.** Suppose we are given a prime p, $|p| = n/k$, chosen uniformly at random in $\mathcal{P}_k$ and that we have access to a $\left(\frac{1}{2} - \frac{\alpha}{n}\right)$-oracle for parity, O_{par}, with $\alpha < k/2$. We can now try to retrieve $x \pmod{p}$ in the following fashion: Choose $a, b \in_U \mathbb{Z}_p$ and assume that $ax + b$ is "small". For instance, assume $ax + b$ is $(2n^c, p)$-small. This happens with probability $\frac{1}{2n^c}$. We now make the following observations: If parity$(ax + b, p) = 0$ then $ax + b \pmod{p}$ is divisible by 2, and $|2^{-1}(ax + b)|_p$ is even smaller than $|ax + b|_p$. On the other hand, If parity$(ax + b, p) = 1$ then $ax + b - 1 \pmod{p}$ is divisible by two and $|2^{-1}(ax + b - 1)|_p$ will be small. Eventually, after $|p|$ parity-tests (if they all gave correct answers), we end up with a representation $y \equiv a''x + b'' \pmod{p}$ with y, a'' and b'' known, and can easily find $x \pmod{p}$. Since we make exactly $|p| = n/k$ parity calls, the probability of getting one (or more) incorrect parity answer is at most $\frac{n}{k}\frac{\alpha}{n} < 1/2$. Finally note that if the initial $ax + b \pmod{p}$ is small and the oracle makes no errors, all successive $a'x + b' \pmod{p}$ will also be small. We have proved:

Lemma 9. *For a randomly chosen prime $p \in_U \mathcal{P}_k$ a $\left(\frac{1}{2} - \frac{\alpha}{n}\right)$-oracle, $\alpha < k/2$, for parity$(ax + b \pmod{p})$ for $(2n^c, p)$-small $ax + b$, can be used to find $x \pmod{p}$ with probability at least $\frac{1}{4n^c}$.*

Comment: There is in fact another way of proving this lemma by using the same gcd-technique as in [1], [3]: Choose $a, b, c, d \in_U \mathbb{Z}_p$ and use the bit-gcd algorithm (which only uses parity tests) to compute $e, f \in \mathbb{Z}_p$ such that $ex + f = \gcd(ax + p \pmod{p}, cx + d \pmod{p})$. With probability greater than $\frac{1}{2}$, $ex + f \equiv 1 \pmod{p}$ and x can easily be found. The same analysis as in [1], [3] show that we make at most $6|p| + 3 = \frac{6}{k}n + 3$ parity calls on an execution of this algorithm. Thus a $\left(\frac{1}{2} - \frac{\alpha}{n}\right)$-oracle with $\alpha < k/18$ would suffice. This can however be accomplished.

Why couldn't Chor et al. use the simpler, first technique? The reason lies in properties of the RSA function. The function $h_{a,b,p}(x) = ax + b \pmod{p}$ is multiplicative: Even if x is unknown we can still use the identity $h_{ac,bc,p}(x) \equiv ch_{a,b,p}(x) \pmod{p}$. We have already mentioned that the RSA function has similar properties. However the first method above uses *additive* properties of $h_{a,b,p}(x)$, namely $h_{a,b,p}(x) \pm c \equiv h_{a,b\pm c,p}(x) \pmod{p}$. The RSA function however, does not have such properties.

Using n^{-c}-oracles for lsb. To allow more erroneous oracles we use a modification of the techniques from [1], [3].

We first set out to describe the parity oracle. As mentioned, we can get a fairly reliable oracle for parity$(ax+b, p)$ using an oracle for lsb$(ax+b \pmod{p})$. The flaw is that this does not work for arbitrary n^{-c}-oracles since we get the phenomenon of "error-doubling" (see [1],[3]) in asking for two points each time. The cure is in these two lemmas, which are slight modifications of those in [1], [3]. For the sake of self containment we sketch them as well as their proofs.

Lemma 10. *It is possible to generate (in polynomial time) a list, P, of $m \in$ poly(n) points of the form $rx + s$ and a set of lists, $L^{(i)}$, $i = 1, 2, \ldots 4m^3$, each $L^{(i)}$ containing m 0-1 elements. The lists satisfy:*

- *The points in P are pairwise independent and uniformly distributed in $\mathbf{Z}_p$.*
- *At least one of the lists $L^{(i)}$ satisfy: $lsb(P_j) = L_j^{(i)}$ for all but a $\frac{2}{\sqrt{m}}$ fraction of the j's.*

Since we now have points of the form $rx + s \pmod{p}$ with "known" lsb, we only need to ask the oracle about $lsb((a + r)x + (b + s) \pmod{p})$ to deduce parity$(ax + b, p)$.

Proof. We go about generating these in the following way: Let $m \in poly(n)$ and divide $\mathbf{Z}_p$ into $m^{3/2}$ intervals of equal length, $I_i = [ip/m^{3/2}, (i + 1)p/m^{3/2})$. Select $r_1, r_2, s_1, s_2 \in_U \mathbf{Z}_p$ and let the j:th point, P_j, be $r_1 x + s_1 + j(r_2 x + s_2)$

$\pmod{p}$, $j = 1, 2, \ldots, m$. It is easy to see that these points are uniformly distributed and pairwise independent.

Suppose we knew intervals I_{i_1}, I_{i_1} (of length $p/m^{3/2}$) such that $y = r_1 x + s_1$

$\pmod{p} \in I_{i_1}$ and $z = r_2 x + s_2 \pmod{p} \in I_{i_2}$. Assume that we in addition knew $lsb(y)$ and $lsb(z)$. Then, since y, z would now be known within $p/m^{3/2}$, $P_j = y + jz$, would be known within $(1 + j)p/m^{3/2} \le 2p/\sqrt{m}$. Then, since $\lfloor P_j/p \rfloor$ would be determined by j, i_1, i_2, $lsb(P_j)$ would be determined by this and $lsb(y), lsb(z)$, unless P_j would happen to lie in an interval of length $2p/\sqrt{m}$ containing a multiple of p. But the later occurs only with probability $\frac{2}{\sqrt{m}}$.

Now, we do not actually "know" all the things assumed above, but there are m^3 possibilities for the pair (i_1, i_2) that determine the intervals and there are 4 choices for the pair $(lsb(y), lsb(z))$. All in all we have a set of $4m^3$ possibilities so we let $L^{(i)}$ consist of our calculated lsb's for each point, based on the i:th possibility for the quadruple $(i_1, i_2, lsb(x), lsb(y))$. Now, exactly one these quadruples is the correct one and hence, the corresponding list will contain m 0-1 elements, the j:th one being equal to $lsb(P_j)$ with probability $1 - \frac{2}{\sqrt{m}}$. $\square$

We can now use these points to get a good parity oracle.

Lemma 11. *Let $\epsilon(n) = n^{-c}$ and let $\alpha > 0$ be any constant. Given an $\epsilon(n)$-oracle for $lsb(ax + b \pmod{p})$ we can in polynomial time construct a set of $4m^3$ oracles, $m \in poly(n)$, such that at least one of them is a $\left(\frac{1}{2} - \frac{\alpha}{n}\right)$-oracle for parity$(ax + b, p)$ for randomly chosen a, x, b, p such that $ax + b$ is $(2\epsilon(n), p)$-small.*

Proof. Same as in [1], [3]. Each oracle gets the sample points P and the j:th oracle gets the list $L^{(j)}$ as created in Lemma 10. In order to get a vote for the parity of some $ax + b \pmod{p}$, the oracle uses a point $P_i = ux + v \pmod{p}$ in P and then only needs to ask the lsb-oracle about $lsb((a + u)x + (b + v)$

$\pmod{p})$. Recall that $lsb(ux + v \pmod{p})$ is already "known" as $L_i^{(j)}$. The oracle that gets the correct choices for $L^{(j)}$ can be shown to be a $\left(\frac{1}{2} - \frac{\alpha}{n}\right)$-oracle for parity$(ax + b, p)$, provided $ax + b$ is $(2\epsilon(n), p)$-small.

The number of sample points, m, needed to make the majority decision reliable enough depends on $\epsilon(n)$, α but can be shown to be polynomial in n, $\epsilon(n)^{-1}$, see [1], [3] for details. ☐

We can now prove the main theorem of this section:

Proof. (Of Theorem 8.) Assume that for some constant $c > 0$ we have a n^{-c}-oracle, O_l, for $\text{lsb}(ax + b \pmod{p})$ for randomly chosen p, a, b, x as in the formulation of the theorem. Choose uniformly at random a set of $4n^{2c+r}$ primes from $\mathcal{P}_k$: $p_1, p_2, \ldots, p_{4n^{2c+r}}$.

For each p_i, use Lemma 11 to convert the lsb-oracle into a set of $4m^3$ parity-oracles, $O_{i,1}, O_{i,2}, \ldots, O_{i,4m^3}$, one of them being a $\left(\frac{1}{2} - \frac{k_i}{2n}\right)$-oracle for parity.

Next, using Lemma 9, for each p_i, use each $O_{i,j}$ to get a list of suggestions for $x \pmod{p_i}$: $z_{i,1}, z_{i,2}, \ldots, z_{i,4m^3}$.

Any set of k *correct* congruences $x \equiv z_{i,j} \pmod{p_i}$ for distinct p_i's will by the Chinese remainder theorem determine $x \pmod{2^n}$. Let X be the random variable that counts the number of i's for which the corresponding list $z_{i,1}, z_{i,2}, \ldots, z_{i,4m^3}$ *does* contain such a correct modular equation. By lemmas 9 and 11, we see that X is binomially distributed with expectance at least n^{c+r} and variance at most n^{2c+r}. By Chebyshevs inequality, we can bound the probability that $X < k$ from above by $O(n^{-r})$. Thus, the probability of retrieving the n-bit string x is at least $1 - n^{-r}$. The result can of course be checked by evaluating f and comparing to $f(x)$.

We do not know which of the parity oracles that is the good one for each p_i and hence neither which of the corresponding $z_{i,j}$'s to use. However, there are $O(n^{k(2c+r)})$ k-subsets in all and for each such there are $4^k m^{3k}$ ways to choose these $z_{i,j}$'s. We have a total of $O(n^{k(2c+r)} m^{3k}) \in poly(n)$ possibilities and each of them can be computed in polynomial time. The entire algorithm is therefore polynomial time. ☐

4.3 Security of other bits

The same reasoning as in [1], [3] also gives

Theorem 12. *Let a, b, p be chosen as in Theorem 8. Let c be any positive constant and let $j \leq c \log |p|$. Then: 1. The j:th least significant bit of $ax + b$ $\pmod{p}$ is n^{-c}-secure. 2. The j least significant bits are simultaneously secure.*

To see 1, note that in the proof of Lemma 11 we can afford to try all possibilities for the j least significant bits. Also, in Lemma 9 we can "guess" the $j - 1$ least significant bits (or assume that they are $000 \cdots 0$) and start the algorithm from the j:th bit.

The simultaneous security follows from 1 and from Yao's unpredictability criterion, [7]: Predicting the j:th bit given the $j - 1$ first bits is equivalent to distinguishing the j least significant bits from random bits.

What about other bits? We first note that it is easy to see that the internal bits are secure with respect to a perfect oracle. Furthermore, we have recently been able to show the following result.

Theorem 13. *For "certain" primes $p \in \mathcal{P}_k$, for any $i = \alpha n/k$, $0 < \alpha < 1$, the i:th bit of $ax + b$ (mod p) for $a, b \in_U \mathbf{Z}_p$ is a family of hard-core predicates for any one-way function.*

The intuitive definition for "certain primes" is primes "sufficiently far from" a multiple of 2^i. At the present time some refinements will have to be made to make this set sufficiently dense among the set of all primes in $\mathcal{P}_k$. We will explore this further in the full paper.

5 Open problems

The proof of Theorem 8 does not hold if we choose $|p|$ too small. However, we ask if it would be possible to improve the result to allow for primes significantly shorter than $\Omega(n)$. For instance, is the theorem still true if $|p| \approx \sqrt{n}$?

The three common examples of UHF's often given in the literature are:

1. Multiplication by a randomly chosen boolean matrix.
2. Linear functions on $GF[2^n]$.
3. Linear functions on $\mathbf{Z}_p$. (Almost universal.)

The results in [4] together with the above results show that all these give a logarithmic number of hard-core bits. The natural conjecture must be that *all* UHF's give hard-core bits. We note that both the proof in [4] as well as the proofs given here rely heavily on the explicit construction of the hash function. Some new technique seems called for to approach the general case.

6 Acknowledgements

I would like to thank my supervisor, Johan Håstad, for many enlightening discussions and for pointing out several errors in the early versions of this paper. Thanks to Christer Berg for further proofreading help.

References

1. W. Alexi, B. Chor, O. Goldreich & C. P. Schnorr: *RSA and Rabin Functions: Certain Parts Are as Hard As the Whole.* SIAM J. on Computing vol 17, no 2 1988, pp. 194–209.
2. J. L. Carter & M. N. Wegman: *Universal Classes of Hash Functions.* JCSS 18 1979, pp. 265–278.
3. B. Chor: *Two Issues in Public Key Cryptography.* An ACM distinguished Dissertation. MIT Press 1985.

4. O. Goldreich & L. A. Levin: *A Hard Core Predicate for any One Way Function.* STOC 1989, pp. 25–32.
5. J. Håstad, R. Impagliazzo, L. A. Levin & M. Luby: *Pseudo Random Number Generators from any One Way Function.* Manuscript 1993. Earlier versions appeared in STOC 1989, 1990.
6. U. V. Vazirani & V. V. Vazirani: *Efficient and Secure Pseudo-Random Number Generation.* FOCS 1984, pp. 458–463.
7. A. C. Yao: *Theory and Applications of Trapdoor Functions.* FOCS 1982, pp. 80–91.

Recycling Random Bits in Composed Perfect Zero-Knowledge

Giovanni Di Crescenzo *

Abstract. In this paper we give techniques for recycling random bits both in the interactive and in the non-interactive model for perfect zero-knowledge proofs. Our first result is a non-interactive perfect zero-knowledge proof system for proving that at least one out of any given polynomial number of statements is true, in which the amount of public random bits used is the same as that for proving a single statement. Our second result is an interactive perfect zero-knowledge proof system for proving any given polynomial number of statements, in which the amount of private random bits used by the prover is, apart from a constant factor, the same as that for proving a single statement. In order to get a randomness-efficient proof system, we also reduce the random string of the verifier by using a multi-bit commitment scheme. The statements considered are of quadratic non residuosity modulo a Blum integer.

1 Introduction

Quantitative aspects of randomness in cryptographic protocols are now emerging as a new interesting research area in cryptology (see, e.g., [5, 13, 25, 26, 23, 1]). In perfect zero-knowledge proof systems the randomness of the prover is crucial to obtain the perfect zero-knowledge property. This paper investigates quantitative aspects of randomness in perfect zero-knowledge proof systems both in the interactive and in the non-interactive model.

Zero-knowledge.

Goldwasser, Micali and Rackoff [19] introduced the concept of interactive proof systems as a method for proving the veridicity of membership of a string to a language. In the same paper, they introduced zero-knowledge proof systems as a method for proving such statements without revealing any additional information. The model for zero-knowledge proofs considers an all-powerful prover interacting with a poly-bounded verifier; moreover, both parties are allowed to flip coins.

Any language having an interactive proof has a computational zero-knowledge proof (see [18, 2, 21]), that is a proof which does not reveal additional information to a poly-bounded verifier. On the other hand, only few languages (basically relying on random self-reducible [27] properties) have been proved to have a perfect zero knowledge proofs (see [19, 18, 17, 27, 3, 12]). Perfect zero-knowledge

---

* Department of Computer Science and Engineering, University of California, San Diego, La Jolla, CA 92093; and Dipartimento di Informatica ed Applicazioni, Università di Salerno, 84081 Baronissi (SA), Italy

L.C. Guillou and J.-J. Quisquater (Eds.): Advances in Cryptology - EUROCRYPT '95, LNCS 921, pp. 367-381, 1995.
© Springer-Verlag Berlin Heidelberg 1995

proofs are proofs which do not reveal additional information even to an infinitely-powerful verifier. It is unlikely that they can be given for NP-complete languages, as this would imply that the polynomial hierarchy collapses to its second level ([16, 7]). Thus giving a perfect zero-knowledge proof for a language allows to give evidence that a language is not NP-complete. Moreover, perfect zero-knowledge proofs do not rely on any unproven hypothesis and really capture the intrinsic properties of the concept of zero-knowledge proof systems. For all these reasons, giving new techniques for perfect zero-knowledge proofs is still an interesting research area.

Randomness.

The soundness of interactive proof systems is strongly based on the unpredictability of the random questions that the verifier makes to the prover. On the other hand, any language having a proof system with a probabilistic prover has one with a deterministic prover: we can choose the prover that maximizes the acceptance probability of the verifier. In [1] techniques for recycling the randomness of the verifier have been given for Arthur-Merlin proof systems. In zero-knowledge proof systems, instead, the prover has to use randomness in computing his messages not to reveal information to the verifer, as he might do in an interactive proof. In [20] the necessity of randomness for provers and verifiers in zero-knowledge proof systems is shown.

In the non-interactive model (see [6, 4]) for zero-knowledge proofs, the prover and the verifier share a random reference string and the proof is a single message sent by the prover to the verifier. One of the main problems of the non-interactive model is that often the size of the public random string bounds the length of the theorem that can be proved. Thus it is very desirable to give non-interactive zero-knowledge proofs for many statements using the same public random string. (see also [4, 15] for discussions). This problem was solved in the case of computational zero-knowledge in [4, 14, 15]. However, in the case of perfect zero-knowledge, a solution to this problem is still unknown. Our first result in this paper is a non-interactive perfect zero-knowledge proof system for proving that at least one out of any polynomial number of statements is true, in which the length of the public random bits is the same as that for proving a single statement.

In the interactive model for perfect zero-knowledge proofs, no technique has been given in order to recycle the randomness of the prover. Thus, the better way for a prover to prove many statements of a certain language in perfect zero-knowledge was using different and independently chosen random strings for each new statement. Our second result in this paper is an interactive perfect zero-knowledge proof system for proving any polynomial number of statements, in which, up to a constant factor, the random string used by the prover is the same as for proving a single statement. To make our proof system randomness-efficient, we also reduce the random string of the verifier by using a multi-bit (weak-to-strong) commitment scheme.

The statements considered are of quadratic non residuosity modulo a Blum integer. Our results are for specific languages, while the result in [15] is given for all languages in NP. On the other hand, non-interactive perfect zero-knowledge proofs have been given so far only for languages that are composition of quadratic non residuosity statements. Also, interactive zero-knowledge proofs for quadratic residuosity are among the most used for cryptographic applications like identification schemes.

In all our proof system the prover runs in probabilistic polynomial time, when given the factorization of the Blum modulus as private input.

Organization of the paper:

In Section 2 we describe some number theoretic properties about quadratic residuosity and Blum integers that will be useful in our protocols, and review the definition of non-interactive perfect zero-knowledge proof systems and the non-interactive perfect zero-knowledge proof system of [4] for the language of quadratic non residuosity. In Section 3 we give our result of recycling public random bits in the non-interactive model for perfect zero-knowledge proofs. In Section 4 we give a multi-bit commitment scheme whose security is based on the difficulty of factoring Blum integers. In Section 5 we give our result of recycling the private random bits of the prover in the interactive model for perfect zero-knowledge proofs.

2 Background and Definitions

2.1 Quadratic residuosity and Blum integers

Quadratic Residuosity. For each integer $x > 0$, the set of integers less than x and relatively prime to x form a group under multiplication modulo x denoted by Z_x^*. We say that $y \in Z_x^*$ is a *quadratic residue* modulo x if and only if there is a $w \in Z_x^*$ such that $w^2 \equiv y \bmod x$. If this is not the case, then y is a *quadratic non residue* modulo x. The quadratic residuosity predicate of an integer $y \in Z_x^*$ can be defined as $Q_x(y) = 0$ if y is a quadratic residue modulo x and 1 otherwise. Define Z_x^{+1} and Z_x^{-1} to be, respectively, the sets of elements of Z_x^* with Jacobi symbol $+1$ and -1 (see [24] for the definition of Jacobi symbol). The Jacobi symbol can be computed in deterministic polynomial time. Also, define the set $QR_x = \{y \in Z_x^* \mid Q_x(y) = 0\}$ of quadratic residues modulo x, and the set $NQR_x = \{y \in Z_x^{+1} \mid Q_x(y) = 1\}$ of quadratic non residues modulo x. The quadratic residuosity predicate defines the following equivalence relation in Z_x^*: $y_1 \sim_x y_2$ if and only if $Q_x(y_1 y_2) = 0$. Thus, the quadratic residues modulo x form a $\sim_x$ equivalence class. If $y \in Z_x^{-1}$, then y is a quadratic non residue modulo x. However, if $y \in Z_x^{+1}$, no efficient algorithm is known to compute $Q_x(y)$. The fastest way known for computing $Q_x(y)$ consists of first factoring x.

Blum integers. In this paper we will consider the special moduli called Blum integers. An integer x is a Blum integer, in symbols $x \in \text{BL}$, if and only if $x = p^{k_1} q^{k_2}$, where p and q are different primes both $\equiv 3 \bmod 4$, and k_1 and k_2 are odd integers. If x is a Blum integer, Z_x^* is partitioned by $\sim_x$ into 4 equally large equivalence classes. Also, $|Z_x^{+1}| = |Z_x^{-1}|$ and Z_x^{+1} is partitioned into 2 equally large equivalence classes, one made of quadratic residues modulo x and the other made of quadratic non residues modulo x. Thus, for this special class of integers we have that for any $y_1, y_2 \in Z_x^*$, $Q_x(y_1) = Q_x(y_2) \implies Q_x(y_1 y_2) = 0$, and $Q_x(y_1) \neq Q_x(y_2) \implies Q_x(y_1 y_2) = 1$. Then a quadratic residue modulo a Blum integers x has exactly four square roots, one in each $\sim_x$ equivalence class, and exactly one of them will be a quadratic residue modulo x. Moreover, if x is a Blum integer, then $-1 \bmod x$ is a quadratic non residue with Jacobi symbol $+1$. This implies that on input a Blum integer x, it is easy to generate a random quadratic non residue in Z_x^{+1}: randomly select $r \in Z_x^*$ and output $-r^2 \bmod x$. Finally, if x is a Blum integer, given its prime factors p, q, it is possible to compute square roots modulo x in deterministic polynomial time.

We refer the reader to [24, 4] for a more formal treatment and for proofs.

2.2 Perfect Zero-Knowledge Proof Systems

Now we review the definition of perfect zero-knowledge proof systems of [19]. Let L be a language and x be an instance to it. Let P a probabilistic Turing machine and V a deterministic Turing machine that runs in time polynomial in the length of its first input.

Definition 1. We say that (P, V) is a Perfect Zero-Knowledge Proof System for the language L if

1. *Completeness.* $\forall x \in L$, $|x| = n$ and for all sufficiently large n,

$$\mathbf{Pr}(P \leftrightarrow V)(x) = \text{ACCEPT} \geq 1 - 2^{-n}.$$

2. *Soundness.* $\forall x \notin L$, $|x| = n$ and for all sufficiently large n, for all probabilistic algorithms P',

$$\mathbf{Pr}(P' \leftrightarrow V)(x) = \text{ACCEPT} \leq 2^{-n}.$$

3. *Perfect Zero Knowledge.* for each V' there exists a probabilistic machine $S_{V'}$ running in expected polynomial time such that $\forall x \in L$, the two probability spaces $S_{V'}(x)$ and $View_{V'}(x)$ are equal, where by $View_V(x)$ we denote the probability space $(R; conv)$, where R is the random tape of V, and $conv$ is the transcript of a conversation between P and V on input x given that R is the random tape of V.

Now we review the definition of non-interactive perfect zero-knowledge proof systems of [4], referring the reader to the original paper for motivations and discussions. Let L be a language and x be an instance to it. Let P a probabilistic Turing machine and V a deterministic Turing machine that runs in time polynomial in the length of its first input.

Definition 2. We say that (P, V) is a Non-Interactive Perfect Zero-Knowledge Proof System for the language L if there exists a positive constant c such that:

1. *Completeness.* $\forall x \in L$, $|x| = n$ and for all sufficiently large n,

$$\mathbf{Pr}(\sigma \leftarrow \{0,1\}^{n^c}; Proof \leftarrow P(\sigma, x) : V(\sigma, x, Proof) = 1) \geq 1 - 2^{-n}.$$

2. *Soundness.* For all probabilistic algorithms *Adversary* giving pairs $(x, Proof)$ as output, where $x \notin L$, $|x| = n$, and all sufficiently large n,

$$\mathbf{Pr}(\sigma \leftarrow \{0,1\}^{n^c}; (x, Proof) \leftarrow Adversary(\sigma) : V(\sigma, x, Proof) = 1) \leq 2^{-n}.$$

3. *Perfect Zero Knowledge.* There exists an efficient simulator algorithm S such that $\forall x \in L$, the two probability spaces $S(x)$ and $View_V(x)$ are equal, where by $View_V(x)$ we denote the probability space $View_V(x) = \{\sigma \leftarrow \{0,1\}^{|x|^c}; Proof \leftarrow P(\sigma, x) : (\sigma, Proof)\}$.

We call the random string σ, input to both P and V, the *reference string*.

2.3 Quadratic non-residuosity

In [4] a non-interactive perfect zero-knowledge proof system for the languages of quadratic non residuosity modulo a *Regular*(2) integer was given (a *Regular*(2) integer is odd, is not a perfect square and has two prime factors). We briefly review it here, as it will be useful to better understand our protocols.

On input an n-bit integer x and an integer $y \in Z_x^{+1}$, the prover takes $2n$ integers $\sigma_1, \ldots, \sigma_{2n}$ from the 20^2-bit long reference string σ such that $\sigma_i \in Z_x^{+1}$, for $i = 1, \ldots, 2n$. Then, for each σ_i, the prover does the following. If σ_i is a quadratic residue modulo x, then he uniformly chooses an integer $r_i \in Z_x^*$ such that $r_i^2 = \sigma_i \bmod x$. On the other hand, if σ_i is a quadratic non residue modulo x, then he uniformly chooses an integer $r_i \in Z_x^*$ such that $r_i^2 = y \cdot \sigma_i \bmod x$. Finally, the prover sends $r_i \in Z_x^*$ to the verifier. The verifier checks that for each $i = 1, \ldots, 2n$, $r_i^2 = y^{b_i} \cdot \sigma_i \bmod x$, for some bit b_i. If so, then he accepts, otherwise he rejects (see [4] for proofs).

If we consider the language of quadratic non residuosity modulo a Blum integer (instead of a *Regular*(2) integer) then we have to add a phase to this protocol, in which it is proved that x is a Blum integer, as in [10]. In this case the length of the reference string is $50n^2$. From now on, we will call this protocol (C,D).

3 OR of many statements on a single random string

In this section we give a non-interactive perfect zero-knowledge proof system for proving the OR of any polynomial number (in the size of the input) of statements of quadratic non residuosity modulo a Blum integer x in which the length of the random reference string used is the same as that of the protocol (C,D) for the language of quadratic non residuosity modulo x.

For simplicity, we consider the language of triples (x, y_1, y_2), such that x is a Blum integer, and at least one of y_1, y_2 is a quadratic non residue modulo x:

$$OR = \{(x, y_1, y_2) \mid x \in BL, y_1, y_2 \in Z_x^{+1}, (y_1 \in NQR_x) \vee (y_2 \in NQR_x)\}.$$

We give a non-interactive perfect zero-knowledge proof system (A,B) for the language OR and then briefly explain how our construction can be easily extended to any polynomial number (in $|x| = n$) of integers y_i.

An informal description. On input (x, y_1, y_2), A writes the reference string σ as the concatenation of two sufficiently long strings σ_1, σ_2. Then, A uses σ_1 to give a non-interactive perfect zero-knowledge proof that x is a Blum integer, for instance, using the proof system given in [10]. Now, we make a simplifying assumption by considering the reference string σ_2 written as the concatenation of $2n$ integers $\sigma_{2i} \in Z_x^{+1}$.[2] The main idea for proving that at least one of y_1, y_2 is a quadratic non residue modulo x is the following. The prover P computes in a careful way (to be specified later) two strings ρ_1, ρ_2 of the same length as σ, such that $\rho_{1i} \cdot \rho_{2i} = \sigma_{2i} \bmod x$, for $i = 1, \ldots, 2n$, and sends a 'proof' that y_1, y_2 are quadratic non residues modulo x, computed using as reference strings ρ_1, ρ_2, respectively. V verifies that the 'proofs' are correctly constructed on the reference strings ρ_1, ρ_2 and that $\rho_{1i} \cdot \rho_{2i} = \sigma_{2i} \bmod x$, for $i = 1, \ldots, 2n$. Of course, P has to convince V even if, say, y_1 is a quadratic residue modulo x. Thus, he computes the two strings ρ_1, ρ_2 in the following way: the string ρ_1 is made of integers of the same quadratic residuosity as y_1, and the string ρ_2 is computed from ρ_1 and σ in order to satisfy $\rho_{1i} \cdot \rho_{2i} = \sigma_{2i} \bmod x$, for $i = 1, \ldots, 2n$. We observe that if y_1 is a quadratic residue, then the string ρ_1 is made of all quadratic residues. Thus P can compute a faked 'proof' of quadratic non residuosity for y_1 using ρ_1 as a reference string. Then, the string ρ_2 is a uniformly distributed string, and if y_2 is a quadratic non residue modulo x, then P can give a non-interactive proof of quadratic non residuosity for y_2 using ρ_2 as a reference string. If B cannot distinguish between integers with the same quadratic residuosity as y_1 and integers with the same quadratic residuosity as y_2, then the faked 'proof' for y_1 and the 'real' proof for y_2 will appear indistinguishable to him. That is, each of the two will appear as a proof of quadratic non residuosity (as in [4]) for y_j using ρ_j as a reference string, for $j = 1, 2$.

Let (E,F) be a non-interactive perfect zero-knowledge proof system for the language BL of Blum integers (see, e.g., [10]). Now we give a formal description of (A,B).

---

[2] Even if this is not true in general, (for instance, in a uniformly distributed random string there may be integers in Z_x^{-1}) in [4, 10] a technique preserving perfect zero-knowledge is given for transforming a uniformly distributed string into a string made of integers in Z_x^{+1}.

Input to A and B: $(x, y_1, y_2) \in \text{OR}$, such that $|x| = n$, and the reference string $\sigma = \sigma_1 \circ \sigma_{21} \circ \cdots \circ \sigma_{2,2n}$, where $\sigma_{2i} \in Z_x^{+1}$, for $i = 1, \ldots, 2n$.

Input to A: x's factorization.

Instructions for A.

  A.1 Prove that x is a Blum integer by running algorithm E on input x and using σ_1 as a reference string. Send E's output Pf to B.

  A.2 If $y_2 \in QR_x$ then set $z_2 = y_1$ and $z_1 = y_2$;
  else set $z_1 = y_1$ and $z_2 = y_2$.

  A.3 For $i = 1, \ldots, 2n$,
  uniformly choose $c_{1i} \in \{0, 1\}$, $r_{1i} \in Z_x^*$;
  set $\rho_{1i} = z_1^{c_{1i}} \cdot r_{1i}^2 \bmod x$ and $\rho_{2i} = \sigma_{2i} \cdot \rho_{1i}^{-1} \bmod x$;
  compute $c_{2i} \in \{0, 1\}$ and a randomly chosen $r_{2i} \in Z_x^*$ such that
  $r_{2i}^2 = z_2^{c_{2i}} \cdot \rho_{2i} \bmod x$.
  if $z_1 = y_1$ then send $(r_{1i}, r_{2i}), (c_{1i}, c_{2i}), (\rho_{1i}, \rho_{2i})$ to B;
  else send $(r_{2i}, r_{1i}), (c_{2i}, c_{1i}), (\rho_{2i}, \rho_{1i})$ to B.

Input to B: The proof Pf and the set $\{(u_{1i}, u_{2i}), (d_{1i}, d_{2i}), (\tau_{1i}, \tau_{2i})\}_{i=1,\ldots,2n}$ sent by A.

Instructions for B.

  B.1 Verify that Pf is a proof that x is a Blum integer by running algorithm F on input x and using σ_1 as a reference string.

  B.2 For $i = 1, \ldots, 2n$,
  verify that $u_{1i}^2 = y_1^{d_{1i}} \cdot \tau_{1i} \bmod x$, and $u_{2i}^2 = y_2^{d_{2i}} \cdot \tau_{2i} \bmod x$;
  verify that $\sigma_{2i} = \tau_{1i} \cdot \tau_{2i} \bmod x$;

  B.3 If all the verifications are successful then accept else reject.

Let (C,D) be the non-interactive perfect zero-knowledge proof system for the language NQR. Now we prove the following

Theorem 3. *(A, B) is a non-interactive perfect zero-knowledge proof system for the language OR, such that the length of the random reference string used by (A,B) is the same as that used by (C,D).*

Proof. *Randomness.* As for (C,D), the random reference string is divided into two parts. The first is $30n^2$-bits long and it is used to prove that x is a Blum integer, as in (C,D). The second is used to prove that at least one of y_1, y_2 is a quadratic non residue modulo x, and is $20n^2$-bits long, that is, exactly as the remaining part of the reference string in (C,D) used to prove the quadratic non residuosity of only one integer.

Completeness. If at least one of y_1, y_2 is a quadratic non residue modulo x, then A, which is given x's factorization as private input, can set z_1 equal to this integer. Also, he can run algorithm E to prove that x is a Blum integer, and finally compute a random square root of $z_2^{c_{2i}} \cdot \rho_{2i} \bmod x$, for some bit c_{2i} and any $i = 1, \ldots, 2n$. Then completeness follows from these observations and the completeness of (E,F).

Soundness. We distinguish two cases. First, assume that x is not a Blum integer. Then from the soundness of (E,F), we have that B accepts with negligible probability. Then, assume that y_1, y_2 are quadratic residues modulo x and that B accepts. Then the two strings ρ_1, ρ_2 can be written as the concatenation of integers that are the product of a quadratic residue and $y_j^{d_{ji}}$, for some bit d_{ji}. Thus each string ρ_j is made of $2n$ quadratic residues modulo x, and so is also σ_2, as B verifies that $\sigma_{2i} = \rho_{1i} \cdot \rho_{2i} \bmod x$, for every $i = 1, \ldots, 2n$. The event that σ_2 is made of all quadratic residues happens with probability $1/2^{2n}$. Then the probability that there exists a modulus x such that B accepts is at most $2^n / 2^{2n} = 1/2^n$, which is negligible.

Perfect Zero-Knowledge. We give a simulator M such that, for each $(x, y_1, y_2) \in$ OR, the output of M on input (x, y_1, y_2) and the view of B in the protocol (A,B) are identically distributed. Let N be the simulator for the non-interactive perfect zero-knowledge proof system (E,F) for the language of Blum integers.

Input to M: $(x, y_1, y_2) \in$ OR, where $|x| = n$.

Instructions for M:

1. Run the algorithm N on input x obtaining as output (σ_1, Pf); set $Proof = Pf$;
2. For $i = 1, \ldots, 2n$,
   uniformly choose $u_{1i}, u_{2i} \in Z_x^*$, and $d_{1i}, d_{2i} \in \{0, 1\}$;
   set $\tau_{1i} = y_1^{-d_{1i}} \cdot u_{1i}^2 \bmod x$ and $\tau_{2i} = y_2^{-d_{2i}} \cdot u_{2i}^2 \bmod x$;
   set $\sigma_{2i} = \tau_{1i} \cdot \tau_{2i} \bmod x$;
   set $Proof = Proof \circ (u_{1i}, u_{2i}) \circ (d_{1i}, d_{2i}) \circ (\tau_{1i}, \tau_{2i})$.
3. Set $\sigma = \sigma_1 \circ \sigma_{21} \circ \cdots \circ \sigma_{2,2n}$ and output $(\sigma, Proof)$.

It is easy to see that the simulator runs in probabilistic polynomial time. Now we prove that the output of M and the view of B in the protocol are equally distributed. First of all let us prove that for each $i = 1, \ldots, 2n$, it holds that d_{1i}, d_{2i} are uniformly distributed bits both in the output of M and in the view of B. This happens as in the output of M both d_{1i} and d_{2i} are uniformly chosen over $\{0, 1\}$; and in the view of B one of them, say d_{1i}, is uniformly chosen over $\{0, 1\}$, and the other, d_{2i}, satisfies $u_{2i}^2 = z_2^{d_{2i}} \cdot \sigma_{2i} \cdot \rho_{1i}^{-1} \bmod x$, where z_2 is a quadratic non residue modulo x. We see from the above equation that the value of d_{2i} depends from the quadratic residuosity of σ_{2i}, and so it is uniformly distributed over $\{0, 1\}$. Then we see that for each $i = 1, \ldots, 2n$, it holds that $u_{1i}, u_{2i}, \tau_{1i}, \tau_{2i}$ are randomly distributed integer in Z_x^* such that $u_{1i}^2 = y_1^{d_{1i}} \cdot \tau_{1i} \bmod x$ and $u_{2i}^2 = y_2^{d_{2i}} \cdot \tau_{2i} \bmod x$ both in the output of M and in the view of B. Finally, also σ_1 is equally distributed in both spaces, from the perfect zero-knowledge of (E,F). $\square$

Now, let us briefly explain how this protocol easily extends to proving an OR of any polynomial number $m = |x|^c$ of quadratic non residuosity modulo x statements. The input is then $(x, y_1, \ldots, y_m)$. Assume that y_m is a quadratic non residue modulo x. Then the prover computes a string τ_j for each y_j in a way much similar as in the algorithm A. More precisely, strings τ_j, for $j = 1, \ldots, m-1$ are computed as ρ_1, and the last string τ_m is computed similarly to ρ_2: by multiplying all elements in previous strings τ_j and the relative element of σ. The remaining parts of the protocol are constructed exactly as in (A,B).

Let $(A,B)_m$ be the above protocol, and let OR_m be the language of $(m+1)$-tuples $(x, y_1, \ldots, y_m)$ such that x is a Blum integer and at least one of $y_1, \ldots, y_m$ is a quadratic non residue modulo x. Also, let (C,D) be the non-interactive perfect zero-knowledge proof system for the language NQR. Then we have the following

Theorem 4. $(A, B)_m$ *is a non-interactive perfect zero-knowledge proof system for the language OR_m, such that the length of the random reference string used by $(A,B)_m$ is the same as that used by (C,D).*

This result improves the protocol to prove an OR of m quadratic non residuosity statements given in [10], which needs a reference string of length $O(m)$ times that of the random string used by (C,D).

4   A multi-bit commitment scheme

In this section we describe a scheme (S,R) in which a sender S can commit to many bits and reveal one of them at each round to a receiver R. The main property of this scheme is that it can be implemented with a small amount of randomness. The scheme will be used to reduce the randomness of the verifier in the construction of a randomness-efficient perfect zero-knowledge proof system for proving any polynomial number of quadratic non residuosity statements. Similar techniques have already been used in literature, e.g. in [9] where efficient weak-to-strong bit commitment schemes are presented, using universal hash functions. Also, in [22] general techniques for efficient weak-to-strong bit commitment schemes have been given. First of all, let us define a multi-bit (weak-to-strong) commitment scheme.

Definition 5. A *(weak-to-strong)*[3] *multi-bit commitment scheme* is a two-phase protocol with two participants: a (weak) sender with probabilistic polynomial-time computing power and a (strong) receiver with unlimited computing power. In the first phase (*the commitment phase*), the sender has m bits $b_1, \ldots, b_m$ and commits to them by computing an $(m+1)$-tuple of "keys" $(Com, Dec_1, \ldots, Dec_m)$ and sends Com (the commitment key) to the receiver. The second phase (*decommitment phase*) can be divided in m subphases. In each of these m subphases

---

[3] Such commitment schemes have been referred in the literature also as *blob schemes*, see, e.g., [8], and *statistically hiding bit commitment schemes*, see, e.g. [9].

the sender reveals the bit b_i along with Dec_i (the i-th decommitment key) to the receiver. A (weak-to-strong) multi-bit commitment scheme has the following two main properties: security and correctness. The *security* property states the following: for each $i = 0, \ldots, m - 1$, from Com and $Dec_1, \ldots, Dec_i$, the receiver cannot guess b_{i+1} with probability significantly better than $1/2$. The *correctness* property states the following: for each $i = 1, \ldots, m$, the receiver obtains a valid decommitment key for a bit c_i, and he is sure that the sender is revealing the same bit b_i to which he committed before.

We observe that given a (weak-to-strong) 1-bit commitment scheme, it is possible to obtain a (weak-to-strong) multi-bit commitment scheme by just using the 1-bit scheme for each of the many bits. In this case, however, the amount of randomness used in an m-bit commitment scheme is m times that used in a 1-bit scheme. Our m-bit commitment scheme is also derived from a 1-bit scheme, but uses an amount of randomness equal to only twice the same amount of the 1-bit commitment scheme. The 1-bit scheme that we use is the following folklore scheme: on input a Blum integer x, and in order to commit to a bit b, the sender uniformly chooses an integer $r \in Z_x^{+1}$ if $b = 0$ and $r \in Z_x^{-1}$ if $b = 1$ and outputs $w = r^2 \bmod x$. In order to reveal bit b, the sender sends r and the receiver sets $b = 0$ if $r \in Z_x^{+1}$, and $b = 1$ if $r \in Z_x^{-1}$. It is easy to see that the receiver obtains a uniformly distributed quadratic residue modulo x for both values of b. Also, if the sender can reveal the commitment in two different ways, then he knows two different square roots modulo x of w and thus he can factor x. Now we extend this commitment scheme to a multi-bit commitment scheme (S,R), using only twice the same amount of randomness. The correctness property is still based on the intractability of factoring Blum integers.

Input to S and R: A Blum integer x and m bits $b_1, \ldots, b_m$.

Commitment Phase

S: Uniformly choose $w_0 \in Z_x^{+1}$ and $s \in Z_x^{-1}$;
for $i = 1, \ldots, m$,
  set $r_i = w_{i-1} \cdot s^{b_i} \bmod x$, $w_i = r_i^2 \bmod x$ and $Dec_i = r_i$;
set $Com = (s, w_m)$ and send Com to R.
R: Verify that $s \in Z_x^{-1}$.

Decommitment Phase

For $i = m, \ldots, 1$,
  S: Send (b_i, Dec_i) to R.
  R: Let $z_i = Dec_i$; verify that $z_i^2 = z_{i+1} \cdot s^{b_i} \bmod x$;
    verify that $(z_i \in Z_x^{+1} \text{ AND } b_i = 0)$ OR $(z_i \in Z_x^{-1} \text{ AND } b_i = 1)$.

The security property of the above scheme follows from the following observation: for any m-tuple of bits $b_1, \ldots, b_m$ input to (S,R), the integers w_m and

$Dec_2, \ldots, Dec_m$ sent by S are uniformly distributed quadratic residues modulo x. The correctness property of the above scheme follows from the following observation: If any sender can correctly reveal two different bits in the i-th decommitment subphase, then he sends two different square root modulo x of a same integer z_i^2, and thus he can factor x. The above discussion informally proves the following

Theorem 6. *If factoring Blum integers is hard, then (S,R) is a (weak-to-strong) multi-bit commitment scheme such that the number of random bits used in (S,R) does not depend on the number of bits committed.*

A formal proof will appear in the final paper.

5 Many statements on a single random string

In this section we give an interactive perfect zero-knowledge proof system (P,V) for proving any polynomial number (in the size of the input) of statements of quadratic non residuosity modulo a Blum integer x, in which the length of the private random string used by the prover is, apart for a small constant factor, the same as that used in [19] for proving a single quadratic residuosity statement.

We consider the language of $(m+1)$-tuples $(x, y_1, \ldots, y_m)$, such that x is a Blum integer, and $y_1, \ldots, y_m$ are quadratic non residues modulo x; formally:

$$\text{AND}_m = \{(x, y_1, \ldots, y_m) \mid x \in \text{BL}, y_i \in Z_x^{+1}, y_i \in NQR_x, \text{for } i = 1, \ldots, m\}.$$

An informal description. The main ideas of this protocol are: 1) a four round interactive perfect zero-knowledge proof by combining a coin-tossing protocol between P and V with a non-interactive proof by P (a similar technique has been used in [11]), and 2) P uses the non-interactive proof of quadratic non residuosity of an integer y_k and V's challenges in order to compute the next non-interactive proof for the integer y_{k+1}, without using any random bits. More precisely, in the first three rounds P and V run a coin-tossing protocol, in which V commits to his random bits using the scheme (S,R) of previous section. After the coin-tossing protocol both parties can compute a random reference string σ. P writes the string σ as $\rho \circ \tau$; then, τ will be used only once to prove that x is a Blum integer using, e.g., the proof system in [11], and ρ will be used to prove that all the y_i's are quadratic non residues modulo x. Now, P proves that the first integer y_1 is a quadratic non residue modulo the Blum integer x, by using a modification of the non-interactive proof system (C,D) of [4] and ρ as a reference string. More precisely, he takes $2n$ integers $\rho_i \in Z_x^{+1}$ from the string ρ; then he randomly chooses a square root $s_{i,1} \in Z_x^{+1}$ of $\rho_i \cdot y_1^{d_{i,1}} \bmod x$, for some bit $d_{i,1}$, and sends $s_{i,1}$ to V. Now, observe that the proof for y_1 constituted by the $s_{i,1}$'s looks very similar to the random integers ρ_i. In fact, both the $s_{i,1}$'s and the ρ_i's are integers in Z_x^{+1}. Now, V will reveal other $2n$ bits $b_{i,2}$ to which he committed in the first round, and P will give a non-interactive proof for y_2 as for y_1, but using as a random reference string the $2n$ integers $u_{i,2} = (-1)^{b_{i,1}} \cdot s_{i,1} \bmod x$. We observe that as -1 is a quadratic non residue modulo x, then the quadratic residuosity of the $u_{i,2}$'s is thus uniformly chosen by V. The proof system (P,V) continues analogously for the other integers y_j. For a better exposition, we avoid to be very formal in our step-by-step description of (P,V).

Input to P and V: $(x, y_1, \ldots, y_m)$ such that $|x| = n$, and $m = n^c$.

Input to P: x's factorization.

(Proving the first statement '$y_1 \in NQR'_x$.)

V.1 Uniformly choose $2nm$ bits $b_{i,k}$ and $10n^2$ bits e_i;
      use algorithm S to commit to them and send Com to P.

P.1 Uniformly choose $50n^2$ bits c_i and send them to V.

V.2 Use algorithm S to reveal all bits d_i to P.

P.2 Use algorithm R to check that V had committed to bits e_i;
      compute σ as the bitwise xor of the c_i's and the e_i's; let $\sigma = \rho \circ \tau$;
      prove that $x \in BL$, using the algorithm E and τ as a reference string;
      for each of the first $2n$ integers $\rho_i \in Z_x^{+1}$ from ρ,
        if $\rho_i \in QR_x$ uniformly choose $s_{i,1} \in Z_x^{+1}$ such that $s_{i,1}^2 = \rho_i \bmod x$;
        if $\rho_i \in NQR_x$ uniformly choose $s_{i,1} \in Z_x^{+1}$ such that $s_{i,1}^2 = y_1 \cdot \rho_i \bmod x$;
        set $u_{i,1} = \rho_i$ and send $s_{i,1}$ to V.

V.3.1 Use algorithm F to verify the proof that x is a Blum integer;
      verify that $s_{i,1}^2 = y_1^{d_{i,1}} \cdot u_{i,1} \bmod x$ for some bit $d_{i,1}$, and $i = 1, \ldots, 2n$;
      use algorithm S to reveal other $2n$ bits $b_{i,2}$ to P.

(Proving the k-th statement '$y_k \in NQR'_x$, for $k = 2, \ldots, m$.)

P.3.k For $i = 1, \ldots, 2n$,
      use algorithm R to check that V had committed to bit $b_{i,k}$;
      set $u_{i,k} = (-1)^{b_{i,k}} \cdot s_{i,k-1} \bmod x$;
      if $u_{i,k} \in QR_x$ uniformly choose $s_{i,k} \in Z_x^{+1}$ such that $s_{i,k}^2 = u_{i,k} \bmod x$;
      if $u_{i,k} \in NQR_x$ uniformly choose $s_{i,k} \in Z_x^{+1}$ such that
      $s_{i,k}^2 = y_k \cdot u_{i,k} \bmod x$;
      send $s_{i,k}$ to V.

V.3.k Verify that $s_{i,k}^2 = y_k^{d_{i,k}} \cdot u_{i,k} \bmod x$ for some bit $d_{i,k}$, and for $i = 1, \ldots, 2n$;
      use algorithm S to reveal other $2n$ bits $b_{i,k}$ and send their Dec_i to P.

V.4 If all the verifications are successful then accept else reject.

We see that our protocol is very randomness-efficient. In fact, the only random bits used by the prover are those chosen in step P.1. Then, the length of the random string used by the prover in order to prove m quadratic non residuosity statements is the same, apart from a constant factor, as in [19] for proving a single quadratic residuosity statement. Also, the length of the random string used by V during a proof of m quadratic non residuosity statements is equal to a $10n^2$-bit initial random string (needed in the proof that x is a Blum integer), and then only $2n$ random bits for each new statement (needed for the soundness in the proof of each new statement). In [19] the verifier uses n random bits for a single quadratic residuosity statement.

For the *completeness* requirement, observe that if x is a Blum integer and all integers $y_1, \ldots, y_m$ are quadratic non residues modulo x, then P, using x's factorization, can run algorithm E, compute the quadratic residuosity of the u_i's and compute the square roots belonging to Z_x^{+1} of the integers $u_{i,k} \cdot y_k^{d_{i,k}} \bmod x$, for some bits $d_{i,k}$. Thus V accepts with overwhelming probability.

For the *soundness* requirement, first of all assume x is not a Blum integer. Then V accepts with negligible probability from the soundness of (E,F). Now, suppose that $y_1, \ldots, y_{k-1}$ are quadratic non residues modulo x and y_k is a quadratic residue, for some $k \in \{1, \ldots, m\}$. We observe that the proof for y_{k-1} sent by P' is made of integers $s_{i,k-1} \in Z_x^{+1}$, whose quadratic residuosity is chosen by P' (we recall that both $s_{i,k-1}$ and $-s_{i,k-1} \bmod x$ are integers in Z_x^{+1} and square roots of a same number). On the other hand, the quadratic residuosity of the $u_{i,k}$'s forming the random string on which P' has to prove y_k is given by the quadratic residuosity of the $s_{i,k-1}$'s sent by P' xored with the new random bits $b_{i,k}$ revealed by V. Then if V accepts the $u_{i,k}$'s associated to y_k are all quadratic residues modulo x, and thus P' has given his $s_{i,k-1}$'s such that the quadratic residuosity of each $s_{i,k-1}$'s is exactly equal to $b_{i,k}$. This implies that P' has guessed $2n$ bits $b_{i,k}$ to which V committed in the first round, but this happens with probability at most $1/2^{2n}$, for the security property of the multi-bit commitment scheme (S,R). Thus, the probability that there exists an n-bit modulus x such that V accepts the proof of any possible $y_k \in QR_x$ is at most $m \cdot 2^n / 2^{2n} = m/2^n$, which is negligible.

For the *perfect zero-knowledge* requirement, we sketch a description of the simulator M on input $(x, y_1, \ldots, y_m) \in \mathrm{AND}_m$. The simulation of the proof of the first statement can be easily derived from that in [11]. Now we describe the simulation of the proof of the k-th statement. Assume that M has successfully simulated the first $k-1$ proofs. This implies that he has learned all the questions $b_{i,j}$ of V' relative to them, and that he has just sent to V' the proof for y_{k-1} consisting in $2n$ integers $s_{i,k-1} \in Z_x^{+1}$. Then M receives from V' other $2n$ bits $b_{i,k}$ to which V' had committed in the first round. After learning bits $b_{i,k}$, M simulates again all P's messages of the proofs of the j-th statements, for $j = 1, \ldots, k-1$, in such a way that he can simulate also the proof of the k-th statement. He does this in the following way: let $d_{i,k}$ be the bits sent by V' relative to the j-th proof. Then M uniformly chooses $2n$ integers $s_{i,k} \in Z_x^{+1}$ and computes $u_{i,k} = y_k^{d_{i,k}} \cdot s_{i,k}^2 \bmod x$, and $s_{i,k-1} = u_{i,k} \cdot (-1)^{b_{i,k}} \bmod x$. M computes the $s_{i,j}$ and $u_{i,j}$ for $j < k$ analogously to the above $s_{i,k-1}$ and $u_{i,k}$. Also, he computes the bits c_i analogously as in the simulation of the proof for y_1. Now M rewinds V' to the state just after his first step and sends the messages just computed to V' in the proper succession in order to simulate P's messages. We observe that if V' does not change any of his decommitted bits, then M succeeds in simulating all proofs of the first k statements. On the other hand, if V' reveals some bits in different way, then by the security property of the multi-bit commitment scheme (S,R), V' sends to M two different square roots of a same quadratic residue modulo x. Thus M can factor x and simulate perfectly the protocol by just running the algorithm of P.

The above discussion informally proves the following

Theorem 7. (P, V) *is a perfect zero-knowledge proof system for the language* AND_m *such that the length of the random string used by P is, up to a small constant factor, the same as in the proof system in [19] for the language QR. Moreover, the number of the random bits used by V is* $2|x|m + O(|x|^2)$.

A formal proof will appear in the final paper.

Acknowledgements

Many thanks go to Alfredo De Santis, Russell Impagliazzo, Markus Jakobsson and Giuseppe Persiano for useful discussions, and to an anonymous referee, whose careful comments have improved the exposition of this paper.

References

1. M. Bellare, O. Goldreich, and S. Goldwasser, *Randomness in Interactive Proof Systems*, Proceedings of the 31th Annual IEEE Symposium on Foundation of Computer Science, 1990, pp. 563–572.
2. M. Ben-Or, O. Goldreich, S. Goldwasser, J. Hastad, S. Micali, and P. Rogaway, *Everything Provable is Provable in Zero Knowledge*, in "Advances in Cryptology – CRYPTO 88", vol. 403 of "Lecture Notes in Computer Science", Springer Verlag, pp. 37–56.
3. J. Boyar, K. Friedl, and C. Lund, *Practical Zero-Knowledge Proofs: Giving Hints and Using Deficiencies*, Journal of Cryptology, n. 4, pp. 185–206, 1991.
4. M. Blum, A. De Santis, S. Micali, and G. Persiano, *Non-Interactive Zero-Knowledge*, SIAM Journal of Computing, vol. 20, no. 6, Dec 1991, pp. 1084–1118.
5. C. Blundo, A. De Santis, and U. Vaccaro, *Randomness in Distribution Protocols*, Proceedings of International Colloquium of Algorithms, Languages and Programming (ICALP) 1994.
6. M. Blum, P. Feldman, and S. Micali, *Non-Interactive Zero-Knowledge and Applications*, Proceedings of the 20th Annual ACM Symposium on Theory of Computing, 1988, pp. 103–112.
7. R. Boppana, J. Hastad, and S. Zachos, *Does co-NP has Short Interactive Proofs ?*, Inf. Proc. Lett., vol. 25, May 1987, pp. 127–132.
8. G. Brassard, C. Crépeau, and M. Yung, *Perfect Zero-Knowledge Computationally Convincing Proofs for NP in Constant Rounds*, Theoretical Computer Science, vol. 84, n. 1 (1991) pp. 23-52.
9. I. Damgaard, T. Pedersen, and B. Pfitzmann, *On the existence of statistically hiding bit commitment schemes and fail-stop signatures*, in "Advances in Cryptology – CRYPTO 93", vol. 773 of "Lecture Notes in Computer Science", Springer Verlag, pp. 250–265.
10. A. De Santis, G. Di Crescenzo, and G. Persiano, *Secret Sharing and Perfect Zero-Knowledge*, in "Advances in Cryptology – CRYPTO 93", vol. 773 of "Lecture Notes in Computer Science", Springer Verlag, pp. 73–84.

11. A. De Santis, G. Di Crescenzo, and G. Persiano, *The Knowledge Complexity of Quadratic Residuosity Languages*, in Theoretical Computer Science, Vol. 132 pp. 291-317 (1994).

12. A. De Santis, G. Di Crescenzo, G. Persiano, and M. Yung, *On Monotone Formula Closure of SZK*, Proceedings of the 35th Annual IEEE Symposium on Foundation on Computer Science, November 1994.

13. Y. Desmedt, C. Goutier, and S. Bengio, *Special Uses and Abuses of the Fiat-Shamir Passport Protocol*, in "Advances in Cryptology – CRYPTO 87", vol. 293 of "Lecture Notes in Computer Science", Springer Verlag.

14. A. De Santis and M. Yung, *Cryptographic Applications of the Non-Interactive Metaproof and Many-Prover Systems*, in "Advances in Cryptology – CRYPTO 90", vol. 537 of "Lecture Notes in Computer Science", Springer Verlag, pp 366-377.

15. U. Feige, D. Lapidot, and A. Shamir, *Multiple Non-Interactive Zero-Knowledge Proofs Based on a Single Random String*, in Proceedings of 22nd Annual Symposium on the Theory of Computing, 1990, pp. 308–317.

16. L. Fortnow, *The Complexity of Perfect Zero-Knowledge*, Proceedings of the 19th Annual ACM Symposium on Theory of Computing, 1987, pp. 204–209.

17. O. Goldreich and E. Kushilevitz, *A Perfect Zero Knowledge Proof for a Decision Problem Equivalent to Discrete Logarithm*, in "Advances in Cryptology - CRYPTO 88", Ed. S. Goldwasser, vol. 403 of "Lecture Notes in Computer Science", Springer-Verlag, pp. 57–70.

18. O. Goldreich, S. Micali, and A. Wigderson, *Proofs that Yield Nothing but their Validity and a Methodology of Cryptographic Design*, Proceedings of 27th Annual Symposium on Foundations of Computer Science, 1986, pp. 174–187.

19. S. Goldwasser, S. Micali, and C. Rackoff, *The Knowledge Complexity of Interactive Proof-Systems*, SIAM Journal on Computing, vol. 18, n. 1, February 1989.

20. O. Goldreich and Y. Oren, *Definitions and Properties of Zero-Knowledge Proof Systems*, Journal of Cryptology, vol. 7, 1994, pp. 1–32.

21. R. Impagliazzo and M. Yung, *Direct Minimum Knowledge Computations* "Advances in Cryptology – CRYPTO 87", vol. 293 of "Lecture Notes in Computer Science", Springer Verlag pp. 40–51.

22. J. Kilian, *A Note on Efficient Zero-Knowledge Proofs and Arguments*, Proceedings of the 24th Annual ACM Symposium on the Theory of Computing, May 1992.

23. E. Kushilevitz and A. Rosen, *A Randomness-Rounds Tradeoff in Private Computations* "Advances in Cryptology – CRYPTO 94", vol. 293 of "Lecture Notes in Computer Science", Springer Verlag pp. 40–51.

24. I. Niven and H. S. Zuckerman, *An Introduction to the Theory of Numbers*, John Wiley and Sons, 1960, New York.

25. T. Okamoto and K. Ohta, *Disposable Zero-Knowledge Authentications and their Applications to Untraceable Electronic Cash* "Advances in Cryptology – CRYPTO 89", vol. 435 of "Lecture Notes in Computer Science", Springer Verlag.

26. T. Okamoto and K. Ohta, *How to Utilize the Randomness of Zero-Knowledge Proof Systems* "Advances in Cryptology – CRYPTO 90", vol. 537 of "Lecture Notes in Computer Science", Springer Verlag.

27. M. Tompa and H. Woll, *Random Self-Reducibility and Zero-Knowledge Interactive Proofs of Possession of Information*, Proc. 28th Symposium on Foundations of Computer Science, 1987, pp. 472–482.

On the Matsumoto and Imai's Human Identification Scheme

Chih-Hung Wang, Tzonelih Hwang, and Jiun-Jang Tsai

Institute of Information Engineering
National Cheng-Kung University
Tainan, Taiwan, R.O.C.

Abstract. At Eurocrypt'91, Matsumoto and Imai presented a human identification scheme for insecure channels, which is suitable for human ability of memorizing and processing a short secret. It prevents an intruder from peeping user in typing password on terminal connected to the central computer. However, in this paper, we are going to propose a new attack, called the *replay challenge attack*, where a malicious terminal pretends to be the host and replays the host's challenges to reveal the secret password. A modified scheme will be proposed to avoid this attack.

1 Introduction

It is very often in the computer systems that an end user has to identify himself to a host. Though human identifications using personal characters, such as fingerprint, voice, or the retinal blood-vessel pattern of a human eye, have been developed and applied actually [5], physical devices for special purpose have to be designed and the cost of these devices are very high.

The design of human identifications without the help of any auxilary device has become an important issue. The password authentication scheme, where a log-in user simply memorizes a short secret and presents it to the host for user authentication, however, suffers both *the peeping attacks* where an intruder stands behind the log-in user to peep the typed password and *the replay attacks* where the intruder intercepts the password from the network and then impersonates the same user by replaying the intercepted password.

In 1991, Matsumoto and Imai proposed a human identification scheme for insecure channel to avoid both replay and peeping attacks by use of a simple challenge-response protocol [3]. Each user and the host are assumed to share a common key. Knowing the common key shared with the user, the verifier (the host) can decide whether an answer replied from the prover (the user) is correct or not. In their scheme, what the user has to do are simply to memorize a short secret and perform very simple operations based on the secret.

In this paper, we are going to study the security of Matsumoto and Imai's human identification scheme by proposing a new attack on it. By this attack, a malicious process first pretends to be the host by replaying a challenge to the login user, and then performs the intercepting or peeping attack to reveal the

L.C. Guillou and J.-J. Quisquater (Eds.): Advances in Cryptology - EUROCRYPT '95, LNCS 921, pp. 382-392, 1995
© Springer-Verlag Berlin Heidelberg 1995

secret by observing the differences in the responses of the login user. Since this kind of attack is particular useful in the environment where the login process is limited to human ability of performing computations, it is valuable to be pointed out here for consideration of constructing even more secure human identification scheme. In addition, we also proposed a modified scheme to avoid this attack.

The structure of this paper is as follows. In Section 2, we review the Matsumoto and Imai's scheme. Then in Section 3. two attacks are proposed to analyze the security of Matsumoto and Imai's scheme. Section 4 proposes a modified scheme to avoid the replay challenge attack and analyzes the security of the newly modified scheme. Finally, some concluding remarks will be given in Section 5.

2 Matsumoto and Imai's Human Identification Scheme

2.1 Notations and Definitions

The following definitions and notations are used in the entire paper.

- $< n >$: the set of all positive integers less than or equal to n.
- $g \circ f$: the composite function of functions f and g.
- Ω : the whole alphabet.
  Q : the question alphabet; a subset of Ω.
- W : the window alphabet; a subset of Q.
- A : the answer alphabet; a subset of Ω.
- $|S|$: the number of elements in the set S. Note: $2 \leq |A| \leq |W| < |Q| \leq |\Omega|$.
- β : the number of blocks.
- α : a threshold value; $1 \leq \alpha \leq \beta$.
- q_j : the jth question block, which is a bijection from $< |Q| >$ to Q.
- a_j : the jth answer block, which is a surjection from $< |Q| >$ onto A.
- SW : a string of secret word, which is a surjection from $< |W| >$ onto A.
- f_j : the window in q_j, which is an injection from $< |W| >$ into $< |Q| >$ such that

$$f_j = sort\left(\{i \in < |Q| > |\ q_j(i) \in W \}\right)$$

, where the sort function is defined below.

Definition 1 *[3] For a totally ordering finite set $(S, \leq)$, the function $sort(S)$ is defined as a bijection, bf, from $< |S| >$ onto S such that*

$$bf(1) \leq bf(2) \leq \cdots \leq bf(|S|)$$

2.2 Matsumoto and Imai's Scheme

The system determines the parameters $\Omega, Q, W, A, \alpha, \beta$ first. A string of secret word SW is known only to P and V. Then, P can identify himself to V by the following steps.

Step1: V generates β question blocks $q_1, q_2, \cdots, q_\beta$, and sends them to the prover P.

Step2: P selects at least α distinct question blocks out from these β blocks to generate the answer blocks by the following substeps.

Step2.1: For each selected question block q_j, P computes the window

$$f_j = sort(\{i \in < |Q| > | q_j(i) \in W \}).$$

Step2.2: P generates the answer blocks

$$a_j\left(f_j(t)\right) = SW(t), \text{for } t = 1, 2, \cdots, |W|.$$

Step2.3 For each r, where $r \in < |Q| >$, and $r \neq f_j(t)$, for $t = 1, 2, \cdots |W|$, P randomly and uniformly selects an elements from A and allocates it to $a_j(r)$.

Step3: For each question block q_j, which is not selected by the prover P at Step2, P allocates a_j with a random block R_j, which is a surjection from $< |Q| >$ onto A.

Step4: P sends the answer blocks $a_1, a_2, \cdots, a_\beta$ to V.

Step5: V verifies the answers as follows.

Step5.1: For each question blocks q_j, V computes the window

$$f_j = sort(\{i \in < |Q| > | q_j(i) \in W \}).$$

Step5.2: If the number of correct answer blocks, i.e., $a_j \circ f_j = SW$, is greater than or equal to α, then V accepts P; otherwise, V rejects P.

2.3 An Example

A simplified example with $\alpha = \beta = 1$ is described here to illustrate Motsumoto and Imai's scheme.

In this example, we use the notations a, q, and f to denote the answer, question and window respectively. Let the question alphabet $Q = \{1, 2, 3, 4, 5, 6, 7, 8\}$, window alphabet $W = \{1, 2, 4, 6\}$ and the answer alphabet $A = \{1, 2, 3, 4\}$. The prover has a string of secret word $SW = 3124$ shared with the verifier. Figure 1 shows the challenge and response between the prover and verifier. The bars in the positions of q show the window positions.

$$W = \{1, 2, 4, 6\}$$
$$SW = 3124$$
$$\text{verify } a \circ f \overset{?}{=} SW$$

Figure 1: An Example of Answer and Question

If an intruder wants to guess the secret word of the prover, he must guess the window alphabet correctly. Since the window size $|W| = 4$ and the question alphbet size $|Q| = 8$, an intruder has to find the window W in $\binom{|Q|}{|W|} = \binom{8}{4}$ trials according to the analysis in [3].

3 The Attacks on Matsumoto and Imai's Scheme

In this section, we propose two attacks on Matsumoto and Imai's scheme . The first attack is a passive attack where the intruder passively observe the login user's responses to a challenge and then intents to guess his password. The second attack, called the *replay challenge attack*, is an active attack where the intruder actively replay the host's challenges to reduce the number of trials in finding out the window W. Both attacks show lower security levels of the scheme than the original proposed one in [3].

3.1 The passive attack

The idea of this attack is based on the following observation: according to the definitions, SW contains all elements of the answer alphabet A at least once, and the other elements in a_j are also from A. Then, one can reduce the number of trials to reveal the password by this observation. Consider the example shown in Fig. 1 again, since the login user's answer contains all elements of $A = \{1, 2, 3, 4\}$ exactly twice, an intruder simply selects one from two of the same elements to guess the window positions and then reveals the user's password based on the window positions. The trials will be reduced to $\binom{2}{1}^4$ in this case which are less than $\binom{8}{4}$ claimed by Matsumoto and Imai.

For simplicity, we consider the case where $\alpha = \beta = 1$. Let $A = \{e_1, e_2, \cdots e_{|A|}\}$; t_i denote the number of e_i in the answer a, for $i = 1, 2, \cdots, |A|$. Then $|Q| = \sum_{i=1}^{|A|} t_i$. s_i is the number of times the elements e_i appear in the secret word SW. Then, $s_1 + s_2 + \cdots + s_{|A|} = |W|$; $1 \le s_i \le Minimum\{|W|, t_i\}$. Thus, the password can be revealed in at most $\sum_{s_1+s_2+\cdots+s_{|A|}=|W|} \binom{t_1}{s_1}\binom{t_2}{s_2} \cdots \binom{t_{|A|}}{s_{|A|}}$ trials.

We will show that the number of trials mentioned above is less than $\binom{|Q|}{|W|}$ in the following theorem.

Theorem 1.

$$\sum_{\substack{s_1+s_2+\cdots+s_{|A|}=|W| \\ 1 \le s_i \le Minimum\{|W|,t_i\}}} \binom{t_1}{s_1}\binom{t_2}{s_2} \cdots \binom{t_{|A|}}{s_{|A|}} < \binom{|Q|}{|W|}.$$

Proof. Since $|Q| = \sum_{i=1}^{|A|} t_i$ and $|W| = \sum_{i=1}^{|A|} s_i$, we have

$$\binom{|Q|}{|W|} = \sum_{\substack{s_1+s_2+\cdots+s_{|A|}=|W| \\ 0 \le s_i \le Minimum\{|W|,t_i\}}} \binom{t_1}{s_1}\binom{t_2}{s_2} \cdots \binom{t_{|A|}}{s_{|A|}}.$$

It is obvious that

$$\sum_{\substack{s_1+s_2+\cdots+s_{|A|}=|W| \\ 0 \le s_i \le Minimum\{|W|,t_i\}}} \binom{t_1}{s_1}\binom{t_2}{s_2}\cdots\binom{t_{|A|}}{s_{|A|}} >$$

$$\sum_{\substack{s_1+s_2+\cdots+s_{|A|}=|W| \\ 1 \le s_i \le Minimum\{|W|,t_i\}}} \binom{t_1}{s_1}\binom{t_2}{s_2}\cdots\binom{t_{|A|}}{s_{|A|}}.$$

Therefore

$$\binom{|Q|}{|W|} > \sum_{\substack{s_1+s_2+\cdots+s_{|A|}=|W| \\ 1 \le s_i \le Minimum\{|W|,t_i\}}} \binom{t_1}{s_1}\binom{t_2}{s_2}\cdots\binom{t_{|A|}}{s_{|A|}}$$

(Q.E.D.)

3.2 The Replay Challenge Attack

In this section, we are going to propose a new attack called the *replay challenge attack* where an attacker impersonates the host to replay an intercepted challenge to the login user and then peeps or intercepts the answers from the user. In an insecure login environment, this attack is considered feasible, since it is not difficult for a malicious node (terminal) to replay a challenge to the end user and then return to the normal state after intercepting the response. For simplicity, we also consider the case where $\alpha = \beta = 1$. By collecting a few answers of the same challenge, the intruder can figure out the secret word of the prover with a high probability . Let $a(i)$, $q(i)$ denote the contents of the ith position in the answer block a and question block q respectively. The positions corresponding to W in the distinct answer blocks for the same question should have the same contents, i.e.,

$$a \circ f = SW = a' \circ f$$

,where a and a' denote these two distinct answer blocks for the same challenge.

Thus, if an intruder discovers that some corresponding positions in the answer blocks a and a' whose contents are distinct, then these positions cannot be the positions of the window W. Thus, we have the following Lemma.

Lemma 1 *Let a and a' be two distinct answer blocks of the same question q . If there exists an i, $i \in< |Q| >$, such that $a(i) \neq a'(i)$, then $q(i) \notin W$.*

Theorem 2. *The window W of Matsumoto and Imai's human identification scheme with $\alpha = \beta = 1$ can be found in $\sum_{k=0}^{|Q|-|W|} \frac{\binom{|Q|-|W|}{k}(|A|-1)^k}{|A|^{|Q|-|W|}} \times \binom{|Q|-k}{|W|}$ expected trials if an intruder replays the same question one time.*

Proof. Let the integer k denote the number of i's such that $a(i) \neq a'(i)$, for $i = 1, 2, \cdots, |Q|$. It is true that $0 \leq k \leq |Q| - |W|$. By Lemma 1, these k positions do not belong to W. In this case, W can be found in $\binom{|Q|-k}{|W|}$ trials. Comparing to the answer a, each of these k positions of a' can only be padded by $(|A|-1)$ possible elements. So, the number of possible permutations for this case is given by $\binom{|Q|-|W|}{k}(|A|-1)^k$. In addition, the other $(|Q| - |W| - k)$ positions of a' must be filled by the same elements of the corresponding positions of a. Thus, the probability for this is $\frac{\binom{|Q|-|W|}{k}(|A|-1)^k}{|A|^{|Q|-|W|}}$ assuming that each element in A is equally likely to be selected by P. Thus, the expected number of trials of finding out W are $\sum_{k=0}^{|Q|-|W|} \frac{\binom{|Q|-|W|}{k}(|A|-1)^k}{|A|^{|Q|-|W|}} \times \binom{|Q|-k}{|W|}$.

$$\textbf{(Q.E.D.)}$$

Corollary 1 *Similar to Theorem 2, an intruder can find the window W in*

$$\sum_{k=0}^{|Q|-|W|} \frac{\binom{|Q|-|W|}{k}(|A|^n - 1)^k}{|A|^{n(|Q|-|W|)}} \times \binom{|Q| - k}{|W|}$$

expected trials if the same question is replayed n times.

Proof. To locate the window positions in the replayed question, the intruder first figures out the positions which do not belong to the window. He replays the same question n times. Then he constructs a matrix using these answer blocks $\{a^0, a^1, \cdots, a^n\}$ as shown in Figure 2. The element at $(x, y) = (row, column)$ is denoted as $a^x(y)$, where $0 \leq x \leq n, 1 \leq y \leq |Q|$. If an intruder finds that the elements in one column, say $\hat{y}$, are not all the same (i.e., $a^0(\hat{y}) = a^1(\hat{y}) = \cdots = a^n(\hat{y})$ is not true), then $q(\hat{y}) \notin W$. The probability of finding k non-window positions is given by

$$\frac{\binom{|Q|-|W|}{k}(|A|^n - 1)^k}{|A|^{n(|Q|-|W|)}}.$$

Therefore, the expected number of trials of finding W are

$$\sum_{k=0}^{|Q|-|W|} \frac{\binom{|Q|-|W|}{k}(|A|^n - 1)^k}{|A|^{n(|Q|-|W|)}} \times \binom{|Q| - k}{|W|}$$

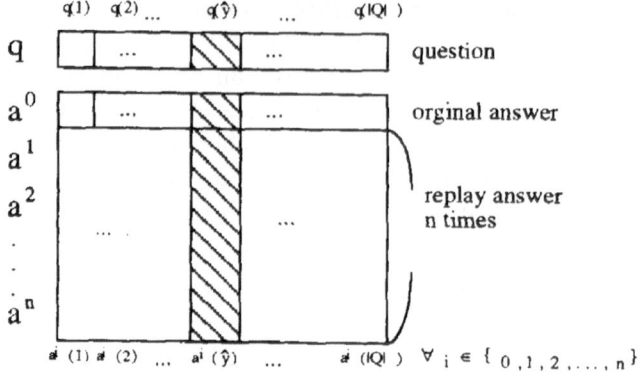

Figure 2. The matrix formed by answer blocks

(Q.E.D.)

Corollary 2 *If the number of times an intruder replays the same question tends to infinity, then the expected number of trials of finding the window W will degrade to* 1

$$(i.e., \lim_{n \to \infty} \sum_{k=0}^{|Q|-|W|} \frac{\binom{|Q|-|W|}{k}(|A|^n-1)^k}{|A|^{n(|Q|-|W|)}} \times \binom{|Q|-k}{|W|} = 1).$$

We use the following examples to show the validity of our attack.

$Test1:|\Omega| = |Q| = 36, |W| = 18, |A| = 2,$ and $\alpha = \beta = 1.$
$Test2:|\Omega| = |Q| = 50, |W| = 10, |A| = 3,$ and $\alpha = \beta = 1.$

Then, the expected number of trials of finding W are as follows

Value of n	Expected trials in Test1	Expected trials in Test2
1	3.08×10^7	3.78×10^6
3	8165.32	174.2
5	65.8	3.328
7	5.3	1.19
9	1.73	1.02

Both attacks discussed previously in this section can be combined into one to improve the probability of revealing the secret word. The first phase of the combined attack is to utilize the replay challenge attack to find the positions of q not belonging to the window W. Then use the passive attack as the the second phase to guess the correct positions of window. Let m $(0 \leq m \leq |W|)$ be the number of positions not belonging to the window W found in first step. The number of elements in these positions of answer a can be subtracted from the corresponding t_i (refer to Section 3.1). Let t_i' be the results of above process and $t_i' \leq t_i$; $\sum_{i=1}^{|A|} (t_i - t_i') = m$. The password can be revealed in at most

$$\sum_{\substack{s_1+s_2+\cdots+s_{|A|}=|W| \\ 1 \leq s_i \leq Minimum\{|W|, t_i'\}}} \binom{t_1'}{s_1}\binom{t_2'}{s_2}\cdots\binom{t_{|A|}'}{s_{|A|}} \text{ trials.}$$

4 Modified Identification Scheme

4.1 The Scheme

A modified identification scheme is devised to avoid the replay challenge attack mentioned in the Section 3. However, the computation required to be executed by the human may be too complicate in the modified scheme thus it may not be a practical human identification scheme.

The answer block is divided into two parts. One is called the *window-padding answer portion* denoted as $a_j(f_j(t))$, for $t = 1, 2, \cdots, |W|$; the other is called the *random-padding answer portion* denoted as $a_j(r)$, $1 \leq r \leq |Q|$, $r \neq f_j(t)$, for $t = 1, 2, \cdots, |W|$. In Matsumoto and Imai's scheme, window-padding answer portion is filled with the secret word SW (i.e., $a_j \circ f_j = SW$). However, as discussed previously, once the question is replayed, the window-padding answer portion of the answers remains the same. In order to avoid this weakness, we introduce a new function Υ_j, which can transfer the scheme from deterministic to probabilistic. In our modified scheme, let $|Q|$ be even and $|W| = \frac{1}{2}|Q|$. Υ_j is defined as follows.

Let the function g_j be an injection from $< |W| >$ into $< |Q| >$, such that

$$g_j = sort\left(\{i \in < |Q| > | \, q_j(i) \notin W\}\right) .$$

Definition 2 *The function $Srand_j$ is a bijection from $< |W| >$ onto $< |W| >$, such that*

(i) *if* $a_j\left(g_j(i)\right) < a_j\left(g_j(i')\right)$, *then* $Srand_j(i) < Srand_j(i')$

(ii) *if* $a_j\left(g_j(i)\right) = a_j\left(g_j(i')\right)$, *and* $i < i'$, *then* $Srand_j(i) < Srand_j(i')$

, where i, $i' \in < |W| >$ *and the order of* a_j *is corresponding to the order of its ASCII code.*

Definition 3 *The function Υ_j is a surjection from $< |W| >$ onto A , such that*

$$\Upsilon_j\left(Srand_j(i)\right) = SW(i)$$

, where $i \in < |W| >$.

Using the above-mentioned definitions, we modify the Matsumoto and Imai scheme on *Step 2.2* and *Step 5.2*. The other steps are the same as those proposed in Matsumoto and Imai's scheme (see also Section 2)

Step2.2: P generates the answer blocks

$$a_j \left(f_j(t) \right) = \Upsilon_j(t), \text{ for } t = 1, 2, \cdots, |W| .$$

. . .

Step5.2: If the number of correct answer blocks, i.e., $a_j \left(f_j(t) \right) = \Upsilon_j(t)$, for $t = 1, 2, \cdots, |W|$, is greater than or equal to α, then V accepts P; otherwise, V rejects P.

. . .

Example 1. Figure 3 illustrates a simplified version (i.e., $\alpha = \beta = 1$) of the newly modified scheme. Let the question alphabet be $Q = \{1, 2, 3, 4, 5, 6, 7, 8\}$, the window $W = \{1, 2, 4, 6\}$, and $A = \{1, 2, 3, 4\}$. The prover shares with the verifier a string of secret word, $SW = 3124$. Figure 3 shows two answers a_1 and a_2 corresponding to the question q.

q =	2	8	5	1	7	3	6	4

a 1=	4	4	3	2	2	1	1	3

$$SW = 3124$$

$$a \circ g = 4321$$

$$Srand = 4321$$

$$\Upsilon = 4213$$

a 2=	1	2	1	4	4	1	3	2

$$SW = 3124$$

$$a \circ g = 2141$$

$$Srand = 3142$$

$$\Upsilon = 1432$$

Figure 3. Example of Answers and Question

4.2 Security Analysis

Here, three attacks will be considered. The first one is the *known-A random attack* proposed in [3] where an attacker knows the function of the protocol and knows the set Ω, Q, $|W|$ and A, but does not know SW and W. The success probability (p_A) of known-A random attack on the modified scheme is given by

$$p_A = \sum_{j=\alpha}^{\beta} \binom{\beta}{j} p^j (1-p)^{\beta-j}$$

, where

$$p = \frac{|A|^{|Q|-|W|}}{|A|^{|Q|} - \sum_{i=1}^{|A|-1} \binom{|A|}{i}(i)^{|Q|}} = \frac{1}{|A|^{|W|}(1 - \sum_{i=1}^{|A|-1} \binom{|A|}{i}(\frac{i}{|A|}))^{|Q|}}$$

This result is equivalent to the security level of Matsumoto and Imai's scheme under the same attack.

The second attack is the passive attack as described in Section 3.1. The intruder has to test at most $\sum_{\substack{s_1+s_2+\cdots+s_{|A|}=|W| \\ 1 \le s_i \le Minimum\{|W|, t_i\}}} \binom{t_1}{s_1}\binom{t_2}{s_2}\cdots\binom{t_{|A|}}{s_{|A|}}$ trials, which are the same as Matsumoto and Imai's scheme to reveal the secret.

The third attack is the *replay challenge attack* as described in Section 3.2. The modified scheme avoids this attack due to the new function Υ_j. For the same question to the end user , the window-padding answer portion of the end user may be changed according to the random-padding answer portion selected by him. It is obvious that the Lemma 1 does not hold in our modified human identification scheme. Therefore, the replay challenge attack will not be successful on the newly modified scheme.

5 Conclusions

Human identification scheme is an important issue for user identifications in the network environment. The scheme proposed by Matsumoto and Imai is a pioneer work . This paper shows an attack, the replay challenge attack, on Matsumoto and Imai's scheme and proposes a modified scheme to avoid this attack. It requires further research to devise secure and practical human identification schemes.

Acknowledgement. We wish to thank Miss Maujy Peng and the referees of this paper for their useful comments.

References

1. Fiat, A. and Shamir, A., "How to Prove Yourself: Practical Solutions to Identical Solutions to Identification and Signature Problems", *Crypto'86*, 1986.
2. Jennifer G. Steiner B. Clifford Neuman, and Jeffrey I. Schiller, " Kerbero: An Authentication Service for Open Network Systems ", *Usenix Conference Proceedings*, pages 183-190, February 1988.
3. Matsumoto, T. and Imai, H., "Human Identification Through Insecure Channel", *Eurocrypt'91*, 1991.
4. Ohta, K. and Okamoto, T., "A Modification of the Fiat-Shamir Scheme", *Crypto'88*, 1988.
5. Davies, D.M. and Price, W.L., *Security for Computer Networks* , Chapter 7, John Wiley & Sons, 1984.
6. Ross, Sheldon M., *Introduction to Probability Models*, Academic Press, Inc., fifth edition.

Receipt-Free Mix-Type Voting Scheme
–A practical solution to the implementation of a voting booth–

Kazue Sako[1] and Joe Kilian[2]

[1] NEC Corporation, 4-1-1 Miyazaki Miyamae, Kawasaki 216, JAPAN
[2] NEC Research Institute, 4 Independence Way, Princeton, NJ 08540, USA

Abstract. We present a receipt-free voting scheme based on a mix-type anonymous channel[Cha81, PIK93]. The receipt-freeness property [BT94] enables voters to hide how they have voted even from a powerful adversary who is trying to coerce him. The work of [BT94] gave the first solution using a *voting booth*, which is a hardware assumption not unlike that in current physical elections. In our proposed scheme, we reduce the physical assumptions required to obtain receipt-freeness. Our sole physical assumption is the existence of a private channel through which the center can send the voter a message without fear of eavesdropping.

1 Introduction

1.1 Receipt-Free Voting Schemes

The ultimate goal of secure electronic voting is to replace physical voting booths. Achieving this goal requires work both on improving the efficiency of current protocols and understanding the security properties that these physical devices can provide. Recently, Benaloh and Tuinstra[BT94] observed that, unlike physical voting protocols, nearly all electronic voting protocols give the voters a receipt by which they can prove how they voted. Such receipts provide a ready means by which voters can sell their votes or another party can coerce a voter. Benaloh and Tuinstra give the first *receipt-free* protocol for electronic voting. In their scheme a trusted center generates for each voter a pair of ballots consisting of a "yes" vote and a "no" vote in random order. Using a trusted beacon and a physical voting booth the center proves to the public that the ballot indeed consists of a well-formed (yes/no) or (no/yes) pair and at the same time proves to the verifier which pair it is. The physical apparatus ensures that by the time the verifier is able to communicate with an outsider, he can forge a proof that the ballot is (yes/no) and also forge a proof that it is (no/yes). Thus, such a proof ceases to provide either proof as a receipt.

Independently, Niemi and Renvall[NR94] tried to solve this problem. They also use a physical voting booth where a voter performs multiparty computation with all the centers.

Both the Benaloh-Tuinstra and the Niemi-Renvall protocols illustrate that receipt-freeness is possible. However, their physical requirements are fairly cumbersome, and are not unlike those faced by participants in physical elections. An important open question is precisely what physical requirements are necessary for achieving receipt-freeness.

L.C. Guillou and J.-J. Quisquater (Eds.): Advances in Cryptology - EUROCRYPT '95, LNCS 921, pp. 393-403, 1995.
© Springer-Verlag Berlin Heidelberg 1995

1.2 Results of This Paper

In this paper we consider how to implement receipt-freeness in a more practical manner. We start with the mix-type protocols of [Cha81, PIK93] and augment them to obtain a protocol which is receipt-free and *universally verifiable*. By universally verifiable we mean that in the course of the protocol the participants broadcast information that allows any voter or interested third party to at a later time verify that the election was properly performed. To make our protocol receipt-free we must by necessity make some physical assumption. We assume the existence of an untappable private channel. Our untappability requirement is physical; cryptographic implementations of untappable channels do not suffice for our purposes. To obtain universal verifiability we develop efficient techniques by which a mixer can prove that they performed correctly, and use the Fiat-Shamir [FS86] technique to make these proofs noninteractive.

1.3 Techniques Used

Chameleon blobs

Brassard, Chaum and Crépeau introduced the concept of zero-knowledge proofs and zero-knowledge bit-commitment schemes[BCC88]. In a zero-knowledge bit commitment scheme the prover commits to b by generating a pair (B, S_b) (B is referred to as a *blob*) and sends B to the verifier. Later, the prover can *open* a blob by sending S_b verifier, who evaluates $\mathsf{open}(B, S_b)$ to obtain $0, 1$ or reject. If the prover behaves properly then $\mathsf{open}(B, S_b) = b$. The distribution on B is independent of b, however a computationally bounded prover cannot generate a triple (B, S_0, S_1) such that $\mathsf{open}(B, S_b) = b$ for $b \in \{0, 1\}$. That is, once a prover has committed to a bit with B, he can open it only one way. A system of *chameleon blobs* is a system with the additional property that the verifier can, on input (B, b) generate S_b such that (B, S_b) evaluates to b. That is, the verifier knows how to open a blob both ways. Furthermore, we require that the conditional distribution on S_b given B be the same as the conditional distribution generated by P. We use chameleon blobs to allow the verifier to forge proofs.

Amortization techniques

In order to achieve universal verifiability, we require the mixers to prove that they are not altering the ballots. These proofs greatly increase the communication complexity of the protocol. To ameliorate this problem, we show how to use techniques similar to those used in [SK94] to reduce the amount of communication and computation necessary to generate, transmit and check the proofs.

1.4 Outline of the Rest of the Paper

In Section 2 we construct a mix-net with the universal verifiability property. In Section 3 we give a receipt-free voting scheme.

2 Universally Verifiable Mix-Net

Mix-net anonymous channels were first proposed by [Cha81]. Subsequently, many voting schemes have been proposed based on this basic technique [FOO92, PIK93]. However, this type of scheme has only *individual verifiability*. That is, a sender can verify whether or not his message has reached its destination, but cannot determine if this is true for the other voters. A disadvantage of this situation is that one has to trust other voters to be vigilant in checking that their vote was counted. Also, one may wish to audit an election, checking that is was fair, without getting back in touch with all of the voters. Thus, universal verifiability is preferable to individual verifiability, provided that it is not too expensive.

In this section, we describe a scheme for mix-net proposed by [PIK93], and give a protocol to make the scheme universally verifiable. Furthermore, we show how to amortize the cost for some of the verification procedures required by our scheme.

2.1 A Scheme with Individual Verifiability

The paper of [PIK93] gives two types of mix-type anonymous channels. Both types of schemes achieve only individual verifiability; we add additional protocols to achieve universal verifiability.

We first outline (with slightly modifications) the anonymous channel protocol proposed in [PIK93]. In this scheme, encrypted messages from the senders are successively processed by the mixing centers until the last center outputs a randomly, untraceably ordered set of unencrypted[3] At a high level, the senders first post their encrypted messages. Center i processes each message posted by Center $i - 1$ (or the senders, when $i = 1$) and posts the results in permuted order. It remains to specify how a message m is initially encrypted by a sender and how Center i processes each message.

In the following, the definition of the "generating element" g is modified from the original scheme, in order to evade an attack proposed by Pfitzmann [Pfi94].

A Mix-Type Anonymous Channel by [PIK93]

Public information : $p = kq + 1$ (p, q prime),
$g = (g')^k \bmod p$ (where g' is a generator mod p)
Public key of center i : $y_i = g^{x_i} \bmod p$
Secret key of center i : $x_i \in Z_q^*$
Message from the sender : m
We define $w_i = y_{i+1}y_{i+2} \cdots y_n$ and $w_n = 1$.

---

[3] That is, the encryptions used for the anonymous channel have been stripped off. Of course, these messages may have been encrypted before being sent through the channel.

Encrypting a message

The sender generates a random number r_0, and posts

$$Z_1 = (G_1, M_1) = (g^{r_0} \bmod p, \quad (w_0)^{r_0} \cdot m \bmod p)$$

for use by Center 1.

Processing a message

On input (G_i, M_i), Center $i(i = 1, \cdots, n - 1)$ generates a random number r_i (independently for each message-pair) and calculates the following using his secret key x_i:

$$
\begin{aligned}
G_{i+1} &= G_i \cdot g^{r_i} \quad \bmod p \\
&= g^{r_0 + \cdots + r_i} \quad \bmod p \\
M_{i+1} &= M_i \cdot w_i^{r_i} / G_i^{x_i} \quad \bmod p \\
&= w_i^{r_0 + \cdots + r_i} \cdot m \quad \bmod p
\end{aligned}
$$

He posts $Z_{i+1} = (G_{i+1}, M_{i+1})$ (permuted with the other processed messages) for use by Center $i + 1$.

Center n recovers m by computing

$$m = M_n / G_n^{x_n} \bmod p.$$

By adding redundancy to the message m, and by having the last center n announce all the received messages (again in permuted order), a sender can check whether or not his message has reached the destination. However, this gives only individual verifiability, as a sender can not directly determine if the other messages have been properly handled. Also, the redundancy can be used as a receipt, precluding the possibility of receipt-freeness.

2.2 Achieving Universal Verifiability

We obtain universal verifiability in the above scheme, and the other scheme discussed in their paper, by requiring each center to prove that they correctly processed their messages. At this time, or later, any interested party can check the resulting proofs to confirm that the messages have all been handled correctly. With this method for achieving universal verifiability there is no need for adding redundancy to the messages. Furthermore, it also helps thwart an attack proposed in [Pfi94].

We first modify the way each center processes the pairs. Given a pair (G_i, M_i), a center computes (G_{i+1}, M_{i+1}), but in the first phase it posts $G_i^{x_i}$. In the second phase it then posts the pairs (G_{i+1}, M_{i+1}) in permuted order. Note that this protocol leaks the value of $G_i^{x_i}$, which was not leaked in the original protocol. We know of no way to exploit this extra information.

A center proves the correctness of each stage separately. We write these protocols in terms of an interactive proof system; they may then be made non-interactive using the Fiat-Shamir technique [FS86].

Proving correctness for the first phase

We can abstract the first phase of the protocol as follows. Given G, the first phase consists of performing decryption and generate $H = G^x \bmod p$. The proof consists of, given $(G, g, y = g^x \bmod p)$, showing that H is generated in this manner from G.

prove DECRYPT

1. The prover uniformly chooses $r \in Z_{p-1}$. Let

$$y' = g^r \bmod p$$
$$G' = G^r \bmod p.$$

   The prover sends (y', G').

2a. With probability $\frac{1}{2}$, the verifier asks the prover to reveal r. The verifier checks that y' and G' are consistent with r.

2b. With probability $\frac{1}{2}$, the verifier asks the prover to reveal $r' = r - x$. The verifier checks that

$$y' = g^{r'} \cdot y \bmod p \text{ and}$$
$$G' = H \cdot G^{r'} \bmod p.$$

Proving correctness for the second phase

We may slightly abstract the second phase as follows.

   Given constants g, w and

$$A = \begin{pmatrix} a_i^{(1)} \\ a_i^{(2)} \end{pmatrix},$$

the second phase consists of generating $r_1, r_2, \ldots$ and a permutation π and generating a set of pairs

$$B = \begin{pmatrix} a_{\pi(i)}{}^{(1)} \cdot g^{r'_{\pi(i)}} \bmod p \\ a_{\pi(i)}{}^{(2)} \cdot w^{r'_{\pi(i)}} \bmod p \end{pmatrix}.$$

Here $a_i^{(1)}$ refer to G's and $a_i^{(2)}$ to M/H's in the first phase. The proof consists of, given (A, B, g, w), showing that B could be generated in this manner from A.

prove SHUFFLE

1. The prover uniformly chooses $t_i \in Z_{p-1}$, random permutation λ and

$$C = \begin{pmatrix} a_{\lambda(i)}{}^{(1)} \cdot g^{t_{\lambda(i)}} \bmod p \\ a_{\lambda(i)}{}^{(2)} \cdot w^{t_{\lambda(i)}} \bmod p \end{pmatrix}.$$

   The prover sends C.

2a. With probability $\frac{1}{2}$, the verifier asks the prover to reveal λ and t_i. The verifier checks that C is consistent with A, λ, t_i in that way.

2b. With probability $\frac{1}{2}$, the verifier asks the prover to reveal $\lambda' = \lambda \circ \pi^{-1}$ and $t'_i = t_i - r'_i$. The verifier checks that C can be generated from B in the following way: For

$$B = \begin{pmatrix} b_i^{(1)} \\ b_i^{(2)} \end{pmatrix},$$

$$C = \begin{pmatrix} b_{\lambda'(i)}^{(1)} \cdot g^{t'_{\lambda'(i)}} \bmod p \\ b_{\lambda'(i)}^{(2)} \cdot w^{t'_{\lambda'(i)}} \bmod p \end{pmatrix}$$

holds.

2.3 Processing Multiple Messages Together

In this section, we show that the centers can process multiple messages together to achieve a reduces amortized cost per message. Instead of executing "prove DECRYPT" protocol for each shuffled component, a center can prove a single statement equivalent to proving that he decrypted all the components correctly in the following way.

We need to show the following equation holds for each component i.

$$H^{(j)} = (G^{(j)})^x \bmod p$$

We can reduce above equations to the following one equation using randomly chosen coefficients c_i.

$$\prod_i (H^{(j)})^{c_i} = \prod ((G^{(j)})^{c_i})^x \bmod p$$

The center can execute above protocol where $G = \prod_i (G^{(j)})^{c_i}$ and $H = \prod (H^{(j)})^{c_i}$. We exploit the fact that if one or more of the original equations is wrong then if the coefficients are chosen randomly the final equation will also be wrong. Note that these coefficients must not be picked by the prover, but should be given by a verifier, beacon or as the output of a suitable hash function.

2.4 Remarks on Vote Duplication

Gennaro [Gen94] has pointed out that in the Sako-Kilian [SK94] voting protocol a malicious voter may copy another voter's vote by simply duplicating the ballot. This attack is readily foiled by a simple modification to the Fiat-Shamir heuristic; see [Gen94] for a number of fixes. However, we note that for some of the mix-type voting schemes proposed in the literature the problem is even more severe. First, in the usual mix-type voting paradigm the ballots have redundancy attached to allow voters to check that their vote has been counted. An attacker may then duplicate a ballot and search the published list of received votes to find the two identical ballots, revealing how that entity voted.

Simple vote duplication may be easily detected by noticing the duplicate ballots in the first stage, and the adversary risks identification as well as detection. A more subtle attack on the [PIK93] scheme (which may therefore also be applied to our current scheme) involves "blinding" a duplicate ballot so that it looks different but eventually yields the same ballot in the end. If the legitimate voter casts an encrypted ballot of the form

$$Z_1 = (G_1, M_1) = (g^{r_0} \bmod p, \quad (y_1 \cdots y_n)^{r_0} \cdot m \bmod p),$$

an attacker can choose $r \in Z_p^*$ at random and cast the ballot

$$Z_1' = (G_1', M_1') = (G_1 \cdot g^{r'} \bmod p, \quad M_1 \cdot (y_1 \cdots y_n)^{r'} \bmod p),$$

giving no obvious relationship between Z_1 and Z_1'. At the end of the anonymous channel protocol, the final center can still detect vote duplication and refuse to reveal these votes, but then it is more difficult for others to be assured that the last center isn't just trying to impede the election.

We note that in our scheme there is no need for extra redundancy. Hence, even a successful vote duplication attack only gives indirect statistical information to the attacker, and is not useful when the number of votes for each choice is large. One approach for evading this problem entirely is to have Z_1 signed and encrypted using the first center's public key. Unless the adversary colludes with the first center, he would not succeed in copying the vote. Even if he does succeed in copying, the inference of a vote may be excluded by omitting the redundancy in each messages. Guarding against a colluding center is more difficult. One approach, is to have a two-step process whereby the voters commit to their first posting and then reveal it. This is somewhat against the spirit of the principle that a voter can vote and then walk away. However, this is not such a big deal when there are a small number of voters, which is precisely the case where such attacks are most troublesome.

3 Proposed Receipt-Free Scheme

In this section, we describe a mix-type receipt-free voting scheme. Subsection 3.1 gives an overview of the scheme and Subsection 3.2 gives a more detailed description. We note that the interactive zero-knowledge proofs can be made non-interactive by again using Fiat-Shamir technique [FS86].

Our assumptions are as follows: First, need a physically untappable means of communication between the mixing centers and the voters. By a standard exclusive-or trick this assumption can be implemented by having a number of communication channels, assuming that the adversary can't simultaneously tap every one of them. Similarly, it would suffice if at some point in the past the verifier shared a random string with the centers. Second, we require that every voter have a discrete-log public-key in which they themselves are guaranteed to know the private key. Note that it doesn't matter if an adversary has coerced a voter to reveal this key.

Subsection 3.3 discusses how to set up a chameleon-blob system with the voters.

3.1 Overview

The scheme takes the following steps. We use freely the techniques of [BT94] and [CY86], adapting them to the mix-type setting.

1. For each voter i, the final counting center posts encrypted 1-votes and 0-votes in random order. He commits to the ordering using chameleon bit commitments. note that the voter can open these commitments arbitrarily. The center executes prove 1-0 vote to prove that he constructed the vote-pairs properly. He decommits the ordering only to the voter through an untappable secure channel.
2. Each centers shuffles the two votes for voter i through the mix-net in reverse order. He commits to how he shuffled using chameleon commitments. Each execute proof SHUFFLE to prove the correctness of his action. He reveals how he shuffled only to the voter i through untappable secure channel.
3. By keeping track of the initial ordering of the pair, and how they were flipped at each stage, each voter knows which vote is which. Each voter submits one of the votes sent down to him.
4. All of the voters' votes are anonymously sent to the counting center using verifiable mix-net described in Section 2. The counting center tallies the votes.

3.2 The Main Protocol

General Constants	: $p = kq + 1$ (p, q prime),
	$g = (g')^k \bmod p$ (where g' is a generator mod$p$)
Center j Public Key	: $y_j = g^{x_j} \bmod p$
Center j Secret Keys	: x_j
Voter i's Public Key	: $\alpha_i = g^{a_i}$
Voter i's Secret Key	: a_i
1-vote	: m_1
0-vote	: m_0

1. The last center n executes the following with each voter i. He first commits a random bit string $\pi^{(i,n)}$ of length $l + 1$ to the voter using public key α_i. For convenience let $\pi_k^{(i,n)}$ denote kth bit of string $\pi^{(i,n)}$. He then generates

$$v^0 = (\bar{G}_n, \bar{M}_n) = (g^{r_{2n}}, m_0 \cdot \bar{y}^{r_{2n}})$$
$$v^1 = (\bar{G}_n{}', \bar{M}_n{}') = (g^{r_{2n-1}}, m_1 \cdot \bar{y}^{r_{2n-1}})$$

where $\bar{y} = \prod y_i$. He places (v^0, v^1) if $\pi_1^{(i,n)} = 0$ and (v^1, v^0) otherwise. He proves that the placed pair is a combination of 1-vote and 0-vote, using prove 1-0 vote (and $\pi^{(i,n)}$) below for l times, where l is a suitable security parameter.
2. The counting center reveals to the voter which is 1-vote by decommitting $\pi^{(i,n)}$.

3. The next center $n - 1$ executes the following with each voter i. He first commits a random bit string $\pi^{(i,n-1)}$ which is $l + 1$ bit long to the voter. For convenience let $\pi_k^{(i,n-1)}$ denote kth bit of string $\pi^{(i,n-1)}$. He calculates

$$(\bar{G}_{n-1}, \bar{M}_{n-1}) = (\bar{G}_n \cdot g^{r_{2(n-1)}}, \bar{M}_n \cdot \bar{y}^{r_{2(n-1)}})$$
$$(\bar{G}_{n-1}{}', \bar{M}_{n-1}{}') = (\bar{G}_n{}' \cdot g^{r_{2(n-1)-1}}, \bar{M}_n{}' \cdot \bar{y}^{r_{2(n-1)-1}})$$

where $(\bar{G}_n, \bar{M}_n)$ and $(\bar{G}_n{}', \bar{M}_n{}')$ are the votes for voter i sent from the previous center. The center $n - 1$ place the votes in this order if $\pi_1^{(i,n-1)} = 0$ and reverse otherwise. He proves that the placed pair is a combination of 1-vote and 0-vote, using **prove** SHUFFLE for l times and for each interaction uses the first unused bit in $\pi^{(i,n-1)}$ for random permutation λ.
4. The center reveals to the voter how he placed by decommitting $\pi^{(i,n-1)}$.
5. Step 3-4 are repeated for center (mixer) $n - 2$ down to the first center.
6. The voter, who can compute which vote of the shuffled pair is a 1-vote and which is a 0-vote, submits the desired vote to the first center of mix, which sends it down to the counting center through the mix-channel described in Section 2.
7. After the last center reveals the permuted votes, anyone can compute the number of 1-votes (m_0) and 0-votes (m_1).

Remark:
We need to choose m_0 and m_1 so that

$$v^0 = (g^{r_{2n}}, m_0 \cdot \bar{y}^{r_{2n}}) \text{ and}$$
$$v^1 = (g^{r_{2n-1}}, m_1 \cdot \bar{y}^{r_{2n-1}})$$

are indistinguishable. The receipt-freeness follows, as with the [BT94] protocol from the fact that the verifier can forge "proofs" that imply that the 1-vote and 0-vote were given in either order.
prove 1-0 vote

1 The prover uniformly chooses $r', r'' \in Z_{p-1}$ and calculates

$$E_0(v^0) = (g^{r'}, m_0 \cdot \bar{y}^{r'})$$
$$E_1(v^1) = (g^{r''}, m_1 \cdot \bar{y}^{r''})$$

send $E_0(v^0), E_1(v^1)$ in the order according to next unused bit in $\pi^{(i,n)}$.
2a. With probability $\frac{1}{2}$, the verifier asks the prover to reveal r' and r''. The verifier checks if $E_0(v^0), E_1(v^1)$ is made consistently.
2b. With probability $\frac{1}{2}$, the verifier asks the prover to reveal $r_{2n} - r'$ and $r_{2n-1} - r''$. The verifier checks the following holds. v^0 and v^1 can be generated from $E_0(v^0), E_1(v^1)$.

For completeness, we summarize the chameleon bit-commitment scheme due to [BKK].

Commitment Sender j commits 0 by g^r and $\alpha_i \cdot g^r$ for 1 to receiver i, who knows a_i satisfying $\alpha_i = g^{a_i}$

Decommitment Sender reveals r. The receiver calculates both g^r and $\alpha_i \cdot g^r$ and obtain the committed bit.

Modified Decommitment Receiver claims he received $r - a_i$ instead of r.

3.3 Confirming the Voter Knows his Secret Key

By adopting non-interactive proofs and chameleon blobs, the voters does not need to interact with centers in a *voting booth* as was needed in [BT94] scheme, if only he can receive messages untapped. The messages must be physically untappable so that the verifier is free to lie about their contents. Also, it is necessary to make sure that voter i in fact knows the discrete logarithm of α_i in order to have the chameleon effect. This is seemingly innocent, but must be stated as a separate assumption. The simplest solution is to assume that the verifier is given a discrete-log at some time in the distant past, such as for a public-key. We note that only one such piece of information must be given for all time. Also, while losing this discrete-log to the center is dangerous, losing it to a coercer does not affect the receipt-freeness of the protocol.

References

[BCC88] G. Brassard, D. Chaum and C. Crépeau. *Minimum Disclosure Proofs of Knowledge.* In *JCSS*, pages 156–189. 1988.

[Ben87] J. Cohen Benaloh. *Verifiable Secret-Ballot Elections.* PhD thesis, Yale University, 1987. YALEU/DCS/TR-561.

[BKK] J. Boyar, M. Krentel and S. Kurtz. A discrete logarithm implementation of perfect zero-knowledge blobs. *Journal of Cryptology*, Vol. 2, No. 2, pp. 63–76, 1990.

[BT94] J. Cohen Benaloh and D. Tuinstra. *Receipt-Free Secret-Ballot Elections.* In *STOC 94*, pages 544–553. 1994.

[Cha81] D. Chaum. Untraceable electronic mail, return addresses, and digital pseudonyms. In *Communications of the ACM*, pages 84–88. ACM, 1981.

[CY86] J. Cohen Benaloh and M. Yung. Distributing the power of a government to enhance the privacy of voters. In *Annual Symposium on Principles of Distributed Computing*, pages 52–62, 1986.

[FOO92] A. Fujioka, T. Okamoto, and K. Ohta. A practical secret voting scheme for large scale elections. In *Advances in Cryptology –Auscrypt '92*, pages 244–251, 1992.

[FS86] A. Fiat and A. Shamir. How to prove yourself: Practical solutions to identification and signature problems. In *Advances in Cryptology –Crypto '86*, pages 186–199. Springer-Verlag, 1986.

[Gen94] R. Gennaro. Using non-interactive proofs to achieve independence efficiently and securely. MIT-LCS Technical Memo 515. 1994.

[NR94] V. Niemi and A. Renvall. How to prevent buying of votes in computer elections. In *ASIACRYPT '94*, pages 141-148. 1994.

[Pfi94]   B. Pfitzmann. Breaking an efficient anonymous channel. In *EUROCRYPT '94*, pages 339-348. 1994.

[PIK93]   C. Park, K. Itoh, and K. Kurosawa. All/nothing election scheme and anonymous channel. In *EUROCRYPT '93*, 1993.

[SK94]   K. Sako and J. Kilian. Secure voting using partially compatible homomorphisms. In *Advances in Cryptology −Crypto '94*, pages 411–424. Springer-Verlag, 1994.

Are Crypto-Accelerators Really Inevitable ?
20 bit zero-knowledge in less than a second on simple 8-bit microcontrollers

David Naccache[1], David M'raïhi[1], William Wolfowicz[2] and Adina di Porto[2]

[1]Gemplus Card International, 1 place de Navarre, F-95208, Sarcelles, FRANCE
[100142.3240 and 100145.2261]@compuserve.com

[2]Fondazione Ugo Bordoni, via Baldassare Castiglione 59, I-00142, Rome, ITALY
cripto@itcaspur.bitnet

Abstract. This paper describes in detail a recent smart-card prototype that performs a 20-bit zero-knowledge identification in less than one second on a simple 8-bit microcontroller without any dedicated crypto-engine aboard.

A curious property of our implementation is its inherent <u>linear</u> complexity : unlike all the other protocols brought to our knowledge, the overall performance of our prover (computation and transmission) is simply proportional to the size of the modulus (and <u>not</u> to its square).

Therefore (as paradoxical as this may seem...) there will always exist a modulus size ℓ above which our software-coded prover will be faster than any general-purpose hardware accelerator.

The choice of a very unusual number representation technique (*particularly* adapted to Fischer-Micali-Rackoff's protocol) combined with a recent modulo delegation scheme, allows to achieve a *complete 20-bit zero-knowledge interaction* in 964 ms (with a 4 MHz clock). The microcontroller (ST16623, the prover), which communicates with a PC via an ISO 7816-3 (115,200 baud) interface, uses only 400 EEPROM bytes for storing its 64-byte keys.

An overhead video-projected demonstration will be done at the end of our talk.

1 Introduction, Context and Basic Bricks

Although crypto-dedicated smart-cards are an industrial reality since several years, the price of these components is still too high for their massive generalization. As a result, the coding of public-key primitives in simple 8-bit microcontrollers is an important commercial issue with a wide gamut of practical applications.

In the past, several software-only implementations were proposed : the first (and probably the best known) is the Fiat-Shamir implementation in the pay-TV system *Videocrypt*. In Eurocrypt'94, Naccache, M'raïhi, Vaudenay and Raphaeli [7] described a DSA variant based on the pre-computation of ready-to-use *signature-coupons*. The NIST has an implementation of the DSA on an 8-bit microcontroller and other remarkable developments in this domain were achieved by Quisquater, Chaum and Fiat's company *Algorithmic Research Limited*.

L.C. Guillou and J.-J. Quisquater (Eds.): Advances in Cryptology - EUROCRYPT '95, LNCS 921, pp. 404-409, 1995.
© Springer-Verlag Berlin Heidelberg 1995

1.1 Fischer-Micali-Rackoff's Protocol

In Eurocrypt'84, Fischer, Micali and Rackoff [4], presented a factoring-based zero-knowledge protocol for proving the knowledge of a secret s (which modular square v is published by the prover).

In itself, this protocol (actually a Fiat-Shamir [3] with $k = 1$) is very simple :

① The prover picks a random r, computes and sends to the verifier $x = r^2 \bmod n$.

② The verifier replies with a random challenge bit b.
  •

③ •   • If $b = 0$, the prover replies with $y = r$
      • If $b = 1$, the prover replies with $y = r\,s \bmod n$

④ and the verifier makes sure that $y^2 \equiv x\, v^b \bmod n$

The security of this *perfect zero-knowledge* protocol is formally established under the sole assumption that factoring n is impossible (we incite the reader to consult [2], [3] and [4] for more details about this method and its numerous generalizations).

1.2 The Brugia-di Porto-Filipponi (BPF) Number Representation System

Denoting by $\{p_i\}$ a set of c co-prime integers, any positive integer $x < g = \prod_{i=1}^{c} p_i$ can be uniquely represented by the list $\{x \bmod p_1, x \bmod p_2, ..., x \bmod p_c\}$.

This representation [1] has the distinct advantage of allowing to perform a multiplication in linear (instead of **square**) complexity : if x and y are represented by the lists $\{x_1, x_2, ..., x_c\}$ and $\{y_1, y_2, ..., y_c\}$ then their product $z = x\, y$ will correspond to :

$$\{z_1 = x_1 y_1 \bmod p_1, z_2 = x_2 y_2 \bmod p_2, ..., z_c = x_c y_c \bmod p_c\}$$

Note that addition (or subtraction) is still linear in this notation (replace the elementwise multiplication by additions or subtractions modulo p_i) but modulo reduction is unfortunately far from being easy (the simplest method seems to be a Chinese remaindering followed by a conventional division). However, a particular property of Fischer-Micali-Rackoff's protocol allows the prover to skip the modular reduction as will be seen later.

Note that if the co-primes are very different from the maximum capacity of the machine words (for instance the first c primes), a considerable redundancy is introduced in each number, a more subtle coding allows to limit this redundancy to it's strict minimum (only 2 bytes are « lost » in each 128-byte value represented on 130 bytes). By choosing $p_1 = 64811$ and $p_c = 6552$ (the biggest possible two-byte prime), most of the most (and this is not a typographic error) significant bits of p_i are ones[1] .

---

[1] with this setting, $g \cong 8.1 \times 10^{312}$ is the hexadecimal number :
b0306dfa10d2bac63339d5fe274fc9ea61d4938dd4c706ea747307fc4ef1465ea49214e30352470531f
de44942461730afeca91c365bc9d867be7e06a46ceeb01ef910a47167592b6b3c8b837f690cc0affcb7
06ac22de64bc8d3f78fc90a3505d10ea547c63e86983f9868d78084b5533044d865c1cd6f40053396e
c2f7783f4d61

After a BPF list-multiplication (scalar product), the terms of the resulting list are reduced modulo p_i with Montgomery's algorithm [5] (this operation, which consists in reducing each 4-byte coordinate modulo a 2-byte p_i requires eight byte-by-byte multiplications per co-ordinate and does not affect the overall time linearity).

In our particular implementation, the Montgomery parasite factors ($2^{-16} \bmod p_i$) are not eliminated by the card as this operation can be trivially sub-contracted to the verifier.

1.3 Randomized Modular Multiplication

In the European patent application EP 91402958.2, Naccache [6] describes how to delegate the modular reduction of the product of two integers to a powerful verifier. In this procedure (the exact parameter sizes and a complete security proof can be found in Shamir's Eurocrypt'94 paper [8]), the sender simply adds to the product a random multiple of the modulus which is eliminated by the receiver. In other words, instead of sending $z = xy \bmod n$, the prover sends $z' = xy + rn$ (where r's role is to mask the non-reduced product xy)[2] and the receiver computes $z = z' \bmod n$.

This technique is applied by our prover in the following way : after the scalar product, n (pre-recorded in the card's EEPROM in BPF format) is multiplied by a random r and added on-the-fly (as there is no carry propagation in BPF format) to the product (r^2 or sr) which is sent to the verifier[3].

2 The Protocol

Given these building bricks, implementation is straightforward :

① The card picks a random r, computes and sends to the PC $x = r^2 + r'n$.

r' is an on-the-fly randomizer and x is represented in BPF format.

② The PC :

      ❶ eliminates the Montgomery constants in each coordinate of x
      ❷ re-codes x in the conventional format (Chinese remaindering)
      ❸ replies with the bit b.

③   • If $b = 0$, the card replies with $y = r$
    • If $b =$, the card replies with $y = rs + r''n$ (r'' is an on-the-fly randomizer).

---

[2] r should normally be bigger than n by about ten bytes.

[3] The generation of a pseudo-random r, both in BPF and such that : $\lceil \sqrt{g} \rceil \le r \le 2^{80} + \lceil \sqrt{g} \rceil$ is done, in $O(\log(n))$, by a proprietary algorithm.

④ The PC :

   ❶ eliminates the Montgomery constants in each coordinate of y

   ❷ re-codes y as a conventional number and checks that $y^2 = xv^b \bmod n$

Note that the verifier can easily check the prover's answer in linear time when $b = 0$ (the prover can then reveal r' which would allow to the verifier to check the responsive in BPF). The linear-time verification of the prover's answer if $b =$ is still an open problem (a positive answer will mean that factoring-based zero-knowledge identification and verification <u>are both</u> feasible in linear time).

Although very low, the complexity attained by our prover is not a provable minimum : it is tempting to imagine that a sub-linear complexity would simply mean that some of the modulus bits are not read by the prover but (surprisingly) such may be the case if the modulus is voluntarily chosen to be redundant (for instance, n can be generated to be compression-suitable). Also, and as paradoxical as this may seem, there exists a modulus size ℓ above which our software-coded verifier will be faster than any general-purpose hardware accelerator.

3 Implementation and Performances

The communication amount requested by our prototype is 285 bytes per round (5700 for jumping over « mystic » 1/1000,000 security level barrier). When transmitted according to the ISO 7816-3 standard at 115,200 bauds this transmission takes 0.6 seconds).

An EEPROM option allows to halve the communication by allocating 128 additional bytes and using 3 different secrets (<u>without</u> multiplying them by each other as done in the Fiat-Shamir : after the commitment phase, the verifier simply « points » the secret key he wants to be used in the round).

The code size (approximately 900 bytes), breaks down as follows :

ROM constants :

```
      compressed prime table          102 bytes
      Montgomery constants            256 bytes
```

ROM code :

```
      communication routines          200 bytes
      BPF multiplication              120 bytes
      Montgomery 4-byte reduction      30 bytes
      randomized multiplication core  200 bytes
```

EEPROM data

n	130 bytes
s	130 bytes
card ID	64 bytes
other « bookkeeping » data	64 bytes

The following table illustrates four of the possible trade-offs of our mask :

number of secrets	*transmission*	*protocol time*
1	5700 bytes	964 ms
3	2850 bytes	658 ms
7	1420 bytes	432 ms
15	710 bytes	314 ms

Table 1 : Possible trade-offs for a security level of 2^{-20} and a 4 MHz clock

4 Further Extensions : Two-Way Authentication

As explained in [6], the randomized multiplication can be applied to Rabin's scheme as well.

This can be applied to implement a very quick two-way authentication (not implemented in our prototype) where the card authenticates the PC as well :

❶ The card picks a random r, and sends $x = r^2 + r'n$ (in BPF) and $y = CRC(r)$.

❷ The PC :
   - remainders x, reduces it modulo n and computes its roots modulo n
   - discards the roots which CRCs do not correspond to y
   - encodes the remaining root w in BPF and sends it to the card

❸ The card compares w and r.

☞   Different ns should be used for sections 2 and 4 and w should not be randomized before being sent to the card...

5 Acknowledgments

The first author would like to acknowledge Jacques Stern's suggestions and improvements regarding the randomized multiplication.

6 Note

This paper does not represent a complete product description, nor does it reflect or represent, in any manner, Gemplus' intention to mass-produce cards using the features herein described. Availability of such final products may depend on commercial agreements between Gemplus and third parties. Gemplus offers no guarantee that such cards will be free of licensing rights belonging to third parties or related to Gemplus' patented technologies.

References

1. O. Brugia, A. di Porto & P. Filiponi, Un metodo per migliorare l'efficienza degli algoritmi di generazione delle chiavi crittografiche basati sull'impiego di grandi numeri primi, Note Recesioni e Notizie, Ministero Poste e Telecommunicazioni, vol. 33, no. 1-2, 1984, pp. 15-22.

2. U. Feige, A. Fiat & A. Shamir, Zero-knowledge proofs of identity, Proc. 19th. ACM Symp. Theory of Computing, 210-217, (1987) and J. Cryptology, 1 (1988), 77-95.

3. A. Fiat & A. Shamir, How to prove yourself: Practical solutions to identification and signature problems, Proc. of Crypto'86, Lecture notes in computer science 263, 181-187.

4. M. Fischer, S. Micali & C. Rackoff, A secure protocol for oblivious transfer, presented at Eurocrypt'84 but missing in the proceedings.

5. P. Montgomery, Modular multiplication without trial division, Mathematics of computation, vol. 44, 1985, pp. 519-521.

6. D. Naccache, Method, sender apparatus and receiver apparatus for modulo operation, European patent application no. 91402958.2, November 5, 1991.

7. D. Naccache, D. M'raihi, S. Vaudenay & D. Raphaeli, Can DSA be Improved ?, Proceedings of Eurocrypt'94, to appear.

8. A. Shamir, How to implement public-key schemes with 16,000 bit moduli on a smart-card with 36 bytes of RAM, presented at the rump session of Eurocrypt'94 (05-10-1994 at 20h11).

Anonymous NIZK Proofs of Knowledge with Preprocessing

Stefano D'Amiano* and Giovanni Di Crescenzo**

Abstract. In this extended abstract we present an unpublished result in [6] which extends a result in [4]. We give a non-interactive zero-knowledge proof system of knowledge with preprocessing, whose main property is that, after executing two preprocessing phases and given a transcript of a proof phase, the verifier is not able to relate the transcript to any of the two preprocessing phases significantly better than random guessing. The technique used has motivated the cash scheme in [3]. Because of this result, only mentioned but used in [3], the main observation of Pfitzmann et al. in [8] against the cash scheme in [3] doesn't hold. We also discuss the other observations of Pfitzmann et al. in [8] against the cash schemes in [3, 5] and show that *all* of them don't hold. As a conclusion, the cash schemes in [3, 5] are not broken at all.

1 The result

A non-interactive zero-knowledge (nizk) proof system of knowledge with preprocessing is a pair of protocols (preprocessing,proof) between two parties P and V in which, after an interactive preprocessing, P can send with a single message to V a zero-knowledge proof that P knows a witness to the truth of a certain statement. Such a system can be used in an electronic cash system by a spender to prove to a shop the knowledge of a signature of a coin released by the bank. Informally, we will say that a nizk proof system with preprocessing is *anonymous* if any verifier V', after running a pair of preprocessing protocols, and given the transcript of any proof protocol, cannot relate the proof protocol to one of the two preprocessing protocols with probability significantly greater than $1/2$.

Our result says that the nizk proof of knowledge with preprocessing given in [4] for all NP languages can be made anonymous. Let us briefly recall the system (A,B) in [4]. Let σ be a public random reference string. In the preprocessing phase the prover A computes a commitment *com* to a string s using coins r and interactively proves the statement H:'I know s and r such that *com* is a commitment to s, using coins r'. The verifier B verifies this proof. Now, let L be any NP-complete language, and let (x, w) be a pair (instance,witness) such that the prover wants to prove that he knows the witness w such that $x \in L$. In order to prove this, A first computes $\beta = w \oplus f_s(x)$, where f_s is a pseudorandom function, and $\oplus$ is the bitwise logical xor operator. Then A gives a nizk proof $proof_T$ (of membership) on the public random reference string σ of the

---

* Computer Science Department, Cornell University, Ithaca, NY, USA
** Department of Computer Science and Engineering, University of California, San Diego, La Jolla, CA, 92093

L.C. Guillou and J.-J. Quisquater (Eds.): Advances in Cryptology - EUROCRYPT '95, LNCS 921, pp. 413-416, 1995
© Springer-Verlag Berlin Heidelberg 1995

statement T: 'there exist strings r, s, w such that com is a commitment to s using coins r, w is a witness for $x \in L$, and $\beta = w \oplus f_s(x)$'. In order to verify this proof, B needs the string com obtained in the preprocessing phase. The view of B in the preprocessing is $(com, H, proof_H)$ and a transcript of the proof phase will be $(\beta, \sigma, T, proof_T)$. Clearly, here the statement T used in the proof phase contains the commitment com used in the preprocessing phase. Thus, after running two commitment phases, given the two views $(com_1, H_1, proof_{H_1})$ and $(com_2, H_2, proof_{H_2})$ and the transcript $(\beta, \sigma, T, proof_T)$ of a proof phase, B can check if the commitment used in T is com_1 or com_2 and thus associate the proof with exactly one of the two preprocessing run.

Our extension is simply stated; our proof system (P,V) is the same as the above (A,B), with the following two modifications. First, in the preprocessing phase V also signs the commitment com and sends the signature sig_{com} to P. Second, in the proof phase, instead of the above statement T, the prover uses the statement T': 'there exist strings r, s, w, com, sig_{com} such that com is a commitment to s using coins r, w is a witness for $x \in L$, $\beta = w \oplus f_s(x)$ and sig_{com} is a correct signature of com'. Then the view of V is: $(com, sig_{com}, H, proof_H)$ in the preprocessing phase, and $(\beta, \sigma, T', proof_{T'})$ in the proof phase. Now the two views are not related, thus a transcript of a proof phase cannot be matched with a view in the preprocessing phase. The formal proof is based on the following idea: assume by contradiction that there is an efficient non-uniform verifier V', which, after running two preprocessing phases $prep_1$ and $prep_2$, and getting a transcript pf of a proof phase, can compute $i \in \{1, 2\}$ such that pf is the proof associated to the preprocessing $prep_i$. Then, it is possible to use V' in order to construct an efficient algorithm A which efficiently opens commitments or contradicts the zero-knowledgeness of the proof system used, or contradicts the pseudo-random property of the function used.

Finally, we remark that this technique has been used in the cash scheme in [3] in the construction of the withdrawing and the spending protocols.

2   The observations by Pfitzmann et al.

In the paper [8], Pfitzmann et al. make some observations against the cash scheme in [2] [3]. Now, we consider all the observations in [8] and show that *all* of them do not hold.

The main untraceability flaw: In their first observation, Pfitzmann et al. state, without proof, that our protocol in [2] doesn't satisfy the untraceability requirement because by receiving a proof of knowledge in a deposit protocol, the bank can link it to the user who run the preprocessing when opening his account. Also, they write '*This is a general problem with NIZKP with preprocessing: At least as long as preprocessing is a 2-party protocol, any reference to it, i.e., any proof, identifies the two parties...*'. In [3] the protocol presented in previous section is

---

[3] This version was the result of a few-days two-paper merging; unclearly, the proceedings version [3] has never been asked.

used, with only a sketch of description, as the same technique extending [4] is used in the construction of the protocols for withdrawing a coin and for spending a coin in the electronic cash scheme. As the above proof system is anonymous, the observation by Pfitzmann et al. is wrong. In particular we stress that it is not true that nizk proofs with preprocessing cannot be used in cash schemes. [4]

Weakness of the definition of untraceability: Here Pfitzmann et al. criticize our definition of untraceability, as it does not require untraceability against payees. We answer by noticing that in our paper, after giving the protocols for withdrawing a coin and spending a coin, we discuss in Section 6 that, by simply applying one of the many secret-sharing based techniques for avoiding the double-spending of a coin, we easily obtain a cash system. This scheme clearly satisfies also untraceability against payees (but it doesn't give transferability!). However, we don't prove such an easy statement, and thus we don't need a definition for it. Thus we investigate transferability of coins and we obtain the first cash scheme in which coins can be transferred without increase in size. This result was not known. Untraceability against payees is not required in it: on the other hand, if it was, then such result would not be possible, as proved in [1].

An untraceability flaw in double-spending detection: Here Pfitzmann et al. write 'Until the double-spender is detected, the bank cannot prove that a coin was double spent, i.e., the users have to believe this. Thus, a dishonest bank could always claim that the coin was deposited twice in order to get the payees to disclose the payers'. This is not true, as in our protocol, the bank discovers that a double-spending has occurred from the fact that she receives two signatures of a same coin; in fact, in Section 6 we write 'the bank broadcasts a message stating that a double-spending of c has occurred ... to prove this, she writes on the public file the two different signatures of c'.

Finding the wrong double-spender: A first observation by Pfitzmann et al. here is: a user 'may have received the coin twice, e.g., from an attacker at different times. In this case, the honest user is likely to be found before the attacker and punished as the double-spender.' This is not true; in fact, when we discuss the single spending requirement, we never say that the first user found to have spent twice the coin is punished. We say that 'the bank uses the signatures received to reconstruct the two paths ... that have been taken ... by the coin c'; this clearly means that the bank will first receive all signatures and then she will reconstruct the whole paths taken by the coin, where the double-spender is the source of these paths. A second observation here is a situation in which a honest user H_1 spends twice the same coin as he received it twice. They say 'H_1 has two completely identical signatures from A_1, so she cannot prove she got the coin twice and is punished'. Here Pfitzmann et al. assume that for signature schemes constructed under general cryptographic tools, two signatures on the same messages are equal. This is clearly not true (see [7] and references therein),

---

[4] Also, when presenting the system in [2], they observe that 'The protocol in [AmCr] is ambiguous about whether the coin must be passed'. In [2] actually it is written 'send $C, Proof$ to U_2' instead of 'send $c, Proof$ to U_2'; however no other occurrences are in the paper for C (except for Chaum).

and actually if it was true, then the results in [7] would prove that nizk proofs of membership are equivalent to general cryptographic tools, i.e., one-way functions (an open problem about nizk proofs).

No divisibility together with transferability: Here it is observed that complexity parameters are not reduced in our divisibility scheme. This is not true, in fact in Section 5 we write '...*to spend a part of c, say, of value* 2^{k-h}, *a user* U_j *gives only the random string used for the commitment at a node at level h'*. We stress that the most expensive complexity parameter used in our construction is just the number of random bits used in commitments, (it is usually one order bigger than the others).

Achieving weak untraceability more easily: Here an alternative solution to obtain untraceability with respect to the bank is suggested by Pfitzmann et al. Unfortunately, the suggested solution doesn't work: a coalition of the user who is depositing the coin and the bank is enough to make the coin traceable.

The system presented at CIAC 94: We also discuss the three observations about the cash scheme in [5]. The first is that the bank cannot announce if it rejects the withdrawal. We have always thought that there are many easy techniques for doing this, all depending from implementation; for instance, the bank can erase the coin published by the withdrawer. The second observation is that there is no untraceability. Of course our protocol does not satisfy a more general definition of untraceability than that given in the paper! This definition is weaker than usual just in order to obtain the non-interactivity of the protocol (which otherwise seems very hard to get). However, we observe that such a definition is not very weak; for instance, real-life cash satisfies it. The third observation seems the only correct so far: the deposit scheme can be attacked by a chosen-message attack. However, it is not clear which utility this attack could give, and also the scheme is easy to repair: the user that deposits the coin just proves to know a signature, instead of giving it (there exist such proofs in literature).

References

1. D. Chaum and T. Pedersen, *Transferred Cash Grows in Size*, Eurocrypt 92.
2. S. D'Amiano and G. Di Crescenzo, *Methodology for Digital Money based on General Cryptographic Tools*, Preproceedings of Eurocrypt 94.
3. S. D'Amiano and G. Di Crescenzo, *Methodology for Digital Money based on General Cryptographic Tools*, to appear on Proceedings of Eurocrypt 94.
4. A. De Santis and G. Persiano, *Communication Efficient Zero-Knowledge Proof of knowledge (with Application to Electronic Cash)*, STACS 92.
5. G. Di Crescenzo, *A Non-Interactive Electronic Cash System*, CIAC 94.
6. G. Di Crescenzo, *Anonymous NIZK Proofs of Knowledge with Preprocessing*, manuscript, December 1993.
7. S. Goldwasser and R. Ostrovsky, *Invariant Signatures and Non-Interactive Zero-Knowledge Proofs are Equivalent*, Crypto 92.
8. B. Pfitzmann, M. Schunter, and M. Waidner, *How to break another provably secure payment system*, Eurocrypt 95.

Author Index